Contemporary Geometry

J.-Q. Zhong Memorial Volume

THE UNIVERSITY SERIES IN MATHEMATICS

Series Editor: Joseph J. Kohn
Princeton University

THE CLASSIFICATION OF FINITE SIMPLE GROUPS
Daniel Gorenstein
VOLUME 1: GROUPS OF NONCHARACTERISTIC 2 TYPE

CONTEMPORARY GEOMETRY (J.-Q. Zhong Memorial Volume)
Edited by Hung-Hsi Wu

**ELLIPTIC DIFFERENTIAL EQUATIONS AND
OBSTACLE PROBLEMS**
Giovanni Maria Troianiello

FINITE SIMPLE GROUPS: An Introduction to Their Classification
Daniel Gorenstein

AN INTRODUCTION TO ALGEBRAIC NUMBER THEORY
Takashi Ono

**INTRODUCTION TO PSEUDODIFFERENTIAL
AND FOURIER INTEGRAL OPERATORS**
François Treves
VOLUME 1: PSEUDODIFFERENTIAL OPERATORS
VOLUME 2: FOURIER INTEGRAL OPERATORS

MATRIX THEORY: A Second Course
James M. Ortega

A SCRAPBOOK OF COMPLEX CURVE THEORY
C. Herbert Clemens

TOPICS IN NUMBER THEORY
J. S. Chahal

Contemporary Geometry

J.-Q. Zhong Memorial Volume

Edited by

Hung-Hsi Wu

University of California
Berkeley, California

Plenum Press • New York and London

Library of Congress Cataloging-in-Publication Data

Contemporary geometry : J.-Q. Zhong memorial volume / edited by Hung
-Hsi Wu.
    p.   cm. -- (The University series in mathematics)
  Includes bibliographical references and index.
  ISBN 978-1-4684-7952-2   ISBN 978-1-4684-7950-8 (eBook)
  DOI 10.1007/978-1-4684-7950-8

  1. Geometry. 2. Zhong, J.-Q. (Jia-Qing), d. 1987. I. Wu, Hung
-Hsi, 1940- II. Series: University series in mathematics (Plenum
Press)
QA446.C66 1991
516--dc20   91-10621
                                                CIP

© 1991 Plenum Press, New York
Softcover reprint of the hardcover 1st edition 1991

A Division of Plenum Publishing Corporation
233 Spring Street, New York, N.Y. 10013

Contributors

Peter Li, Department of Mathematics, University of Arizona, Tucson, Arizona 85721

Qi-Keng Lu, Institute of Mathematics, Academia Sinica, Beijing 100080, People's Republic of China

Yum-Tong Siu, Department of Mathematics, Harvard University, Cambridge, Massachusetts 02138

Andrejs Treibergs, Department of Mathematics, University of Utah, Salt Lake City, Utah 84112

Hung-Hsi Wu, Department of Mathematics, University of California, Berkeley, California 94720

Preface

Early one morning in April of 1987, the Chinese mathematician J.-Q. Zhong died unexpectedly of a heart attack in New York. He was then near the end of a one-year visit in the United States. When news of his death reached his Chinese–American friends, it was immediately decided by one and all that something should be done to preserve his memory. The present volume is an outgrowth of this sentiment. His friends in China have also established a Zhong Jia-Qing Memorial Fund, which has since twice awarded the Zhong Jia-Qing prizes for Chinese mathematics graduate students. It is hoped that at least part of the reasons for the esteem and affection in which he was held by all who knew him would come through in the succeeding pages of this volume.

The three survey chapters by Li and Treibergs, Lu, and Siu (Chapters 1–3) all center around the areas of mathematics in which Zhong made noteworthy contributions. In addition to putting Zhong's mathematical contributions in perspective, these articles should be useful also to a large segment of the mathematical community; together they give a coherent picture of a sizable portion of contemporary geometry. The survey of Lu differs from the other two in that it gives a firsthand account of the work done in the People's Republic of China in several complex variables in the last four decades. Although it represents of necessity only one man's view of an area not previously open to the West, it is hoped that this effort will lead to similar surveys in other areas of mathematics. The 14 papers by Zhong collected in this volume give a good account of his mathematical work. With but three exceptions (Chapters 13–15), all the papers have been translated from Chinese especially for this volume, and are therefore made available in English for the first time. Finally, the biographical essay may be of some interest to those who wish to understand China; through the life of one mathematician, one gets an intimate, if restricted, view of that nation's social life in the last 25 years.

A very important part of my duty as editor of this volume was to coax others into doing the job on my behalf. In this regard, I must count myself a great success. I am indebted to Professors P. Li, Q.-K. Lu, Y.-T. Siu, and A. E. Treibergs for writing the survey articles. In particular, Professors Li and Treibergs undertook this task at my request at the eleventh hour, and for this I am especially grateful. Much less obvious is the debt I owe a group of mathematicians in China who made the translations of Zhong's papers possible: Professor Lo Yang for coordinating the translation project, Professors Zhi-Hua Chen, Yi Hong, Xiao-Jiang Tan, and Wei-Ping Yin for actually doing the work, and Professors Qi-Keng Lu, He Shi, Xian-Rou Sun, Qi-Ming Wang (who has since regretfully passed away), and Mei-Juan Wu (Mrs. Zhong) for providing the all-important editorial assistance. I have also received invaluable help from Shiu-Yuen Cheng, Robert E. Greene, Jiang-Hua Lu, and Mei-Juan Wu in ways too diverse to enumerate. Professor J. J. Kohn made possible the publication of this volume in the University Series in Mathematics, and Mr. L. S. Marchand of Plenum Publishing Corporation has been most helpful and understanding throughout the long and tortuous process in bringing this volume to completion. Last but not least, the editorial staff of Plenum performed miracles with a manuscript that was in places all but undecipherable. I am happy to avail myself of this opportunity to extend to them my most sincere appreciation.

H. Wu

Berkeley, California

Contents

PART I: SURVEYS OF CONTEMPORARY GEOMETRY

1. **Applications of Eigenvalue Techniques to Geometry**
   Peter Li and Andrejs E. Treibergs

PART II: SELECTED PAPERS OF ZHONG JIA-QING

Zhong Jia-Qing (1937–1987)

Hung-Hsi Wu

Zhong Jia-Qing[1] was born on December 4, 1937, in Anhui Province, which lies in the east coastal region of China. He died unexpectedly of a heart attack in the early morning of April 12, 1987, in New York while he was a visiting professor at Columbia University. He is survived by his wife, Wu Mei-Juan, who is a professor of physics at Qinghua University in Beijing, a daughter Zhong Wen (born 1967), and a son Zhong Ning (born 1970).

Zhong's talent in mathematics manifested itself in his high school days, when he walked away with practically all the awards and honors in mathematics that his school had to offer. In 1956, he entered Peking University in the department of mathematics.[2] After graduation in 1962,[3] he was accepted by Hua Lo-Keng as a graduate student[4] in the Institute of Mathematics in Beijing.[5] When Hua left for The China University of Science and Technology[6] in 1964 to take up the position of vice-president, Zhong followed him there. However, a year later, when Zhong was barely 28, external circumstances beyond his control were to shatter his mathematical life for the next 12 years: the Cultural Revolution (1966-1976) was about to ravage China and to claim among its victims much more than just a generation of young Chinese mathematicians. During 12 years that would have been crucial to his maturation as a mathematician, Zhong, like so many of his compatriots, was compelled at various times to do manual

Research for this chapter was partially supported by the National Science Foundation.

Hung-Hsi Wu • Department of Mathematics, University of California, Berkeley, California 94720.

1

labor on farms, machine work in factories, or tutorial work for the benefit of the so-called *mainstays of the nation,* namely, the workers, peasants, and soldiers. Even so, Zhong managed to sneak in some research work in secret, often at great personal risk. But the obvious stress and the extremely adverse physical conditions took their toll, and his health was never the same again.

In 1978, Zhong rejoined the Institute of Mathematics. That was the time when scientific contact with the outside world was beginning to be reestablished in China, and Zhong's mathematical work entered a new phase. The remaining 10 years of his life were full of frantic activity, both in absorbing the mathematics currently done in the West and, at the same time, in contributing to it. Meanwhile, he was promoted to associate professor in 1979, and eventually to professor in 1986. He visited the United States on three separate occasions: 1980-1981 at Stanford, 1983-1984 at The Institute for Advanced Study, and 1986-1987 at (successively) Harvard and Columbia. These visits resulted in some of his best papers. In the years immediately preceding his death, his mathematical work was gaining momentum (see, for example, the nine posthumous papers [23-31][7]), and his contributions in real and complex geometry were beginning to attract the attention they deserved. At the beginning of 1987, he was notified of having been selected as the first recipient of the Shiing-Shen Chern award in mathematics given by the Chinese Mathematical Society. In order to attend the award ceremony, he decided to cut short his visit at Columbia and return home a month ahead of schedule. But a few days before his departure from New York, he was struck down by the heart condition that was a legacy of his years in the Cultural Revolution. He was not quite 50 years old.

The life of a typical mathematician in the West is predictably bland and uneventful. Even the life of a great mathematician such as Poincaré would seem to have little to offer in terms of human interests and drama.[8] Not so the life of a Chinese mathematician who lived through the Cultural Revolution. This cataclysmic event, so much talked about and written about in bits and pieces, but never documented in full to produce anything resembling a global perspective, may yet prove to be beyond normal comprehension in its entirety.[9] Fortunately, we are here only dealing with that part of the Cultural Revolution that is within the confines of the mathematical world, and, in this narrow context, some understanding may yet be possible. To be sure, Zhong survived the Cultural Revolution, but not before his health and his mathematical career had been permanently damaged. Let us first go back to the beginning of this career, which can be clearly pinpointed at the time he took a course in several complex variables from Lu Qi-Keng in 1959. Lu was a student of Hua Lo-Keng and has spent all his life at the

Institute of Mathematics. In 1959, he was, however, a visiting professor at Peking University,[10] where Zhong was an undergraduate at the time. The main thrust of Lu's lectures was the latest research by Hua and Lu himself on the geometry and function theory of the classical domains; in spirit those lectures were quite close to the by now well-known work of Hua on the subject[11] (for more details, see Chapter 2 of this volume). This approach to complex analysis puts a premium on one's computational facility, particularly with matrices. Right from the beginning Zhong displayed an affinity for this sort of computation, and this ability proved to be an asset in all his subsequent publications. When it was time to write his graduation thesis, it was a foregone conclusion that Zhong would choose a topic related to the classical domains. It turned out to be the determination of the largest automorphism group of the classical domain of type III [1] (joint work with Yin Wei-Ping), that of the other three types having already been determined in the meantime.

From the beginning, Zhong was regarded as the best student in his age group, and he further confirmed his superiority by achieving an almost perfect score in the entrance examination of the Institute of Mathematics in 1962. As mentioned earlier (cf. note 4), he wanted to study with Hua, and in return Hua duly took note of the arrival of his latest star student. However, even in 1962, the demands on Hua's time were already enormous,[12] and one can well imagine that his duties as a thesis advisor were honored more in the breach than the observance. Fortunately, Lu was available, and Zhong obviously saw more of Lu than of Hua.[13] It was therefore to be expected that when Hua was called upon to become Vice-President of The China University of Science and Technology in 1963 (see note 6), Zhong would request that he be allowed to stay behind at the Institute of Mathematics to work with Lu rather than follow Hua across town. However, the request was eventually denied, so in 1964, Zhong was formally transferred to The China University of Science and Technology.

At the new institution, Zhong found the intellectual atmosphere to be less stimulating than what he was used to at the Institute of Mathematics. Nevertheless, he put the next year and a half to good use by studying intensively Hermann Weyl's *The Classical Groups*. The knowledge he gained in representation theory out of this period of hard work was to stand him in good stead in his later works (cf. [3, 4, 9, 16]). Unfortunately, Zhong's mathematical career was disrupted at this point by the "Four Purifications Movement,"[14] which was soon to culminate in the Cultural Revolution. Zhong was sent to a village in the suburb of Beijing in 1965 as an inspector of the communes and the cadres. Tens of thousands of intellectuals all over China were sent to the countryside at this time for the purpose of acquainting themselves with the lives of the peasants and workers. During the next 10

years, Zhong together with all the other university professors were essentially
cut off from any access to research (those were the days when Euclid's
axiomatic system was officially criticized for being based on idealistic
precepts rather than on Marxist-Leninist materialism).[15] It is a testimony
to the tenacity of the human spirit that Zhong could get as far as he did in
mathematics after his comeback in 1977.

People in the West are generally under the mistaken impression that
social turmoil in China began and ended with the Cultural Revolution,
whereas in fact that nation has been racked every two to three years by one
upheaval or another since the mid-1950s. What is true is that the Cultural
Revolution disrupted all phases of life in China to a far greater degree than
any of the other upheavals before or since. After 1966, all intellectual or
normal social activities came to a dead halt. Thus, although Zhong was
officially graduated from The China University of Science and Technology
in 1967, in reality he and almost all the other university students spent
1966–1968 traveling around the country. Indeed, since academic work was
considered reactionary, and since trains and lodging were free for one and
all, there was clearly no better way to while away the time. But one good
thing did come out of this period. In February of 1967, Zhong married Wu
Mei-Juan, whom he had known since 1956. At the time, Mrs. Zhong[16] was
teaching at Qinghua University, which is also in Beijing and quite near
Peking University. Zhong, on the other hand, was still officially a member
of the mathematics department of The China University of Science and
Technology.[17] When the latter was moved from Beijing to Hefei at the end
of 1969 (see note 6), it was decided that Zhong would join the faculty of
the mathematics department at Qinghua University so that the family could
stay together[18]; for by then, the Zhongs already had a two-year-old daughter
and a second child was on its way. However, this arrangement was achieved
at a steep price: before the Zhongs could settle down at Qinghua University,
the whole family was required to first move to a farm near Nanchang in
Jiangxi Province,[19] where many other Qinghua professors had already been
sent to "reform themselves through manual labor." There they were to stay
until 1971. Those two years completely changed the Zhongs' lives.

To understand these proceedings, one should be aware of the fact that
during the Cultural Revolution, Peking University and Qinghua University
were the base of the ultraleftists. In fact, Qinghua University was under the
direct supervision of Madame Mao (Jiang Qing, head of the "Gang of
Four"). Under the circumstances, for Qinghua University to send its own
faculty members to farms to perform manual labor was merely par for the
course. However, the official in charge of this policy was apparently one
of the most vicious characters in the Cultural Revolution, and his choice
of that particular farm in Jiangxi Province was for a particular purpose:

the water there was infested with parasitic flukes (a blood-sucking leechlike flatworm) whose bite was well-known to transmit schistosomiasis, an incurable parasitic disease that leads to irreversible damage of the internal organs, especially the liver and the heart.[20] To this official, if all the faculty members working there did contract schistosomiasis, it would be no more than fitting punishment for these lowly intellectuals.[21] Thus it came to pass that all the faculty members of Qinghua University at the farm did contract schistosomiasis, including Zhong and his wife.

When I first met Zhong, in 1978, he told me that his liver and his heart had been quite severely damaged. As we now know, this heart condition was eventually to claim his life 10 years later. But it was an irony of fate that Zhong *really could have avoided contracting this hideous disease* had he not disobeyed orders from the authorities. For as soon as the Zhongs arrived at the farm, Zhong was appointed the group leader of the "untouchables,"[22] consisting mainly of university professors and other intellectuals who had been found guilty of being "reactionary academic authorities," "counterrevolutionaries," "ex-rightists," etc.[23] The leadership at the farm informed Mrs. Zhong that they were extending her husband a special favor in the hope that he could "turn around" these reactionary intellectuals and stay on to make a substantial contribution to the Great Cultural Revolution. By being a group leader, Zhong would have been in a position to assign the most onerous chores to the others while he occupied himself with merely supervisory duties. In particular, there would have been no need for Zhong to do any work in the water and run the risk of contracting schistosomiasis.

The leadership had their own designs in giving Zhong this special treatment. Knowing realistically that they could not hope to wipe out the intellectuals, they needed to win over some of these lowly characters in order to "carry on the revolution." Zhong was young, was already a graduate of one of the most prestigious universities under a famous teacher, and was obviously admired and well liked by his peers. To be able to land a person of this caliber would certinly be a major coup. These cultural revolutionaries, who understood mankind only through its baser instincts, were caught off guard from time to time by genuine manifestations of human decency. In this case, Zhong chose to show his appreciation of the special treatment he received from the leadership by identifying himself with all his charges. Rather than sitting back to order them around, he himself took over the most undesirable assignments and in particular, joined them in the field and, hence, in the water. He also got along with them famously. At a time when, given a choice, everyone would avoid them like the plague, the "untouchables" were obviously moved by Zhong's quiet heroism. To them, Zhong and his wife were above reproach. After Zhong's death, an old Qinghua University professor reminisced about Zhong and said: "At the farm, Zhong took over from me all the strenuous chores. I will never forget him."

Not surprisingly, this kind of behavior from someone they were trying to court did not sit well with the leadership. They sought out Mrs. Zhong and told her to remind her husband that "the line between classes must be clearly drawn" and that "class struggles" must be kept foremost on his mind at all times. When Zhong learned of this, he reassured his understandably apprehensive wife that he knew what he was doing and that he had the situation under control. He was right on the first point, but on the second he would soon realize how wrong he was.

As mentioned earlier, Mrs. Zhong was already five months pregnant with her second child when she arrived at the farm. Shortly thereafter, her health came under the double threat of malnutrition and the strain of extended working hours. Each meal at the farm consisted of rice and turnip strips, both of the lowest quality, and nothing else. Moreover, she (like everybody else) had to work each day from four o'clock in the morning until dark. When the child was born in April of 1970, it probably would not have survived the first few months without the extraordinary effort put out by the other people in the farm. They managed to catch eels in the fields to supplement her diet, and their children climbed trees to snatch eggs from birds' nests for her consumption. Nevertheless, six months after giving birth, Mrs. Zhong found her child's physical development to be alarmingly subnormal and, after exhausting all other avenues, appealed to the leadership to let her send the child to her mother's care in another city. But Zhong's behavior being what it was, the leadership naturally found it easy to turn down the request. A big argument ensued, and, afterward, the Zhongs decided that the situation was desperate enough to justify their taking the matter into their own hands. They would take their child to her mother the very next morning, although they had been warned that this would lead to their being dubbed "counterrevolutionaries," a most serious charge during the Cultural Revolution. But that evening, the leadership, for whatever reason, relented and arrived at a compromise with the Zhongs: they would grant the Zhongs a leave of absence for this purpose in a month, and in exchange the Zhongs would raise no more fuss about this issue. After this, the rest of the Zhongs' stay at the farm was relatively uneventful until the farm was dissolved in 1971. Thereupon, they returned to Qinghua University in Beijing.

In 1972, Zhong was only 35, but the two years at the farm had aged him. And, even though he had been allowed to return to Qinghua University, he, like other professors, was still not allowed to resume research. Moreover, Qinghua University at that point was exclusively devoted to the education of the workers, peasants, and soldiers. This was the result of an official policy aimed at redressing, overnight, the traditional underrepresentation of these three groups in the universities. Recognizing the fact that these students had not qualified for this eminent institution on intellectual grounds, the government ordered the professors not just to teach but *to teach until the students learned.* Obviously, such an order was easier to give than to implement, especially in an abstract subject such as mathematics. Some students did not understand, for example why $1/2 + 1/2 = 1$, because it seemed more reasonable to have

$$\frac{1}{2} + \frac{1}{2} = \frac{1+1}{2+2} = \frac{2}{4}.$$

The addition law for arbitrary fractions presented even graver problems. When it came to decimals, they wondered why $0.5 = 1/2$. If they had already learned about $1/2$, why were they asked to learn another way of writing the same thing? Could it be that the professor purposely concocted this 0.5 to torment them? If so, this professor had to be *criticized.*[24] Under the circumstance, teaching could not possibly be confined to the classroom; the professors must follow the students around the whole day and try to explain the material at all hours until enlightenment finally dawned on them. Zhong once told me of such an experience when I first met him in 1978. At one point he had to teach a certain student the formula $(a + b)^2 = a^2 + 2ab + b^2$. Apparently the student would much prefer that $(a + b)^2 = a^2 + b^2$. After doggedly following him for several days, Zhong finally convinced him of the correctness of the more cumbersome formula. Just to make sure, Zhong asked the student the morning after the breakthrough what $(x + y)^2$ was equal to. The student stared at him in amazement and said he had no idea. Zhong then patiently went over once more their common struggle of the past few days with $(a + b)^2$, and thereupon the student exclaimed: "You only taught me $(a + b)^2$, but you never said anything about $(x + y)^2$!" So day after day in 1972–1976, Zhong devoted all his available hours to education of this type. Only after everybody had gone to sleep could he spend some time looking at mathematical journals. But he never gave up his conviction that soon everything would be back to normal. Sadly as we now know with hindsight, things did *not* immediately go back to normal even after the downfall of the "Gang of Four" in October of 1976, and certainly not in Qinghua University, the one-time bastion of the ultraleftists. Totally discouraged, Zhong asked to be reinstated at the Institute of Mathematics. With the help of Lu, this was finally accomplished in 1978. By then, Zhong was already 40 years old.

During the years 1965–1976, Zhong kept up his research by working in secrecy and isolation. All or part of the papers [3–9] were based on the work done in that period, although they were all published after 1976. But by 1978, contact with the West had finally been reestablished, and the Chinese mathematicians of Zhong's generation were faced with a crisis of major proportions. It did not take them long to realize that the isolation of China from 1950 to 1978 had kept them out of some important mathematical developments. This was especially true of several complex variables and symmetric spaces, the two areas that have direct impact on the study of classical domains. Like his compatriots, Zhong was faced with the choice of either remaining in mathematical isolation by continuing with his old works or making a leap into the unknown by picking up the new mathematics at an advanced age, mathematically speaking. Zhong chose the latter. At a time when the works of S.-T. Yau and Y.-T. Siu were beginning to command worldwide attention, one can imagine that Zhong's choice of his new field of specialization was in fact quite easy; it was to be the geometric aspects of several complex variables. This was an area not far from what he had been working on, and now Zhong had the opportunity to learn directly from the top practitioners of the art, as both Siu and Yau went back to China to give extended lectures in 1979. The rest of Zhong's development as a mathematician is well documented in the Western journals, and in any case the survey chapters by Li and Treibergs and Siu in this volume (Chapters 1 and 3) have obviated the need of any comments by me in this regard. Instead, I will make a few remarks about Zhong the man.

I never got to know Zhong very well on a personal level. By all accounts, he was a loving father and husband. I learned about this part of his life only after Zhong's death, indirectly through friends who knew him better than I did. I also got a glimpse of his family life from the commemorative essay written by Mrs. Zhong and from the moving speech by Zhong Ning in accepting, on his father's behalf, the first Shiing-Shen Chern award in mathematics in May of 1987. In that speech, Zhong Ning recalled, sometimes by directly quoting from his father's letters, what Zhong had told the family about his personal aspirations and his attitude toward Chinese mathematics. What he said in the intimacy of his home was entirely consistent with his public persona. He had no illusions about either his own achievements or the present state of Chinese mathematics, but his dedication to the future of Chinese mathematics was so complete that it left no room for complacency or despair. However, one could detect that, through it all, there was an underlying current of regret over how much more he could have done had he not been stripped of those 12 years surrounding the Cultural Revolution.

Those of us who knew him and his ability have also come to share the same sense of regret.

My association with Zhong was, on the whole, a professional one, although near the end there was reason to believe that a real friendship was about to emerge. Among the Chinese mathematicians that I know, he probably came closest to approximating the Chinese ideal of a scholar. A gentle and unassuming man, he nevertheless had a seemingly inexhaustible reserve of inner stability and will. Substance would always take precedence over form. Money, fame, and power affected him almost not at all. I remember being once incensed over the situation concerning his promotion (or rather, *non*-promotion) at the Institute of Mathematics and offering to give the authorities a piece of my mind, but Zhong just good-naturedly declined the offer with a smile. In the years before he passed away, he was increasingly relied on by many of the Chinese–American mathematicians for advice concerning Chinese mathematics. To my knowledge, his well-considered opinions were always accepted without question.

There was one incident I know of where Zhong's integrity was put to the test, although this incident did not come about as a result of any person's design. In the early part of 1985, there was a plan afoot to draft Zhong to teach in the summer school of '86. This is a summer school, sponsored by the Chinese Ministry of Education, devoted exclusively to mathematics. Each year since 1984, such a summer school has been held by turns in one of the universities in China during the summer months. Foreign and Chinese mathematicians are carefully chosen to give intensive courses of current interest. Under normal circumstances, an invitation to teach there would be considered quite an honor. In the case of Zhong, however, the invitation required maneuvers of great delicacy. The reason was that the summer school was born amidst an intense controversy involving Chinese and Chinese–American mathematicians on both sides of the Pacific. It was a group of Chinese–American mathematicians who first proposed the idea of this summer school, and, for reasons we need not go into here, some Chinese mathematicians as well as another group of Chinese–American mathematicians found this idea objectionable. From the very beginning, Zhong maintained a position of total neutrality on this issue. It may be conjectured that he in fact knew very well that the summer school was good for Chinese mathematics, but he also happened to have strong ties to those who voiced objections to the summer school. Because human relations are taken much more seriously in China than in the West,[25] Zhong had no choice but to stay out of the controversy in order to avoid stepping on the toes of his friends and superiors. On the other hand, Zhong was already at that point of his career a leading figure in Chinese mathematics, and the summer school could not hope to be a success without getting him involved

in it. In June of 1985, I happened to be in Beijing, and I took it upon myself
to sound him out on this matter. I knew about his commitment to educating
the next generation of Chinese mathematicians, but would he perhaps allow
his personal considerations to override his better judgment? I had expected
a long and drawn out discussion, but when I finally put the question to
him, it took him only a few moments to collect himself and say that he
would be inclined to accept if the position were offered. I well remember
the moment when in October later that year, words reached me that Zhong
had officially accepted. In the land with the one-time official policy that
"politics reigns over all," some can and *do* rise above politics after all.

Although I had never heard any of Zhong's lectures, I had heard
reliable reports that he was an excellent lecturer.[26] So it occurred to me
that his talent as a teacher was essentially wasted in the Institute of Mathematics. For, regardless of its other virtues, that institution simply does not
have a functional graduate program to train its own graduate students;
among other things, its faculty has a zero teaching load. Knowing how
strongly he felt about mathematics education, I asked him (also in June of
1985) whether he had ever considered transferring to a regular teaching
university such as the Peking University (his alma mater). Again to my
surprise, he said with no hesitation that indeed he had, and had even made
preliminary explorations of this option with the officials of the Peking
University. However, he was held back from pursuing it further because
there was still a personal obligation that he was duty bound to fulfill. I
knew very well what that obligation was, and, like him, I could not find a
way out of the predicament. But it was an understanding between us that
in a few years this obstacle would be a thing of the past and then he would
be free to realize this ambition. Alas, he was never granted those few extra
years.

One of the most poignant moments of my stay in China came during
that same meeting with Zhong. Zhong mentioned to me that from his contact
with the Chinese-American mathematicians in the past few years, he discovered that some of them seemed to care more about China than did the
younger generation of Chinese mathematicians themselves. We discussed
the problem of the nonreturning Chinese students abroad,[27] and he made
some pointed remarks about some of those whom we both knew personally.
For someone so totally committed to staying behind in his homeland to
fight the good fight, the realization that he might be the lone warrior in his
chosen battleground must have been devastating. Although I had detected
not a little bitterness during the course of the conversation, I also knew
that there was nothing I could say to lessen his pain.

When I first met Zhong in 1978, for a few days we had the opportunity
to talk in some depth. It was through those conversations that I had my

first personal encounter with the terrible waste of human resources during the Cultural Revolution. It was also during one of those conversations that he made the wistful remark that, for him to start all over again at his age, doing truly outstanding work in mathematics was no longer a realistic goal. He thought that his main contribution to Chinese mathematics should lie in the direction of passing the torch to the next generation. If I say that among the many people who had made a similar remark to me Zhong was almost alone in remaining true to his words to the end of his life, it is not to pass judgment on the others. We all know how our altruistic impulses will eventually taper off in the face of reality, and if the reality turns out to be particularly unforgiving, then there is all the more reason for us to be compassionate. But by the same token, for those who do remain constant and unflagging in the pursuit of their ideals, our admiration and affection are that much more intense. Zhong was such a man. We miss his forthrightness and his integrity. Above all, we miss this man who, when confronted with an issue affecting the common good, would rise to the occasion and hold his self-interest in check.

Is the present social climate of China conducive to the production of such a person?

Postscriptum (September 1, 1989)

In view of the momentous events of April–July 1989, in Beijing, I am impelled to point out that the preceding essay was written in January of 1989. In particular, pages 9–11 must be read with these dates in mind. How would Zhong have reacted to these events? The readers of the preceding pages doubtlessly already know the answer.

Acknowledgment

This article could not have been written without the information supplied by Professor Wu Mei-Juan, and I wish to thank her warmly. However, I am afraid I must take sole responsibility for all the opinions and the interpretations of the facts.

Notes

1. Zhong is the last name, but in Chinese the last name comes first. In his English publications, Zhong himself followed the usual practice in English and signed himself as Jia-Qing Zhong.

2. All college students in China must declare their majors at the time of their application for admission.

3. In those days, Peking University adopted a six-year system, and the last two years were the equivalent of a master's program in the United States.

4. Unlike the graduate schools in the United States, which accept graduate students in a general subject and ask them to choose an area of specialization and a thesis advisor only later in their course of study, a prospective graduate student in China must pick his or her thesis advisor (and therewith the area of specialization) at the time of application for admission.

5. The Institute of Mathematics is part of the Academia Sinica, but, contrary to what the name might suggest, it did, and still does, accept students for graduate study.

6. At the time, still in Beijing. During the Cultural Revolution, it was moved to Hefei, Anhui Province, where it has remained to this day.

7. Numbers in brackets refer to "Publications of Zhong Jia-Qing," pp. 15–17 of this volume.

8. That both David Hilbert and his student Richard Courant happened to have led lives colorful enough to merit two biographical volumes must be regarded as the exception rather than the rule.

9. Given the absence of *any* reliable source of information or record keeping all through the Cultural Revolution, in fact all through the first 40 years of the People's Republic of China, can we ever hope to know with certainty what exactly transpired during those fateful years? For over 2000 years, China had a proud tradition of always keeping the historical record straight, but it may be no more.

10. The two institutions are within walking distance of each other.

11. See Ref. 79 of Chapter 2.

12. Of course, this was nothing compared with the years of the Cultural Revolution when Hua started to get involved in mathematical education for the masses. During that time, Hua would be sent all over China to give lectures to audiences numbering tens of thousands.

13. According to Lu, Zhong was nevertheless Hua's student in the true sense of the word, because Hua was the person who exerted the greatest mathematical impact on Zhong.

14. Purification of politics, of thoughts, of organization, and of economics; it was started in 1964.

15. Apparently, a few professors in some of the universities were granted special permission to keep the research going. Those were the "designated researchers," so to speak.

16. In China, Professor Wu Mei-Juan is always known by her maiden name, because the Chinese—quite rightly I think—do not believe that the exchange of marriage vows should automatically confer on the husband the divine right to strip his wife of her identity. However, the phonetic transcription of Chinese names being difficult to read at best, I have decided to follow the American custom for the sake of intelligibility.

17. For the benefit of those readers shell-shocked from past job-hunting experiences, let me point out that in China every university automatically keeps its own graduates on the faculty. Thus a typical Chinese mathematician would get *all* his degrees and go through *all* the teaching ranks to reach full professorship in the *same* department. Some attempts have been made in recent years to break this pattern of inbreeding, but for complicated sociological reasons there has been very little success so far.

18. Without entering into the details of what Zhong had to go through to get this approved, let me just say that such an arrangement is far from trivial in China, then as now, and in fact usually impossible. What we have come to refer to in a jocular fashion as "the two-body problem" in the United States—husband and wife living in two different cities—has always been a fact of life in China.

19. The farm is about 900 miles directly south of Beijing; Jiangxi Province borders the southern part of Anhui Province.
20. In my many visits to China, I was told about this official and his motive by a great many people on different occasions, and all the versions I heard are entirely consistent.
21. During the Cultural Revolution, the Chinese people were divided into classes according to their worth on the revolutionary scale. Workers, peasants, and soldiers were at the top of the order, and intellectuals were at the very bottom. The latter fact was a reflection of Mao's utter contempt for the intelligentsia.
22. The literal translation of this Chinese term is "oxlike ghosts and snakelike spirits."
23. Those were the people who could think straight but were unwise enough to have expressed themselves clearly at one time or another.
24. This is a technical term in the lexicon of the Chinese communists. In this context, it means that the professor should be roundly reviled and abused in a public forum until he or she abjectly confessed his or her mistakes. The slightest sign of resistance or attempt at self-defense would lead to a longer session and a higher level of abuse.
25. Just ask any American businessman who has ever attempted to get anything done in China.
26. Indeed, had he been otherwise, he would not have been invited to teach in the summer school.
27. This problem has attracted even more attention now, in 1989, both from the Chinese government and from the American media.

Publications of Zhong Jia-Qing

1. The largest automorphism group of the skew-symmetric hyperbolic spaces, *J. Peking Univ.* **8**, 226-244 (1962) (with Yin Wei-Ping) (in Chinese).
2. On the group of motions of the extension spaces of the classical domains of type I, *J. China Univ. Sci. Technol.* **1**, 229-238 (1965) (in Chinese).
3. A class of integral determinants and their applications to the representation theory of groups, *Acta Math. Sinica* **19**, 88-106 (1976).
4. Harmonic analysis on rotation groups—Abel summability, *J. China Univ. Sci. Technol.* **9**, 31-43 (1979) (in Chinese) (Chapter 4 of this volume).
5. Coxeter-Killing transformations of simple Lie algebras, *Acta Math. Sinica* **22**, 291-302 (1979) (in Chinese with English summary) (Chapter 5 of this volume).
6. Dimensions of the rings of invariant differential operators on bounded homogeneous domains, *Chinese Ann. Math.* **1**, 261-272 (1980) (in Chinese with English summary) (Chapter 6 of this volume).
7. On prime ideals of the ring of differential operators, *Chinese Ann. Math.* **1**, 359-374 (1980) (in Chinese with English summary) (Chapter 7 of this volume).
8. On the sum of class functions of Weyl groups, *Acta Math. Sinica* **23**, 836-850 (1980) (in Chinese) (Chapter 8 of this volume).
9. The trace formula of the Weyl group representation of the symmetric group, *Acta Math. Sinica* **23**, 836-850 (1980) (in Chinese) (Chapter 9 of this volume).
10. The realization of affine homogeneous cones, *Acta Math. Sinica* **24**, 116-142 (1981) (with Lu Qi-Keng) (in Chinese).

11. Some types of nonsymmetric homogeneous domains, *Acta Máth. Sinica* **24**, 587–613 (1981) (with Yin Wei-Ping) (in Chinese) (Chapter 10 of this volume).

12. The extension spaces of nonsymmetric classical domains, *Acta Math, Sinica* **24**, 614–640 (1981) (with Yin Wei-Ping) (in Chinese) (Chapter 11 of this volume).

13. Cohomology of extension spaces for classical domains, *Acta Math. Sinica* **24**, 931–944 (1981) (in Chinese) (Chapter 12 of this volume).

14. The Siegel–Godement transformations on positive Hermitian cones, *Acta Match. Sinica* **25**, 236–243 (1982) (in Chinese).

15. Pinching theorems for the first eigenvalue on positively curved manifolds, *Invent. Math.* **65**, 221–226 (1981) (with P. Li).

16. A note on Schubert calculus, *Proc. 1980 Beijing Symp. Diff. Geom. and Diff. Eqs.*, S. S. Chern et al. eds., Vol. 3, Science Publishers, Beijing, and Gordon and Breach, New York, 1981, pp. 1697–1708.

17. The degree of strong nondegeneracy of the bisectional curvature of exceptional bounded symmetric domains, *Several Complex Variables* (Proc. 1981 Hangzhou Conf.), J. J. Kohn et al. eds., Birkhauser, Boston, 1984, pp. 127–139 (Chapter 13 of this volume).

18. On the estimate of the first eigenvalue of a compact Riemannian manifold, *Zhong Guo Kexue Ser. A* **9** 812–820 (1983) (with Yang Hong-Cang) (in Chinese).

19. On the estimate of the first eigenvalue of compact Riemannian manifolds, *Sci. Sinica Ser. A* **27**, 1265–1273 (1984) (with Yang Hong-Cang) (Chapter 14 of this volume).

20. Variétés kähleriennes d'Einstein compactes de courbure bisectionnelle semi-positive, *C. R. Acad. Sci. Paris* **299** 473–475 (1984).

21. Curvature characterization of compact Hermitian symmetric spaces, *J. Diff. Geom.* **23**, 15–67 (1986) (with N. Mok) (Chapter 15 of this volume).

22. An estimate of the gap of the first two eigenvalues in Laplace operator, *Trans. Amer. Math. Soc.* **294**, 341–349 (1986) (with Yu Qi-Hang).

23. The Buseman function on the classical domains, *Acta Math. Sinica*, to appear (in Chinese).

24. Schubert calculus and Schur functions, *Zong Guo Kexue*, Ser. A, No. 8, 819–827 (1989) (in Chinese) (Chapter 16 of this volume).

25. An expansion in Schur functions and its applications in enumerative geometry, *Zhong Guo Kexue*, Ser. A, No. 10, 1018–1029 (1989) (in Chinese) (Chapter 17 of this volume).

26. On some problems in complex geometry, *Some New Advances in Modern Mathematics*, Anhui Science and Technology Press, to appear (in Chinese).

27. Curvature invariants on Kähler-Einstein manifolds, to appear (in Chinese).
28. Sturm theory and Jacobi fields, to appear (in Chinese).
29. A remark on Szëgo's calculation of the first two eigenvalues, to appear (in Chinese).
30. An estimate of the gap of the first two eigenvalues of the Schrödinger operator, to appear (in Chinese).
31. Kähler-Einstein surfaces of negative curvature, to appear (in Chinese).
32. Compactifying complete Kähler-Einstein manifolds of finite topological type and bounded curvature, *Ann. Math.* **129**, 427-470 (1989) (with N. Mok).

Part I

Surveys of Contemporary Geometry

1

Applications of Eigenvalue Techniques to Geometry

Peter Li and Andrejs Treibergs

Analysis has always been a powerful tool in the study of the geometry of manifolds. This fact has been reconfirmed by the recent developments in connection with the application of geometric analysis to various fields such as differential topology, algebraic geometry, mathematical physics, number theory, etc. Crucial to the application of partial differential equations to geometric problems are techniques developed in studying the Laplace operator and its eigenvalues. This is because in most cases when a partial differential equation occurs in geometry, the linearized equation has a principal term given by the Laplacian. The question of inverting the linearized operator naturally leads to the studying of the spectrum of the Laplacian. Among all the eigenvalues, the first nonzero eigenvalue plays the most important role. This is due to the fact that it occurs in the Poincaré inequality (see Section 1.4), which is one of the most powerful inequalities in the theory of nonlinear analysis.

Even though Jiaquing Zhong studied eigenvalues only for a short while, his contributions are valuable. One of these is the sharp lower bound for the first eigenvalue of a compact nonnegatively Ricci curved manifold, which he obtained in collaboration with Yang. We restrict our look at this

Both authors are partially supported by National Science Foundation grant DMS-8700783.

Peter Li • Department of Mathematics, University of Arizona, Tucson, Arizona 85721. *Andrejs Treibergs* • Department of Mathematics, University of Utah, Salt Lake City, Utah 84112.

enormous area of geometry by considering only cases that are related to Zhong's work. We hope that some of the exposition will simplify and unify some known results. In this chapter we will describe some recent research in the area of spectrum of the Laplacian, in particular, about the first nonzero eigenvalue. We will discuss some of the ways in which studying eigenvalues leads to geometric theorems other than in the analysis of the elliptic PDEs that occur in geometric problems, although this is one of the most important applications. Few proofs will be given, with the exception of the lower bound of Zhong and Yang. This is partly because their original article was published in a journal which is not universally accessible. We also take this opportunity to highlight and simplify the key points of the argument which were less transparent.

This small survey of results about eigenvalues of manifolds and applications to geometry cannot be anything but a sampler reflecting the whim of the authors. By now there are many books and articles surveying and covering the subject from various vantage points. The original text of Berger, Gauduchon, and Mazet [7] is still the first place for the interested reader to start. There are the works of Polya and Szegö [72], Payne [69], Protter [73], and Sperb [78], which are excellent surveys for applications of eigenvalue methods to various estimates in PDEs. Geometric applications are described in Bandle [1], Bérard [4], and Chavel [17].

1.1. Eigenvalues of the Laplacian

Let (M^n, g_{ij}) be an n-dimensional compact Riemannian manifold with or without boundary. The Laplace operator acting on functions is defined by

$$\Delta f = \frac{1}{\sqrt{g(x)}} \sum_{ij} \frac{\partial}{\partial x^i} \left\{ \sqrt{g(x)} g^{ij}(x) \frac{\partial f}{\partial x^j} \right\},$$

where

$$(g^{ij}) = (g_{ij})^{-1} \quad \text{and} \quad g = \det(g_{ij}).$$

Eigenvalues of the Laplacian are constants λ which satisfy

$$\Delta u + \lambda u = 0$$

for some not identically zero function u taken from an appropriate space. For compact manifolds under either the Dirichlet boundary conditions

$$u(x) = 0 \quad \text{for all } x \in \partial M$$

or the Neumann boundary conditions (including $\partial M = \varnothing$)

$$\frac{\partial u}{\partial \nu}(x) = 0 \qquad \text{for all } x \in \partial M,$$

it is known that each of these problems has a discrete spectrum. For the Dirichlet problem,

$$\text{Spec}_D(M) = \{\lambda_1 < \lambda_2 \leq \lambda_3 \cdots \to \infty\}.$$

For the Neumann problem, or for a compact manifold without boundary,

$$\text{Spec}_N(M) = \{\lambda_0 < \lambda_1 \leq \lambda_2 \leq \cdots \to \infty\}.$$

This notation varies among authors. We distinguish these because for the Neumann problem, $0 = \lambda_0$, so λ_1 is the first nontrivial eigenvalue in both cases. Denote the eigenfunction corresponding to λ_l by ϕ_l. The eigenspace E_λ of eigenfunctions corresponding to λ may have multiplicity $m_l = \dim(E_{\lambda_l})$, which is finite. However, the eigenfunctions are chosen to form a complete orthonormal set in $L^2(M)$ satisfying the appropriate boundary conditions.

Let us collect some facts about geometry which will be applied later. Given a Riemannian manifold M, in a neighborhood of a point we choose an orthonormal frame $\{e_1, \ldots, e_n\}$ of unit vector fields and the corresponding coframe of one-forms $\{\omega^i\}$ such that $\omega^i(e_j) = \delta^i{}_j$. The connection one-forms $\omega_i^j = \omega_j^i$ and the curvature tensor $R_{i\,ml}^j$ are defined from the structure equations

$$d\omega^i = \omega^j \wedge \omega_j^i, \tag{1.1}$$

$$d\omega_i^j = \omega_i^m \wedge \omega_m{}^j - \tfrac{1}{2} R_{i\,ml}^j \omega^m \wedge \omega^l. \tag{1.2}$$

Given a function $u \in C^2(M)$, we may compute its gradient u_i and Hessian $u_{ij} = u_{ji}$ by

$$u_i \omega^i = du,$$

$$u_{im}\omega^m = du_i - u_m \omega_i{}^m. \tag{1.3}$$

The Laplacian of u is then given by

$$\Delta u = \sum_m u_{mm}.$$

The third covariant derivative is defined by

$$u_{ijm}\omega^m = du_{ij} - u_{im}\omega_j^m - u_{mj}\omega_i^m.$$

Differentiating (1.2) again and using (1.3) gives the commutation formula (Ricci identity)

$$u_{ijm} = u_{imj} + u_p R_i^{\ p}{}_{mj}. \tag{1.4}$$

Hence, taking the trace gives

$$\Delta(u_i) = (u_i)_{mm} = u_{mim} = (\Delta u)_i + u_p R_m^{\ p}{}_{mi} = (\Delta u)_i + u_p R_{pi}, \tag{1.5}$$

where $R_{ij} = R_m^{\ i}{}_{mj}$ is the Ricci curvature.

With this notation we present a theorem of Lichnerowicz [62] (see also Ref. 10).

THEOREM 1.1. *Let M^n be a compact manifold without boundary whose Ricci curvature satisfies $R_{ij} \geq (n-1)k\delta_{ij}$ for some constant $k > 0$. Then the first nontrivial eigenvalue satisfies $\lambda_1 \geq kn$.*

PROOF. Consider the length squared $v = |Du|^2$ of the gradient of the first eigenfunction. Differentiating and using (1.4), we have

$$\tfrac{1}{2}v_i = u_m u_{mi},$$

and

$$\tfrac{1}{2}\Delta v = u_{mi}u_{mi} + u_m u_{mii} = u_{mi}u_{mi} - \lambda v + k(n-1)v. \tag{1.6}$$

From the Schwarz inequality,

$$u_{mi}u_{mi} \geq \frac{1}{n}(\Delta u)^2 = -\frac{\lambda}{n}u\,\Delta u.$$

Integrating (1.6) over M yields

$$0 \geq \int_M u_{ij}u_{ij} - \lambda v + k(n-1)v$$

$$\geq \int_M -\frac{\lambda}{n}u\,\Delta u - \lambda v + k(n-1)v$$

$$= \left\{\frac{\lambda}{n} - \lambda + k(n-1)\right\}\int_M v,$$

proving the result.   □

Obata [67] considered the case of equality.

THEOREM 1.2. *Suppose that M^n is a compact manifold whose Ricci curvature satisfies $R_{ij} \geq (n - 1)k\delta_{ij}$ for some constant $k > 0$. Suppose that $\lambda_1(M) = \lambda_1(S^n) = kn$, where S^n is the standard round sphere of constant sectional curvature k. Then M is isometric to S^n.*

There are corresponding statements for manifolds with boundary. For the Dirichlet problem the result was achieved by Reilly [74], for the Neumann problem by Escobar [37].

THEOREM 1.3. *Suppose M^n is a compact manifold with boundary such that the Ricci curvature satisfies $R_{ij} \geq (n - 1)k\delta_{ij}$ for some constant $k > 0$. Assume that either (1) λ_1 is the first Dirichlet eigenvalue and ∂M has nonnegative mean curvature in M with respect to the interior normal, or (2) λ_1 is the first Neumann eigenvalue and ∂M is convex. Then $\lambda_1 \geq nk$ and equality holds if and only if M is isometric to a great hemisphere of the standard sphere of constant sectional curvature k.*

The eigenvalues can also be characterized by variational principles. The Poincaré minimum principle for eigenvalues for the Dirichlet problem is

$$\lambda_{l+1} = \inf_{\substack{u \in H_{1,2}(M) \\ u \perp \phi_j, j=1,\cdots,l \\ u = 0 \text{ on } \partial M}} \frac{\int_M |Du|^2}{\int_M u^2}. \tag{1.7}$$

If $\Omega \subset M$, then an eigenfunction of Ω can be extended by zero to M and used as a test function. This gives the monotonicity of the first Dirichlet eigenvalue: $\lambda_1(M) \leq \lambda_1(\Omega)$. There are other related variational principles. The interested reader should refer to Refs. 33, 46, and 82.

1.2. Upper Bounds via Conformal Geometry

The variational characterization of eigenvalues is particularly useful in obtaining upper bounds. In general, this can be achieved by carefully choosing test functions and substituting them into a variational principle. If the manifold is of dimension 2, then it is well known that the Dirichlet integral is invariant under conformal transformations. The conformal geometry therefore provides a powerful method for studying eigenvalues.

By abstracting the idea of an argument of Szegö [79], Hersch [47] gave an upper bound for the first eigenvalue depending only on the area.

THEOREM 1.4. *Let* $(\mathbf{S}^2, ds^2)$ *be a sphere with an arbitrary metric. Then*

$$\frac{1}{\lambda_1} + \frac{1}{\lambda_2} + \frac{1}{\lambda_3} \geq \frac{3 \, \text{Area} \, (\mathbf{S}^2)}{8\pi},$$

where $\text{Area}(\mathbf{S}^2)$ *is the total area of the sphere with respect to the metric* ds^2. *Moreover, equality holds if and only if the metric* ds^2 *is the standard metric.*

This type of estimate was extended to an arbitrary oriented Riemann surface by Yang and Yau [87].

THEOREM 1.5. *Let* (Σ_g^2, ds^2) *be a compact oriented Riemann surface of genus g with an arbitrary metric. Then*

$$\frac{1}{\lambda_1} + \frac{1}{\lambda_2} + \frac{1}{\lambda_3} \geq \frac{3 \, \text{Area}(\Sigma)}{8\pi(g+1)}.$$

Observing that $\lambda_1 \leq \lambda_2 \leq \lambda_3$, we have, of course,

$$\lambda_1(\Sigma_g) \leq \frac{8\pi(g+1)}{\text{Area} \, (\Sigma)}. \tag{1.8}$$

It is interesting to point out that for higher-dimensional n-spheres it was shown by Berger [6] that the standard metric does not yield the maximum value for the sum of the reciprocals of the first $n + 1$ eigenvalues. However, the first eigenvalue cannot be bounded by the volume alone. Urakawa [81(a)] gave a family of metrics with unit volume on $\mathbf{S}^3$ whose first eigenvalue tends to infinity. Li and Yau [60] proved a corresponding version to Theorem 1.4 for the real projective space $\mathbf{RP}^2$.

THEOREM 1.6. *For any metric* ds^2 *on* $\mathbf{RP}^2$,

$$\lambda_1 \leq \frac{12\pi}{\text{Area} \, (\mathbf{RP}^2)}.$$

Equality holds if and only if ds^2 is the standard metric of $\mathbf{RP}^2$.

In the process of proving this theorem, the notion of conformal volume was formulated by Li and Yau [60] (also see Ref. 84). This was used to study a conjecture of Willmore [83], which asserts that among all embedded surfaces in $\mathbf{R}^3$ with genus 1 the Clifford torus has the smallest integral squared mean curvature, $\int_M H^2$.

Let M^n be a compact manifold which admits a branched conformal immersion $\phi: M \to S^p$. Let ds^2 be the metric on the sphere and ds_0^2 the standard metric on S^p. Let $\mathscr{C}$ denote the group of conformal diffeomorphisms of S^p. The *p-conformal volume of ϕ* can be defined to be

$$V_c(p, \phi) = \sup_{g \in \mathscr{C}} \int_M dV_g,$$

where dV_g is the volume element $\phi^* g^* \, ds_0^2$. The *p-conformal volume of M* is defined to be

$$V_c(p, M) = \inf_{\phi} V_c(p, \phi),$$

where ϕ runs through branched conformal immersions of M into S^p. The conformal volume of a surface is related to its first eigenvalue by the following estimate, which was proved in Ref. 60.

THEOREM 1.7. *Let M be a compact surface. Then*

$$\lambda_1(M) \, \text{Area}\,(M) \le 2V_c(p, M) \tag{1.9}$$

for all p such that $V_c(p, M)$ is defined. Equality implies M is a minimal surface of S^p whose immersion is given by $\{\psi_1, \ldots, \psi_{p+1}\} \in S^p \subset R^{p+1}$ consisting of eigenfunctions for the eigenvalue $\lambda_1(M)$.

In some interesting geometric cases it is also possible to estimate the conformal volume from above. In fact, the conformal volume of the real projective space $\mathbf{RP}^2$ is 6π, which yields an equality in Theorem 1.7. On the other hand, the relationship of the conformal volume to the Willmore problem is given by the following lemma.

LEMMA 1.8. *Let M^2 be a compact surface without boundary in R^p. Then*

$$V_c(p, M) \le \int_M |H|^2.$$

Furthermore, equality implies M is the image of a minimal surface in $\mathbf{S}^p$ under some stereographic projection.

A consequence of Theorem 1.7 and Lemma 1.8 is Theorem 2.6.

THEOREM 1.9. *Let M^2 be a compact surface in $\mathbf{R}^p$ homeomorphic to $\mathbf{RP}^2$. Then*

$$6\pi \le \int_M |H|^2.$$

Equality holds if and only if M is the stereographic projection of the Veronese surface in $\mathbf{S}^4$.

By a similar type of argument, Willmore's question is also partially answered in Ref. 60 for the 2-torus.

THEOREM 1.10. *Suppose M is an oriented surface of genus 1 in $\mathbf{R}^p$. Suppose that M is conformally equivalent to a flat torus with lattice generated by $\{(1,0),(x,y)\}$, where $0 \le x \le 1/2$, $y > 0$, and $(1 - x^2)^{1/2} \le y \le 1$. Then*

$$2\pi^2 \le \int_M |H|^2.$$

Equality implies M is conformally equivalent to the square torus and is the image of a stereographic projection of a minimal torus in $\mathbf{S}^3$.

This result was later improved by Montiel and Ros (65).

THEOREM 1.11. *Suppose M is an oriented surface of genus 1 in $\mathbf{R}^p$. Suppose that M is conformally equivalent to a flat torus with lattice generated by $\{(1,0),(x,y)\}$, where $0 \le x \le 1/2, 0 < y$, and $(x - 1/2)^2 + (y - 1)^2 \le 1/4$. Then*

$$2\pi^2 \le \int_M |H|^2.$$

Finally, the idea of estimating the first eigenvalue from above using the conformal geometry has been carried through to algebraic submanifolds of arbitrary dimension of complex projective spaces by Bourguignon, Li, and Yau [14]. In that case, the group of holomorphic transformations is utilized.

1.3. Upper Bounds Using Cheng's Comparison Theorem

In this section we consider another method for obtaining upper bounds for the eigenvalues of Riemannian manifold. Cheeger [18] has shown that for compact manifolds M^n with nonnegative sectional curvature there holds

$$\lambda_1(M) \le \frac{c}{d(M)^2}$$

where c depends only on the dimension n and $d(M)$ denotes the diameter of M. Later, Cheng [23] succeeded in providing more general results. The first theorem is the comparison theorem for the first Dirichlet eigenvalues of balls.

THEOREM 1.12. *Let M^n be a Riemannian manifold whose Ricci curvature is bounded below by $R_{ij} \ge k(n-1)\delta_{ij}$ for some constant $k \in \mathbf{R}$. Denote the geodesic ball by $B_r(x) = \{y \in M: d(x, y) < r\}$. Then the first Dirichlet eigenvalue satisfies*

$$\lambda_1(B_r(x)) \le \lambda_1(\bar{B}_r), \qquad (1.10)$$

where $\bar{B}_r$ is a geodesic ball in the model spaceform $\bar{M}$ of constant sectional curvature k. Moreover, equality holds if and only if $B_r(x)$ is isometric to $\bar{B}_r$.

The method of proof is to transplant the first eigenfunction of the spaceform to M by the exponential map and substitute it into the Rayleigh quotient (1.7). Using the Poincaré minimum principle, Cheng [23] applied this theorem to give upper bounds on higher eigenvalues as well.

THEOREM 1.13. *Let M^n be a compact manifold without boundary. Suppose the Ricci curvature is bounded below by $R_{ij} \ge (n-1)k\delta_{ij}$, where $k \in \mathbf{R}$ is any constant. Then*

$$\lambda_l(M) \le \bar{\lambda}_1(\bar{B}_{d/2l}),$$

where d is the diameter of M and $\bar{\lambda}_1$ is the first Dirichlet eigenvalue of a ball in the model space of constant sectional curvature k. In particular, by estimating the first eigenvalue of a geodesic ball in a spaceform, this implies that for $n \ge 2$,

$$\lambda_l \le \begin{cases} \dfrac{2l^2 n(n+4)}{d^2}, & \text{if } k \ge 0, \\[3mm] \dfrac{n^2|k|}{4} + \dfrac{66l^2 2^{2n}}{d^2}, & \text{if } k < 0. \end{cases}$$

Gage [41] sharpened and generalized this upper bound to include rotationally symmetric models.

We end this section with a purely geometric application of Cheng's, generalizing the Toponogov sphere theorem.

THEOREM 1.14. *Suppose M^n is a manifold with Ricci curvature bounded below by $R_{ij} \geq (n-1)k\delta_{ij}$ for some constant $k > 0$, and the diameter is bounded below by $d \geq \pi/\sqrt{k}$. Then M is isometric to the standard sphere of constant curvature k.*

1.4. Lower Bounds by Gradient Estimates

Other than the fact that the first nonzero eigenvalue carries substantial geometric information, it also plays an important role in the theory of partial differential equations. This is clearly demonstrated by the variational characterization of λ_1, which is generally referred to as the Poincaré inequality. Precisely, it asserts that

$$\lambda_1 \int_M u^2 \leq \int_M |Du|^2$$

for all functions $u \in H_{1,2}$ satisfying $\int_M u = 0$. In order to use this inequality, it is clear that a lower bound for λ_1 is required.

In this section we describe how to get such bounds using a priori estimates. The idea of using the maximum principle to estimate solutions of partial differential equations has been successful for many years. The reader should consult some of the many surveys covering these methods, for example, Payne [69], Protter [73], and Sperb [78].

In view of Cheng's upper bound of λ_1, Yau [88] sought a lower bound in terms of similar geometric quantities. He succeeded in estimating λ_1 from below in terms of the lower bound of the Ricci curvature, the upper bound of the diameter, and the lower bound of the volume of the manifold. In the same article, he conjectured that a lower bound should depend only on the lower bound of the Ricci curvature and the upper bound of the diameter alone. This conjecture was first verified by Li [54] when the Ricci curvature is nonnegative. Shortly thereafter, Li and Yau [59] successfully verified the conjecture in the general case by using the maximum principle technique.

THEOREM 1.15. *Let M^n be a compact manifold. Suppose d denotes the diameter of M and the Ricci curvature of M has a lower bound $R_{ij} \geq k(n-1)\delta_{ij}$ for some constant $k \leq 0$. Then the first nonzero eigenvalue λ_1 of M must satisfy*

$$\lambda_1 \geq \frac{1}{2(n-1)d^2} \exp\{-1 - \sqrt{1 + 4(n-1)^2 d^2 |k|}\}. \tag{1.11}$$

Independently, Gromov [44] also proved Yau's conjecture by estimating an isoperimetric inequality. The main difference in the two approaches is that the maximum principle is more versatile in the sense that it applies to manifolds with boundaries and other types of differential equations. In fact, estimates of λ_1 for the Dirichlet and the Neumann problem were also derived in Ref. 59. However, the isoperimetric inequality approach, in some instances, provides sharper estimates when the manifold is positively curved.

In case the manifold has nonnegative Ricci curvature, the estimate of Li and Yau [59, 56] yields a lower bound which is only half the first eigenvalue of all the observed examples. Hence, they were led to conjecture that the correct lower bound should be double the one they obtained. This was finally confirmed by Zhong and Yang [90]. Their key observation to sharpen the previous estimates was to appreciate the importance of the invariant $a(u)$, the eigenfunction midrange first exploited by Li and Zhong [61].

Let λ_1 be the least nontrivial eigenvalue of a compact manifold, and let ϕ be the corresponding eigenfunction. By multiplying by a constant, one can arrange that

$$a - 1 = \inf_M \phi, \qquad a + 1 = \sup_M \phi, \tag{1.12}$$

where $0 \le a(\phi) < 1$ is the midrange of ϕ.

LEMMA 1.16. *Suppose M^n is a compact manifold boundary whose Ricci curvature is nonnegative. Then the first nontrivial eigenvalue satisfies*

$$\lambda_1 \ge \frac{\pi^2}{(1 + a)d^2}, \tag{1.13}$$

where d is the diameter of M and $a = a(\phi)$ is the midrange of the normalized first eigenfunction.

PROOF. For $\lambda = \lambda_1$ and $u = \phi - a$ the equation becomes

$$\Delta u = -\lambda(u + a).$$

Let $P = |Du|^2 + cu^2$, where $c = \lambda(1 + a) + \varepsilon$ for some $\varepsilon > 0$. Let $x_0 \in M$ be the point where P is maximum. If $|Du(x_0)| \ne 0$, we may rotate the frame so that $u_1(x_0) = |Du(x_0)|$. Differentiating gives

$$\tfrac{1}{2}P_i = u_m u_{mi} + cuu_i,$$

so at x_0,

$$0 = u_1(u_{11} + cu)$$

and

$$u_{ij}u_{ij} \geq u_{11}^2 = c^2u^2. \tag{1.14}$$

Differentiating, using the commutation formula (1.14) the definition of P, and evaluating at x_0 give

$$\begin{aligned}
0 \geq \tfrac{1}{2}\Delta P &= u_{mi}u_{mi} + u_m u_{mii} + cu_1^2 + cu\,\Delta u \\
&\geq c^2u^2 + u_m(\Delta u)_m + R_{mp}u_m u_p + cu_1^2 - c\lambda u(u + a) \\
&\geq c^2u^2 - \lambda u_1^2 + cu_1^2 - c\lambda u(u + a) \\
&= (c - \lambda)(u_1^2 + cu^2) - ac\lambda u \\
&\geq (a\lambda + \varepsilon)P(x_0) - ac\lambda. \tag{1.15}
\end{aligned}$$

Therefore, by letting $\varepsilon \to 0$, for all $x \in M$, we have

$$|Du(x)|^2 \leq \lambda(1 + a)(1 - u(x)^2). \tag{1.16}$$

Also (1.16) is trivially satisfied if $Du(x_0) = 0$.

Let γ be the shortest geodesic from the minimizing point of u to the maximizing point. The length of γ is at most d. Integrating the gradient estimate (1.16) along this segment with respect to arclength gives

$$d\sqrt{\lambda(1 + a)} \geq \sqrt{\lambda(1 + a)} \int_\gamma ds \geq \int_\gamma \frac{|Du|\,ds}{\sqrt{1 - u^2}} \geq \int_{-1}^{1} \frac{du}{\sqrt{1 - u^2}} = \pi.$$

In order to sharpen this estimate, Zhong and Yang had to squeeze the maximum principle and make a judicious choice of test function. $\square$

LEMMA 1.17. *Suppose M is a compact manifold without boundary whose Ricci curvature is nonnegative. Assume that a nontrivial eigenfunction ϕ corresponding to the eigenvalue λ is normalized so that, for $0 \leq a < 1$, $a + 1 = \sup \phi$ and $a - 1 = \inf_M \phi$. With the substitution $u = \phi - a$, the gradient satisfies the estimate*

$$|Du|^2 \leq \lambda(1 - u^2) + 2a\lambda z(u) \tag{1.17}$$

where

$$z(u) = \frac{2}{\pi}\{\arcsin(u) + u\sqrt{1 - u^2}\} - u. \tag{1.18}$$

PROOF. We will first prove the estimate (1.17) for $u = \theta(\phi - a)$, where $\theta < 1$ is sufficiently close to 1. The lemma will follow by letting $\theta \to 1$. By the definition of u, we have

$$\Delta u = -\lambda(u + \theta a)$$

with $|u| \leq \theta$. By (1.16) we may assume $0 < a < 1$. Using calculus we verify

$$2|z(u)| \leq 1 - u^2. \tag{1.19}$$

Consider the function

$$Q = |Du|^2 - c(1 - u^2) - 2a\lambda z(u), \tag{1.20}$$

where by (1.16) and (1.19) we can choose c large enough so that $\sup_M Q = 0$. The lemma follows if $c \leq \lambda$. Hence we may assume that $c > \lambda$.

Let the maximizing point of Q be x_0. We claim that $|Du(x_0)| > 0$, since otherwise $Du(x_0) = 0$ and, by (1.19),

$$0 = Q(x_0) = -c(1 - u^2)(x_0) - 2a\lambda z(x_0) \leq -(c - a\lambda)(1 - \theta^2),$$

which is a contradiction. Differentiating, we have

$$\tfrac{1}{2}Q_i = u_m u_{mi} + cuu_i - a\lambda \dot{z}u_i. \tag{1.21}$$

At x_0, rotating the frame so that $u_1(x_0) = |Du(x_0)|$ and using $Q_i = 0$, we have

$$u_{mi}u_{mi} \geq u_{11}^2 = (cu - a\lambda\dot{z})^2. \tag{1.22}$$

Differentiating again, using the commutation formula, $Q(x_0) = 0$, (1.19), (1.21), and (1.22), we get

$$0 \geq \tfrac{1}{2}\Delta Q(x_0)$$

$$= u_{mi}u_{mi} + u_m u_{mii} + cu_1^2 + cu\,\Delta u - a\lambda\ddot{z}u_1^2 - a\lambda\dot{z}\,\Delta u$$

$$= u_{mi}u_{mi} + u_m(\Delta u)_m + R_{mi}u_m u_i + (c - a\lambda\ddot{z})u_1^2 + (cu - a\lambda\dot{z})\,\Delta u$$

$$\geq (cu - a\lambda\dot{z})^2 + (c - \lambda - a\lambda\ddot{z})[c(1 - u^2) + 2a\lambda z]$$

$$\quad - (c\lambda u - a\lambda^2\dot{z})(u + a\theta)$$

$$= -ac\lambda\{(1 - u^2)\ddot{z} + u\dot{z} + u\} + a^2\lambda^2\{-2z\ddot{z} + \dot{z}^2 + \theta\dot{z}\}$$

$$\quad + a\lambda(c - \lambda)\{-u\dot{z} + 2z + 1\} + ac\lambda(1 - \theta)u + (c - \lambda)(c - a\lambda). \tag{1.23}$$

The function z was chosen to satisfy

$$(1 - u^2)\ddot{z} + u\dot{z} + u = 0.$$

We also check that

$$\dot{z} \leq \frac{4}{\pi^2}(1 - u^2),$$

$$-2z\ddot{z} + \dot{z}^2 + \dot{z} \geq \frac{8}{3\pi}\left(1 - \frac{2}{\pi}\right)(1 - u^2),$$

$$-u\dot{z} + 2z + 1 \geq 0.$$

We conclude that

$$0 \geq a^2\lambda^2\left\{\frac{8}{3\pi}\left(1 - \frac{2}{\pi}\right) - \frac{4}{\pi^2}(1 - \theta)\right\}(1 - u^2)$$

$$+ ac\lambda(1 - \theta)u + (c - \lambda)(c - a\lambda).$$

The first term is nonnegative if $3\theta \geq 7 - 2\pi$, and the rest is positive if $ac\lambda(1 - \theta) < (c - \lambda)(c - a\lambda)$. For such θ we have a contradiction; therefore $c \leq \lambda$. Taking $\theta \to 1$ concludes the proof. $\square$

THEOREM 1.18. *Suppose M is a compact manifold without boundary whose Ricci curvature is nonnegative. Let $a \geq 0$ be the midrange of a normalized first eigenfunction $a + 1 = \sup \phi$, $a - 1 = \inf \phi$, and let d be the diameter. Then the first nontrivial eigenvalue satisfies*

$$d^2\lambda_1 \geq \pi^2 + \frac{6}{\pi}\left(\frac{\pi}{2} - 1\right)^4 a^2 \geq \pi^2(1 + 0.02a^2).$$

PROOF. Arguing with $u = \phi - a$ as before, let γ be the shortest geodesic from the minimizing point of u to the maximizing point with length at most d. Integrating the gradient estimate (1.17) along this segment with respect to arclength and using oddness, we get

$$d\lambda^{1/2} \geq \lambda^{1/2} \int_{\gamma} ds$$

$$\geq \int_{\gamma} \frac{|Du|\ ds}{\sqrt{1 - u^2 + 2az(u)}}$$

$$\geq \int_0^1 \left\{ \frac{1}{\sqrt{1 - u^2 + 2az}} + \frac{1}{\sqrt{1 - u^2 - 2az}} \right\} du$$

$$\geq \int_0^1 \frac{1}{\sqrt{1 - u^2}} \left\{ 2 + \frac{3a^2 z^2}{1 - u^2} \right\} du$$

$$\geq \pi + 3a^2 \left(\int_0^1 \frac{z}{\sqrt{1 - u^2}} \right)^2$$

$$= \pi + \frac{3a^2}{\pi^2} \left(\frac{\pi}{2} - 1 \right)^4. \tag{1.24}$$

The technique applies to manifolds with boundary. Let M^n be a compact manifold with smooth boundary whose Ricci curvature is nonnegative. Suppose that the second fundamental form of ∂M is nonnegative. Then the first nontrivial eigenvalue of the Laplacian with Neumann boundary conditions also satisfies the inequality (1.17). The proof runs the same as Lemma 1.17 except that the possibility of the maximum of the test function Q at the boundary must be handled. In fact, the boundary convexity assumption implies that the maximum of Q cannot occur on the boundary. This method has been used for various estimates of the eigenvalues of the Dirichlet and Neumann boundary problems [59]. One of the most general estimates for the Neumann boundary problem without assuming convexity of the boundary was recently proved by Meyer [64] and Chen [21].

Recently, Yang [85, 86] proved a lower bound for λ_1 which unifies the estimates of Li and Yau [59] and of Zhong and Yang [90] for compact manifolds with or without boundary. In particular, for a compact manifold without boundary, he showed that under the assumptions of Theorem 1.15,

$$\lambda_1 \geq \frac{\pi^2}{d^2} \exp\left\{ -C_n \sqrt{(n-1)|k|d^2} \right\},$$

where $C_n = \max\{\sqrt{n-1}, \sqrt{2}\}$.

We give an application to physics accomplished by using a priori estimation of eigenfunctions. It was first reported by Singer *et al.* [77] for convex domains of Euclidean space. The problem was to estimate the mass gap, the energy jump from the lowest to the second lowest eigenstate of a physical system satisfying the Schrödinger equation. The estimate was sharpened by Yu and Zhong [89].

THEOREM 1.19. *Let* $\Omega \subset \mathbf{R}^n$ *be a smooth strictly convex domain. Let* $V(x) \in C^2(\Omega)$ *be a nonnegative, convex function. Let* $0 < \nu_1 < \nu_2 \leq \cdots$ *be the eigenvalues of the Schrödinger operator for the Dirichlet problem*

$$\Delta u - Vu = -\nu u \qquad \text{on } \Omega,$$

$$u = 0 \qquad \text{on } \partial\Omega.$$

Then

$$\nu_2 - \nu_1 \geq \frac{\pi^2}{d^2},$$

where d is the diameter of Ω.

The generalization to compact manifolds without boundary was made by Chen [20]. In that case the mass gap is estimated in terms of diameter, curvature $\|R^j_{i\,kl}\|_{C^1(M)}$, and potential $\|V\|_{C^2(M)}$. The key to using eigenvalue estimation for this result is an equation involving the gap. Let u_1 and u_2 be eigenfunctions corresponding to the first two eigenvalues. Then $q = u_2/u_1$ satisfies

$$\Delta q = \nu_1 - \nu_2 - \langle Dq, D \log u_1 \rangle.$$

The proof proceeds by estimating $|Dq|^2$.

Obata's theorem leads one to wonder if there is a stable version: Suppose the manifold has some curvature condition and the first eigenvalue is just *near* the first eigenvalue of the sphere. Does it follow that the manifold itself is near the sphere? The early work made many geometric hypotheses for the problem. Li and Zhong [61] realized that one had to prove a sharp lower bound for the diameter of a manifold and then apply the diameter pinching theorem of Grove and Shiohama [45]: If the sectional curvature of a manifold exceeds 1 and the diameter strictly exceeds $\pi/2$, then the manifold is homeomorphic to the sphere. The eigenvalue pinching theorem was proved by Li and Zhong for $n \leq 3$, by Li and Treibergs [58] for $n = 4$, and by Croke [35] for all n.

THEOREM 1.20. *Suppose M^n is a compact manifold whose sectional curvature $K^M \geq 1$. Then there is a constant $c(n) > 0$ such that if the first nonzero eigenvalue of M satisfies $\lambda_1 \leq n + c(n)$, then M is homeomorphic to the sphere.*

Li and Zhong's idea was to show under the weaker assumption $R_{ij} \geq (n-1)\delta_{ij}$ that the gradient of the first eigenfunction has a much better inequality than it would for nonnegative Ricci curvature, and, therefore, the corresponding inequality like (1.24) could be read as an improved lower bound for the diameter. Croke, however, used an isoperimetric inequality to obtain his diameter estimate.

1.5. Isoperimetric Inequalities

Analytic inequalities are equivalent to isoperimetric inequalities on a manifold. The Poincaré inequality (the lower bound of the spectrum) is no exception. Through isoperimetric inequalities, which are essentially entirely geometric, one can estimate analytic quantities. In this section we illustrate the relationship and give an application of Brooks. There are many excellent surveys and examples of this manner of interaction between geometry and analysis [4, 76].

We begin by defining isoperimetric constants. Let $p > 0$. Then for Dirichlet boundary data [43], the p-isoperimetric constant is defined to be

$$I(p)(M) = \inf_{\substack{\Omega \subset M \\ \partial\Omega \cap \partial M = \varnothing \\ \partial\Omega = \text{hypersurface}}} \frac{\text{vol}_{n-1}(\partial\Omega)}{\text{vol}_n(\Omega)^{1/p}}. \tag{1.25}$$

For Neumann data or an empty boundary, the definition is

$$I(p)(M) = \inf_{\substack{S \subset M \\ S = \text{hypersurface} \\ M = \Omega_1 \cup S \cup \Omega_2 \\ S = \partial\Omega_1 = \partial\Omega_2}} \frac{\text{vol}_{n-1}(S)}{\min\{\text{vol}_n(\Omega_1)^{1/p}, \text{vol}_n(\Omega_2)^{1/p}\}}, \tag{1.26}$$

where Ω_i are nodal domains. $I(1)$ is known as the Cheeger constant [19] because he found the basic inequality which we describe. It is a manifold version of Faber and Krahn's [38, 51, 1] result that the ball is the unique Euclidean domain with the smallest Dirichlet first eigenvalue among all domains with fixed volume.

THEOREM 1.21. *Let M^n be a compact manifold with or without boundary. Then the first nontrivial eigenvalue satisfying either the Neumann or Dirichlet boundary condition satisfies*

$$\lambda \geq \tfrac{1}{4} I(1)^2(M).$$

Cheeger showed how to estimate the Rayleigh quotient by using $I(1)$. In fact, Yau [88] showed that a lower bound on $I(1)$ is equivalent to the L^1-Poincaré inequality. By the Schwarz inequality, this implies an L^2-Poincaré inequality or a λ_1 bound. He was able to show the following L^1-Poincaré inequality, which gave the best estimate at the time.

THEOREM 1.22. *Let M be a compact manifold without boundary whose Ricci curvature is bounded below by $(n-1)k$. If $\partial M = \varnothing$, there is a constant $c > 0$ depending on diameter, volume, and k so that for all Lipschitz f,*

$$c \inf_{\beta \in \mathbf{R}} \int_M |f - \beta| \leq \int_M |Df|. \tag{1.27}$$

There is a corresponding statement for manifolds with boundary, but the constant c also depends on the volume of an included ball.

Yau went on to show that the isoperimetric constant $I(n/(n-1))$ is equivalent to the Sobolev inequality, and similarly for all $1 < p < n/(n-1)$ (see also Bombieri [11].) However, $I(n/(n-1))$ is just the classical isoperimetric constant. A Hölder inequality applied to the Sobolev inequality then implies bounds on eigenvalues [55].

THEOREM 1.23. *Let M^n be a compact manifold with or without boundary. Let $I(p)$ be the isoperimetric constant as in (1.25) or (1.26) with $1 \leq p \leq n/(n-1)$. Then for all Lipschitz functions u satisfying the corresponding Neumann or Dirichlet boundary conditions there holds*

$$I(p) \inf_{\beta \in \mathbf{R}} \left(\int_M |u - \beta|^p \right)^{1/p} \leq \int_M |Du|. \tag{1.28}$$

PROOF. Let β be chosen so that $F = u - \beta$ satisfies $\mathbf{vol}_n(M_0) \leq V/2$ and $\mathbf{vol}_n(N_0) \leq V/2$, where $V = \mathbf{vol}_n(M)$, $N_0 = \{x : f(x) < 0\}$, and $M_0 = \{x : f(x) > 0\}$. Let $M_t = \{x \in M_0 : f(x) > t\}$. Arguing M_0 first, by the coarea formula and a calculus inequality for $\text{vol}_n(M_t)$ nonincreasing [11], we get

$$\int_{M_0} |Df| = \int_0^\infty \text{vol}_{n-1}(\partial M_t)\, dt$$

$$\geq I(p) \int_0^\infty \text{vol}_n (M_t)^{1/p}\, dt$$

$$\geq I(p) \left(p \int_0^\infty t^{p-1} \text{vol}_n (M_t)\, dt \right)^{1/p}$$

$$= I(p) \left(\int_{M_0} f^p\, dx \right)^{1/p}.$$

N_0 is estimated similarly, and (1.28) follows from Minkowski's inequality.
$\square$

The L^1 Sobolev inequality implies the L^p. One can also deduce the Poincaré inequality from the Sobolev inequality [55]. Finally, Croke [34] has given general estimates on the isoperimetric constant $I(p)$, thereby making another route to bounding eigenvalues from below.

THEOREM 1.24. *Let M^n be a compact manifold with or without boundary. If $\partial M = \varnothing$ or for Dirichlet data, or if M is geodesically convex with Neumann data, then there is a lower bound for the isoperimetric constant $I(n/(n-1))$ in terms of the volume, diameter, and lower bound of Ricci curvature of M.*

This has been recently generalized by Gallot [42, 43]. For example, for a compact manifold without boundary, there is a lower bound for the isoperimetric constant $I(p)$ in terms of the volume, diameter, n, p, and an *integral* lower bound of the negative part of the Ricci curvature $\|k_-\|_{L^{1/(2-2p)}(M)}$ provided that this is small enough, where $R_{ij}(x) \geq k(x)\delta_{ij}$ and $k_- = \max\{-k(x), 0\}$. There has been much significant progress in isoperimetric estimates lately, and we encourage the reader to seek further references [4].

There is another type of isoperimetric inequality which is more suitable for manifolds with positive curvature. This was first considered by P. Levy for subsets of the sphere. Suppose M^n is a compact manifold without boundary. Define

$$\check{I}_\alpha = \inf_{\substack{S \subset M \\ S = \text{hypersurface} \\ M - S = \Omega_1 \cup \Omega_2 \\ \text{vol}_n \Omega_1 = \alpha \, \text{vol}_n M}} \frac{\text{vol}_{n-1}(S)}{\text{vol}_n(\Omega_1)}. \tag{1.29}$$

where the infimum is taken over all hypersurfaces dividing M into two disjoint domains Ω_i. Because $I(1) = \inf_\alpha \breve{I}_\alpha$, estimates on either isoperimetric constant are equivalent. This isoperimetric constant has been estimated by Gromov [44] for his lower bound of λ_1 for compact manifolds in terms of a lower bound on the Ricci curvature and an upper bound on the diameter; see also Ref. 16.

THEOREM 1.25. *Let M^n be a compact manifold without boundary such that the Ricci curvature is bounded below $R_{ij} \geq (n-1)k\delta_{ij}$. Then there is a positive lower bound of $\breve{I}_\alpha$ in terms of α, k, and d, the diameter of M. If $k \leq 0$ and $\alpha \leq 1/2$, this takes the form*

$$\breve{I}_\alpha \geq \left\{ \int_0^d \cosh^n \left(\sqrt{-k}\tau \right) d\tau \right\}^{-1}. \tag{1.30}$$

If $k > 0$, letting $\bar{M}^n$ be the sphere of sectional curvature k and $\bar{\Omega}_1 \subset \bar{M}$ be a spherical cap so that $\mathrm{vol}_n (\bar{\Omega}_1) = \alpha \, \mathrm{vol}_n (\bar{M})$, we have

$$\breve{I}_\alpha \geq \frac{\mathrm{vol}_{n-1} (\partial \bar{\Omega}_1)}{\mathrm{vol}_n (\bar{M})}. \tag{1.31}$$

Equality holds in (1.31) if and only if M is isometric to $\bar{M}$ and Ω_1 is isometric to $\bar{\Omega}_1$.

Note that (1.30) is an estimate on Cheeger's constant and thus another lower bound for $\lambda_1(M)$. Bérard and Meyer [5] used this to prove Lichnerowicz's Theorem 1.1. There have been recent extensions of this, assuming integral bounds on Ricci as well [43]. Croke gave a version to prove the pinching theorem [35].

The proof of Theorem 1.25 uses Fermi coordinates around an area-minimizing separating hypersurface, which has constant mean curvature, and a Jacobian comparison result. Another application of this method is Schoen's estimate [75], not involving diameter but making stronger curvature assumptions.

THEOREM 1.26. *Let M^n be a compact manifold of dimension $n \geq 3$ whose sectional curvature is negatively pinched, $-1 \leq K_M \leq k < 0$. Then the first nontrivial eigenvalue $\lambda_1(M)$ has a lower bound in terms of k, n, and $\mathrm{vol}_n M$.*

This result fails for Riemann surfaces. There are examples [76, 15] of fixed genus and area but with arbitrarily small eigenvalue.

Cheeger's isoperimetric constant can also give *upper* bounds on the first eigenvalue by a theorem of Buser [16].

THEOREM 1.27. *Let M^n be a compact manifold without boundary whose Ricci curvature $R_{ij} \geq (n-1)k\delta_{ij}$ is bounded below by a negative constant*

$k \leq 0$. *Then*

$$\lambda_1 \leq c(I(1)^2 - kI(1)),$$

where c is a constant depending only on n and $I(1) = I(1)(M)$ is Cheeger's isoperimetric constant defined in (1.26).

A notion of the first Dirichlet eigenvalue may be extended to complete manifolds M^n as follows. Let

$$\sigma_1(M) = \inf_{f \in H_0^1(M)} \frac{\int_M |Df|^2}{\int_M f^2}. \qquad (1.32)$$

Bounds were obtained by McKean [63], Cheng and Yau [27], and Cheng [22].

THEOREM 1.28. *Let M^n be a complete manifold.*

1. *If M is simply connected and the sectional curvature satisfies $K_M \leq k < 0$, then $\sigma_1 \geq (1/4)(n-1)^2|k|$.*
2. *If M has polynomial volume growth, then $\sigma_1 = 0$.*
3. *If $R_{ij} \geq (n-1)k\delta_{ij}$, $k \leq 0$, then $\sigma_1 \leq (1/4)(n-1)^2|k|$.*

The vanishing of σ_1 determines the behavior of the fundamental group. We give an application of Brooks [12].

THEOREM 1.29. *Let M^n be a compact or noncompact manifold. Then $\sigma_1(\tilde{M}) = 0$ if and only if $\pi_1(M)$ is amenable.*

A countably generated discrete group G is called amenable if it admits a finitely additive left invariant mean μ on G; that is, there is a bounded linear functional $\mu \in L^\infty(G) \to \mathbf{R}$ so that $\inf_{g \in G} f(g) \leq \mu(g) \leq \sup_{g \in G} f(g)$ and $\mu(g \circ f) = \mu(f)$, where $g \circ f(x) := f(g^{-1}x)$. Groups with subexponential growth are amenable, such as the finite or abelian, and groups containing the free subgroup on two generators are nonamenable.

COROLLARY 1.30. *Let N be a simply connected Riemannian manifold and Γ a group of isometries of N such that N/Γ is a compact manifold. Then Γ is amenable if and only if any other group of isometries satisfying the same condition is amenable.*

The connection between amenability and eigenvalues occurs because of the following proposition.

PROPOSITION 1.31. *Let M^n be a compact manifold and $\tilde{M}$ the universal cover. Let F be a fundamental domain in $\tilde{M}$ obtained by taking the union of one n-simplex in the lift of each n-simplex in a triangulation of M. Then $\pi_1(M)$*

is amenable if and only if for every such F and $\varepsilon > 0$ there is a finite subset $E \subset \pi_1(M)$ such that

$$H = \bigcup_{g \in E} gF$$

satisfies $\mathbf{vol}_{n-1}(\partial H) < \varepsilon \, \mathbf{vol}_n(H)$.

An upper bound of σ_1 is obtained by approximating the characteristic function of H. By the Cheeger lower bound, Theorem 1.21, one can find a sequence of domains so that the ratio of area to volume is arbitrarily small and then approximate by a union of fundamental domains.

It follows from Theorem 1.28(1) and Theorem 1.29 that the fundamental group of a compact manifold with negative sectional curvature is not amenable. Brooks [13] also looked at towers of coverings that have non-vanishing eigenvalue.

1.6. A Lower Bound of Choi and Wang and Its Application to Minimal Surfaces

Let us now consider the case when the manifold M^n is a minimal submanifold of the standard m-sphere S^m. It is known that minimality implies that the embedding coordinate functions of M into $S^m \subset \mathbf{R}^{m+1}$ are eigenfunctions corresponding to the eigenvalue $n = \dim M$. Clearly, in this case estimates of eigenvalues are more directly related to the structure of M. For example, an eigenvalue comparison theorem of Cheng, Li, and Yau [25] yields a gap theorem for the volume of minimal submanifolds in spheres.

THEOREM 1.32. *Suppose $M^n \subset S^{n+l}$ is a compact minimally immersed submanifold of the standard unit sphere of codimension l. Suppose that the immersion is full; that is, M is not contained in any proper linear subspace of $\mathbf{R}^{n+l+1}$. Then there exists a constant $c_n > 0$ depending only on dimension so that*

$$\mathbf{vol}_n(M) \geq (1 + c_n l) \, \mathbf{vol}_n(S^n).$$

For embedded minimal surfaces in S^3, Yau conjectured that the coordinate functions are the first nonconstant eigenfunctions. This is the same as saying that $\lambda_1 \geq 2$. The result of Choi and Wang [30] gives an estimate of λ_1 which is independent of the geometry of M. This can be viewed as supporting evidence for Yau's conjecture. The method was inspired by a result of Reilly [74], which gives an alternative proof of Alexandrov's theorem.

THEOREM 1.33. *Let M^n be a compact, embedded, minimal hypersurface in a compact manifold N^{n+1} whose Ricci curvature satisfies $R_{AB}^N \geq kn\delta_{AB}$ for some constant $k > 0$. Then*

$$\lambda_1(M) \geq \tfrac{1}{2}kn.$$

In particular, if N is the standard $(n + 1)$-sphere with constant sectional curvature 1, the estimate is

$$\lambda_1(M) \geq \frac{n}{2}.$$

Applying this estimate for compact minimal surfaces and Theorem 2.2 of Yang and Yau, we conclude the following.

COROLLARY 1.34. *Let M^2 be an embedded compact minimal surface in N^3, a compact manifold with positive Ricci curvature, $R_{AB}^N \geq 2k\delta_{AB}$ for some constant $k > 0$. Then*

$$\text{Area}\,(M) \leq \frac{4\pi}{k}\left(\frac{4}{|\pi_1(N)|} - \chi(M)\right),$$

where $\chi(M)$ is the Euler characteristic and $|\pi_1(N)|$ is the order of the fundamental group.

This area estimate is then utilized by Choi and Schoen [29] to obtain the following important consequence about the compactness of the space of minimal surfaces.

THEOREM 1.35. *Let N^3 be a compact manifold with positive Ricci curvature. Suppose M is an embedded compact minimal surface in N. Then there is a constant C depending only on N and the Euler characteristic of M such that*

$$\sup_M |h_{ij}| \leq C,$$

where h_{ij} is the second fundamental form of M. Hence, the space of compact, minimal, embedded surfaces $M \subset N$ of a fixed topological type is compact in the C^j topology for any $j \geq 2$. Furthermore, if N is real analytic, then the space is a compact, finite-dimensional, real analytic variety.

1.7. Nodal Domains and Multiplicity Bounds

Given an eigenfunction u on M, the nodal domains are the components of $M - u^{-1}(0)$ given by $\Omega_1 \cup \Omega_2 \cup \cdots \cup \Omega_\nu$. Courant's nodal domain theorem [33] asserts the following.

LEMMA 1.36. *Let M^n be a compact manifold without boundary. Then the number of nodal domains of a λ_l eigenfunction is at most $l + 1$.*

This has been sharpened by Bérard and Meyer [5], generalizing a result of Peetre [70] and Pleijel [71].

THEOREM 1.37. *Let M^n be a compact manifold with boundary. Let ν_l denote the number of nodal domains of the lth Dirichlet eigenfunction. Then*

$$\limsup_{l\to\infty} \frac{\nu_l}{l} \le \frac{2^{n-2}n^2\Gamma(n/2)^2}{j^n_{(n-2)/2}} < 1,$$

where j_s is the first positive zero of the Bessel function J_s.

There is no increasing lower bound on the number of nodal domains. Indeed, Lewy [53] has shown that on the round 2-sphere, for each l, there is a spherical harmonic of degree l, an eigenfunction in $E_{l(l+1)}$, which has two or three nodal domains depending on if l is odd or even, respectively. Lemma 1.36 is used to bound the multiplicity $m_l = \dim E_{\lambda_l}$ of the eigenspace by Cheng [24], Besson [8, 9], and Colin de Verdière [32].

THEOREM 1.38. *Let M^2 be an orientable and N^2 a nonorientable compact connected surface with genus g and arbitrary metric. Then the multiplicity of the lth eigenfunction satisfies*

$$m_l(M) \le \begin{cases} 4g + 2l + 1, & \text{for all } (g, l) \ne (1, 1), \\ 6, & \text{if } (g, l) = (1, 1) \quad (M \cong T^2); \end{cases}$$

$$m_l(N) \le \begin{cases} 4g + l + 3, & \text{for all } (g, l) \ne (1, 1), \\ 5, & \text{if } (g, l) = (0, 1) \text{ or } (1, 1) \quad (N \cong \mathbf{K}^2 \text{ or } N \cong \mathbf{RP}^2). \end{cases}$$

Besson also showed that there is a sphere other than the round one with $m_1(S^2) = 3$. Colin de Verdière [32] showed that there is a metric on a 2-torus with $m_1 = 6$ and one on a real projective 2-space with $m_1 = 5$. When the dimension of the manifold is greater than 2, Colin de Verdière [31] constructed examples of arbitrary multiplicity for a fixed topological type. However, Uhlenbeck [81] has shown that generically eigenspaces have multiplicity 1.

1.8. Stability of Minimal Surfaces

The techniques arising from eigenvalue studies can be applied, for example, to stability of minimal surfaces. We would like to explain the connection and illustrate this with several applications.

Let $M^2 \subset N^3$ be an immersed minimal surface. The second variation formula yields the stability operator

$$L = \Delta - K_M + R^N + \tfrac{1}{2}|A|^2,$$

where K_M is the Gaussian curvature of M, R^N is the scalar curvature of N, and $|A|^2 = \sum h_{ij}^2$ is the square length of the second fundamental form. That is, if e_1, e_2 are an orthonormal frame tangent to M, η is a compactly supported function of M, and the variation vector field is $\dot{X}|_{t=0} = \eta e_3$, where $X(t)$ is the position vector for a parameterized family of surfaces $M(t)$, then

$$\frac{d^2 \text{ Area } (M_t)}{dt^2}\bigg|_{t=0} = \int_M |D\eta|^2 - (R^N + \tfrac{1}{2}|A|^2 - K_M)\eta^2 \, da$$

$$= -\int_M \eta L\eta \, da.$$

The eigenvalues of L may be defined by a variational characterization for compact subdomains $\Omega \subset M$:

$$\mu_{l+1}(\Omega) = \inf_{\substack{u \in H_0^1(\Omega) \\ u \perp \eta_j, \, j=1,\cdots,l}} \frac{\int_M |Du|^2 - (R^N + (1/2)|A|^2 - K_M)u^2}{\int_M u^2},$$

where $\mu_1 < \mu_2 \leq \cdots \to \infty$ are the eigenvalues of L and η_l is an eigenfunction corresponding to μ_l. The *index* of Ω, denoted ind (Ω), is a number l such that $\mu_1, \ldots, \mu_l < 0 \leq \mu_{l+1}$, which corresponds to the number of directions which decrease area. We say that Ω is *stable* if ind $(\Omega) = 0$ or, equivalently,

$$\frac{d^2 \text{ Area}(M_t)}{dt^2}\bigg|_{t=0} \geq 0.$$

Eigenvalue monotonicity implies index monotonicity: $\Omega' \subset \Omega \Rightarrow$ ind $(\Omega') \leq$ ind (Ω). If M is complete, as in (1.32), the index of M is defined to be

$$\text{ind } (M) = \lim_{r \to \infty} \text{ind } (B_r(p_0)).$$

If $N = \mathbf{R}^3$, the study of the index can be formulated as an eigenvalue problem for the Laplacian of a new metric. Assuming M is orientable with a global normal unit vector field $\mathbf{e}_3$, the Gauss map $\mathscr{G}: M \to S^2$ is defined to be $\mathscr{G}(x) = \mathbf{e}_3(x)$. It is known that the Gauss map of a minimal surface is a conformal map, and the pull-back of the standard metric ds_0^2 of S^2 is given by

$$\mathscr{G}^*(ds_0^2) = -K_M \, ds^2,$$

where ds^2 is the intrinsic metric of M. Since $R^N = 0$, we can rewrite the stability operator, using the Gauss curvature equation, as

$$L = \Delta - 2K_M.$$

The index is defined to be the number of negative eigenvalues for the Rayleigh quotient:

$$\frac{\int_M (|Du|^2 + 2K_M u^2)}{\int_M u^2} = \left(\frac{\int_M K_M u^2}{\int_M u^2}\right)\left(\frac{\int_M |Du|^2}{\int_M K_M u^2} + 2\right).$$

This is equal to the number of negative eigenvalues for the Rayleigh quotient:

$$\frac{\int_M |Du|^2}{\int_M K_M u^2} + 2,$$

which is associated with the operator $\Delta_0 - 2$, defined with respect to the metric $\mathscr{G}^*(ds_0^2) = -K_M \, ds^2$. Hence, the index is just the number of eigenvalues with value less than 2 for the Laplacian Δ_0 with respect to the pull-back metric. We will now discuss a few examples which are related to this point of view.

The first application is a theorem of Barbosa and do Carmo [2, 3].

THEOREM 1.39. *Let M be an orientable minimal surface in* $\mathbf{R}^3$ *and* $\Omega \subset M$ *a piecewise* C^1 *precompact subdomain. If the area of the Gauss image of* Ω *satisfies* Area$(\mathscr{G}(\Omega)) < 2\pi$, *then* Ω *is stable.*

This result was used by Nitsche [66] to prove a uniqueness theorem for minimal surfaces: A regular analytic Jordan curve $\Gamma \subset \mathbf{R}^3$ whose total curvature satisfies $\int_\Gamma \kappa \, ds \leq 4\pi$ bounds precisely one solution of Plateau's problem (a minimal disk with given boundary curve.)

The next result was obtained by Fischer-Colbrie and Schoen [40] and by do Carmo and Peng [36].

THEOREM 1.40. *A complete, stable minimal surface in* $\mathbf{R}^3$ *must be a plane.*

Fischer-Colbrie and Schoen [40] actually considered complete, stable minimal surfaces in a three-dimensional manifold of nonnegative scalar curvature. Fischer-Colbrie [39] later proved that the total curvature is related to index.

THEOREM 1.41. *Let* M *be a complete minimal surface in* $\mathbf{R}^3$. *Then* M *has finite index if and only if it has finite total curvature.*

When a complete surface has finite total curvature, by a theorem of Huber [48] it must be conformally equivalent to a Riemann surface $\Sigma - \{p_1, \ldots, p_m\}$ with finitely many punctures. If the surface is a minimal surface of $\mathbf{R}^3$, then by the holomorphicity of the Gauss map, Osserman [68] observed that it can be extended to a holomorphic branched cover of $\mathscr{G}: \Sigma \to \mathbf{S}^2$ of finite degree d. Therefore, the index can then be viewed as the number of eigenvalues less than 2 for the Laplacian of the metric $\mathscr{G}^*(ds_0^2)$. Due to the presence of branch points, this metric, in general, has isolated degeneracy.

In view of the theorem of Fischer-Colbrie, it is natural to ask if there is a formula relating to the index and the total curvature. Again, by the fact that $G^*(ds_0^2) = -K_M ds^2$, the total curvature is given by the degree of the Gauss map times a factor of 4π, which is the area of $\mathbf{S}^2$. With this interpretation of the total curvature, the question becomes, is there a formula relating the index to the geometry of the Gauss map—for example, the total branching order, the degree, and the genus of Σ.

If $M \subset \mathbf{R}^3$ is the catenoid or Enneper's surface, $\Sigma = \mathbf{S}^2$, $\mathscr{G}$ is a diffeomorphism and there is only one eigenfunction of $\mathbf{S}^2$ with eigenvalue less than 2; hence, ind $(M) = 1$. Cheng and Tysk [26] showed that the only complete oriented minimal surface with embedded ends and index 1 is the catenoid. Without assuming embeddedness, Choe [28] showed that index 1 implies a catenoid or Enneper's surface. Choe also estimated the index of many examples by counting eigenvalue multiplicity.

THEOREM 1.42. *The index of the following minimal surfaces has a lower bound*:

1. *The Jorge-Meeks* [50] *surface with p-ends* $M_p \subset \mathbf{R}^3$ *has* ind $(M_p) \geq 2p - 3$.
2. *The Costa-Hoffman-Meeks* [49] *minimal surface* $Q_g \subset \mathbf{R}^3$ *of genus g has* ind $(Q_g) \geq 2g + 1$.
3. *Lawson's* [52] *minimal surface* $\xi_{m,p} \subset \mathbf{S}^3$ *of genus mp has* ind $(\xi_{m,p}) \geq$ max $(2m + 1, 2p + 1)$.
4. *Any complete nonorientable minimal surface* $U \subset \mathbf{R}^3$ *with finite total curvature and conformally equivalent to the projective plane with finitely many punctures has* ind $(U) \geq 2$.

Tysk [80] and Cheng and Tysk [26] have estimated upper bounds on the index in terms of the total curvature of $M^2 \subset \mathbf{R}^3$.

THEOREM 1.43. *Let $M^2 \subset \mathbf{R}^3$ be a complete, oriented minimal surface. There is an absolute constant $c \approx 0.6133 \ldots$ such that the index of M satisfies*

$$\mathrm{ind}\,(M) \leq - c \int_M K_M \, da,$$

where K_M is the Gauss curvature of M.

The estimate of the index boils down to counting eigenvalues on a branched cover of the sphere and relating them to the sphere through a heat equation argument.

Finally, using the maximum principle for eigenvalues, Li and Tam [57] showed that in some special cases there is a relation between the degree and the index.

THEOREM 1.44. *Let M be a complete, orientable minimal surface in $\mathbf{R}^3$ such that the image of the branch point set lies in a great circle of $\mathbf{S}^2$. Then $\mathrm{ind}\,(M) = 2d - 1$, where d is the degree of the Gauss map.*

Thus Choe's bound is sharp for the case of the catenoid, trinoid, and the whole Jorge-Meeks family of surfaces.

References

1. C. Bandle, *Isoperimetric Inequalities and Applications*, Academic Press, New York (1981).
2. J. Barbosa and M. do Carmo, On the size of a stable minimal surface in $\mathbf{R}^3$, *Am. J. Math.* **98**, 515–528 (1976).
3. J. Barbosa and M. do Carmo, Stability of minimal surfaces and eigenvalues of the Laplacian, *Math. Z.* **173**, 13–28 (1980).
4. P. Bérard, *Spectral Geometry: Direct and Inverse Problems*, Lecture Notes in Math., Vol. 1207, Springer-Verlag, Berlin (1986).
5. P. Bérard and D. Meyer, Inégalités isopérimétriques et applications, *Ann. Sci. École Norm. Sup. Paris* **15**, 513–542 (1982).
6. M. Berger, Sur les premières valeurs propres des variétés riemanniennes, *Compositio Math.* **26**, 129–149 (1973).
7. M. Berger, P. Gauduchon, and E. Mazet, *Le Spectre d'une Variété Riemanniennes*, Lecture Notes in Math., Vol. 194, Springer-Verlag, New York (1974).
8. G. Besson, Sur la multiplicité de la première valeur propre des surfaces riemanniennes, *Ann. Inst. Fourier* **30**, 109–128 (1980).
9. G. Besson, On the multiplicity of the eigenvalues of the Laplacian, in *Geometry and Analysis on Manifolds*, (T. Sunada, ed.), Lecture Notes in Math., Vol. 1339, pp. 32–53, Springer-Verlag, New York (1988).
10. S. Bochner, Vector fields and Ricci curvature, *Bull. Am. Math. Soc.* **52**, 776–797 (1946).

11. E. Bombieri, Theory of minimal surfaces and a counterexample to the Bernstein conjecture in high dimensions, Lecture notes.

12. R. Brooks, The fundamental group and the spectrum of the Laplacian, *Comment. Math. Helv.* **56**, 581–598 (1981).

13. R. Brooks, The spectral geometry of a tower of coverings, *J. Diff. Geom.* **23**, 97–107 (1986).

14. J. P. Bourguignon, P. Li, and S. T. Yau, An upper bound for the first eigenvalue of submanifolds of a complex projective space, preprint.

15. P. Buser, Riemannsche Flächen mit Eigenwerten in $(0, 1/4)$, *Comment. Math. Helv.* **52**, 25–34 (1977).

16. P. Buser, A note on the isoperimetric constant, *Ann. Sci. École Norm. Sup. Paris* **15**, 213–230 (1982).

17. I. Chavel, *Eigenvalues in Riemannian Geometry*, Academic Press, Orlando (1984).

18. J. Cheeger, The relation between the Laplacian and the diameter for manifolds of non-negative curvature, *Arch. Math* **19**, 558–560 (1968).

19. J. Cheeger, A lower bound for the smallest eigenvalue of the Laplacian, in *Problems in Analysis, a Symposium in Honor of S. Bochner*, Princeton University Press, Princeton (1970).

20. R. Chen, Eigenvalue estimate on a compact riemannian manifold, *Am. J. Math.* **111**, 769–781 (1989).

21. R. Chen, Neumann eigenvalue estimate on a compact Riemannian manifold, *Proc. Amer. Math. Soc.* **108**, 961–970 (1990).

22. S. Y. Cheng, *Eigenfunctions and Eigenvalues of the Laplacian*, Proc. Symp. Pure Appl. Math., Vol. 27, pp. 184–193, American Mathematical Society, Providence, RI (1975).

23. S. Y. Cheng, Eigenvalue comparison theorems and its geometric application, *Math Z.* **143**, 289–297 (1975).

24. S. Y. Cheng, Eigenfunctions and nodal sets, *Comment. Math. Helv.* **51**, 43–55 (1976).

25. S. Y. Cheng, P. Li, and S. T. Yau, Heat equations on minimal submanifolds and their applications, *Am. J. Math.* **106**, 1033–1065 (1984).

26. S. Y. Cheng and J. Tysk, An index characterization of the catenoid and index bounds for minimal surfaces in $\mathbf{R}^4$, *Pacific J. Math.* **134**, 251–260 (1988).

27. S. Y. Cheng and S. T. Yau, Differential equations on Riemannian manifolds and their geometric applications, *Comm. Pure Appl. Math.* **28**, 333–354 (1975).

28. J. Choe, Index, vision number and stability of complete minimal surfaces, *Arch. Rat. Mech. Anal.* **109**, 195–212 (1990).

29. H. I. Choi and R. Schoen, The space of minimal embeddings of a surface into a three dimensional manifold of positive Ricci curvature, *Invent. Math.* **81**, 387–394 (1985).

30. H. I. Choi and A. N. Wang, A first eigenvalue estimate for minimal hypersurfaces, *J. Diff. Geom.* **18**, 559–562 (1983).

31. Y. Colin de Verdière, Sur la multiplicité de la première valeur propre non nulle du laplacien, *Comment. Math. Helv.* **61**, 254–270 (1986).

32. Y. Colin de Verdière, Construction de laplaciens dont une partie finie du spectre esp donnée, *Ann. Sci. École Norm. Sup. Paris* **20**, 599–615 (1987).

33. R. Courant and D. Hilbert, *Methods of Mathematical Physics*, Vol. 1, Interscience, New York (1953).

34. C. Croke, Some isoperimetric inequalities and eigenvalue estimates, *Ann. Sci. École Norm. Sup.* **13**, 419–435 (1980).

35. C. Croke, An eigenvalue pinching theorem, *Invent. Math.* **68**, 253–256 (1982).

36. M. do Carmo and C. K. Peng, Stable complete minimal surfaces in $\mathbf{R}^3$ are planes, *Bull. Am. Math. Soc.* **1**, 903–906 (1979).

37. J. Escobar, Sharp constant in a Sobolev trace inequality, *Indiana Math. J.* **37**, 687–698 (1988).

38. C. Faber, Beweiss, dass unter allen homogenen Membrane von gleicher Fläche und gleicher Spannung die kreisförmige die tiefsten Grundton gibt, *Sitzungsber.-Bayer. Akad. Wiss, Math.-Phys. München* 169-172 (1923).

39. D. Fischer-Colbrie, On complete minimal surfaces with finite Morse index in three manifolds, *Invent. Math.* **82**, 121-132 (1985).

40. D. Fischer-Colbrie and R. Schoen, The structure of complete stable minimal surfaces in 3-manifolds of nonnegative scalar curvature, *Comm. Pure Appl. Math.* **33**, 199-211 (1980).

41. M. Gage, Upper bounds for the first eigenvalue of the Laplace-Beltrami operator, *Indiana U. Math. J.* **29**, 877-912 (1980).

42. S. Gallot, Isoperimetric inequalities based on integral norms of Ricci curvature, *Soc. Math France Astérisque* **157-158**, 191-216 (1988).

43. S. Gallot, Inégalités isopérimetriques et analytiques sur les variétés riemanniennes, *Soc. Math. France Astérisque* **163-164**, 31-91 (1988).

44. M. Gromov, Paul Levy's isoperimetric inequality, *Inst. Hautes Etudes Sci. Publ. Math.* (1980) (preprint).

45. K. Grove and K. Shiohama, A generalized sphere theorem, *Ann. Math.* **106**, 201-211 (1977).

46. J. Hersch, Caractérisation variationelle d'une somme de valeurs propres consécutives; généralisation d'inégalités de Pólya-Schiffer et de Weyl, *C. R. Acad. Sci. Paris* **252**, 1714-1716 (1961).

47. J. Hersch, Quatre propriétés isopérimétriques de membranes sphériques homogènes, *C. R. Acad. Sci. Paris* **270**, 1645-1648 (1970).

48. A. Huber, On subharmonic functions and differential geometry, *Comment. Math. Helv.* **32**, 13-72 (1957).

49. D. Hoffman and W. Meeks III, Complete embedded minimal surfaces of finite total curvature, *Bull. Am. Math. Soc.* **12**, 134-136 (1985).

50. L. Jorge and W. Meeks III, The topology of complete minimal surfaces with finite total Gaussian curvature, *Topology* **22**, 203-221 (1983).

51. E. Krahn, Über eine von Rayleigh formulierte Minimaleigenschaft des Kreises, *Math. Ann.* **94**, 97-100 (1925).

52. H. B. Lawson, Complete minimal surfaces in S^3, *Ann. Math.* **92**, 335-374 (1970).

53. H. Lewy, On the minimum number of domains in which nodal lines of spherical harmonics divide the sphere, *Comm. PDE* **2**, 1233-1244 (1977).

54. P. Li, A lower bound for the first eigenvalue for the Laplacian on compact manifolds, *Indiana U. Math. J.* **28**, 1013-1019 (1979).

55. P. Li, On the Sobolev constant and the p-spectrum of a compact Riemannian manifold, *Ann. Sci. École Norm. Sup.* **13**, 451-468 (1980).

56. P. Li, Poincaré inequalities on Riemannian manifolds, in Seminar on Differential Geometry (S. T. Yau, ed.), Ann. of Math., Studies, Vol. 102, pp. 73-83, Princeton University Press, Princeton (1982).

57. P. Li, F. Stenger, L. F. Tam, and A. Treibergs, On the index of the Costa-Hoffman-Meeks surface, to be published.

58. P. Li and A. Treibergs, Pinching theorem for the first eigenvalue on positively curved four manifolds, *Invent. Math.* **66**, 35-38 (1982).

59. P. Li and S. T. Yau, *Eigenvalues of a Compact Riemannian Manifold.*, Proc. Symp. Pure Math., Vol. 36, pp. 205-239, American Mathematical Society, Providence, RI (1980).

60. P. Li and S. T. Yau, A new conformal invariant and its application to the Willmore conjecture and the first eigenvalue of compact surfaces, *Invent. Math.* **69**, 269-291 (1982).

61. P. Li and J. Q. Zhong, Pinching theorem for the first eigenvalue on positively curved manifolds, *Invent. Math.* **65**, 221-225 (1981).

62. A. Lichnerowicz, *Géometrie des Groupes de Transformations*, Dunod, Paris (1958).

63. H. P. McKean, An upper bound to the spectrum of Δ on a manifold of negative curvature, *J. Diff. Geom.* **4**, 359–366 (1970).

64. D. Meyer, Minoration de la première valeur propre non nulle du problème de Neumann sur les variétés riemanniennes a bord, *Ann. Inst. Fourier, Grenoble* **36**, 113–125 (1986).

65. S. Monteil and A. Ros, Minimal immersions of surfaces by the first eigenfunctions and conformal area, *Invent. Math.* **83**, 153–166 (1986).

66. J. Nitsche, A new uniqueness theorem for minimal surfaces, *Arch. Rat. Mech. Anal.* **52**, 319–329 (1973).

67. M. Obata, Certain conditions for a Riemannian manifold to be isometric to the sphere, *J. Math. Soc. Japan* **14**, 333–340 (1962).

68. R. Osserman, Global properties of complete minimal surfaces in E^3 and E^n, *Ann. Math.* **80**, 340–364 (1964).

69. L. Payne, Isoperimetric inequalities and their applications, *SIAM Rev.* **9**, 453–488 (1967).

70. J. Peetre, Estimates on the number of nodal domains, in *XIII Congr. Math. Scand.*, 1957, pp. 198–201.

71. A. Pleijel, Remarks on Courant's nodal line theorem, *Comm. Pure Appl. Math.* **9**, 543–550 (1956).

72. G. Polya and G. Szegö, *Isoperimetric Inequalities in Mathematical Physics*, Ann. of Math. Studies, no. 27, Princeton University Press, Princeton (1951).

73. M. Protter, The maximum principle and eigenvalue problems, in *Beijing Symp. on Differential Geometry and Partial Differential Equations* (1980).

74. R. Reilly, Applications of the Hessian operator in a Riemannian manifold, *Indiana U. Math. J.* **26**, 459–472 (1977).

75. R. Schoen, A lower bound for the first eigenvalue of a negatively curved manifold, *J. Diff. Geom.* **17**, 233–238 (1982).

76. R. Schoen, S. Wolpert, and S. T. Yau, Geometric bounds on the low eigenvalues of a compact surface, in *Geometry of the Laplace Operator*, Proc. Symp. Pure Math., Vol. 36, pp. 279–285, American Mathematical Society, Providence, RI (1980).

77. I. Singer, B. Wong, S. S. T. Yau, and S. T. Yau, An estimate of the gap of the first two eigenvalues of the Schrödinger operator, *Ann. Scuola Norm. Pisa, Cl. Sci. IV* **12**, 319–333 (1985).

78. R. P. Sperb, *Maximum Principles and Their Applications*, Academic Press, New York (1981).

79. G. Szegö, Inequalities for certain eigenvalues of a membrane of given area, *J. Rat. Mech. Anal.* **3**, 343–356 (1954).

80. J. Tysk, Eigenvalue estimates with applications to minimal surfaces, *Pacific J. Math.* **128**, 361–366 (1987).

81. K. Uhlenbeck, Generic properties of eigenfunctions, *Am. J. Math.* **98**, 1059–1078 (1976). (a) H. Urakawa, On the least positive eigenvalue of the Laplacian for compact group manifolds, J. Math. Soc. Jpn. **31**, 209–226 (1979).

82. H. Weinberger, *Variational Methods for Eigenvalue Approximation*, CBMS, Society for Industrial and Applied Mathematics, Philadelphia (1974).

83. T. J. Willmore, Note on embedded surfaces, *An. St. Univ. Isai, s. I. a. Mat.* **11B**, 493–496 (1965).

84. T. J. Willmore, *Total Curvature in Riemannian Geometry*, Ellis Horwood, Chichester (1982).

85. H. C. Yang, Estimates of the first eigenvalue for a compact Riemann manifold, *Science in China Series* A**33**, 39–51 (1990).

86. H. C. Yang, Estimates of the first eigenvalue of compact Riemannian manifolds with Dirichlet boundary condition, preprint (1990).

87. P. Yang and S. T. Yau, Eigenvalues of the Laplacian on compact Riemannian surfaces and minimal submanifolds., *Ann. Scuola Norm. Sup. Pisa* **7**, 55–63 (1980).

88. S. T. Yau, Isoperimetric constants and the first eigenvalue of a compact Riemannian manifold, *Ann. Sci. École. Norm. Sup.* **8**, 62–507 (1975).

89. Q. H. Yu and J. Q. Zhong, Lower bounds on the gap between the first and second eigenvalues of the Schrödinger operator, *Proc. Am. Math. Soc.* **294**, 341–349 (1986).

90. J. Q. Zhong and H. C. Yang, On the estimate of first eigenvalue of a compact Riemannian manifold, *Sci. Sinica Ser. A* **27**, 1265–1273 (1984).

2

The Theory of Functions of Several Complex Variables in China from 1949 to 1989

Qi-Keng Lu

Thirty years ago a paper with a similar title was written surveying the development of the same subject in China from 1949 to 1959 (Lu [100]). Since then, a number of younger mathematicians have come of age. Professor L.-K. Hua, the founder of several complex variables in China, died in 1985. Professor J.-Q. Zhong, the most distinguished Chinese mathematician of his generation, died in 1987 before his fiftieth birthday. It is time to review again what we have done in the past 40 years. With the help of my colleagues, I have done my best to collect in the bibliography all obtainable papers or titles of papers written by the Chinese in this field. Special thanks are due to Professors Z.-H. Chen, Y.-C. Su, J.-H. Zhang, and T.-D. Zhong for their cooperation. But I am afraid that this bibliography is still far from complete; for example, I did not have at hand all the journals published by the hundreds of universities in China.

Let me recall the brief history about the development of several complex variables in China since 1949. In the early 1950s I was the only student studying several complex variables under the guidance of Professor Hua. In 1952 to 1953 there was a seminar in Peking University for studying B. A. Fuks's book, *Theory of Analytic Functions of Several Complex Variables*

Qi-Keng Lu ● Institute of Mathematics, Academia Sinica, Beijing 100080, People's Republic of China.

(in Russian), which was published in 1948 and was brought to China by Hua from Moscow. There were about 10 participants in the seminar, including some famous Chinese mathematicians, among them Professors Cheng Ming-De, Chuang Qi-Tai, Min Si-He, Xu Bao-Lu. I learned a lot about the Bergman metric and the Bergman kernel from this book and from Hua's works on constructing the orthonormal systems of functions in the classical domains. But no one in Peking University went on to study several complex variables after this seminar. In 1954, T.-D. Zhong came from Xiaman University to the Institute of Mathematics, Academia Sinica, as a visiting scholar for two years, and S. Gong moved later from Fudan University in Shanghai to Beijing and worked in our institute. Then we had a group of four persons. In 1954 to 1955 we had a seminar on unitary geometry and several complex variables, the lecture notes of which were eventually published (Lu [93]). Unitary geometry is now called Hermitian geometry. But we mainly discussed Kähler geometry, especially the geometry of the Bergman metric. Two teachers, Dong Wai-Yuan and Chen Jie, from Peking University participated in this seminar, but they no longer work in this field. Till the end of the decade the number of people studying several complex variables remained four.

Beginning in 1959, I gave a series of courses to 10 graduate students in Peking University. Zhong Jia-Qing was one of them. The content of these courses was published in two books (Lu [101, 103]). Most of the 10 students continued their study of several complex variables after they graduated in 1962 and became the principal Chinese mathematicians in this field. At nearly the same time Zhong Tong-De gave a course almost each year on this subject at Xiaman University. A group was established there. Another group headed by Gong Sheng was formed at the University of Science and Technology of China. But I can say nothing about the subject during the Cultural Revolution from 1966 to 1978. The study of several complex variables began to recover after the summer of 1978. Since then a number of graduate students began working in this field. Now the number is about 50. They reside in various universities. Some of them are still working for their Ph.D.'s in the United States and Europe. Most of the older specialists have visited these countries for a period from one month to two years. On the other hand, many foreign mathematicians have visited our country, such as S. S. Chern, H. Wu, J. Kohn, S.-Y. Cheng, Y.-T. Siu, S.-T. Yau, P. A. Griffiths, A. Borel, E. M. Stein, K. Stein, R. Remmert, S. Kobayashi, F. Hirzebruch, H. Grauert, L. Carleson, E. Vesentini. They gave talks on related subjects. In particular, H. Wu, Y.-T. Siu, and S.-T. Yau gave systematic lectures. All these activities brought some fresh air to the study of several complex variables in China. This can be seen from the later works of J.-Q. Zhong.

2.1. The Classical Domains

In 1935, H. Cartan [11] proved that the group of holomorphic automorphisms of a bounded domain in $\mathbb{C}^n$ is a real Lie group. Immediately É. Cartan [10] proved that if the bounded domain is symmetric, then the Lie group must be semisimple, and he classified the bounded symmetric domains into four main types and two exceptional ones of complex dimensions 16 and 27, respectively. C. L. Siegel [151] and L.-K. Hua [67] realized these four main types of bounded symmetric domains by matrices and studied their geometrical and analytic properties. These four types are

$$R_{\mathrm{I}}(m, n) = \{Z \in \mathbb{C}^{m \times n} \mid I - Z\bar{Z}' > 0\},$$

$$R_{\mathrm{II}}(p) = \{Z \in \mathbb{C}^{p \times p} \mid I - Z\bar{Z}' > 0, Z = Z'\},$$

$$R_{\mathrm{III}}(q) = \{Z \in \mathbb{C}^{q \times q} \mid I - Z\bar{Z}' > 0, Z = -Z'\},$$

$$R_{\mathrm{IV}}(N) = \{Z \in \mathbb{C}^{N} \mid 1 + |zz'|^2 - 2z\bar{z}' > 0, 1 - |zz'|^2 > 0\},$$

where $Z \in \mathbb{C}^{m \times n}$ means that A is an $m \times n$ matrix, the elements of which belong to $\mathbb{C}$, and Z' is the transposed matrix of Z. $H > 0$ means that H is a positive definite Hermitian matrix.

In the bounded symmetric domain, the Riemannian metric used by É. Cartan is the Bergman metric, which is defined from the Bergman kernel

$$K(z, \bar{z}) = \sum_{\nu=0}^{\infty} \phi_\nu(z)\overline{\phi_\nu(z)}, \tag{2.1}$$

where $\{\phi_\nu(z)\}_{\nu=0,1,2,\ldots}$ is a complex orthonormal system of the space $H^2(D)$, the space of all holomorphic and absolutely square integrable functions in a bounded domain D. Hua [72, 73, 76] in the first half of the 1950s used group representation theory to construct the complete orthonormal system of $H^2(R_A)$ $(A = \mathrm{I, II, III, IV})$. At first he established two wonderful algebraic identities, by which he could decompose a representation into irreducible ones. He proved that the components of the irreducible representations formed a complete orthogonal system. In order to normalize the orthogonal system, he calculated a series of matrix integrals, such as the volumes of the classical domains, various complex subgroups of a unitary group, and various compact quotient spaces. This shows that Hua was proficient in group representation theory and had marvelous tricks in matrix and integral operations. Then he summed up the complete orthonormal systems as in

(2.1) and obtained the Bergman kernels of all classical domains. They are

$$K_{R_I(m,\,n)}(Z, \bar{Z}) = [\, V(R_I) \det (I - Z\bar{Z}')^{m+n}]^{-1},$$

$$K_{R_{II}(p)}(Z, \bar{Z}) = [\, V(R_{II}) \det (I - Z\bar{Z}')^{p+1}]^{-1},$$

$$K_{R_{III}(q)}(Z, \bar{Z}) = [\, V(R_{III}) \det (I - Z\bar{Z}')^{q-1}]^{-1}, \qquad (2.2)$$

$$K_{R_{IV}(q)}(Z, \bar{Z}) = [\, V(R_{IV})(1 + |zz'|^2 - 2z\bar{z}')^N]^{-1},$$

where $V(R_A)$ is the Euclidean volume of R_A. From (2.2) we can obtain from the definitions the Bergman metrics (Hua [67], Lu [93])

$$ds^2_{R_A} = \partial\bar{\partial} \log K_{R_A}(Z, \bar{Z}) = \alpha(R_A)\, \mathrm{tr}\,[(I - Z\bar{Z}')^{-1}\, dZ\, (I - \bar{Z}'Z)^{-1}\, d\bar{Z}'], \tag{2.3}$$

where $\alpha(R_I(m, n)) = m + n$, $\alpha(R_{II}(p)) = p + 1$, $\alpha(R_{III}(q)) = 1 - q$, and

$$ds^2_{R_{IV}} = \frac{2N}{(1 + |zz'|^2 - 2z\bar{z}')^2}[(1 + |zz'|^2 - 2z\bar{z}')\, dz\, \overline{dz}' + 4z\bar{z}'|z\, dz'|^2$$
$$- 2(\overline{zz}'z\, dz'\, z\, \overline{dz}' + zz'\overline{z\, dz}'\, \bar{z}\, dz') - 2(|z\, dz'|^2 - (z\, \overline{dz}')^2)]. \tag{2.4}$$

On the other hand, one can define the Bergman kernel of a domain D without knowing the complete orthonormal system (Bergman [5]); that is,

$$K(z, \bar{z}) = \sup_{f \in H_1^2(D)} |f(z)|^2, \tag{2.5}$$

where

$$H_1^2(D) = \left\{ f \in H^2(D)\, \middle|\, \int_D |f(z)|^2 \dot{z} \le 1 \right\}$$

and $\dot{z}$ is the Euclidean volume element. If one can get the Bergman kernel in some way, then, conversely, one can construct a complete orthonormal system from the Bergman kernel (Lu [67]). Let $t = (t_1, \ldots, t_n)$ be a fixed point of D, and let

$$a_{m_1,\ldots,m_n}(z) = \frac{\partial^m K(z, \bar{t})}{\partial \bar{t}_1^{m_1} \cdots \partial \bar{t}_n^{m_n}}, \qquad m_1 + \cdots + m_n = m, \quad m = 0, 1, 2, \ldots.$$

Arrange the indices $(m_1, \ldots, m_n)$ in a natural way into a single index ν such that $\{a_{m_1,\ldots,m_n}(z)\} = \{a_\nu(z)\}_{\nu=0,1,2,\ldots}$. Let

$$a_{\alpha\bar{\beta}} = \int_D a_\alpha(z)\overline{a_\beta(z)}\dot{z},$$

$$A^{(\nu)} = \begin{vmatrix} a_{0\bar{0}} & a_{0\bar{1}} & \cdots & a_{0\bar{\nu}} \\ a_{1\bar{0}} & a_{1\bar{1}} & \cdots & a_{1\bar{\nu}} \\ \cdots & \cdots & \cdots & \cdots \\ a_{\nu\bar{0}} & a_{\nu\bar{1}} & \cdots & a_{\nu\bar{\nu}} \end{vmatrix},$$

and let $A^{(\nu)}_{\alpha\bar{\beta}}$ be the cofactor corresponding to the element $a_{\alpha\bar{\beta}}$ of the determinant $A^{(\nu)}$. It was proved (Lu [67]) that the functions

$$\phi_\nu(z) = \frac{\overline{A^{(\nu)}_{0\bar{\nu}}}a_0(z) + \cdots + \overline{A^{(\nu)}_{\nu\bar{\nu}}}a_\nu(z)}{(A^{(\nu)}_{\nu\bar{\nu}}A^{(\nu)})^{1/2}}, \qquad \nu = 0, 1, 2, \ldots, \qquad (2.6)$$

form a complete orthonormal system of $H^2(D)$.

Suppose now D is transitive under the group Aut (D) of holomorphic automorphisms (i.e., D is a bounded homogeneous domain); then since $K(z, \bar{z}) dz_1 \wedge \cdots \wedge dz_n \wedge \overline{dz_1} \wedge \overline{dz_n}$ is an invariant volume element under the group Aut (D), one can easily prove (Hua and Lu [87]) that

$$K(z, \bar{z}) = c \left| \det \frac{\partial \sigma_z(\zeta)}{\partial \zeta} \right|_{\zeta=z}, \qquad (2.7)$$

where $\sigma_z \in$ Aut (D) satisfies $\sigma_z(z) = t$ (a fixed point of D) and $\partial\sigma_z(\zeta)/\partial\zeta$ means the Jacobian matrix of σ_z. Hence, if we know σ_z explicitly, then we can obtain $K(z, \bar{z})$ explicitly. Since the explicit expression of Aut (R_A) is well known (see Siegel [151], Hua [67, 69], Hua and Lu [87]), we can derive (2.2) directly from σ_z of Aut (R_A). In Hua and Lu's formulation, any element of Aut (R_A) for $A =$ I, II, III can be written in the form

$$\sigma_z(T) = A(T - Z)(I - \bar{Z}'T)^{-1}D^{-1}, \qquad T, Z \in R_A, \qquad (2.8)$$

where A and D are any solutions of the matrix equations

$$\bar{A}'A = (I - Z\bar{Z}')^{-1}, \qquad \bar{D}'D = (I - \bar{Z}'Z)^{-1}. \qquad (2.9)$$

As for the group Aut (R_{IV}),

$$\sigma_z(\zeta) = \left\{ \left[\left(\frac{1 + \zeta\zeta'}{2}, \frac{1 - \zeta\zeta'}{2i} \right) - \zeta X' \right] A \binom{1}{i} \right\}^{-1} \left[\zeta - \left(\frac{1 + \zeta\zeta'}{2}, \frac{1 - \zeta\zeta'}{2} \right) X \right] D,$$

$$(2.10)$$

where

$$X = \frac{-1}{1 - |zz'|^2} \left(\frac{(\overline{zz'} - 1)z + (zz' - 1)\bar{z}}{i(zz' + 1)\bar{z} - i(\overline{zz'} + 1)z} \right),$$

$$AA' = (I - XX')^{-1}, \qquad DD' = (I - X'X)^{-1}. \tag{2.11}$$

Xu Yi-Chao [167] also used (2.7) to construct the Bergman kernels of bounded homogeneous domains, and proved that such domains are L-domains; i.e., the kernel $K(z, \bar{t})$ possesses no zero when $(z, t) \in D \times D$.

Professor Hua [70] introduced an invariant metric for a bounded homogeneous domain D in the following manner. We can assume after a holomorphic linear transformation that D contains $0 \in \mathbb{C}^n$, and when $\sigma_0 \in$ Aut (D), $\sigma_0(0) = 0$, the matrix

$$\left[\frac{\partial \sigma_0(\zeta)}{\partial \zeta} \right]_{\zeta=0}$$

is unitary. All such Jacobians form a compact subgroup of the unitary group $U(n)$. It is isomorphic to the isotropy subgroup $\text{Aut}_0 (D)$ of Aut (D) [11]. Since D is transitive under Aut (D), any point $z \in D$ corresponds to at least one $\sigma_z \in$ Aut (D) such that $\sigma_z(z) = 0$. Hua's metric is defined by

$$d\sigma^2 = dz \left[\frac{\partial \sigma_z(\zeta)}{\partial \zeta} \right]_{\zeta=z} \left[\frac{\overline{\partial \sigma_z(\zeta)}}{\partial \zeta} \right]_{\zeta=z}' \overline{dz'}, \tag{2.12}$$

which is invariant under Aut (D). Since the classical domain R_A is an irreducible global symmetric space, Hua's metric differs from Bergman's in (2.3)-(2.4) only by a constant factor. But for reducible classical domains these two metrics are different in general. If $z = (u, v) \in D_1 \times D_2$, where D_1 and D_2 are bounded domains in $\mathbb{C}^p$ and $\mathbb{C}^q$, respectively, it was proved in 1956 (Lu [93]) that the Bergman kernels satisfy

$$K_{D_1 \times D_2}(z, \bar{z}) = K_{D_1}(u, \bar{u}) K_{D_2}(v, \bar{v}), \tag{2.13}$$

and then the Bergman metrics satisfy

$$ds^2_{D_1 \times D_2} = ds^2_{D_1} + ds^2_{D_2}. \tag{2.14}$$

Lu and Xu [123] proved that the connected component of the Lie group Aut $(D_1 \times D_2)$ is the product of the connected components of Aut (D_1) and Aut (D_2). Hence Hua's metric possesses a similar property:

$$d\sigma^2_{D_1 \times D_2} = d\sigma^2_{D_1} + d\sigma^2_{D_2}. \tag{2.15}$$

But we cannot conclude that $d\sigma^2_{D_1 \times D_2}$ is proportional to $ds^2_{D_1 \times D_2}$ when $d\sigma^2_{D_1}$ is proportional to $ds^2_{D_1}$ and $d\sigma^2_{D_2}$ to $ds^2_{D_2}$. Moreover, Hua's metric cannot be defined for any bounded domain in $\mathbb{C}^n$. However, it can be defined on any homogeneous Riemannian manifold.

If D is a Riemannian manifold of real dimension m and its Riemannian metric in local coordinates $x = (x^1, \ldots, x^m)$ is

$$ds^2 = \sum_{j,k=1}^{m} g_{jk} \, dx^j \, dx^k,$$

the Laplace-Beltrami operator is defined by

$$\Delta = \frac{1}{\sqrt{g}} \sum_{j,k=1}^{m} \frac{\partial}{\partial x^j} \left(\sqrt{g} \, g^{jk} \frac{\partial}{\partial x^k} \right), \tag{2.16}$$

where (g^{jk}) is the inverse matrix of (g_{jk}) and $g = \det(g_{jk})$. When D is a Hermitian manifold of complex dimension n and the Hermitian metric in complex local coordinates is

$$ds^2 = \sum_{\alpha,\beta=1}^{m} h_{\alpha\bar\beta} \, dz^\alpha \, \overline{dz^\beta},$$

then

$$\Delta = \sum_{\alpha,\beta=1}^{m} h^{\bar\beta\alpha} \frac{\partial^2}{\partial z^\alpha \, \partial \bar z^\beta}, \tag{2.17}$$

where $(h^{\bar\beta\alpha})$ is the inverse matrix of $(h_{\alpha\bar\beta})$. A real- or complex-valued function u is called harmonic in D if $\Delta u = 0$.

In a classical domain R_A, Hua called the following subset of its boundary ∂R_A the characteristic manifold of $\bar R_A$ (the closure of R_A):

$$S_A = \{Z \in \bar R_A \mid I - Z\bar Z' = 0\}, \qquad \text{for } A = \text{I, II, III} \tag{2.18}$$

and

$$S_{\text{IV}} = \{z \in \bar R_{\text{IV}} \mid z = e^{i\theta}x, \ x \in \mathbb{R}^N, \ xx' = 1, \ 0 \le \theta < \pi\} \tag{2.18}_1$$

$$= \{z \in \bar R_{\text{IV}} \mid 1 + |zz'|^2 - 2z\bar z' = 0, \ 1 - |zz'|^2 = 0\}.$$

As we shall see, the characteristic manifold S_A is exactly the Šilov boundary of holomorphic functions and of harmonic functions in R_A. Then it is natural to consider the following Dirichlet problem:

$$\Delta u = 0 \text{ in } R_A, \qquad u|_{S_A} = \phi \text{ in } S_A,$$

where ϕ is a given continuous function on S_A.

By a clue due to Mitchell [129] and Hua [77], Hua and Lu [87] solved the Dirichlet problem by constructing the Poisson kernel of R_A. Even as the Bergman kernel can be considered to be the coefficient of an invariant volume form, the Poisson kernel can be defined as the coefficient of a differential form

$$P_A(Z, U)\omega_1(U) \wedge \cdots \wedge \omega_M(U), \qquad (Z, U) \in R_A \times S_A, \quad M = \dim(S_A),$$
$$(2.19)$$

which is invariant under $\text{Aut}(R_A)$, and $\omega_1(U), \ldots, \omega_M(U)$ is a coframe of the characteristic manifold S_A. Then

$$P_A(Z, U) = \left| \det \frac{\partial \sigma_Z(U)}{\partial U} \right|. \qquad (2.20)$$

Since σ_z is known according to (2.8) and (2.10), the Jacobian of σ_z acting on S_A is

$$P_A(Z, U) = [\det(I - Z\bar{Z}')|\det(I - Z\bar{U}')|^{-2}]^{\beta(R_A)} \qquad (A = \text{I, II, III}), \qquad (2.21)$$

where

$$\beta(R_I(m, n)) = \min\{m, n\},$$

$$\beta(R_{II}(p)) = \frac{p+1}{2}, \qquad (2.22)$$

$$\beta(R_{III}) = \text{integer}\left(\frac{q-1}{2}\right) + 1,$$

and

$$P_{IV}(z, u) = [(1 + |zz'|^2 - 2z\bar{z}')|1 + uu'z\bar{z}' - 2u\bar{z}'|^{-2}]^{N/2}. \qquad (2.23)$$

It can be proved that for every $U \in S_A$,

$$\Delta P_A(Z, U) = 0. \tag{2.24}$$

Then

$$u(Z) = \frac{1}{V(S_A)} \int_{S_A} \phi(U) P_A(Z, U) \omega_1(U) \wedge \cdots \wedge \omega_M(U) \tag{2.25}$$

is a solution of the Dirichlet problem because one can prove, by studying the boundary behavior of the Poisson integral, that, for any $V \in S_A$,

$$\lim_{Z \to V} u(Z) = \phi(V). \tag{2.26}$$

For the proof of the uniqueness of the solution defined by the Poisson integral (2.26), one must study carefully the geometric structure of the boundary B_A of R_A and establish a maximal principle, which states that a function harmonic in $\overline{R_A}$ must take its maximum and minimum on the characteristic manifold S_A. The structure of B_A is much more complicated. It can be decomposed into a chain of slit spaces (Lu [98, 103]).

The Cauchy formulas of classical domains were originally established by Hua [79]. They came from the construction of the Szegö kernels in harmonic analysis. After the solution of the Dirichlet problem, these formulas can be deduced from the Poisson formulas. For any function $u(Z)$ harmonic in $\overline{R_A}$ we have from the uniqueness of the solution the Poisson formula

$$u(Z) = \frac{1}{V(S_A)} \int_{S_A} u(U) P_A(Z, U) \dot{U}, \qquad \dot{U} = \omega_1(U) \wedge \cdots \wedge \omega_M(U). \tag{2.27}$$

Notice that from (2.21) and (2.23) the Poisson kernel can be written as

$$P_A(Z, U) = \frac{|H_A(Z, \bar{U})|^2}{H_A(Z, Z)} \tag{2.28}$$

where

$$H_A(Z, \bar{U}) = \det (I - Z\bar{U}')^{-\beta(R_A)} \qquad (A = \text{I, II, III}),$$

$$H_{IV}(z, \bar{u}) = (1 + zz'\overline{uu}' - 2z\bar{u}')^{-N/2}, \tag{2.29}$$

each of which is holomorphic with respect to $Z \in R_A$. If $f(Z)$ is holomorphic on $\bar{R}_A$, then substituting $u(Z) = f(Z)[H_A(Z, W)]^{-1}$ ($W \in R_A$) into (2.27) and setting $Z = W$ we have

$$f(Z) = \frac{1}{V(S_A)} \int_{S_A} f(U) H_A(Z, \bar{U}) \dot{U}. \tag{2.30}$$

This is the Cauchy formula, and $H_A(Z, \bar{U})$ is the Szegö kernel of R_A (Hua and Lu [82]).

From (2.30) one can easily prove that

$$|f(Z)| \leq \sup_{U \in S_A} |f(U)|. \tag{2.31}$$

This shows that the characteristic manifold S_A is exactly the Šilov boundary of R_A.

In the proof of (2.24) for $A = I$, Hua [77] noted that the Poisson kernel $P_I(Z, U)$ of $R_I(m, n)$ ($m \leq n$) satisfies not only the Laplace-Beltrami equation but also a set of differential equations

$$\Delta_{j\bar{k}} P_I(Z, U) = 0,$$

where

$$\Delta_{j\bar{k}} = \sum_{\alpha,\beta=1}^{n} \left(\delta_{\alpha\beta} - \sum_{l=1}^{m} z_{l\alpha} \bar{z}_{l\beta} \right) \frac{\partial^2}{\partial z_{j\alpha} \, \partial \bar{z}_{k\beta}}.$$

Then by the uniqueness of the Dirichlet problem, a function u harmonic in R_A must satisfy the set of equations $\Delta_{j\bar{k}} u = 0$. Moreover, if $P(x_{11}, x_{12}, \ldots, x_{mm})$ is a polynomial, then $P(\Delta_{11}, \Delta_{12}, \ldots, \Delta_{mm}) u = 0$. Hua proposed the following problem: how "large" is the ring R of linear differential operators such that $L \in R$ iff $LP_I(Z, U) = 0$? J.-Q. Zhong [207] partially answered this question. In the ring of linear differential operators $I(D)$ on a domain D, he denoted by $(T_1, \ldots, T_m)$ the left ideal generated by $T_1, \ldots, T_m \in I(D)$ and $F = \{f \in C^\infty(D) \mid T_1 f = 0, \ldots, T_m f = 0\}$. Let $H(T_1, \ldots, T_m) = \{T \in I(D) \mid Tf = 0, f \in F\}$. He called $(T_1, \ldots, T_m)$ prime if $H(T_1, \ldots, T_m) = (T_1, \ldots, T_m)$. He proved that when $D = R^n$ or $R_I(1, n)$, (Δ) is a prime ideal.

Hua, in his classic treatise on classical domains [79], introduced matrix polar coordinates. In $R_I(m, n)$ $(m \leq n)$, for example, any $Z \in R_I$ can be written in the form

$$Z = U \Lambda V, \quad \Lambda = \begin{pmatrix} \lambda_1 & & \\ & \ddots & \\ & & \lambda_m \end{pmatrix}, \quad 1 > \lambda_1 \geq \cdots \geq \lambda_m \geq 0,$$

$$U \in U(n), \quad V \in \mathbb{C}^{m \times n}, \quad V\bar{V}' = I. \tag{2.32}$$

When the unitary matrices U and V are fixed, the locus of Z is an m-dimensional real plane section in R_I. The induced metric from (2.3) is

$$d\sigma^2 = \text{tr}\,[(I - \Lambda\Lambda')^{-1}\,d\Lambda\,(I - \Lambda'\Lambda)^{-1}\,d\Lambda'] = \sum_{i=1}^{m} \left(d \log \frac{1 + \lambda_i}{1 - \lambda_i} \right)^2.$$

This a Euclidean metric, so its curvature must be zero. Later we can see that this plane section is a totally geodesic submanifold of R_I.

In fact, the geodesic between two points Z_0 and Z of R_A with respect to Hua's (or the Bergman) metric was known (Lu [93, 103]). Especially in R_I there is a $\sigma_{Z_0} \in \text{Aut}\,(R_I)$ such that $\sigma_{Z_0}(Z_0) = 0$ and $\sigma_{Z_0}(Z) = \Lambda(I, 0)$, where Λ appeared in the polar coordinate (2.32) of Z. Then the geodesic passing through the points Z_0 and Z is

$$\Gamma(s) = \sigma_{Z_0}^{-1}(\Lambda(s)(I, 0)),$$

where

$$\Lambda(s) = \begin{pmatrix} \tanh a_1 s & & \\ & \ddots & \\ & & \tanh a_m s \end{pmatrix},$$

$$a_k = \log \frac{1 + \lambda_k}{1 - \lambda_k} \bigg/ \left[\sum_{i=1}^{m} \left(\log \frac{1 + \lambda_i}{1 - \lambda_i} \right)^2 \right]^{1/2}. \tag{2.33}$$

The geodesic distance between Z_0 and Z is

$$\sigma_I(Z_0, Z) = \frac{1}{2} \text{tr} \left[\log^2 \frac{I + Q^{1/2}(Z_0, Z)}{I - Q^{1/2}(Z_0, Z)} \right]^{1/2}, \tag{2.34}$$

where

$$Q(Z_0, Z) = (Z - Z_0)(I - \bar{Z}_0'Z)^{-1}(\bar{Z}' - \bar{Z}_0')(I - Z_0\bar{Z}')^{-1}. \tag{2.35}$$

J.-Q. Zhong [214] proved that the Buseman function of R_I is

$$\beta_\Gamma(T) = \lim_{s \to \infty} [\sigma_I(T, \Gamma(s)) - \sigma_I(Z_0, \Gamma(s))] = - \sum_{k=1}^{m} a_k \log d_{kk}(T), \quad (2.36)$$

where $d_{jk}(T)$ are the elements of the matrix

$$U'(I - T\bar{U}'_\infty)^{-1}(I - T\bar{T}')^{-1}(I - U_\infty \bar{T}')^{-1} \bar{U}' \quad (2.37)$$

with $U_\infty = \lim_{s \to \infty} \Gamma(s)$.

In 1927, Carathéodory [7] defined a distance function of two points a and b in a bounded domain D of $\mathbb{C}^n$ by

$$C_D(a, b) = \frac{1}{2} \sup_{f \in H_1^\infty(D)} \log \frac{|1 - f(a)\overline{f(b)}| + |f(a) - f(b)|}{|1 - f(a)\overline{f(b)}| - |f(a) - f(b)|}, \quad (2.38)$$

where $H^\infty(D)$ is the set of functions bounded and holomorphic in D and

$$H_1^\infty(D) = \{f \in H^\infty(D) | |f| < 1\}.$$

To see what the Carathéodory metric is in a classical domain, it is convenient to define the classical domains by a single function. Let $\lambda(H)$ be the maximal characteristic root of a Hermitian matrix H. Then R_A can be defined by the inequality

$$\rho(Z) := \lambda(Z\bar{Z}') < 1, \quad (2.39)$$

where $Z \in \mathbb{C}^{m \times n}$ for $A = I$, $Z \in \mathbb{C}^{p \times p}$, $Z = Z'$ for $A = II$ and $Z \in \mathbb{C}^{q \times q}$, $Z = -Z'$ for $A = III$. For $A = IV$, R_{IV} is defined by

$$\rho(z) := z\bar{z}' + [(z\bar{z}')^2 - |zz'|^2]^{1/2} < 1, \quad z \in \mathbb{C}^n. \quad (2.40)$$

It can be proved that $\rho(Z)$ is plurisubharmonic in R_A. In 1962 Wang Da-Ming [159] proved that the Carathéodory metric is

$$C_{R_A}(0, Z) = \frac{1}{2} \log \frac{1 + \rho^{1/2}(Z)}{1 - \rho^{1/2}(Z)}. \quad (2.41)$$

Moreover, he changed Z to $\varepsilon \, dZ$ and obtained

$$dC_{R_A}(0, dZ) := \frac{1}{\varepsilon} \lim_{\varepsilon \to 0} C_{R_A}(0, \varepsilon \, dZ) = \sqrt{\rho(dZ)}. \quad (2.42)$$

Then he noted that $dC_{R_A}(0, dZ)$ is a Finsler metric, though the Carathéodory differential metric was not introduced until 1965, by Reiffen [135].

2.2. The Classical Manifolds

The method used above to handle the geometric and analytic problems of the classical domains can be generalized to handle the nonexceptional global symmetric spaces. The first step is, of course, to find a matrix realization of such spaces. We will do it first for the nonexceptional compact Hermitian symmetric spaces, which are most related to the classical domains and are called the classical manifolds.

At the beginning of the book by Behnke and Thullen [4], they mentioned the extension spaces (*erweiterten Raume*) of $\mathbb{C}^n$. It can be summarized as a space such that

1. By adding points of infinities of $\mathbb{C}^n$, it becomes a compact complex space $\bar{\mathbb{C}}^n$ of the same dimension.
2. The general points of $\bar{\mathbb{C}}^n$ form a homogeneous complex manifold.

However, they noted that there are at least two extension spaces of $\mathbb{C}^2$: the topological product of two Riemannian spheres and the two-dimensional complex projective space. Hua's [69] discussion of extension spaces imposed one more condition:

3. There is a bounded homogeneous domain D called "absolute" in $\mathbb{C}^n$ such that the subgroup of $\text{Aut}(\bar{\mathbb{C}}^n)$ that leaves D invariant is the group $\text{Aut}(D)$.

If we choose D to be the bidisk D^2 of $\mathbb{C}^2$, then $\bar{\mathbb{C}}^2$ must be $\mathbb{CP}^1 \times \mathbb{CP}^1$. If we choose D to be the ball B^2 of $\mathbb{C}^2$, then $\bar{\mathbb{C}}^2$ must be $\mathbb{CP}^2$. Since D^2 and B^2 are the only bounded homogeneous domains in $\mathbb{C}^2$ [10], $\mathbb{CP}^1 \times \mathbb{CP}^1$ and $\mathbb{CP}^2$ are the only compact Hermitian symmetric spaces dual to D^2 and B^2.

Now let the bounded domain be $R_1(m, n)$ $(m \le n)$. The dual compact Hermitian symmetric space is the complex Grassmannian manifold $G_1(m, n)$, which can be realized in the following manner (Lu [102, 103]). Let $E(m, n)$ be the space of all $m \times (m + n)$ complex matrices of rank m. Two matrices, $\mathfrak{Z}_1$ and $\mathfrak{Z}_2$, of $E(m, n)$ are said to be equivalent iff there is a nonsingular complex matrix P such that $\mathfrak{Z}_1 = P\mathfrak{Z}_2$. The space $G_1(m, n)$ of equivalent classes is the complex Grassmannian manifold. There is a natural map $\pi: E(m, n) \to G_1(m, n)$.

Let $\mathfrak{Z} = (\mathfrak{Z}_1, \dots, \mathfrak{Z}_{m+n}) \in E(m, n)$, where $\mathfrak{Z}_1, \dots, \mathfrak{Z}_{m+n}$ are $m \times 1$ matrices, and $V(\alpha_1, \dots, \alpha_m)$ be the subset of $E(m, n)$ such that $(\mathfrak{Z}_{\alpha_1}, \dots, \mathfrak{Z}_{\alpha_m})$ is a nonsingular matrix $(1 \le \alpha_1 < \cdots < \alpha_m \le m + n)$.

Denote by $\{\alpha_{m+1}, \ldots, \alpha_{m+n}\}$ the complement of $\{\alpha_1, \ldots, \alpha_m\}$ in the set $\{1, 2, \ldots, m + n\}$. In the set $\mathfrak{V}(\alpha_1, \ldots, \alpha_m) = \pi(V(\alpha_1, \ldots, \alpha_m))$ of $G_{\mathrm{I}}(m, n)$ one can introduce local coordinates

$$Z = (\mathfrak{Z}_{\alpha_1}, \ldots, \mathfrak{Z}_{\alpha_m})^{-1}(\mathfrak{Z}_{\alpha_{m+1}}, \ldots, \mathfrak{Z}_{\alpha_{m+n}})$$

corresponding to the point $\pi(\mathfrak{Z}) \in \mathfrak{V}(\alpha_1, \ldots, \alpha_m)$. $\mathfrak{Z}$ is called the homogeneous coordinate (function), Z the canonical local coordinate (function) of the point $\pi(\mathfrak{Z})$, and $\mathfrak{V}(\alpha_1, \ldots, \alpha_m)$ the canonical coordinate neighborhood. The classical domain $R_{\mathrm{I}}(m, n)$ can be considered as a hyperball defined by the matrix inequality

$$\mathfrak{Z}\begin{pmatrix} 0 & I \\ -I & 0 \end{pmatrix}\bar{\mathfrak{Z}}' > 0 \tag{2.43}$$

in the neighborhood $\mathfrak{V}(1, 2, \ldots, m)$. $R_{\mathrm{I}}(m, n)$ can serve as the "absolute" such that $G_{\mathrm{I}}(m, n)$ is the extension space of $\mathbb{C}^{mn}$.

Let $G_{\mathrm{II}}(p)$ be the complex submanifold of $G_{\mathrm{I}}(p, p)$ defined by the equations in homogeneous coordinates

$$\mathfrak{Z}\begin{pmatrix} 0 & I \\ -I & 0 \end{pmatrix}\mathfrak{Z}' = 0. \tag{2.44}$$

Then the classical domain $R_{\mathrm{II}}(p) = R_{\mathrm{I}}(p, p) \cap G_{\mathrm{II}}(p)$, and $G_{\mathrm{II}}(p)$ is the dual compact Hermitian symmetric space of $R_{\mathrm{II}}(p)$ or the extension space with respect to the absolute $R_{\mathrm{II}}(p)$.

Let $G_{\mathrm{III}}(q)$ be the complex submanifold of $G_{\mathrm{I}}(q, q)$ defined by

$$\mathfrak{Z}\begin{pmatrix} 0 & I \\ I & 0 \end{pmatrix}\mathfrak{Z}' = 0. \tag{2.45}$$

Then $R_{\mathrm{III}}(q) = R_{\mathrm{I}}(q, q) \cap G_{\mathrm{III}}(q)$, and $G_{\mathrm{III}}(q)$ is the dual compact Hermitian symmetric space of $R_{\mathrm{III}}(q)$ or the extension space with respect to the absolute $R_{\mathrm{III}}(q)$.

Denote by $G_{\mathrm{IV}}(N)$ the submanifold of $G_{\mathrm{I}}(1, N + 2)$ $(=\mathbb{CP}^{N+1})$ defined by

$$\mathfrak{Z}\begin{pmatrix} I^{(2)} & 0 \\ 0 & -I^{(N)} \end{pmatrix}\mathfrak{Z}' = 0. \tag{2.46}$$

Then the classical domain $R_{\mathrm{IV}}(N)$ is the hyperball

$$\mathfrak{Z}\begin{pmatrix} I^{(2)} & 0 \\ 0 & -I^{(N)} \end{pmatrix}\bar{\mathfrak{Z}}' > 0$$

in $G_{IV}(N)$, which is the dual compact Hermitian symmetric space of $R_{IV}(N)$ or the extension space with respect to the absolute $R_{IV}(N)$.

Let $\mathfrak{Z}_1$ and $\mathfrak{Z}_2$ be the homogeneous coordinates of two points of $G_I(m, n)$. A distance function can be defined in $G_I(m, n)$; that is (Lu [106]),

$$\sigma(\mathfrak{Z}_1, \mathfrak{Z}_2) = \text{arc cos} \frac{|\det \mathfrak{Z}_1 \bar{\mathfrak{Z}}_2'|}{[\det \mathfrak{Z}_1 \bar{\mathfrak{Z}}_1' \det \mathfrak{Z}_2 \bar{\mathfrak{Z}}_2']^{1/2}}. \tag{2.47}$$

It is remarkable that Sun [155, 156] could use such a metric in numerical analysis.

If both of the two points are in the same canonical coordinate neighborhood and their canonical coordinates are Z_1 and Z_2, respectively, then

$$\sigma(\mathfrak{Z}_1, \mathfrak{Z}_2) = \text{arc cos} \frac{\det (I + Z_1 \bar{Z}_2')}{[\det (I + Z_1 \bar{Z}_1') \det (I + Z_2 \bar{Z}_2')]^{1/2}}. \tag{2.48}$$

Taking $Z_1 = Z$ and $Z_2 = Z + \varepsilon \, dZ$ in (2.48) and dividing it by ε, we have, when $\varepsilon \to 0$, the differential metric

$$d\sigma_I^2 = \text{tr} \, [(I + Z\bar{Z}')^{-1} \, dZ \, (I + \bar{Z}'Z)^{-1} \, d\bar{Z}'] = \partial\bar{\partial} \log \det (I + Z\bar{Z}'),$$

which is a Kähler metric and also a Hua metric. When $m = n$ and this metric is restricted to the submanifold $G_{II}(n)$ or $G_{III}(n)$, we obtain the invariant metric of $G_{II}(n)$ or $G_{III}(n)$. As for $G_{IV}(N)$, the metric is

$$d\sigma_{IV}^2 = \tfrac{1}{2} \partial\bar{\partial} \log [1 + |z_1^2 + \cdots + z_N^2|^2 + 2(|z_1|^2 + \cdots + |z_N|^2)]. \tag{2.49}$$

By such a matrix realization of compact Hermitian symmetric space, many geometric results can be obtained by a more direct and simpler way, though some of them have been proved by other methods.

Ye [175] studied the geodesics of the complex Grassmannian manifold $G_I(m, n)$. Let O be a fixed point with homogeneous coordinate $(I, 0)$. The geodesic starting from O in the canonical coordinate of $\mathfrak{H}(1, 2, \ldots, m)$ is

$$Z(s) = U\Lambda(s)V, \qquad \Lambda(s) = \begin{pmatrix} \tan a_1 s & & 0 \\ & \ddots & \\ 0 & & \tan a_m s \end{pmatrix}, \tag{2.50}$$

where $U\bar{U}' = I$, $V\bar{V}' = I$, and $\sum_{j=1}^{\infty} a_j^2 = 1$. He proved that the point Q with homogeneous coordinates $(O^{(m)}, I^{(m)}, O^{(m, n-m)})$ is a conjugate point of O and the set of shortest geodesics from O to Q is isomorphic to the unitary group $U(m)$. Moreover, the index of every nonshortest geodesic from O to Q is not less than $2m(n - m + 1)$. Then by Morse theory (Milnor [128]) the homotopy group $\pi_j(G_I(m, n)) \cong \pi_{j-1}(U(m))$ for $j \le 2m(n - m + 1)$.

J.-Q. Zhong [211] used the theory of holomorphic vector fields of Carrell and Lieberman [8, 9] for Kähler manifolds to find a base of cohomology groups of G_A ($A = $ II, III). He constructed, according to the matrix formulation of G_A, a holomorphic vector field

$$v = \text{tr}\left(\mathfrak{Z}\Lambda\frac{\partial}{\partial\mathfrak{Z}'}\right), \tag{2.51}$$

where Λ is a constant diagonal matrix. Then he obtained the Poincaré polynomials (the coefficient of t^k is the kth Betti number)

$$P_{R_{II}(p)}(t) = (1 + t^2)(1 + t^4) \cdots (1 + t^{2p}),$$

$$P_{R_{III}(q)}(t) = (1 + t^2)(1 + t^4) \cdots (1 + t^{2q-2}).$$

It seems that a simpler and more general method to calculate the Betti numbers of a compact nonexceptional symmetric space is again to use the Morse theory. For example, if we restrict the projection $\pi: E(m, n) \to G_I(m, n)$ to the Stiefel manifold

$$S(m, n) = \{\mathfrak{Z} \in E(m, n) \mid \mathfrak{Z}\bar{\mathfrak{Z}}' = I\}, \tag{2.52}$$

then the function

$$\phi = \text{tr}(\mathfrak{Z}\Lambda\bar{\mathfrak{Z}}') \tag{2.53}$$

defined on $S(m, n)$ induces a function on $G_I(m, n)$. This ϕ is a Morse function, and one can calculate the index of the critical points of ϕ. By Morse theory (Milnor [128]) one would obtain the Poincaré polynomial of G_I. Since G_A ($A = $ II, III, IV) is a complex submanifold of a certain R_I, one can use the restriction of ϕ to the submanifold as its Morse function and calculate the Poincaré polynomial of it. It is possible that this method can be applied to other compact symmetric spaces. For instance, the function

$$\psi = \text{tr}(\mathfrak{X}\Lambda\mathfrak{X}'), \tag{2.54}$$

where $\mathfrak{X}$ is an $m \times (m + n)$ real matrix satisfying $\mathfrak{X}\mathfrak{X}' = I$, is a well-defined function on the real Grassmannian manifold.

Now let us list the matrix realizations of the nonexceptional and irreducible Riemannian global symmetric spaces (NIRGSS) according to an unpublished work (Lu [104]). The symbols and notations of symmetric spaces and classical groups used here are the same as those used in Chapter IX of Helgason's book [60].

Table 2.1. NIRGSS of Type III

Quotient Space	Matrix Realization
$SL(n, R)/SO(n)$	$S > 0$, $\det S = 1$, $S^{(n)} = S' = \bar{S}$
	or
	$I - TT' > 0$, $T^{(n)} = T' = \bar{T}$,
	$\det(I + T) = \det(I - T)$
$SU^*(2n)/Sp(n)$	$H > 0$, $\det H = 1$, $H^{(2n)} = \bar{H}'$, $\bar{H}J = JH$,
	$J = \begin{pmatrix} 0 & I^{(n)} \\ -I^{(n)} & 0 \end{pmatrix}$
	or
	$I - T\bar{T}' > 0$, $T^{(2n)} = -T'$, $\bar{T}J = JT$,
	$\det(J + T) = \det(J - T)$
$SU(p, q)/S(U(p) \times U(q))$	$I - ZZ' > 0$, $Z^{(p, q)}$ being complex matrix
$SO(p, q)/S(O(p) \times O(q))$	$I - XX > 0$, $X^{(p, q)}$ being real matrix
$SO^*(2n)/U(n)$	$I - Z\bar{Z}' > 0$, $Z^{(n)} = -Z'$ being complex
$Sp(n, R)/U(n)$	$I - Z\bar{Z}' > 0$, $Z^{(n)} = Z'$ being complex
$Sp(p, q)/Sp(p) \times Sp(q)$	$I - Z\bar{Z}' > 0$, $Z^{(2p, 2q)}$ being complex,
	$K_0^{(2p)}Z = \bar{Z}K_0^{(2q)}$, $K_0^{(2p)} = I^{(p)} \times \begin{pmatrix} 0 & 1 \\ -1 & 0 \end{pmatrix}$
	or
	$I - QQ^* > 0$, Q being a $p \times q$ matrix of quaternions, Q^* its conjugate and transposed matrix

Here $A \dotplus B$ of two matrices A and B means the matrix $\left(\begin{smallmatrix} A & 0 \\ 0 & B \end{smallmatrix}\right)$, and

$$A \cdot \times B = \begin{pmatrix} a_{11}B & \cdots & a_{1n}B \\ \cdots & \cdots & \cdots \\ a_{m1}B & \cdots & a_{mn}B \end{pmatrix},$$

where $A = (a_{jk})$.

The NIRGSS of the group spaces $SU(n + 1)$, $SO(2n)$, $SO(2n + 1)$, and $Sp(n)$ are themselves matrices.

Table 2.2. NIRGSS of Type I

Quotient Space	Matrix Realization
$SU(n)/SO(n)$	$S^{(n)} = S'$, $S\bar{S}' = I$, det $S = 1$
$SU(2n)/Sp(n)$	$K^{(2n)} = -K'$, $K\bar{K}' = I$, det $K = 1$
$SU(p + q)/S(U(p) \times U(q))$	The space $G_{\mathrm{I}}(p, q)$ (the complex Grassmannian manifold) formed from all $p \times (p + q)$ complex matrices of rank p as its homogeneous coordinates
$SO(p + q)/S(O(p) \times O(q))$	The space $G^R(p, q)$ (the real Grassmannian manifold) formed from all $p \times (p + q)$ real matrices of rank p as its homogeneous coordinates
$SO(2n)/U(n)$	The space $G_{\mathrm{III}}(n)$ formed from all $n \times 2n$ complex matrices $\mathfrak{Z}$ of rank n as its homogeneous coordinates and satisfying $\mathfrak{Z}\mathfrak{Z}' = 0$
$Sp(n)/U(n)$	The space $G_{\mathrm{II}}(n)$ formed from all $n \times 2n$ complex matrices $\mathfrak{Z}$ of rank n as its homogeneous coordinates and satisfying $\mathfrak{Z}J\mathfrak{Z}' = 0$, $$J = \begin{pmatrix} 0 & I^{(n)} \\ -I^{(n)} & 0 \end{pmatrix}$$
$Sp(p + q)/Sp(p) \times Sp(q)$	The space $G^Q(p, q)$ formed from all $2p \times (2p + 2q)$ complex matrices $\mathfrak{Z}$ of rank $2p$ as its homogeneous coordinates and satisfying $$\mathfrak{Z}K_0^{(2p+2q)} = K_0^{(2p)}\mathfrak{Z}, \quad K_0^{(2p)} = I^{(p)} \times \begin{pmatrix} 0 & 1 \\ -1 & 0 \end{pmatrix}$$ (the quaternionic Grassmannian manifold)

Even the exceptional bounded symmetric domains of 16 and 27 dimensions can be realized by matrices, and the geometry and analysis on these domains can be studied by their matrix formulations (Wu and Xu [162], Yin [180, 181]).

By the above matrix realizations the theory of harmonic functions can be generalized to real NIRGSS of type III (Hua and Lu [87], Sun [153]). It seems that one can apply the Morse theory to the NIRGSS of type I and calculate the index of a Morse function or a geodesic in it to get some information on the homology groups or the periodicity of the homotopy groups.

Table 2.3. NIRGSS of Type IV

Quotient Space	Matrix Realization
$SL(n + 1, C)/SU(n + 1)$	$H > 0$, $H^{(n+1)} = \bar{H}'$, $\det H = 1$
$SO(2n + 1, C)/SO(2n + 1)$	All $(2n + 1) \times (2n + 1)$ matrices $$H = RP_0\left[\begin{pmatrix} e^{-\lambda_1} & 0 \\ 0 & e^{\lambda_1} \end{pmatrix} \dotplus \cdots \dotplus \begin{pmatrix} e^{-\lambda_n} & 0 \\ 0 & e^{\lambda_n} \end{pmatrix} \dotplus 1\right]\bar{P}_0'R'$$ where R is any real orthogonal matrix, $$P_0 = I^{(n)} \cdot \times \begin{pmatrix} 2^{-1/2} & 2^{-1/2} \\ i2^{-1/2} & -i2^{-1/2} \end{pmatrix} \dotplus 1 \text{ and}$$ $\lambda_1 \geq \cdots \geq \lambda_n \geq 0$
$Sp(n, C)/Sp(n)$	All $2n \times 2n$ matrices $$H = U\begin{pmatrix} \Lambda & 0 \\ 0 & \Lambda^{-1} \end{pmatrix}\bar{U}', U \text{ being unitary,}$$ $$\bar{U}J = JU, J = \begin{pmatrix} 0 & I^{(n)} \\ -I^{(n)} & 0 \end{pmatrix},$$ $$\Lambda = \begin{pmatrix} \lambda_1 & & \\ & \ddots & \\ & & \lambda_n \end{pmatrix}, \text{ and } \lambda_1 \geq \cdots \geq \lambda_n > 0$$
$SO(2n, C)/SO(2n)$	All $2n \times 2n$ matrices $$H = RP_0\left[\begin{pmatrix} e^{-\lambda_1} & 0 \\ 0 & e^{\lambda_1} \end{pmatrix} \dotplus \cdots \dotplus \begin{pmatrix} e^{-\lambda_n} & 0 \\ 0 & e^{\lambda_n} \end{pmatrix}\right]\bar{P}_0'R'$$ where R is a real orthogonal matrix, $$P_0 = I^{(n)} \cdot \times \begin{pmatrix} 2^{-1/2} & 2^{-1/2} \\ i2^{-1/2} & -i2^{-1/2} \end{pmatrix}, \text{ and}$$ $\lambda_1 \geq \cdots \geq \lambda_n \geq 0.$

Though it is somewhat different from several complex variables, it is interesting to state some of the results for illustration. In Table 2.1, the space

$$\mathcal{R}(m, n) = \{X \in \mathbb{R}^{m \times n} \mid I - XX' > 0\} \tag{2.55}$$

is a NIRGSS. It is not hard to see that the Hua metric in this space is

$$ds^2 = \text{tr}\left[(I - XX')^{-1} dX (I - X'X)^{-1} dx'\right] \tag{2.56}$$

(Hua and Lu [6]) and the corresponding Laplace-Beltrami operator is

$$
\Delta = \sum_{j,k=1}^{m} \left(\delta_{jk} - \sum_{\gamma=1}^{n} x_{j\gamma} x_{j\gamma} \right)
$$

$$
\cdot \left[\sum_{\alpha,\beta=1}^{n} \left(\delta_{\alpha\beta} - \sum_{l=1}^{m} x_{l\alpha} x_{l\beta} \right) \frac{\partial^2}{\partial x_{j\alpha} \partial x_{k\beta}} - 2 \sum_{\alpha=1}^{n} x_{j\alpha} \frac{\partial}{\partial x_{k\alpha}} \right]. \tag{2.57}
$$

Then the Poisson integral (when $m \leq n$)

$$
u(X) = \frac{1}{V(\mathscr{S})} \int_{\Gamma\Gamma'=I} \phi(\Gamma) \frac{\det (I - XX')^{(n-1)/2}}{\det (I - X\Gamma')^{n-1}} \dot{\Gamma} \tag{2.58}
$$

is the unique solution of the Dirichlet problem

$$
\Delta u(X) = 0 \quad \text{and} \quad \lim_{X \to \Gamma_0} u(X) = \phi(\Gamma_0), \qquad \Gamma_0 \in \mathscr{S} = \{\Gamma \in \mathbb{R}^{m \times n} | \Gamma\Gamma' = I\} \tag{2.59}
$$

for any given function $\phi(\Gamma)$ continuous on the characteristic manifold $\mathscr{S}$.

In particular, when $m = 1$, $n = 2$, and $X = (x, y)$, the equation

$$
\Delta u := (1 - x^2 - y^2) \left[(1 - x^2) \frac{\partial^2 u}{\partial x^2} - 2xy \frac{\partial^2 u}{\partial x \partial y} + (1 - y^2) \frac{\partial^2 u}{\partial y^2} - 2x \frac{\partial u}{\partial x} - 2y \frac{\partial u}{\partial y} \right]
$$

$$
= 0 \tag{2.60}
$$

is a differential equation of mixed type in the entire (x, y)-plane. Hua [80, 81] studied this equation in the (x, y)-plane and inspired a series of works in China on mixed-type equations.

In the theory of functions of one complex variable, one of the most important theorems is the uniformization theorem related to Riemann surfaces. The universal covering space of any Riemann surface is a global Hermitian symmetric space, which is either the complex plane $\mathbb{C}$ with zero curvature or the sphere with positive constant curvature or the unit disk with negative constant curvature. Therefore, whether a Kähler manifold is a Hermitian symmetric space is important for several complex variables. Mok and Zhong [130, 131] proved that when M is a compact Kähler-Einstein manifold of nonnegative holomorphic bisectional curvature and positive Ricci curvature, M must be isometric to a compact Hermitian symmetric space. Since this result is reviewed by Siu in Chapter 3, it will not be discussed here any further.

2.3. Homogeneous Manifolds

É. Cartan [10] asked whether every bounded homogeneous domain is symmetric. Pyatetski-Shapiro [133] gave the first counterexamples in 1959. Later Lu and Xu [123] gave counterexamples to a conjecture of Hua—the curvature of a bounded homogeneous domain is nonpositive. However, Pyatetski-Shapiro still called the numerous bounded homogeneous domains the classical domains. So, correspondingly, a manifold acted on transitively by a classical group of matrices will be called a classical manifold in this generalized sense.

Pyatetski-Shapiro [134] proved that every bounded homogeneous domain must be holomorphically equivalent to a Siegel domain. However, not every Siegel domain is a homogeneous domain. Lu [102, 103] wished to prove that every bounded homogeneous domain could be realized by a hyperball in an extended space. This is true to a certain extent (Lu and Zhong [125]), but the realization is too complicated to be reviewed here.

Xu [164–168] realized the homogeneous Siegel domains by a set of matrices that satisfy a set of matrix equations and called them N-Siegel domains. The exact definition of an N-Siegel domain is omitted here, because it is hard to write it clearly in less than two pages. The reader should refer to Xu's original papers. However, Xu proved that any N-Siegel domain is homogeneous and that any homogeneous Siegel domain is linearly equivalent to an N-Siegel domain. The construction of the Bergman kernel $K(z, \bar{t})$ and the Cauchy-Szegö kernel $H(z, \bar{u})$ was known (Gindikin [32], Xu [167]). Xu [170] proved that any bounded homogeneous domain is an L-domain (the so-called Lu Qi-Keng domain)—that is, a domain D whose Bergman kernel $K(z, \bar{t})$ does not vanish for all $(w, t) \in D \times D$. Moreover, Wu and Xu [162] estimated the lower and upper bounds of the unitary curvature of a bounded homogeneous domain and gave the exact values of the lower bounds for the exceptional bounded symmetric domains R_V of 16 dimensions and R_{VI} of 27 dimensions; that is, $L(R_V) = -1/6$ and $L(R_{VI}) = -1/9$.

The Šilov boundary S of a homogeneous Siegel domain D can be defined. Let the Poisson kernel be formally defined by

$$P(z, u) = \frac{|H(z, \bar{u})|^2}{H(z, \bar{z})}, \qquad (z, u) \in D \times S$$

[cf. (1.28)]. Then, contrary to the usual situation, R.-Q. Lu [127] gave the first example of a homogeneous Siegel domain such that $\Delta P(z, u) \neq 0$, where Δ is the Laplace-Beltrami operator of the given domain. Then the method used by Hua and Lu [87] to solve the Dirichlet problem by the

Poisson integral fails. Xu [167] proved that a necessary and sufficient condition that $P(z, u)$ satisfy the Laplace-Beltrami equation corresponding to the Bergman metric is that the homogeneous Siegel domain be symmetric.

Of course, the exceptional domains R_V and R_{VI} are among the N-Siegel domains. Xu exhibits them explicitly, and J.-Q. Zhong also knew this fact.

As already noted in Section 2.1, Hua's metric is proportional to the Bergman metric for irreducible symmetric domains, and this is not true for reducible domains in the case of bounded symmetric domains. But for nonsymmetric bounded domains Hua's metric is, in general, not proportional to the Bergman metric, even if the homogeneous domain is irreducible, because the isotropy group of such a domain is, in general, "small"; hence, there may be many different invariant metrics. The corresponding Laplace-Beltrami operators are also different. It is interesting to see how many invariant differential operators there are in a given homogeneous domain.

Denote by $I(D)$ the ring of linear differential operators invariant under the connected component of the Lie group Aut (D). When D is a global symmetric space, it is known (for example, see Helgason [60]) that $I(D)$ is generated by $r = \text{rank}(D)$ invariant differential operators. Gong Hui-Sheng and Ye Fang-Cao [33] constructed explicitly a set of invariant generators of $I(D)$ when D is nonexceptional. Furthermore, Ye [176] constructed the generators of $I(D)$ for all NIRGSS in Tables 2.1-2.3. The number of generators or the dimension of the ring $I(D)$ depends on the group Aut (D): the "smaller" the group Aut (D), the bigger the number of generators. If D is a bounded homogeneous domain, Selberg's result [136] implies that the dimension of $I(D)$ is finite. J.-Q. Zhong [207] explored a method to calculate this dimension by the theory of group representations. Denote by K the connected component of the isotropy group $\text{Aut}_0(D)$. It may be assumed to be a compact subgroup of $U(n)$ [11]. Let $I_{k,l}(D)$ be the subset of $I(D)$ such that $L \in I_{k,l}(D)$ means that its highest nonvanishing coefficient is a (k, l)-type contravariant tensor. He obtained the following recurrence formula:

$$\dim \left[\frac{I_{k,l}(D)}{\sum_{(k_1, l_1) < (k, l)} I_{k_1, l_1}(D)} \right] = \int_K \chi_{(k, 0, \ldots, 0)}(K) \chi_{(l, 0, \ldots, 0)}(K) \dot{K},$$

where $\chi_{(k, 0, \ldots, 0)}$ is the character of the irreducible representation of $U(n)$ corresponding to the index $(k, 0, \ldots, 0)$ (cf. Hua [79]).

In 1962 to 1963 Lu [102, 103] constructed a class of homogeneous complex manifolds $G(r_1, \ldots, r_p; s_1, \ldots, s_p)$, which are submanifolds of the Grassmannian manifold $G_1(r_1 + \cdots + r_p, s_1 + \cdots + s_p)$ such that the closure of such a submanifold is a compact complex space. Then suitable hyperballs were introduced in $G(r_1, \ldots, r_p; s_1, \ldots, s_p)$ such that they are homogeneous domains, most of which are not symmetric but holomorphically equivalent to a bounded domain in $\mathbb{C}^N$. J.-Q. Zhong and Yin [217] generalized this construction. They started with the construction of homogeneous domains whose underlying convex cones (in the sense of Pyatetski-Shapiro [133]) are not self-conjugate. (Most of the underlying cones of the homogeneous domains constructed by Pyatetski-Shapiro are self-conjugate.) Then they studied the unitary geometry of such domains and constructed their extension spaces. It should be noted that the works of Hua's school in studying several complex variables (even in some other fields) are constructive and intuitive but sometimes complicated.

2.4. Integral Representations and Their Boundary Values

In 1957, Q.-K. Lu and T.-D. Zhong [126] studied the boundary value of the Bochner-Martinelli integral. Let D be a bounded domain in $\mathbb{C}^n$ with C^2 boundary ∂D, and $\phi(u)$ a function satisfying the Hölder condition on ∂D. Then the principal value (p.v.) of the Cauchy integral

$$\Phi(z) := \int_{\partial D} \phi(u)B(z, u)$$

exists, where $B(z, u)$ is the Bochner-Martinelli kernel

$$B(z, u) = \frac{(n-1)!}{(2\pi d)^n} \sum_{k=1}^{n} (-1)^{k-1} \frac{u_k - z_k}{|u - z|^{2n}}$$

$$\times du_1 \wedge \cdots \wedge d\hat{u}_k \wedge \cdots \wedge du_n \wedge d\bar{u}_1 \wedge \cdots \wedge d\bar{u}_n.$$

Moreover, if $v \in \partial D$, then

$$\Phi^+(v) := \lim_{\substack{z \to v \\ z \in D}} \Phi(z) = \text{p.v.} \int_{\partial D} \phi(u)B(v, u) + \tfrac{1}{2}\phi(v),$$

$$(2.61)$$

$$\Phi^-(v) := \lim_{\substack{z \to v \\ z \in \mathbb{C}^n \setminus \bar{D}}} \Phi(z) = \text{p.v.} \int_{\partial D} \phi(u)B(v, u) - \tfrac{1}{2}\phi(v).$$

Eighteen years later, Harvey and Lawson [58] rediscovered this formula in an appendix and called it the Plemelj formula. When $n > 1$, $B(z, u)$ is, in general, not holomorphic with respect to $z \in D$ when $u \in \partial D$. But for special domains there are various Cauchy–Szegö kernels $H(z, u)$ that are holomorphic with respect to z. It is interesting to see whether the Plemelj formula holds for such kernels. The first attempt was made by Gong and Sun [52] in 1965 by using Hua's Cauchy–Szegö kernel in a ball $R_I(1, n)$ [see (2.29)]. They proved that various principal values can be defined. Given any two nonnegative constants α and β ($\alpha + \beta \neq 0$), they defined a principal value such that, for $v \in \partial R_I(1, n)$,

$$\lim_{z \to v} \int_{u \bar{u}' = 1} \frac{\phi(u)}{(1 - z\bar{u}')^n} \, \dot{u} = \text{p.v.} \int_{u \bar{u}' = 1} \frac{\phi(u)}{(1 - v\bar{u}')^n} \, \dot{u} + \frac{1}{2} \left(\frac{2\beta}{\alpha + \beta} \right)^{n-1} \phi(v).$$

A similar Plemelj formula holds even for the Henkin integral for a strictly pseudoconvex domain (Gong and Shi [47]).

After the Plemelj formula of a certain kind of Cauchy integral has been established, the further problem is to study the singular integral equations of several complex variables. This is reduced to establishing three formulas:

1. The transposed formula of a singular double integral
2. The composition formula of singular integrals
3. The inverse formula of a singular integral

All these formulas can be found in the books of Gong [41] and T.-D. Zhong [227].

Some authors (S.-T. Chen [15–18], Li [90], Yao [174], L.-Y. Liu [92], J.-X. Chen [14], He [59], etc.) constructed other Cauchy integrals and studied their boundary behavior.

As noted by Norguet [132], most of the various Cauchy formulas can be derived from the Cauchy–Fantappiè formula established by Leray [89]. In 1966, Lu [108] proved that Hua's Cauchy formulas for classical domains can also be derived from the Cauchy–Fantappiè formula. Since then, the publication of the books by Aizenberg and Juzakov [1] and by Henkin and Leiterer [61] in 1979 and 1984, respectively, reconfirmed the importance of the Cauchy–Fantappiè formula.

2.5. The Schwarz Lemma

Let D be a bounded domain in $\mathbb{C}^n$ and

$$T(z, \bar{z}) = \left(\frac{\partial^2 \log K(z, \bar{z})}{\partial z_\alpha \, \partial \bar{z}_\beta} \right)_{1 \leq \alpha, \beta \leq n},$$

the matrix of the Bergman metric tensor. If $f = (f_1, \ldots, f_n): D \to \mathbb{C}^n$ is holomorphic and bounded by a constant M (i.e., $f\bar{f}' \le M^2$), then Lu [94] proved in 1957 that

$$\frac{\partial f}{\partial z} \frac{\partial \bar{f}'}{\partial \bar{z}} \le M^2 T(z, \bar{z}). \qquad (2.62)$$

Especially when $f = (f_1, 0, \ldots, 0)$ and $M = 1$, (2.62) obviously implies

$$\sup_{f_1 \in H_1^\infty(D)} |df_1|^2 \le ds^2, \qquad (2.63)$$

where ds^2 is the Bergman metric and the left-hand side is exactly the Carathéodory differential metric defined eight years later by Reiffen [135]. Twenty years later Hahn [57] reproduced the proof of (2.62) and noted that the Carathéodory differential metric is not greater than the Bergman metric.

If D is homogeneous and Hom (D) is the set of all holomorphic maps of D into D, then there is a constant $k(D)$ such that (Lu [94])

$$f^* ds^2 \le k^2(D) ds^2 \qquad \text{for all } f \in \text{Hom } (D). \qquad (2.64)$$

Obviously $k^2(D)$ is an analytic invariant. If D is symmetric and irreducible, $k^2(D)$ is equal to the rank of D (Lu [94] for nonexceptional domains and Koranyi [88] for general symmetric domains). Since the Hua metric is proportional to the Bergman metric for irreducible symmetric domains, (2.64) holds for the Hua metric too. Moreover, for Hua metrics,

$$k^2(D_1 \times D_2) = k^2(D_1) + k^2(D_2) \qquad (2.65)$$

for any two symmetric domains D_1 and D_2. If we denote by $L(R_A)$ and $U(R_A)$ the lower and upper bounds of the holomorphic sectional curvature of the classical domains, then

$$k^2(R_A) = \frac{L(R_A)}{U(R_A)}, \qquad A = \text{I, II, III, IV.} \qquad (2.66)$$

This revealed for the first time the relation between the Schwarz lemma and curvature in several complex variables, but in one complex variable it was due to Ahlfors. Both $L(R_A)$ and $U(R_A)$ are analytic invariants and satisfy the relations

$$L(R_A \times R_B) = \min\{L(R_A), L(R_B)\}, \qquad \frac{1}{U(R_A \times R_B)} = \frac{1}{U(R_A)} + \frac{1}{U(R_B)}.$$

(2.67)

Since $k^2(R_A)$ and $L(R_A)$ are known for all irreducible bounded symmetric domains, it is easy to prove that these two invariants form a complete set of analytic invariants that characterize such domains.

The Schwarz lemma can be regarded as an estimate of the first derivatives of a bounded holomorphic map $f: D \to C^N$ by the Bergman metric tensor. The derivatives of higher orders of a map f can also be estimated by the Bergman metric tensor and the quantities related to the unitary curvature tensor $R_{\bar{p}jk\bar{q}}$ (Lu [110, 112]). When D is a classical domain and $f \in \mathrm{Hom}\,(D)$, then (Lu [97])

$$\sum_{j,k,p,q=1}^{n} R_{\bar{p}jk\bar{q}}(f(z)) \, \overline{df^p} \, df^j \, df^k \, \overline{df^q}$$

$$\geq k^2(D) \sum_{j,k,p,q=1}^{n} R_{\bar{p}jk\bar{q}}(z) \, \overline{dz^p} \, dz^j \, dz^k \, \overline{dz^q}.$$

(2.68)

Lu [113] generalized the Schwarz inequality to quartic forms. Let D be a Kähler manifold with a complex structure J. If its sectional curvature is nonpositive or nonnegative at a point, then at this point

$$\langle R(X, Y)Y, X \rangle^2 \leq \langle R(X, JX)JX, X \rangle \langle R(Y, JY)JY, Y \rangle.$$

(2.69)

One can use this inequality to calculate the lower or upper bounds of the sectional curvature of R_A.

When D is a complete Kähler manifold, Chen, Cheng, and Lu [22] proved the following Schwarz lemma, which is a modified form of S.-T. Yau's Schwarz lemma. Let M be a complete Kähler manifold whose holomorphic sectional curvature is bounded from below by K_1 and whose Riemannian sectional curvature is also bounded from below. Let N be a Hermitian manifold whose holomorphic sectional curvature is bounded from above by $K_2 < 0$. If $f: M \to N$ is a nonconstant holomorphic map, then

$$f^* \, ds_N^2 \leq \frac{K_1}{K_2} \, ds_M^2.$$

(2.70)

Chen and Yang [25] generalized this theorem to a Hermitian manifold M by adding a condition to its holomorphic bisectional curvature and to the torsion. However, the method they used is the same as that in Chen *et al.* In a later work (Chen and Yang [26]) they developed a method that they called holomorphic variation. It can be roughly described in terms of local coordinates as follows. Given a geodesic $\gamma(t)$ in a Hermitian manifold M with Hermitian metric $ds^2 = \sum_{j,k=1}^m g_{j\bar{k}}\, dz^j\, \overline{dz^k}$, they vary the geodesic $\gamma(t)$ $(0 \le t \le \rho)$ in the following manner:

$$z^\alpha(t, w) = \gamma^\alpha(t) + w\xi^\alpha(t), \qquad |w| < \varepsilon,$$

where w is a complex parameter and $\xi^\alpha(0) = \xi^\alpha(\rho) = 0$. Let

$$L(w) = \int_0^\rho \left[\sum_{j,k=1}^m g_{j\bar{k}}\, dz^j\, \overline{dz^k} \right]^{1/2}.$$

Then they calculated $\partial L(w)/\partial w\, \partial \bar{w}|_{w=0}$, the complex second variation, and obtained some comparison theorems in terms of holomorphic sectional, bisectional curvatures and quantities related to the torsion; in particular, they derived by this method the Schwarz lemmas of Chen *et al.* and others.

In 1969, Chern, Levine, and Nirenberg [31] introduced a metric defined by one-dimensional homology classes on a complex manifold M. Bedford and Taylor [2] generalized the definition to higher-dimensional homology classes; that is, if $\gamma \in \dot{H}_*(M, \mathbb{R})$, this norm is defined by

$$N(\gamma) = \begin{cases} \sup_{u \in \mathscr{F}} \inf_{T \in \gamma} |T(d^c u \wedge (dd^c u)^{k-1})|, & \text{if } \dim \gamma = 2k - 1, \\[2mm] \sup_{u \in \mathscr{F}} \inf_{T \in \gamma} |T(d^c u \wedge (dd^c u)^{k-1} \wedge du)|, & \text{if } \dim \gamma = 2k, \end{cases}$$

where $\mathscr{F}$ is the family of C^2 plurisubharmonic functions u such that $0 < u < 1$, and T is the current that represents γ. The norm $N(\partial\Omega')$ can be computed when $M = \Omega' - \Omega_1$, where Ω' and Ω_1 are concentric balls. J.-H. Zhang [196] studied the case when M is a domain homeomorphic to a ball and Ω', Ω_1 are

$$\Omega' = \{z \in M \,||f_j| < a_j, a_j > 1, j = 1, \ldots, n\},$$

$$\Omega_1 = \{z \in M \,||f_j| < 1, j = 1, \ldots, n\},$$

where $f_1, \ldots, f_n$ are holomorphic in M. He proved that

$$N(\partial\Omega') \ge \frac{(2\pi)^n}{(n-1)! \log a_1 \cdots \log a_n},$$

and equality holds when $f_j = z_j$.

2.6. Harmonic Analysis in Classical Domains
and on Their Characteristic Manifolds

Hua [79] noted that the characteristic manifold S_I of the classical domain $R_I(m, n)$, when $m = n$, is the unitary group $U(n)$. The Poisson kernel can be developed into a series

$$\frac{1}{c} P_I(Z, U) = \sum_{\nu=0}^{\infty} \psi_\nu(Z)\phi_\nu(U), \tag{2.71}$$

where $c = V(S_I)$ and $\{\phi_\nu(U)\}_{\nu=0, 1, 2,\ldots}$ is a complete orthonormal system, which can be chosen as the elements from all irreducible representations of $U(n)$ except for normalizing factors, and $\psi_\nu(Z)$ is harmonic in R_I such that $\psi_\nu(rV) = \rho_\nu(r)\phi_\nu(V)$ for $V \in U(n)$. The series (2.91) is uniformly convergent when Z is on any compact set in R_I and V is in $U(n)$. For a continuous function $\phi(U)$ defined on $U(n)$, the series

$$\sum_{\nu=0}^{\infty} a_\nu\phi_\nu(U) \quad \text{with } a_\nu = \int_{S_I} \phi(V)\overline{\phi_\nu(U)}\dot{U}$$

is called the Fourier series of $\phi(U)$, and it is, in general, not convergent in the usual sense. However, the series

$$\sum_{\nu=0}^{\infty} \rho_\nu(r)a_\nu\phi_\nu(U) = \frac{1}{c}\int_{S_I} \phi(V)P_I(rU, V)\dot{V} \tag{2.72}$$

is convergent, and by (2.26) it tends to $\phi(U)$ when $r \to 1$. Hua [78] called (2.72) the Abel summability of the Fourier series of $\phi(U)$. Gong [34–38] determined all $\rho_\nu(r)$ by a series of complicated calculations. J.-Q. Zhong [204] introduced a determinant of integrals

$$L(A; f_1,\ldots,f_n; g_1,\ldots,g_n) = \det\left(\frac{1}{2\pi i}\int_\Gamma \frac{f_j(z)g_k(z)}{A(z)}\,dz\right)_{1\le j,\, k\le n},$$

where Γ is a simple closed curve in the complex plane and $A(z), f_j(z), g_j(z)$ are functions holomorphic in the open set bounded by Γ and continuous on the closure. Taking $A(z) = \prod_{j=1}^{n}(z - z_j)$, where z_j are points inside Γ, Zhong proved various algebraic identities that are useful for group representations; in particular, he calculated by a simpler method the functions $\rho_\nu(r)$.

It is surprising that J.-Q. Zhong [212, 215] was able to apply the theory of group representations to the Schubert calculus, and he used the aforementioned algebraic identities to calculate the product of two Schubert cycles.

In the complex Grassmannian manifold $G_1(m, n)$, denoted by σ_a the Schubert cycle corresponding to the index $a = (a_1, \ldots, a_m)$ $(n \geq a_1 \geq \cdots \geq a_m \geq 1)$. The product $\sigma_a \cdot \sigma_b$ of two Schubert cycles σ_a and σ_b means their intersection, which can again be expressed as a linear combination of Schubert cycles σ_c,

$$\sigma_a \cdot \sigma_b = \sum_c \delta(a, b; c)\sigma_c,$$

where the $\delta(a, b; c)$'s are integers. He proved that when $a = (a_1, \ldots, a_k)$, $b = (b_1, \ldots, b_l)$, and $c = (c_1, \ldots, c_{k+l})$,

$$\delta(a, b; c) = \int_{U(k) \times U(l)} \sigma_c(t_1, \ldots, t_{k+l})\sigma_a(t_1, \ldots, t_k)\sigma_b(t_1, \ldots, t_l)\dot{U}(k)\dot{U}(l),$$

where

$$\sigma_{(a_1, \ldots, a_m)}(t_1, \ldots, t_m) = \frac{\det (t_j^{a_k+k-j})_{1 \leq j, k \leq m}}{\prod_{j \neq k} (t_j - t_k)}$$

is the character of the irreducible representation of $U(m)$ corresponding to the index a. The details of the proof will appear in Zhong [215].

In the Poisson integral (2.58) of a real symmetric domain, Zhong [205] noted that in the special case $m = n$ it is sufficient to take the integral over the real orthogonal group $SO(n)$ instead of $O(n)$. He developed the Poisson kernel $P(rI, \Gamma)$ in (1.58) into a series

$$P(rI, U) = \sum_m \rho_m(r)\sigma_m(\Gamma), \qquad \Gamma \in SO(n),$$

where $\sigma_m(\Gamma)$ are the characters of all the irreducible representations of $SO(n)$. He calculated all the $\rho_m(r)$ and obtained correspondingly an Abel summability for Fourier series of a continuous function on $SO(n)$.

Since the symmetric space $Sp(p, q)/Sp(p) \times Sp(q)$ in Table 2.1 can be realized in a matrix form similar to the classical domains, Sun [153], as mentioned before, obtained the Poisson kernel of such a space. A corresponding Abel summability could work for the Fourier series on the symplectic unitary group $Sp(n)$ (Chen and He [13]).

For further development in China of harmonic analysis on a compact classical group, one could consult the book by Gong [42].

Another kind of harmonic analysis is the study of the eigenfunctions of the Laplace-Beltrami operator; i.e.,

$$\Delta u + \lambda u = 0,$$

where u is called the eigenfunction and λ the eigenvalue. G.-X. Chen [12] noted that for any complex constant c the Poisson kernel of R_I satisfies

$$[P_I(Z, U)]^c = \lambda(c)[P_I(Z, U)]^c,$$

where $\lambda(c)$ is a constant depending only on c. This shows that Δ has a continuous spectrum. This is much different from a compact manifold, where the eigenvalues are discrete and nonnegative. Zhong and Yang [216] estimated the first positive eigenvalue and gave a sharp result: if M is a compact Riemannian manifold with nonnegative Ricci curvature, then the first eigenvalue $\lambda_1 \geq (\pi/d)^2$, where d is the diameter of M. This result was reviewed in Chapter 1 of this volume.

2.7. Pseudoconvex Domains

In 1966, Lu [107] proved that if D is a bounded domain in $\mathbb{C}^n$ such that the Bergman metric is complete and its unitary (holomorphic sectional) curvature is constant, then there exists a unique biholomorphic map f:

$$w_j = f_j(z_1, \ldots, z_n), \qquad j = 1, \ldots, n,$$

such that $B = f(D)$ is a ball and f maps a given point $t \in D$ into the center of B and

$$\left. \frac{\partial w_j}{\partial z_k} \right|_{z=t} = \delta_{jk};$$

this map can be expressed as

$$f_j = \sum_{k=1}^{n} T^{kj}(t, \bar{t}) \frac{\partial}{\partial \bar{t}^k} \log \frac{K(z, \bar{t})}{K(z, t)}, \tag{2.73}$$

where $K(z, \bar{t})$ is the Bergman kernel of D and $T^{kj}(t, \bar{t})$ is the inverse of the matrix $[\partial^2 \log K(t, \bar{t})]/\partial t_j \partial \bar{t}_k$.

The map (2.73) can be defined for any bounded domain D if $K(z, \bar{t})$ possesses no zero in $(z, t) \in D \times D$. Lu said "it seems that no one has proved that $K(z, \bar{t})$ possesses no zero in D, though a number of examples show that it is true." Skwarziński [152] called $K(z, \bar{t})$ possessing no zero in D the Lu Qi-Keng conjecture, but in the meantime he gave a counterexample to it. Then he called the domain in which the Bergman kernel possesses no zero a Lu Qi-Keng domain. Since then a number of authors have studied this conjecture. But papers that gave counterexamples were much more numerous than papers that gave positive answers. R. E. Greene and H. Wu asked in 1979 whether this conjecture is true for a bounded simply connected and strictly pseudoconvex domain. But even then Boas [6] gave counter-examples to it. It remains to see whether one can give a counterexample for a (geometrically) convex domain.

Yin [186–188] studied Reinhardt domains $D(\alpha) := \{z \in \mathbb{C}^n \mid |z_1|^{2\alpha} + |z_1|^2 + \cdots + |z_n|^2 < 1\}$ $(\alpha > 0)$. This is a pseudoconvex domain, but the boundary is not real analytic except when α is an integer. Yin proved that $D(\alpha)$ possesses a Kähler-Einstein metric, and the unitary curvature with respect to any Aut (D)-invariant Kähler metric is constant on an $(n-1)$-subspace of the tangent space $T^{1,0}(D)$.

A local holomorphic function on the closure $\bar{D}$ of a bounded domain D is a function holomorphic either on an open set $V \subset D$ or on $V \cap D$ and smooth up to $V \cap \partial D$. Zhang Jin-Hao [199] denoted by $\mathfrak{G}$ the sheaf of germs of local holomorphic functions and obtained such a vanishing theorem: if D is a bounded pseudoconvex domain with smooth boundary and for every point of D the subelliptic estimate of the $\bar{\partial}$-Neumann problem for (p, q)-forms holds for $q > 0$, then $H^q(\bar{D}, \mathfrak{G}) = 0$.

Let D' be a Stein neighborhood of D and V' a subvariety of D' and $V = V' \cap D$. V' and ∂D are said to be regularly separated if for any point $z \in V' \cap \partial D$, there exists a neighborhood U and $m > 0$ such that

$$\text{dist}\,(z, U \cap V' \cap \partial D)^m < c\,\text{dist}\,(z, U \cap V'), \qquad c > 0.$$

Denote by $\mathfrak{G}V$ the sheaf of germs of local holomorphic functions vanishing on V and by $\Gamma(\bar{D}, \mathfrak{G})$ the sections of $\mathfrak{G}_V$. Applying the foregoing theorem, Zhang [201] proved that when D satisfies the same condition of the theorem, the subvarieties $(V' \cap \partial D) = 0$, and V' and D are regularly separated, then $\Gamma(\bar{D}, \mathfrak{G}_V)$ can be generated by functions $f_1, \ldots, f_k$ of $\Gamma(D, \mathfrak{G})$, with $k = \text{codim } V$.

References

The papers of Chinese authors listed in the following are written in Chinese except those marked "in English."

1. L. A. Aizenberg and A. P. Juzakov, *Integral Representation and Residues in Multidimensional Complex Analysis*, Nauka, Sigirsk. otdel., Novosibirsk (1979) (in Russian).

2. E. Bedford and B. A. Taylor, The Dirichlet problem for a complex Mongé-Ampère equation, *Invent. Math.* **37**, 1–44 (1976).

3. E. Bedford and B. A. Taylor, Variational properties of the complex Monof-Ampère equation and intrinsic norms, *Am. J. Math.* **7**, 249–279 (1979).

4. H. Behnke and P. Thullen, *Theorie der Funktionen Mehreren Komplexer Veränderlichen*, Ergebnisse Math., Springer, Berlin (1933).

5. S. Bergman, *Sur les Fonctions Orthogonales de Plusieurs Variables Complexes avec les Applications à la Theorie des Fonctions*, Gauthier-Villars, Paris (1947).

6. H. Boas, Counterexample to the Lu Qi-Keng conjecture, *Proc. Am. Math. Soc.* **97**, 36–37 (1986).

7. C. Carathéodory, Ueber das Schwarz Lemma bei analytischen Funktionen von mehreren Veränderlichen, *Math. Ann.* **97**, 76–98 (1927).

8. J. Carrell and D. Lieberman, Holomorphic vector fields and Kähler manifolds, *Invent. Math.* **21**, 303–309 (1973).

9. J. Carrell and D. Lieberman, Chern classes of the Grassmannians and Schubert calculus, *Topology* **17**, 177–183 (1978).

10. É. Cartan, Sur les domaines bornés homogènes de l'éspace de n variables complexes, *Hamburg Univ. Math. Sem. Abh.* **11**, 106–162 (1936).

11. H. Cartan, *Sur les Groupes de Transformations Analytiques*, Actualités Sci. Ind., Exposés Math., IX, Paris (1935).

12. Guang-Xiao Chen, On the fundamental solutions of Poisson equation of matrix classical domain, *Sci. Sinica Ser. A* **8**, 813–825 (1987) (in English).

13. Guang-Xiao Chen and Jao-Qi He, Dissertation, Beijing (1981).

14. Jian-Xin Chen, The Cauchy type integral of $(p, n-1)$-forms, *J. Xiaman Univ.* **22**, 398–405 (1983).

15. Shu-Jin Chen, The integral representation on a convex domain in $\mathbb{C}^n$, *Acta Math. Sinica* **22**, 743–750 (1979).

16. Shu-Jin Chen, An integral representation in $\mathbb{C}^n$, *Acta Math. Sinica* **24**, 538–544 (1981).

17. Shu-Jin Chen, The Leray-Stokes formula of a polyhedron in $\mathbb{C}^n$, *Kexue Tongbao* **21**, 1292–1295 (1982).

18. Shu-Jin Chen, The Sakotski-Plemelj formula on a class of Cauchy-Fantappié type integral, *Kexue Tongbao* **23**, 267–273 (1984).

19. Shu-Jin Chen, The integral representation of holomorphic functions in $\mathbb{C}^n$, *J. Univ. Xiaman* **25**, 1–9 (1986).

20. Zhi-Hua Chen, A note on Kähler immersion, *Kexue Tongbao, Math. Special Ser.* **25**, 111–114 (1980).

21. Zhi-Hua Chen, Some theorems on the totally geodesic maps, *Sci. Sinica Ser. A* **2**, 97–103 (1985).

22. Zhi-Hua Chen, Shiu-Yuen Cheng, and Qi-Keng Lu, On the Schwarz lemma for complete Kähler manifolds, *Sci. Sinica* **12**, 1238–1247 (1979) (in English).

23. Zhi-Hua Chen and Hong-Cang Yang, A Schwarz lemma for a compact manifold whose Ricci curvature is bounded from below, *Acta Math. Sinica* **24**, 945–952 (1981).

24. Zhi-Hua Chen and Hong-Cang Yang, Estimation of decreasing coefficients of K-dilatation harmonic maps, *Kexue Tongbao, Foreign Land Ed.* **28**, 979–982 (1983).

25. Zhu-Hua Chen and Hong-Cang Yang, On Schwarz lemma for complete Hermitian manifolds, in *Several Complex Variables* (Proc. 1981 Hangzhou Conf.), pp. 99–116, Birkhauser (1984) (in English).

26. Zhi-Hua Chen and Hong-Cang Yang, Estimation of the upper bound on the Levi form of the distance function on Hermitian manifolds and its applications, *Acta Math. Sinica* **27**, 631–643 (1984).

27. Zhi-Hua Chen and Hong-Cang Yang, The Schwarz lemma of quasi-conformal harmonic mappings, *Progress in Math.* **13**, 311–317 (1984).

28. Zhi-Hua Chen and Hong-Cang Yang, Torsion and Steinness, *Sci. Sinica Ser. A.* **27**, 457–482 (1984) (in English).

29. Zhi-Hua Chen and Hong-Cang Yang, A class of Liouville theorems, *Acta Math. Sinica* **28**, 218–232 (1985).

30. Zhi-Hua Chen and Hong-Cang Yang, A note on Steinness, *Chinese Ann. Math. Ser. A* **7**, 119–122 (1986).

31. S. S. Chern, H. I. Levine, and L. Nirenberg, Intrinsic norms on a complex manifold, in *Global Analysis, Papers in Honor of K. Kodaira*, pp. 119–139, Princeton University Press, Princeton (1969).

32. S. G. Gindikin, Analysis in homogeneous domains, *Uspehi Mat. Nauk* **19**, 3–92 (1964); English transl., *Russian Math. Surveys* **19**, 1–89 (1964).

33. Hui-Sheng Gong and Fang-Cao Ye, The invariant differential operators of classical symmetric domain, *J. Xiaman Univ.* **11**, 55–67 (1960).

34. Sheng Gong, Fourier analysis on unitary group. I, *Acta Math. Sinica* **10**, 239–261 (1960).

35. Sheng Gong, Fourier analysis on unitary group. II, *Acta Math. Sinica* **12**, 17–31 (1962).

36. Sheng Gong, Fourier analysis on unitary group. III, *Acta Math. Sinica* **13**, 152–161 (1963).

37. Sheng Gong, Fourier analysis on unitary group. IV, *Acta Math. Sinica* **13**, 323–331 (1963).

38. Sheng Gong, Fourier analysis on unitary group. V, *Acta Math. Sinica* **15**, 305–325 (1965).

39. Sheng Gong, Partial sum of Fourier series on a rotation group, *J. China Univ. Sci. Technol. in China* **9**, 25–30 (1979).

40. Sheng Gong, A remark on the integral of Cauchy type on strictly pseudo-convex domains, in *Proc. 1980 Symp. Differential Geometry and Differential Equations*, vol. III, pp. 1183–1190, Science Press, Beijing (1982) (in English).

41. Sheng Gong, *Singular Integral in Several Complex Variables*, Shanghai Science and Technology Press, Shanghai (1982).

42. Sheng Gong, *Harmonic Analysis on Classical Group*, Science Press, Beijing (1983).

43. Sheng Gong, Schwarz lemma on Einstein space of several complex variables, *Acta Math. Sinica* **7**, 471–476 (1957).

44. Sheng Gong, A remark on Möbius transformations for higher dimensions, in *Proc. 1981 Shanghai-Hofei Symp. on Differential Geometry and Differential Equations*, pp. 75–94 (in English) Science Press, Beijing (1989).

45. Sheng Gong, On Reinhardt domains. I, *Kexue Tongbao* **31**, 433–438 (1986).

46. Sheng Gong and S. S. Miller, Partial differential subordinations in inequalities defined on complete circular domains, *Comm. Partial Diff. Eq.* **11**, 1243–1255 (1986) (in English).

47. Sheng Gong and Ji-Huai Shi, A note on the Poisson formula for the transitive domain of several complex variables, *J. Univ. Sci. Technol. in China* **2** 16–20 (1966).

48. Sheng Gong and Ji-Huai Shi, Singular integrals in several complex variables. I: Henkin integrals of strictly pseudoconvex domains, *Chinese Ann. Math.* **3**, 473–502 (1982).

49. Sheng Gong and Ji-Huai Shi, Singular integrals in several complex variables. II: Hadamard principal value on a sphere, *Chinese Ann. Math. Ser. B* **4**, 307–318 (1983).

50. Sheng Gong and Ji-Huai Shi, Singular integrals in several complex variables. III: Cauchy integrals of classical domains, *Chinese Ann. Math. Ser. B* **4**, 467–484 (1983).

51. Sheng Gong and Ji-Huai Shi, Singular integrals in several complex variables. IV: The derivatives of Cauchy integrals on balls, *Chinese Ann. Math. Ser. B* **5**, 21–36 (1984).

52. Sheng Gong and Ji-Guang Sun, The integral representation of holomorphic functions in a circular domain, *J. Univ. Sci. Technol. in China* **1**, 219–226 (1965).

53. Sheng Gong and Ji-Guang Sun, The Cauchy type of several complex variables. I: The Cauchy type integral of a ball, *Acta Math.* **15**, 431–443 (1965).

54. Sheng Gong and Ji-Guang Sun, The Cauchy type integral of several complex variables. II: The Cauchy type integral of the hyperbolic sphere of Lie spheres, *Acta Math. Sinica* **15**, 775–799 (1965).

55. Sheng Gong and Ji-Guang Sun, The Cauchy type integral of several complex variables. III: The Cauchy type integral of hyperbolic space of matrices, *Acta Math. Sinica* **15**, 800–811 (1965).

56. Sheng Gong and Ji-Guang Sun, On the singular integral equation on a sphere, *Acta Math. Sinica* **16**, 194–210 (1966).

57. K. T. Hahn, Inequality between the Bergman metric and Carathéodory differential metric, *Proc. Am. Math. Soc.* **68**, 193–194 (1978).

58. F. R. Harvey and H. B. Lawson, On boundaries of complex analytic varieties. I, *Ann. of Math.* **102**, 223–290 (1975).

59. Wen-Hua He, The singular integral and the singular integral equations on a sphere, *Math. J.* **4**, 127–144 (1984).

60. S. Helgason, *Differential Geometry and Symmetric Spaces*, Academic Press, New York (1962).

61. G. M. Henkin and J. Leiterer, *Theory of Functions on Complex Manifolds*, Akademie-Verlag, Berlin (1984).

62. Zu-Qi Ho and Guang-Xiao Chen, The harmonic analysis on unitary group. I: The criterion of the convergency of a Fourier series, *Math. Res. Rev.* **1**, 29–41 (1981).

63. Yi Hong, Ding-Zhu Du, and Hong-Cang Yang, An inequality for convex functions, *Kexue Tongbao, Foreign Land Ed.* **27**, 1266–1270 (1982).

64. Yi Hong and Hong-Cang Yang, Hardy's inequalities for a compact Riemann manifold, *Kexue Tongbao* **28**, 1351–1354 (1983).

65. Yi Hong and Hong-Cang Yang, On the estimate of the first eigenvalue of a compact Riemannian manifold, *Sci. Sinica Ser. A* **27**, 1265–1273 (1984) (in English).

66. Yi Hong and Hong-Cang Yang, Some generalization of Hardy's inequality, *Research Rep. Math. Sci.* **IMT-18** (1985).

67. Loo-Keng Hua, On the theory of automorphic functions of a matrix variable. I: Geometrical basis, *Am. J. Math.* **66**, 470–488 (1944) (in English).

68. Loo-Keng Hua, On the theory of automorphic functions of a matrix variable. II: The classification of hypercircles under the symplectic group, *Am. J. Math.* **66**, 531–563 (1944) (in English).

69. Loo-Keng Hua, On the extended space of several complex variables I: The space of complex spheres, *Quart. J. Math. Oxford Ser.* **17**, 214–222 (1946) (in English).

70. Loo-Keng Hua, Theory of Fuchsian functions of several complex variables, *Ann. Math.* **47**, 167–191 (1946) (in English).

71. Loo-Keng Hua, Introduction of the theory of vector modular forms, *Akad. Nauk. Azerbaidzan SSSR Trudy Inst. Fiz. Mat.* **3**, 32–43 (1948) (in Russian).

72. Loo-Keng Hua, Theory of functions of several complex variables: I. A complete orthonormal system in the hyperbolic space of matrices, *Acta Math. Sinica* **4**, 288–323 (1954).

73. Loo-Keng Hua, Theory of functions of several complex variables. II: A complete orthonormal system in hyperbolic space of hyperspheres, *Acta Math. Sinica* **5**, 1–25 (1955).

74. Loo-Keng Hua, On the estimation of the unitary curvature of the space of several variables, *Sci. Sinica* **4**, 1–26 (1955) (in English).

75. Loo-Keng Hua, On non-continuable domain with constant curvature, *Sci. Sinica* **4**, 27–32 (1955) (in English).

76. Loo-Keng Hua, On the theory of functions of several complex variables. A complete orthonormal system in the hyperbolic space of symmetric and antisymmetric matrices, *Dokl. Akad. Nauk SSSR* **101**, (1955) (in Russian).

77. Loo-Keng Hua, On a system of partial differential equations, *Sci. Record* (*N.S.*) 1, 1-4 (1957).
78. Loo-Keng Hua, A convergence theorem in the space of continuous functions on a compact group, *Sci. Record* (*N.S.*) 2, 280-284 (1958) (in English).
79. Loo-Keng Hua, *Harmonic Analysis of Functions of Several Complex Variables in Classical Domains*, Science Press, Beijing (1958); rev. ed., 1965; Russian transl., Izd. Lit., Moskva (1959); English transl., American Mathematical Society, Providence, RI (1963).
80. Loo-Keng Hua, On Lavrentiev's partial differential equation of mixed type, *Sci. Sinica* 13, 1755-1762 (1964).
81. Loo-Keng Hua, *Starting with the Unit Circle*, Science Press, Beijing (1977). English transl., Springer-Verlag, New York (1981).
82. Loo-Keng Hua and Qi-Keng Lu (=K. H. Look), On Cauchy formula for space of skew-symmetric matrices of odd order, *Sci. Record* (*N.S.*) 2, 19-22 (1958) (in English).
83. Loo-Keng Hua and Qi-Keng Lu (=K. H. Look), Boundary properties of the Poisson integral of Lie sphere, *Sci. Record* (*N.S.*) 2, 77-80 (1958) (in English).
84. Loo-Keng Hua and Qi-Keng Lu (=K. H. Look), Theory of harmonic functions of classical domains. I: Harmonic functions in the hyperbolic space of matrices, *Acta Math. Sinica* 8, 531-547 (1958).
85. Loo-Keng Hua and Qi-Keng Lu (=K. H. Look), Theory of harmonic functions of classical domains. II: Harmonic functions in the hyperbolic space of the symmetric matrices, *Acta Math. Sinica* 9, 259-306 (1959).
86. Loo-Keng Hua and Qi-Keng Lu (=K. H. Look), Theory of harmonic functions of classical domains. III: Harmonic functions in the hyperbolic space of skew-symmetric matrices, *Acta Math. Sinica* 9, 306-314 (1959).
87. Loo-Keng Hua and Qi-Keng Lu, Theory of harmonic functions in classical domains, *Sci. Sinica* 8, 1031-1094 (1959) (in English).
88. Adam Koranyi, Analytic invariants of bounded symmetric domains, *Proc. Am. Math. Soc.* 19, 279-284 (1968).
89. J. Leray, Le calcul différentiel et intégral sur une variété analytique complexe (problem de Cauchy III), *Bull. Soc. Math. France* 87, 81-180 (1959).
90. Lun-Huan Li, The boundary properties of the integral representation of a non-degenerate Weil polyhedron, *J. Xiaman Univ.* 25, 117-127 (1986).
91. Peter Li and Jia-Qing Zhong, Pinching theorem for the first eigenvalue on positively curved manifolds, *Trans. Am. Math. Soc.* 294, 341-349 (1986).
92. Liang-Yu Liu, On the boundary properties of Cauchy type integral on piece-wise smooth manifolds, *Acta Math. Sinica* 31, 547-557 (1988).
93. Qi-Keng Lu (=K. H. Look), The unitary geometry and the theory of functions of complex variables, *Progress in Math.* 2, 587-662 (1956).
94. Qi-Keng Lu (=K. H. Look), Schwarz lemma in the theory of functions of several complex variables, *Acta Math. Sinica* 7, 370-420 (1957).
95. Qi-Keng Lu (=K. H. Look), An analytic invariant and its characteristic properties, *Acta Math. Sinica* 8, 243-252 (1958).
96. Qi-Keng Lu (=K. H. Look), Schwarz lemma and analytic invariants, *Sci. Sinica* 7, 453-504 (1958) (in English).
97. Qi-Keng Lu (=K. H. Look), A theorem about the inner holomorphic maps of classical domains, *J. Xiaman Univ.* 4, 8-17 (1957).
98. Qi-Keng Lu (=K. H. Look), Slit space and extremal principle, *Sci. Record* (*N.S.*) 3, 289-294 (1959) (in English).
99. Qi-Keng Lu (=K. H. Look), Explicit formula of the infinitesimal connection for a continuous representation of GL(*n, R*), *Sci. Record* (*N.S.*) 3, 296-300 (1959) (in English).

100. Qi-Keng Lu (=K. H. Look), A study of the theory of functions of several complex variables in China during the last decade, *Sci. Sinica* **8**, 1230-1237 (1959) (in English).
101. Qi-Keng Lu (=K. H. Look), *An Introduction to the Theory of Functions of Several Complex Variables*, Science Press, Beijing (1961).
102. Qi-Keng Lu (=K. H. Look), On a class of homogeneous complex manifolds, *Sci. Sinica* **11**, 723-751 (1962) (in English).
103. Qi-Keng Lu (=K. H. Look), *The Classical Manifolds and Classical Domains*, Shanghai Press of Science and Technology, Shanghai (1963).
104. Qi-Keng Lu (=K. H. Look), The matrix representation of non-exceptional and irreducible Riemann global symmetric spaces (unpublished, 1963).
105. Qi-Keng Lu (=K. H. Look), The solution of a quadratic matrix equation and its geometrical meaning, *J. Xiaman Univ.* **10**, 89-134 (1953).
106. Qi-Keng Lu (=K. H. Look), The elliptic geometry of extended spaces, *Acta Math. Sinica* **13**, 49-62 (1963).
107. Qi-Keng Lu (=K. H. Look), On the complete Kähler manifolds of constant unitary curvature, *Acta Math. Sinica* **16**, 269-281 (1966); English transl., Chinese Mathematics (1966).
108. Qi-Keng Lu (=K. H. Look), On the Cauchy-Fantappié formula, *Acta Math. Sinica* **16**, 344-363 (1966); English transl., Chinese Mathematics (1966).
109. Qi-Keng Lu (=K. H. Look), Gauge theory and the principal fibre bundle, *Acta Phys. Sinica* **23**, 241-263 (1974).
110. Qi-Keng Lu (=K. H. Look), The estimation of the intrinsic derivatives of the holomorphic mappings of bounded domains, *Sci. Sinica* **22**, 1-17 (1979).
111. Qi-Keng Lu (=K. H. Look), *Differential Geometry and Its Application to Physics*, Science Press, Beijing (1982).
112. Qi-Keng Lu (=K. H. Look), On the Schwarz lemma of higher order, in *Proc. 1980 Beijing Symp. on Differential Geometry and Differential Equations* vol. III, pp. 1308-1318, Science Publishers and Gordon & Breach, Beijing and New York (1982) (in English).
113. Qi-Keng Lu (=K. H. Look), A note on the extremum of the sectional curvature of a Kähler manifold, *Sci. Sinica Ser. A* **27**, 367-371 (1983) (in English).
114. Qi-Keng Lu (=K. H. Look), On the representative domains, in *Proc. 1981 Hangzhou Conf. on Several Complex Variables*, Birkhauser, pp. 199-210 (1984) (in English).
115. Qi-Keng Lu (=K. H. Look), The classification of the curvature tensors of the Kähler manifolds, *Chinese Quart. J. Math.* **2**, 15-32 (1986) (in English).
116. Qi-Keng Lu (=K. H. Look), The Green forms and the Cauchy formulas on a Kähler manifold, *Chinese Quart. J. Math.* **2**, 1-11 (1987) (in English).
117. Qi-Keng Lu (=K. H. Look), The Green forms with respect to the intrinsic metric of a ball, *Sci. Sinica* (to appear).
118. Qi-Keng Lu (=K. H. Look), The heat kernel of a ball in $\mathbb{C}^n$, *Research Memorandum*, no. 18, vol. 3, *Inst. of Math.*, *Acad. Sinica* (1987).
119. Qi-Keng Lu (=K. H. Look), The various kernels of classical domains and classical manifolds, in *Reports in the Symposium dedicated to Professor L. K. Hua, Beijing* (1988).
120. Qi-Keng Lu, Han-Ying Guo, and Ke Wu, A formulation of nonlinear gauge theory and its applications, *Comm. Theor. Phys. (Beijing, China)* **2**, 1029-1038 (1983).
121. Qi-Keng Lu and Yi Hong, The heat kernel of the classical domain $R_I(m, n)$, preprint.
122. Qi-Keng Lu and Yong-Shi Wu, The right translation invariant metric and variational principles on a principal fibre bundle, in *Proc. 1980 Quangzhou Conf. of Theoretical Particle Physics*, pp. 985-1005.
123. Qi-Keng Lu and Yi-Chao Xu, A note on transitive domains, *Acta Math. Sinica* **11**, 11-23 (1961).

124. Qi-Keng Lu and Wei-Ping Yin, The solution of the Cauchy problem of a wave equation with variable coefficients, *Chinese Ann. Math.* **1**, 115–128 (1980).

125. Qi-Keng Lu and Jia-Qing Zhong, The realization of the homogeneous cone, *Acta Math. Sinica* **24**, 117–142 (1981).

126. Qi-Keng Lu and Tong-De Zhong, The generalization of the Privalov theorem, *Acta Math. Sinica* **7**, 144–165 (1957).

127. Ru-Qian Lu, On the harmonic functions on a class of non-symmetric transitive domains, *Sci. Sinica* **14**, 315–324 (1965) (in English).

128. J. Milnor, *Morse Theory*, Princeton University Press, Princeton, 1963.

129. J. Mitchell, Potential theory in the geometry of matrices, *Trans. Am. Math. Soc.* **19**, 401–422 (1955).

130. Ngai-Ming Mok and Jia-Qing Zhong, Variétés compactes kähleriennes d'Einstein de courbure bisectionnelle semipositive, *C. R. Acad. Sci. Paris, Ser. I, No. 12* **299** (1984).

131. Ngai-Ming Mok and Jia-Qing Zhong, Curvature characterization of compact Hermitian symmetric spaces, *J. Diff. Geom.* **23**, 15–67 (1986). (Chapter 15 of this volume).

132. F. Norguet, Problem sur les formes differentielle et les courant, *Ann. Inst. Fourier Grenoble* **11**, 1–82 (1961).

133. I. I. Pyatetski-Shapiro, On a problem proposed by É. Cartan, *Dokl. Akad. Nauk SSSR* **124**, 272–273 (1959).

134. I. I. Pyatetski-Shapiro, *The Geometry of Classical Domains and Theory of Automorphic Functions*, Fitzmatgiz, Moskow (1961).

135. H. J. Reiffen, Die Carathéodory Distanz und Ihre Zugehoerige Differential Metrik, *Math. Ann.* **161**, 315–324 (1965).

136. A. Selberg, Harmonic analysis and discontinuous groups in weakly symmetric Riemannian spaces with application to Dirichlet series, *J. Ind. Math. Soc.* **20**, 47–87 (1956).

137. Ji-Huai Shi, On the Cauchy integrals for the hypersphere, *J. Univ. Sci. Technol. in China* **10**, 1–9 (1980).

138. Ji-Huai Shi, A new proof of the polydisc's analysis nonequivalence to any ball, *Kexue Tongbao* **26**, 952–953 (1981) (in English).

139. Ji-Huai Shi, Approximating univalent analytic mappings of several complex variables by chains of subordinate polynomial mappings, *J. Univ. Sci. Technol. in China* **11**, 7–15 (1981).

140. Ji-Huai Shi, On the bound of convexity of univalent analytic maps of the ball, *Kexue Tongbao* **27**, 473–476 (1982) (in English).

141. Ji-Huai Shi, Some applications of the Ryll-Wojtaszyk polynomial, *Acta Math. Sinica* (*N.S.*) **1**, 359–365 (1985) (in English).

142. Ji-Huai Shi, Two results of Bloch functions of several complex variables, *J. Univ. Sci. Technol. in China* **16**, 147–152 (1986).

143. Ji-Huai Shi, On the rate of growth of the mean M_p of holomorphic and pluriharmonic functions on bounded symmetric domains of C^n, *J. Math. Anal. Appl.* **126**, 161–175 (1987).

144. Ji-Huai Shi, Representations of linear functions on Lipschitz spaces Λ_ϕ of functions holomorphic in the unit ball of C^n, *Chinese Ann. Math. Ser. A* **8**, 189–198 (1987).

145. Ji-Huai Shi, Hadamard products of functions holomorphic in the ball of C^n, *Chinese Ann. Math. Ser. A* **9**, 49–59 (1988).

146. Ji-Huai Shi, The spaces $H^p(B_n)$, $0 < p < 1$, and $B_{pq}(B_n)$, $0 < p < q < 1$, are not locally convex, *Proc. Am. Math. Soc.* **103**, 69–74 (1988).

147. Ji-Huai Shi, Coefficient multipliers of H^p and B_{pq} spaces in the unit ball of C^n, in *Complex Variables Theory and Applications* (to appear).

148. Ji-Huai Shi, Hardy-Littlewood theorem on bounded symmetric domains of C^n, *Sci. Sinica* **31**, 916–926 (1988).

149. Ji-Huai Shi, Mackey topologies of Hardy spaces and Bergman spaces of several complex variables, *Kexue Tongbao* **33**, 881–884 (1988).

150. Ji-Huai Shi, Cesaro means of power series of several complex variables, *Chinese Quart. J. Math.* (to appear).
151. C. L. Siegel, Symplectic geometry, *Am. J. Math.* **65**, 1-85 (1943).
152. M. Skwarzinski, The distance in theory of pseudo-conformal transformations and the Lu Qi-Keng conjecture, *Proc. Am. Math. Soc.* **22**, 305-310 (1969).
153. Ji-Guang Sun, The harmonic functions on a class of bounded symmetric domains, *J. Univ. Sci. Technol. in China* **2**, 35-43 (1973).
154. Ji-Guang Sun, The singular integral equation on a closed smooth manifold, *Acta Math. Sinica* **22**, 675-692 (1979).
155. Ji-Guang Sun, The regulation of singular integral equations on a sphere, *J. Xiaman Univ.* **18** (1979).
156. Ji-Guang Sun, Perturbation theorems for generalized singular value, *J. Comput. Math.* **1**, 233-242 (1983).
157. Ji-Guang Sun, Perturbation analysis for the generalized singular value problem, *SIAM J. Numer. Anal.* **20**, 611-625 (1983) (in English).
158. H. C. Vinberg, Homogeneous cones, *Dokl. Akad. Nauk SSSR, No.* 1 **133**, 9-12 (1960).
159. Da-Ming Wang, *The Carathéodory Metric of Classical Domains*, Dissertation, Beijing (1962).
160. H. C. Wang, Closed manifolds with homogeneous complex structure, *Amer. J. Math.* **76**, 1-33 (1954).
161. Shi-Kun Wang and Dao-Zhen Dong, Harmonic analysis on rotational group. I: The criterion of the convergency of a Fourier series, *Chinese Ann. Math.* **4**, 195-206 (1983).
162. Lanfang Wu and Yi-Chao Xu, Holomorphic sectional curvatures of homogeneous bounded domains, *Kexue Tongbao* **28**, 393-396 (1983).
163. Yi-Chao Xu, On the group of holomorphic automorphisms of bounded positive (m, p)-circular domains, *Acta. Math. Sinica* **13**, 419-432 (1963).
164. Yi-Chao Xu, On the group of holomorphic automorphisms of bounded homogeneous domains, *Acta Math. Sinica* **19**, 169-191 (1976).
165. Yi-Chao Xu, On the holomorphic isomorphisms of bounded homogeneous domains, *Acta Math. Sinica* **19**, 248-266 (1976).
166. Yi-Chao Xu, On the Siegel domain of first kind over a square cone, *Acta Math. Sinica* **21**, 1-17 (1978).
167. Yi-Chao Xu, On the Bergman kernel functions of homogeneous bounded domains, *Sci. Sinica Special Issue II*, 80-90 (1979) (in English).
168. Yi-Chao Xu, The classification of square type domains, *Sci. Sinica* **22**, 375-392 (1979) (in English).
169. Yi-Chao Xu, The invariant differential operators of order 2 in an N-Siegel domain, *Acta Math. Sinica* **25**, 340-352 (1982).
170. Yi-Chao Xu, On homogeneous bounded domains, *Sci. Sinica Ser. A* **26**, 25-34 (1983) (in English).
171. Yi-Chao Xu, On the classification of homogeneous bounded domains, 1983 (unpublished).
172. Yi-Chao Xu, Classification of a class of homogeneous Kähler manifolds, *Sci. Sinica* **29**, 449-463 (1986) (in English).
173. Yi-Chao Xu and De-Ling Wang, The classification of bounded semi-circular domains in C^2, *Acta Math. Sinica* **23**, 372-384 (1980).
174. Zong-Yuan Yao, On the B-M integral representation of first kind in C^n, *J. Xiaman Univ.* **26**, 390-398 (1987).
175. Fang-Cao Ye, The conjugate points of the complex Grassmann manifold, *Acta Math. Sinica* **21**, 367-374 (1978).

176. Fang-Cao Ye, The invariant differential operators of NIRGSS. I-III, *J. Xiaman Univ.* (to appear).
177. Wei-Ping Yin, The explicit solutions of the Cauchy problem for some hyperbolic equations, *Acta Math. Sinica* 23, 102–117 (1980).
178. Wei-Ping Yin, The Riemann functions of a class of hyperbolic equations, *J. China Univ. Sci. Technol.* 10, 117–128 (1980).
179. Wei-Ping Yin, A note to Poisson kernel, *Kexue Tongbao* 26, 581–583 (1981).
180. Wei-Ping Yin, Two problems on bounded homogeneous domains, *J. Univ. Sci. Technol. in China* 16, 241–247 (1986).
181. Wei-Ping Yin, Two problems on Cartan domains, *J. Univ. Sci. Technol. in China* 16, 130–156 (1986).
182. Wei-Ping Yin, A special function on non-self-dual cones, *J. Univ. Sci. Technol. in China* 16, 367–384 (1986).
183. Wei-Ping Yin, A characterization of bounded symmetric domains, *Kexue Tongbao* 31, 9–11 (1987).
184. Wei-Ping Yin, The curvature of nonsymmetric Siegel domain of the first kind, *J. Univ. Sci. Technol. in China* 17, 1–16 (1987).
185. Wei-Ping Yin, Invariant metrics and functions of Reinhardt domains, *Kexue Tongbao* 33, 329–332 (1987).
186. Wei-Ping Yin, The Kähler geometry of Reinhardt domains, *Kexue Tongbao* 32, 876–877 (1987).
187. Wei-Ping Yin, Some remarks on certain Reinhardt domain, *Progress in Math.* 17, 206–207 (1988).
188. Wei-Ping Yin, The curvature and group invariant functions of a class of domains, *Sci. Sinica Ser. A* 12, 1245–1257 (1988).
189. Wei-Ping Yin, The unitary geometry of nonsymmetric homogeneous Siegel domains of the first type (I), (II), (III), *Acta Math. Sinica* 24, 753–764, 764–779, 879–891 (1981).
190. Wei-Ping Yin, Some notes on extension spaces, *J. Univ. Sci. Technol. of China* 12, 13–19 (1982).
191. Wei-Ping Yin, Some results on the nonsymmetric Siegel domains and explicit solutions of some partial differential equations, in: *Proc. 1980 Beijing Symp. on Differential Geometry and Differential Equations*, pp. 1681–1694, Science Press, Beijing (1982) (in English).
192. Wei-Ping Yin, On the group of analytic automorphisms of Cartan domains, *J. Univ. Sci. Technol. of China* 17, 291–302 (1987).
193. Wei-Ping Yin, The Aut (D)-invariant harmonic functions, *Chinese Ann. Math.*, to appear.
194. Wei-Ping Yin, Some new phenomena on bounded nonhomogeneous domains, *Acta Math. Sinica*, to appear.
195. Qi-Huang Yu and Jia-Qing Zhong, Lower bounds of the gap between the first and second eigenvalues of the Schroedinger operator, *Trans. Am. Math. Soc.* 296, 342–349 (1986).
196. Jin-Hao Zhang, Coefficient problem of holomorphic mappings on Reinhardt domains, *Kexue Tongbao* 25, 103–108 (1980).
197. Jin-Hao Zhang, On the limit of coefficient problem on star-like mapping, *J. Fudan Univ.* 24, 65–72 (1985).
198. Jin-Hao Zhang, The S metric on holomorphy domain, *Kexue Tongbao* 33, 353–356 (1986).
199. Jin-Hao Zhang, Subelliptic estimates and vanishing theorems of some cohomology groups, *Acta Math. Sinica New Ser.* 2, 57–72 (1986) (in English).
200. Jin-Hao Zhang, Chern-Levine-Nirenberg norm on homology class, *Northeastern Math. J.* 3, 429–438 (1987) (in English).
201. Jin-Hao Zhang, Cohomology vanishing theorem and finitely generated ideals in functional algebras, *Acta Math. Sinica New Ser.* 4, 1–4 (1988) (in English).

202. Jia-Qing Zhong, The group of automorphisms of the hyperbolic space of skew-symmetric matrices, *J. Peking Univ.* **8**, 226–244 (1962).
203. Jia-Qing Zhong, On the group of motions of the extension spaces of the classical domains of type I, *J. Univ. Sci. Technol. in China* **1**, 229–238 (1965).
204. Jia-Qing Zhong, A class of integral determinants and their applications to the representation theory of groups, *Acta Math. Sinica* **19**, 88–106 (1976).
205. Jia-Qing Zhong, Harmonic analysis on rotation groups, Abel summability, *J. Univ. Sci. Technol. in China* **7**, 31–43 (1979) (Chapter 4 of this volume).
206. Jia Qing Zhong, Coxeter-Killing transformations of simple Lie algebras, *Acta Math. Sinica* **22**, 291–302 (1979) (Chapter 5 of this volume).
207. Jia-Qing Zhong, Dimensions of the rings of invariant differential operators on bounded homogeneous domains, *Acta Math. Sinica* **23**, 261–272 (1980).
208. Jia-Qing Zhong, On prime ideals of the ring of differential operators, *Chinese Ann. Math.* **1**, 259–374 (1980) (Chapter 7 of this volume).
209. Jia-Qing Zhong, On the sum of class functions of Weyl groups, *Acta Math. Sinica* **23**, 695–711 (1980) (Chapter 8 of this volume).
210. Jia-Qing Zhong, The trace formula of the Weyl group representations of the symmetric group, *Acta Math. Sinica* **23**, 836–850 (1980) (Chapter 9 of this volume).
211. Jia-Qing Zhong, The Siegel-Godement transformations on positive Hermitian cones, *Acta Math. Sinica* **25**, 236–243 (1982).
212. Jia-Qing Zhong, A note on Schubert calculus, in *Proc. 1980 Beijing Symp. on Differential Geometry and Differential Equations*, vol. III, pp. 1697–1708, Science Publishers, Beijing (1981) (in English).
213. Jia-Qing Zhong, The degree of strong non-degeneracy of the bisectional curvature of the exceptional bounded symmetric spaces, in *Several Complex Variables*, pp. 127–139 (Proc. 1981 Hangzhou Conf.), Birkhauser (1984) (in English) (Chapter 13 of this volume).
214. Jia-Qing Zhong, The Busemann function of the classical domains, *Acta Math. Sinica*, to appear.
215. Jia-Qing Zhong, The Schubert calculus and the theory of group representations, *Sci. Sinica*, to appear.
216. Jia-Qing Zhong and Hong-Cang Yang, On the estimate of the first eigenvalue of a compact Riemann manifold, *Sci. Sinica* **27**, 1265–1273 (1984) (in English) (Chapter 14 of this volume).
217. Jia-Qing Zhong and Wei-Ping Yin, Some classes of non-symmetric homogeneous domains, *Acta Math. Sinica* **24**, 587–613 (1981).
218. Jia-Qing Zhong and Wei-Ping Yin, The extension spaces of non-symmetric homogeneous domains, *Acta Math. Sinica* **24**, 931–947 (1981) (Chapter 11 of this volume).
219. Tong-De Zhong, On pseudo-analytic vector fields on a pseudo-hermitian manifold, *J. Xiaman Univ.* **1**, 93–102 (1957).
220. Tong-De Zhong, On the boundary properties of the Cauchy type integral of several complex variables, *Acta Math. Sinica* **15**, 227–241 (1965).
221. Tong-De Zhong, The composite formula of singular integral equation on the characteristic manifold, *J. Xiaman Univ.* **17**, 1–13 (1978).
222. Tong-De Zhong, The singular integral equation of a smooth manifold, *J. Xiaman Univ.* **18**, 1–8 (1979).
223. Tong-De Zhong, The transposed formula of the singular integral of Bochner-Martinelli kernel, *Acta Math. Sinica* **23**, 538–550 (1980).
224. Tong-De Zhong, The regulation of the singular integral equation on a characteristic manifold, *Kexue Tongbao* **23**, 667–668 (1981); *J. Xiaman Univ.* **20**, 1–6 (1981).
225. Tong-De Zhong, Singular integral equations with Bochner-Martinelli kernel on smooth orientable manifolds, in *Proc. 1980 Beijing Symp. on Differential Geometry and Differential Equations*, vol. III, pp. 1709–1711, Science Press, Beijing (1982) (in English).

226. Tong-De Zhong, Some applications of Bochner-Martinelli integral representation, in *Several Complex Variables*, 217–225 (Proc. 1981 Hangzhou Conf.), Birkhauser (1984).

227. Tong-De Zhong, *The Integral Representation and the Singular Integral Equation of Multidimensions*, Xiaman Univ. Press (1986).

228. Tong-De Zhong, The integral representation of holomorphic functions on a Stein manifold, *J. Xiaman Univ.* **26**, 385–389 (1987).

229. Tong-De Zhong, The Andreotti-Norguet formula on a Stein manifold and its generalization, *J. Xiaman Univ.* **26**, 641–643 (1987).

230. Tong-De Zhong, Holomorphic extension on Stein manifold, Research Report No. 10, Mittag-Leffler Institute (1987).

3

Uniformization in Several Complex Variables

Yum-Tong Siu

In this chapter we survey the current state of the theory of uniformization in several complex variables. In the case of one complex variable the uniformization theorem says that a simply connected Riemann surface is either the Riemann sphere $\mathbb{P}_1$ or the open unit disk Δ or the Gaussian plane $\mathbb{C}$. The theory of uniformization in the higher-dimensional case attempts to find a suitable analog of it in the case of several complex variables. In the case of several complex variables the analog of $\mathbb{P}_1$ is the Hermitian symmetric manifold of compact type. That means the complex projective space $\mathbb{P}_n$, the hyperquadric $\sum_{\nu=0}^{n+1} z_\nu^2 = 0$ in $\mathbb{P}_{n+1}$, the Grassmannians, etc. The analog of Δ is the bounded symmetric domain. That means the open ball in $\mathbb{C}^n$, the set of all $m \times n$ matrices A with $I - AA^t$ negative definite (with no symmetry conditions on A, or with the additional condition of skew-symmetry or symmetry on A), etc. The analog of $\mathbb{C}$ is simply $\mathbb{C}^n$. The problem is the following: when is a given complex manifold biholomorphic to a Hermitian symmetric manifold of compact type, a bounded symmetric domain, or the complex Euclidean space? In the case of Riemann surface the property of simple connectedness is sufficient to narrow it down to $\mathbb{P}_1$, Δ, or $\mathbb{C}$. The topological condition of compactness separates $\mathbb{P}_1$ from Δ and

Partially supported by a grant from the National Science Foundation.

Yum-Tong Siu ● Department of Mathematics, Harvard University, Cambridge, Massachusetts 02138.

C. To distinguish between Δ and $\mathbb{C}$, one needs conditions such as curvature assumptions. In the case of higher dimension is it possible to characterize the symmetric spaces by topological or curvature or other conditions? Such a characterization is also known by the name of rigidity or strong rigidity, which means that it is not possible to deform the manifold or find a different one under such conditions.

3.1. Characterization of Compact Type by Topological Conditions

3.1.1. With Kähler Condition

For the case of complex projective space Hirzebruch and Kodaira [32] proved the following result. Suppose M is a compact Kähler manifold that is topologically the same as the complex projective space. Then M is biholomorphic to the complex projective space if the complex dimension of M is odd. When the complex dimension of M is even, the same result holds with the additional condition that the anticanonical line bundle is not negative.

The idea of the proof of the Hirzebruch–Kodaira result for the complex projective space is to use the line bundle L on M corresponding to the generator of $H^2(M, \mathbb{Z})$, which (after replacing it by its dual if necessary) is positive because of the existence of the Kähler metric and because the second Betti number is 1. By using the theorem of Riemann-Roch and the vanishing theorem of Kodaira, they showed that $\Gamma(M, L)$ is generated by $n + 1$ sections, and these $n + 1$ sections define a biholomorphic map from M to $\mathbb{P}_n$. The nonnegativity of the anticanonical line bundle is needed when the Riemann-Roch theorem is used.

Yau [87] applied the theory of Kähler–Einstein metrics, developed by Calabi [15–17], Aubin [3, 4], and Yau [88], to remove the additional assumption of the nonnegativity of the anticanonical line bundle in the case of even-dimensional complex projective space in the following way. If the anticanonical line bundle is negative, there exists a Kähler–Einstein metric on the manifold. Once it is known that a Kähler–Einstein metric of negative scalar curvature exists, then one can use an earlier known result (Chen and Ogiue [21, Theorem 2]) that, for an n-dimensional compact Kähler–Einstein manifold,

$$(-1)^n c_1^n \le (-1)^n \frac{2(n + 1)}{n} c_2 c_1^{n-2},$$

with equality only for a compact quotient of the ball or the complex projective space, where c_ν is the νth Chern class of the manifold. (For a compact Kähler-Einstein manifold of complex dimension 2, the Chern number inequality $c_1^2 \leq 3c_2$, with equality only in the case of the ball and the complex projective space, was pointed out in 1952 by Guggenheimer [30].) One can now conclude that the manifold is a compact quotient of the ball, contradicting its simple connectedness.

Brieskorn [13] generalized the Hirzebruch-Kodaira result from the complex projective space to the hyperquadric. Whether the additional assumption of the nonnegativity of the anticanonical line bundle for the even-dimensional hyperquadric can be removed is still unknown.

CONJECTURE 3.1. *Suppose M is a compact Kähler manifold of even dimension $n \geq 4$. If M is homeomorphic to the complex hyperquadric Q_n, then M is biholomorphic to Q_n.*

In the case of complex dimension 2 the hyperquadric is simply the product $\mathbb{P}_1 \times \mathbb{P}_1$ of two complex lines. Hirzebruch [31] discovered a class of compact Kähler surfaces that are diffeomorphic to $\mathbb{P}_1 \times \mathbb{P}_1$ but not biholomorphic to it. Hirzebruch's class of surfaces is now known as the Hirzebruch surfaces, and they are defined as follows. For any nonnegative integer n take the line bundle L over $\mathbb{P}_1$ whose first Chern class is n. The projective line bundle corresponding to L is a Hirzebruch surface Σ_n. The Hirzebruch surfaces for nonzero even integers n are diffeomorphic to $\mathbb{P}_1 \times \mathbb{P}_1$ but are not biholomorphic to it. As a matter of fact, for $n \geq 2q$ the Hirzebruch surface Σ_{n-2q} can be deformed to the Hirzebruch surface Σ_n in the following way.

Let $[z_0, z_1]$ be the homogeneous coordinates of the base manifold $\mathbb{P}_1$, and let $[w_0, w_1, w_2]$ be the homogeneous coordinates of the fiber manifold $\mathbb{P}_2$. For $t \in \mathbb{C}$ let M_t be the hypersurface in $\mathbb{P}_1 \times \mathbb{P}_2$ defined by $z_0^n w_0 - z_1^n w_1 - tz_0^{n-q} z_1^q w_2 = 0$. Then M_0 is biholomorphic to Σ_n, and M_t is biholomorphic to Σ_{n-2q} for $t \neq 0$. One can see this by the following argument, which computes the Chern class of the line bundle associated with the $\mathbb{P}_1$-bundle $M_t \to \mathbb{P}_1$ whose projection is induced by the natural projection $\mathbb{P}_1 \times \mathbb{P}_2 \to \mathbb{P}_1$.

Let us show that for $t \neq 0$ the surface $z_0^n w_0 - z_1^n w_1 - tz_0^{n-q} z_1^q w_2 = 0$ in $\mathbb{P}_1 \times \mathbb{P}_2$ is Σ_{n-2q}, by producing two nonintersecting holomorphic sections of $M_t \to \mathbb{P}_1$. The first one is E_∞, given by the two equations $z_0^n w_0 - tz_1^q w_2 = 0$ and $w_1 = 0$. The second one is E_0, given by the two equations $z_1^{n-q} w_1 - tz_0^{n-q} w_2 = 0$ and $w_0 = 0$. The subvariety E_∞ defines a section, because $\{z_0 = 0\} \cap E_\infty$ gives the three equations $z_0 = 0$, $tz_1^q w_2 = 0$, and $w_1 = 0$, and the only solution for these three equations is the point given by $[z_0, z_1] = [0, 1]$ and $[w_0, w_1, w_2] = [1, 0, 0]$. Likewise the subvariety E_0 defines a section. These two sections do not intersect, because at $E_0 \cap E_\infty$ we have the four equations $w_0 = 0$, $w_1 = 0$, $tz_1^q w_2 = 0$, and $tz_0^{n-q} w_2 = 0$. Since one of z_0, z_1 is nonzero, we have $w_2 = 0$, which is not possible because one of w_0, w_1, w_2

must be nonzero. By using E_0 as the zero section and E_∞ as the section at infinity, we conclude that M_t is biholomorphic to the projective line bundle associated with a line bundle over $\mathbb{P}_1$. We now determine the Chern class of this line bundle. Take a third section E given by $w_2 = 0$. Consider $E \cap E_\infty$, which is defined by the three equations $z_0^q w_0 = 0$, $w_1 = 0$, and $w_2 = 0$. Since w_0 cannot be zero, we have $z_0^q = 0$. So the only point of intersection of E and E_∞ is a q-tuple point at $([z_0, z_1], [w_0, w_1, w_2]) = ([0, 1], [1, 0, 0])$. Likewise $E \cap E_0$ is defined by the three equations $z_1^{n-q} w_1 = 0$, $w_0 = 0$, and $w_2 = 0$. The only point of intersection of E and E_0 is an $(n - q)$-tuple point at $([z_0, z_1], [w_0, w_1, w_2]) = ([1, 0], [0, 1, 0])$. So the Chern class of the line bundle is $(n - q) - q = n - 2q$. (For a more general discussion of such deformations see Brieskorn [14, p. 355], in which he discussed the more general case of holomorphic $\mathbb{P}_n$-bundles over $\mathbb{P}_1$.)

The conjecture for the complex dimension 2 case after taking into account the Hirzebruch surfaces is the following.

CONJECTURE 3.2. *Suppose M is a compact Kähler surface that is homeomorphic to* $\mathbb{P}_1 \times \mathbb{P}_1$. *Then M is biholomorphic to a Hirzebruch surface* Σ_n *for some even integer n.*

Some partial result of this conjecture was obtained by Andreotti [2]. He proved that the Hirzebruch surfaces with even n are the only compact complex algebraic surfaces M that are diffeomorphic to $\mathbb{P}_1 \times \mathbb{P}_1$ and whose 3-genus $\dim_\mathbb{C} \Gamma(M, K_M^3)$ is not equal to 25, where K_M is the canonical line bundle of M.

There is also a conjecture for general irreducible compact Hermitian symmetric manifolds.

CONJECTURE 3.3. *Suppose M is a compact Kähler manifold that is homeomorphic to an irreducible compact Hermitian symmetric manifold N. Then M is biholomorphic to N.*

3.1.2. Without the Kähler Condition

Another unsolved problem concerning this characterization of the complex projective space and the hyperquadric is whether the Kähler condition can be removed. In the case of complex dimension 2, the Kähler condition is equivalent to the topological condition of the evenness of the first Betti number (see Ref. 72) and can thus be easily removed.

The removal of the Kähler condition is related to the question of the nonexistence of complex structure on the 6-sphere. There are two papers (Adler [1], Hsiung [33]) giving proofs of the nonexistence of the complex structure on the 6-sphere. Adler also gave a proof that any holomorphic deformation of $\mathbb{P}_n$ is biholomorphic to $\mathbb{P}_n$ in that paper. However, there are still reservations concerning the completeness of the proofs in these two papers.

The 6-sphere is obtained by adding one infinity point to $\mathbb{R}^6$. The complex projective space $\mathbb{P}_3$ of complex dimension 3 is obtained by adding the infinity hyperplane $\mathbb{P}_2$ to $\mathbb{C}^3$. Since $\mathbb{C}^3$ is the same as $\mathbb{R}^6$ topologically, if the 6-sphere carries a complex structure one can blow up the infinity point and obtain a compact complex manifold that is topologically the same as $\mathbb{P}_3$. If one can prove that any compact complex manifold that is topologically the same as $\mathbb{P}_3$ must be biholomorphic to $\mathbb{P}_3$, then one can blow down to a point some complex hypersurface in $\mathbb{P}_3$ that is biholomorphic to $\mathbb{P}_2$. We know that we cannot blow down to a point such a complex hypersurface in $\mathbb{P}_3$, because there are many nonconstant holomorphic functions on the complement of any complex hypersurface S in $\mathbb{P}_3$, and if some complex hypersurface can be blown down to a point then any holomorphic function on $U - S$ for some open neighborhood U of S can be extended across S, which makes the existence of any nonconstant holomorphic function on $\mathbb{P}_3 - S$ impossible. So we know that the 6-sphere does not carry any complex structure if we can prove that any compact complex manifold that is topologically the same as $\mathbb{P}_3$ must be biholomorphic to it.

Hsiung's proof uses only local differential geometric methods and does not seem to need any global properties of the 6-sphere. Adler considers the standard almost complex structure of S^6 defined by the Cayley numbers and the standard metric on S^6. Though the standard almost complex structure is not integrable on S^6, the standard metric satisfies equations that are reduced to equations characterizing a Kähler metric if the torsion tensor for the almost complex structure vanishes. If there is an integrable almost complex structure on S^6, it can be joined to the standard one by a family of almost complex structures parametrized by t in the unit interval $[0, 1]$ so that $t = 0$ corresponds to the standard almost complex structure and $t = 1$ corresponds to the integrable almost complex structure. He then shows by a continuity argument, starting from $t = 0$, that for all t, including $t = 1$, there exist metrics satisfying equations that are reduced to equations characterizing a Kähler metric if the torsion tensor for the almost complex structure vanishes. The result for $t = 1$ would give us a Kähler metric on an integrable complex structure on S^6, which is impossible topologically. The reservations concerning his arguments involve the closedness of the continuity method.

3.1.3. With the Moishezon Condition

A compact complex manifold of complex dimension n is Moishezon if the transcendence degree d of the field of meromorphic functions is equal to n. In general, the transcendence degree d is always less than or equal to

n by a theorem of Thimm [78]. A Moishezon manifold is projective algebraic if and only if it is Kähler [44].

Peternell [61] and Nakamura [57] showed that a compact complex manifold M that is topologically the same as $\mathbb{P}_3$ is biholomorphic to $\mathbb{P}_3$ if M is Moishezon. Unfortunately the additional assumption of the Moishezon property makes impossible any application to the problem of the non-existence of complex structure on the 6-sphere. Peternell [62] and Nakamura [58, 59] also showed that a compact complex manifold M that is topologically the same as the three-dimensional hyperquadric Q_3 is biholomorphic to Q_3 if M is Moishezon. Nakamura's method needs the assumption that there is no nontrivial holomorphic section for any positive power of the canonical line bundle.

It is not yet proved for general n whether a compact complex manifold M that is topologically the same as $\mathbb{P}_n$ is biholomorphic to $\mathbb{P}_n$ if M is Moishezon.

3.1.4. The Case of Deformation

An easier problem is the question of global deformation of $\mathbb{P}_n$ posed by Kodaira and Spencer [39]. The problem is the following. Suppose $\pi : M \to \Delta$ is a holomorphic family of compact complex manifolds parametrized by the unit disk Δ in $\mathbb{C}$. If for every nonzero t the fiber $M_t := \pi^{-1}(t)$ is biholomorphic to $\mathbb{P}_n$, does it follow that the fiber M_0 at $t = 0$ is also biholomorphic to $\mathbb{P}_n$?

An affirmative answer to this question is given in Ref. 74, which builds on the work of Tsuji [81] and uses Ref. 38. We sketch the proof, which is not exactly the same as that of Ref. 74 because the last step (p. 218, line 27) contains a gap, which was pointed out by Mabuchi and Tsuji. However, the method given here follows the same line as that of Ref. 74 and uses the counting of the zeros of holomorphic vector fields.

Kuhlmann also wrote a series of preprints on this problem (one of which is Ref. 40). The case of a general irreducible Hermitian symmetric manifold of compact type remains unknown.

CONJECTURE 3.4. *Suppose* $\pi : M \to \Delta$ *is a holomorphic family of compact complex manifolds parametrized by the unit disk* Δ *in* $\mathbb{C}$. *If for every nonzero* t *the fiber* $M_t := \pi^{-1}(t)$ *is biholomorphic to an irreducible Hermitian symmetric manifold, then the fiber* M_0 *at* $t = 0$ *is also biholomorphic to the irreducible Hermitian symmetric manifold.*

In all cases the fiber M_0 is always Moishezon by the argument of the semicontinuity of $\Gamma(M_t, K_{M_t}^{-\nu})$ in t, where K_{M_t} is the canonical line bundle of M_t. For the case of $\mathbb{P}_n$ and Q_n, once one knows that the fiber M_0 is

Kähler, the conclusion for the case of the complex projective and the hyperquadric follows from the theorem of Hirzebruch and Kodaira [32] and the theorem of Brieskorn [13]. In the case of a general irreducible Hermitian symmetric manifold of compact type, one has the following easier conjecture when one assumes that M_0 is Kähler.

CONJECTURE 3.5. *Suppose* $\pi: M \to \Delta$ *is a holomorphic family of compact complex manifolds parametrized by the unit disk* Δ *in* $\mathbb{C}$. *If for every nonzero* t *the fiber* $M_t := \pi^{-1}(t)$ *is biholomorphic to an irreducible Hermitian symmetric manifold and if the fiber* M_0 *at* $t = 0$ *is Kähler, then* M_0 *is also biholomorphic to the irreducible Hermitian symmetric manifold.*

In Ref. 37 Kilambi sketched a proof for the case in which the fibers M_t are assumed only to be homogeneous compact complex manifolds but the Kähler metric of M_t is assumed to be extendible to a continuous Hermitian metric on M_0. There he also sketched a proof that a homogeneous compact complex manifold that is diffeomorphic to an irreducible Hermitian symmetric manifold must be biholomorphic to it.

We now sketch the proof of the global nondeformability of the complex projective space. Let $\pi: M \to \Delta$ be a holomorphic family of compact complex manifolds parametrized by the unit disk Δ in $\mathbb{C}$ such that, for every nonzero t, the fiber $M_t := \pi^{-1}(t)$ is biholomorphic to $\mathbb{P}_n$. We want to show that the fiber M_0 at $t = 0$ is also biholomorphic to $\mathbb{P}_n$.

There is a line bundle L over M whose restriction to M_t is the hyperplane section line bundle of M_t for $t \neq 0$. Choose a holomorphic section s_0 of L over M and choose a holomorphic section τ of $M - \{s_0 = 0\}$ over Δ (after shrinking Δ if necessary). For $t \neq 0$ there is an Euler vector field on M_t that fixes the "hyperplane at infinity" $M_t \cap \{s_0 = 0\}$ and the "origin" $\{\tau(t)\}$. One seeks to extend the holomorphic $\mathbb{C}^*$ action of this Euler vector field to a holomorphic $\mathbb{C}^*$ action of M_0. Suppose such an extension is possible. Then the component $M \cap \{s_0 = 0\}$ of the fixed-point set of the family of the $\mathbb{C}^*$ actions forms a family of compact complex manifolds of complex dimension $n - 1$ over Δ. It follows from induction on n that the component $M_0 \cap \{s_0 = 0\}$ of the fixed-point set of the $\mathbb{C}^*$ action on M_0 is $\mathbb{P}_{n-1}$. By extending the holomorphic sections of L over $M \cap \{s_0 = 0\}$ to M, we can find holomorphic sections $s_1, \ldots, s_n$ of L over M so that $s_0, \ldots, s_n$ have no common zeros and they map M_0 biholomorphically onto $\mathbb{P}_n$. This approach of using Euler vector fields for the problem of nonformability of the complex projective space was first introduced by Tsuji [81].

The main difficulty is that the Euler vector field on M_t for $t \neq 0$ approaches a holomorphic vector field on M_0 only after multiplication by a positive power k of t, and one merely gets a $\mathbb{C}$ action instead of a $\mathbb{C}^*$ action from the limit vector field. The principal part of the proof is to overcome this difficulty.

First observe that there is a Zariski open subset G of M_0 such that $M_0 - \{s_0 = 0\}$ cannot contain any irreducible compact complex curve that intersects G, because for l sufficiently large, for a basis $\{u_\nu\}$ of $\Gamma(M_0, \Gamma^l)$ the function $\Sigma_\nu |u_\nu/s_0^l|^2$ is plurisubharmonic on $M_0 - \{s_0 = 0\}$ and is strictly plurisubharmonic $G - \{s_0 = 0\}$ for some Zariski open subset G of M_0.

We now suppose that for each holomorphic path of origins $\tau(t)$ and for each holomorphic family of infinity hyperplane sections we have the Euler vector fields X_t on M_t and a nonidentically zero holomorphic vector field $X^* = \lim_{t \to 0} t^k X_t$ on M_0 for some *positive* integer k. We want to derive a contradiction. Consider the set Σ of all holomorphic paths $\tau(t)$ of origins. We use the same family of infinity hyperplane sections $\{s_0 = 0\}$ for each path of origins. For each member τ of Σ we denote X^* by $X(\tau)$. The Lie bracket $[X(\tau_1), X(\tau_2)]$ of $X(\tau_1)$ and $X(\tau_2)$ is zero for any two members τ_1 and τ_2 of Σ because of the positivity of the integers k.

Let q be the positive integer such that there exist members $\tau_1, \ldots, \tau_q$ of Σ such that the vector fields $X(\tau_1), \ldots, X(\tau_q)$ are linearly independent at some point P_0 of G and the vector fields $X(\tau_1'), \ldots, X(\tau_{q+1}')$ are linearly dependent at every point of M_0 for any $q + 1$ members $\tau_1', \ldots, \tau_{q+1}'$ of Σ.

Since the Lie bracket of any two of the vector fields $X(\tau_1), \ldots, X(\tau_q)$ is identically zero, it follows that we can integrate them at the same time and get a submanifold V in $M_0 - \{s_0 = 0\}$ containing P_0 such that there is a biholomorphic map Ψ from $\mathbb{C}^q$ to V with $\Psi(0) = P_0$. For $a_1, \ldots, a_q$ not all zero we let $X(a) = \sum_{\nu=1}^q a_\nu X(\tau_\nu)$ and denote by $C(a)$ the topological closure of its orbit containing P_0.

Let us first look at the case $q > 1$. When we have a and a' that are not nonzero multiples of each other, we let $W(a, a')$ be the closure of the union of orbits of $X(a')$ that intersect $C(a)$. Since the Lie bracket of $X(a)$ and $X(a')$ is zero, it follows that both $X(a)$ and $X(a')$ are tangential to $W(a, a')$. Moreover, $W(a, a') - \{s_0 = 0\}$ is biholomorphic to $\mathbb{C}^2$, and $X(a)/s_0$ and $X(a')/s_0$ are holomorphic.

We make the following observation. Suppose we have a compact complex surface W, a divisor D in W, and two holomorphic vector fields X and Y on W, all of whose one-dimensional orbits have complex curves as their topological closure. Further assume that $W - D$ is biholomorphic to $\mathbb{C}^2$, with X and Y corresponding to the partial differentiation with respect to the coordinates of $\mathbb{C}^2$, and that the closure of every orbit of X and Y in $W - D$ contains a common zero of X and Y. Then the manifold obtained by blowing down successively the exceptional curves of the first kind in W is biholomorphic to $\mathbb{P}_2$.

By a holomorphic vector field on a local singularity we mean a holomorphic vector field on some ambient manifold that annihilates the local defining functions for the local singularity. For any method of *canonical* resolution of local singularities (in the sense that resolutions of biholomorphic local singularities are naturally biholomorphic), any holomorphic vector field on the local singularities can be lifted holomorphically to the resolution.

Let $\tilde{W} = W(a, a')$, and take a canonical desingularization W of $\tilde{W}$ with the map $\xi: W \to \tilde{W}$. We apply the above observation to the case $D = \xi^{-1}(\{s_0 = 0\})$ with X and Y equal to the lifting of $X(a)$ and $X(a')$. When we lift the vector fields to the desingularization, we divide by s_0 first. A generic orbit of X or Y in $\xi^{-1}(\tilde{W} - \{s_0 = 0\})$ intersects $\xi^{-1}(G)$, and therefore must intersect $\xi^{-1}(\{s_0 = 0\})$; and contains a common zero of X and Y. So every orbit of X or Y in $\xi^{-1}(\tilde{W} - \{s_0 = 0\})$ contains a common zero of X and Y. One can blow down W to $\mathbb{P}_2$ by blowing down successively exceptional curves of the first kind. Moreover, the image of $\xi^{-1}(\{s_0 = 0\})$ in $\mathbb{P}_2$ is the infinity line. To recover W from $\mathbb{P}_2$, we look at the first time a point P_1 in $\mathbb{P}_2$ is blown up to a curve Γ. The point P_1 must be in the common zero set of the two vector fields corresponding to X and Y, or else their liftings X and Y cannot be holomorphic. This zero set is the infinity line. From the explicit description of the blowup we know that Γ is not contained in the common zero set of the vector fields corresponding to X and Y. We ignore the finite number of points $Q_1, \ldots, Q_l$ of Γ that are to be blown up later. We know that $\Gamma - \{Q_1, \ldots, Q_l\}$ cannot be mapped to $\{s_0 = 0\}$ by ξ, or it would be in the common zero set of X and Y. Take Q in $\Gamma - \{Q_1, \ldots, Q_l\}$, which is not mapped to $\{s_0 = 0\}$ by ξ. We can take a projective line E in $\mathbb{P}_2$ containing P_1 such that its proper transform $\tilde{E}$, for the process of blowing up P_1, contains Q and $\xi(\tilde{E})$ intersects G. The curve $\xi(\tilde{E})$ is an irreducible compact complex curve in $M_0 \cap \{s_0 = 0\}$ that intersects G, contradicting the nonexistence of such a curve. So we know that there is no blowdown of W. Hence, W is biholomorphic to $\mathbb{P}_2$, and $\xi^{-1}(\{s_0 = 0\})$ is the infinity line of $\mathbb{P}_2$. Thus $\xi: W \to W(a, a')$ is a normalization of $W(a, a)$. Let Z be any nonzero linear combination of $X(a)$ and $X(a')$. The lifting of Z/s_0 to W is a $\xi^*(L)^{-1}$-valued holomorphic vector field on W tangential to D, and its zero set consists only of a single point in D. It follows that ξ must be a biholomorphism and $W(a, a')$ is biholomorphic to $\mathbb{P}_2$. It follows from the arbitrariness of a and a' that the biholomorphism Ψ can be extended to a biholomorphism from $\mathbb{P}_q$ to the topological closure V^- of V, as one can easily see by considering the moduli space of all complex curves C in V^- with $\Psi^{-1}(C)$ tangential to the Euler vector field of $\mathbb{C}^q$ at 0.

We now consider the case $q = 1$. Now V is simply the integral curve of $X(\tau_1)$ through P_0. By counting the number of zeros of the lifting $\tilde{X}$ of

$X(\tau_1)/s_0$ to the normalization of V^-, we conclude that V^- intersects $\{s_0 = 0\}$ at an only point of intersection P_1. Moreover, $X(\tau_1)/s_0$ intersects $\{s_0 = 0\}$ normally at P_1 and must vanish at P_1, or else some neighboring orbits are disjoint from $\{s_0 = 0\}$ and intersect G. The counting of zeros of $\tilde{X}$ now shows that V^- must be regular and biholomorphic to $\mathbb{P}_1$.

We assume that $q < n$, because we are done if $q = n$. We denote V^- by $S(\tau_1, \ldots, \tau_q, P_0)$. For a path τ of origins, we denote $X(\tau)/s_0$ by $Y(\tau)$. We now distinguish two cases.

Case 1. The zero set Z of $Y(\tau_1) \wedge \cdots \wedge Y(\tau_q)$ is of codimension at least 2 in M_0. Take a member τ of Σ so that the origin $\tau(0)$ for $t = 0$ lies in V. By Cramer's rule and the codimension of Z we can express $Y(\tau)$ as a linear combination of $Y(\tau_1), \ldots, Y(\tau_q)$ with constant coefficients, making it impossible for $Y(\tau)$ to vanish at the point $\tau(0)$ where $Y(\tau_1), \ldots, Y(\tau_q)$ are linearly independent.

Case 2. $Y(\tau_1) \wedge \cdots \wedge Y(\tau_q)$ vanishes on some hypersurface defined by $\rho = 0$ in M_0, where ρ is a holomorphic section of some power L^l of L with $l > 0$. Thus, $(1/\rho) Y(\tau_1) \wedge \cdots \wedge Y(\tau_q)$ is holomorphic on M_0. We know that $(1/\rho) Y(\tau_1) \wedge \cdots \wedge Y(\tau_q)$ cannot be nowhere zero on $S(\tau_1, \ldots, P_0)$; otherwise the integration of the distribution $(1/\rho) Y(\tau_1) \wedge \cdots \wedge Y(\tau_q)$ would make $S(\tau_1, \ldots, P_0)$ disjoint from $S(\tau_1, \ldots, \tau_q, P)$ for P close to P_0, and the normal bundle of $S(\tau_1, \ldots, \tau_q, P_0)$ in M_0 is trivial. Then for any $a \in \mathbb{P}_{q-1}$ the Chern class of the normal bundle of $C(a)$ in M_0 equals its Chern class of the normal bundle of $C(a)$ in $S(\tau_1, \ldots, \tau_q, P_0)$ and is $q - 1$. This yields a contradiction, because the anticanonical line bundle of M_0 is L^{n+1} and $L|C(a)$ has Chern class 1, forcing the Chern class of the normal bundle of $C(a)$ in M_0 to be $n - 1$. Let $m \geq 1$ be the degree of the hypersurface that is the zero set of $(1/\rho) Y(\tau_1) \wedge \cdots \wedge Y(\tau_q)$ in $S(\tau_1, \ldots, \tau_q, P_0)$. Then the degree of the zero set of $X(\tau_1) \wedge \cdots \wedge X(\tau_q)$ in $S(\tau_1, \ldots, \tau_q, P_0)$ equals $q + l + m$. On the other hand, the restriction of $X(\tau_1) \wedge \cdots \wedge X(\tau_q)$ to $S(\tau_1, \ldots, \tau_q, P_0)$ is a holomorphic section of the anticanonical line bundle of $S(\tau_1, \ldots, \tau_q, P_0)$, and its zero set in $S(\tau_1, \ldots, \tau_q, P_0)$ must be a hypersurface of degree $q + 1$, contradicting $l \geq 1$ and $m \geq 1$.

3.2. Characterization of Noncompact Type by Topological Conditions

3.2.1. Results of Strong Rigidity

Results on the characterization of noncompact type by topological conditions also go by the name of strong rigidity. The concept of strong rigidity has its origin in Weil's papers [83–85] on local rigidity, in which he proved that a cocompact lattice Γ in a semisimple Lie group G has no deformation if the Lie group has no compact or three-dimensional factors. More precisely, any homomorphism from Γ to G sufficiently close to the inclusion map $\Gamma \subset G$ can be obtained from the inclusion map by conjugation on the target G by an element of G. (A lattice Γ in a semisimple Lie group G means that Γ is a discrete subgroup of G and G/Γ has finite G-invariant measure. A cocompact lattice Γ means a lattice satisfying the additional assumption that G/Γ is compact.)

Then Mostow [53] proved the following important result on strong rigidity. The fundamental group of a compact locally symmetric Riemannian manifold of nonpositive curvature determines the manifold up to an isometry and a choice of normalizing constants if the manifold admits no closed one- or two-dimensional geodesic submanifolds that are locally direct factors. (An equivalent statement is the following: if G and G' are semisimple Lie groups without center and compact factors and if Γ and Γ' are discrete cocompact subgroups of G and G', respectively, then any isomorphism between Γ and Γ can be extended to an isomorphism between G and G' if there is no homomorphism from G to $PSL(2, \mathbb{R})$ with the image of Γ discrete.) In particular, two compact quotients of an irreducible bounded symmetric domain of complex dimension ≥ 2 with isomorphic fundamental groups are either biholomorphic or antibiholomorphic.

Margulis extended Mostow's result from a cocompact lattice to a general lattice [42].

Then in Ref. 67 it was discovered that, in the category of compact Kähler manifolds, to show that two compact Kähler manifolds M and N of the same topology type are biholomorphic (or antibiholomorphic), it suffices to assume that only one of M and N is a compact quotient of an irreducible symmetric domain of complex dimension ≥ 2.

One can reformulate this result in the terminology of strong rigidity. A compact Kähler manifold M is said to be *strongly rigid* (in the category of compact Kähler manifolds) if any compact Kähler manifold N that is of the same homotopy type as M is either biholomorphic or antibiholomorphic to M. The result then says that any compact quotient of an irreducible bounded symmetric domain of complex dimension at least 2 is strongly rigid.

The idea involved in the proof of this strong rigidity result is rather simple. It consists of representing the topological equivalence by harmonic maps and applying the Bochner–Kodaira technique to the differential of type $(0, 1)$ of the harmonic map.

3.2.2. Harmonic Maps and the Bochner–Kodaira Technique

A differential map f from a compact Riemannian manifold N to another Riemannian manifold M is called harmonic if the integral over N of the pointwise square norm of the differential of f as a functional on the set of all differential maps from N to M is critical at the map f. By the Euler-Lagrange equation an equivalent definition is that the Laplacian of f defined by using the connections of M and N is identically zero. A theorem of Eells and Sampson [23] tells us that if the sectional curvature of M is seminegative, then every homotopy class of maps from N to M can be represented by a harmonic map.

The Bochner-Kodaira technique is applied in the following way. Suppose M and N are compact Kähler manifolds and the sectional curvature of M is nonnegative with complex dimension n of N at least 2. Take a harmonic map f from N to M. Let $h_{\alpha\bar{\beta}}$ be the Kähler metric of M and ω_N the Kähler form of N. We multiply both sides of the following equation by ω_N^{n-2} and integrate over N:

$$\partial\bar{\partial}(h_{\alpha\bar{\beta}}\,\bar{\partial}f^\alpha \wedge \overline{\partial f^\beta}) = -h_{\alpha\bar{\beta}}D\,\bar{\partial}f^\alpha \wedge \bar{D}\,\overline{\partial f^\beta} - R_{\alpha\bar{\beta}\gamma\bar{\delta}}\,\partial f^\gamma \wedge \overline{\partial f^\delta} \wedge \bar{\partial}f^\alpha \wedge \overline{\partial f^\beta}.$$

(The summation convention is being used.) Here D is the covariant version of the ∂ operator, and $R_{\alpha\bar{\beta}\gamma\bar{\delta}}$ is the curvature tensor, which in normal coordinates equals $-\partial_\gamma\partial_{\bar{\delta}}h_{\alpha\bar{\beta}}$. For the first term we have, with respect to normal coordinates for both manifolds (when $\omega_N = \sqrt{-1}\sum_{i=1}^n dz^i \wedge \overline{dz^i}$),

$$h_{\alpha\bar{\beta}}D\,\bar{\partial}f^\alpha \wedge \bar{D}\,\overline{\partial f^\beta} \wedge \omega_N^{n-2} = \frac{-1}{n(n-1)}\sum_{\alpha=1}^n \left(\sum_{i,j=1}^n |\partial_i\partial_{\bar{j}}f^\alpha|^2 - \left|\sum_{i=1}^n \partial_i\partial_{\bar{i}}f^\alpha\right|^2\right)\omega_N^n.$$

$$= \frac{-1}{n(n-1)}|D\bar{\partial}f|^2\omega_N^n.$$

For the second term we have, with respect to normal coordinates of N,

$$R_{\alpha\bar{\beta}\gamma\bar{\delta}}\,\partial f^\gamma \wedge \overline{\partial f^\delta} \wedge \bar{\partial}f^\alpha \wedge \overline{\partial f^\beta} \wedge \omega_N^{n-2} = \frac{1}{n(n-1)}\left(\sum_{i,j=1}^n R_{\alpha\bar{\beta}\gamma\bar{\delta}}\zeta_{ij}^{\alpha\bar{\delta}}\zeta_{ij}^{\bar{\beta}\gamma}\right)\omega_N^n,$$

where $\zeta_{ij}^{\alpha\bar{\beta}}\,\overline{dz^i} \wedge \overline{dz^j} = 2\bar{\partial}f^\alpha \wedge \overline{\partial f^\beta}$ (in other words, $\zeta_{ij}^{\alpha\bar{\beta}} = \partial_{\bar{i}}f^\alpha\,\partial_{\bar{j}}f^\beta - \partial_{\bar{j}}f^\alpha\,\partial_{\bar{i}}f^\beta$).

The important point of the formula is that the Ricci curvature of the domain manifold does enter into the formula. This formula belongs to the same class of formulas that Kodaira used to prove his vanishing theorem. Here we use the bundle $f^*T_M^{1,0}$ and the $f^*T_M^{1,0}$-valued $(0, 1)$-form $\bar{\partial}f$. Because the bundle $f^*T_M^{1,0}$ depends on the form $\bar{\partial}f$, our Bochner formula is not linear but quasilinear.

The reason why the Ricci curvature does not enter into the formula is that this formula corresponds to the Bochner formula for $*(\bar{\partial}f)$ (where $*$ is the star operator), which is an $f^*(T_M^{1,0})^* \otimes K_N$-valued $(0, n-1)$-form. Because the canonical line bundle K_N of N occurs as a factor, the Ricci curvature that occurs normally in such Bochner formulas gets canceled from the curvature of K_N. Because the cotangent bundle $(T_M^{1,0})^*$ is used, one needs the positivity of the cotangent bundle (rather than the tangent bundle) of M for the vanishing of $\bar{\partial}f$ or ∂f. Equivalently one can derive the formula by using the divergence of $*(\bar{\partial}f)$ in the same way that Kodaira derived the formula for his vanishing theorem.

3.2.3. Holomorphicity from the Bochner–Kodaira Formula

From this Bochner–Kodaira-type formula one gets the holomorphicity (or antiholomorphicity) of harmonic maps into compact target manifolds satisfying certain negative curvature conditions. The formulation is as follows. For any positive integer s, the bisectional curvature of a Kähler manifold is said to be strongly s-nondegenerate at a point if it is not possible to find two nonzero complex subspaces V and W of the tangent space such that the bisectional curvature in the direction of ξ and η vanishes for $\xi \in V$ and $\eta \in W$ and $\dim_C V + \dim_C W > s$. Suppose M is a compact Kähler manifold, the cotangent bundle of M is semipositive in the sense of Nakano [60], and the bisectional curvature of M is strongly s-nondegenerate. (The curvature tensor $\Omega_{\alpha\bar{\beta}i\bar{j}}$ for a vector bundle is semipositive in the sense of Nakano if the quadratic form $(\zeta^{\alpha i}) \to \Sigma\Omega_{\alpha\bar{\beta}i\bar{j}}\zeta^{\alpha i}\overline{\zeta^{\beta j}}$ is semipositive, where α, β are indices for fiber coordinates and i, j are indices for coordinates of the base manifold.) If N is a compact Kähler manifold, $f: N \to M$ is a harmonic map, and $\text{rank}_R \, df > 2s + 1$, then f is either biholomorphic or antibiholomorphic. As a consequence, a compact Kähler manifold M is strongly rigid if the cotangent bundle of M is semipositive in the sense of Nakano, and the bisectional curvature of M is strongly s-nondegenerate for some $s < \dim_C M$.

3.2.4. Strong Rigidity for Locally Symmetric Manifolds

It follows from the curvature property of bounded symmetric domains that the compact quotient of an irreducible bounded symmetric domain of complex dimension at least 2 is strongly rigid [67, 70].

We define the degree of strong nondegeneracy of the bisectional curvature to be the smallest positive integer s so that the bisectional curvature is strongly s-nondegenerate at every point. To verify that the degree s of strong nondegeneracy of the bisectional curvature of an irreducible bounded symmetric domain of dimension ≥ 2 is less than the dimension, one can use an explicit expression of the curvature tensor or one can use the Lie algebra structure of the automorphism group of the irreducible bounded symmetric domain. The use of the explicit expression of the curvature tensor works easily for the four classical types, and such a computation of the degree of strong nondegeneracy of the bisectional curvature was carried out in Ref. 67. For the two exceptional domains it is easier to use the Lie algebra structure of the automorphism group of the irreducible bounded symmetric domain. In Ref. 70 the Lie algebra structure was used to show that the degree s of strong nondegeneracy of the bisectional curvature is less than the dimension for the two exceptional domains. However, the exact values of s for the two exceptional domains were not calculated in Ref. 70, but Zhong [89] calculated them to be 6 and 11.

The formulation of the curvature condition in terms of Lie algebra structure given in Ref. 70, p. 865, is as follows. Let $\mathfrak{g} = \mathfrak{k} + \mathfrak{p}$ be the Cartan decomposition of the Lie algebra of the automorphism group of the bounded symmetric domain. The complexification $\mathfrak{p}_C$ of $\mathfrak{p}$ breaks up into two summands $\mathfrak{p}^{1,0}$, $\mathfrak{p}^{0,1}$ corresponding to the vector space of type $(1, 0)$ and type $(0, 1)$, respectively. Let $\{e_\alpha\}$ be a basis for $\mathfrak{p}^{1,0}$. For $\zeta_{ij}^{\alpha\bar\beta} = \partial_{\bar i} f^\alpha \, \partial_j f^\beta - \partial_{\bar j} f^\alpha \, \partial_i f^\beta$,

$$\sum_{i,j=1}^n R_{\alpha\bar\beta\gamma\bar\delta}\zeta_{ij}^{\alpha\bar d}\,\zeta_{ij}^{\beta\bar\gamma} = -\sum_{i,j}\|[\partial_{\bar i}f^\alpha e_\alpha, \overline{\partial_j f^\beta e_\beta}] - [\partial_{\bar j}f^\alpha e_\alpha, \overline{\partial_i f^\beta e_\beta}]\|^2$$

$$= -\sum_{i,j}\|[\partial_{\bar i}f^\alpha e_\alpha + \overline{\partial_i f^\alpha e_\alpha}, \partial_{\bar j}f^\beta e_\beta + \overline{\partial_j f^\beta e_\beta}]\|^2.$$

Since

$$R_{\alpha\bar\beta\gamma\bar\delta}\xi^\alpha\overline{\xi^\beta}\,\eta^\gamma\overline{\eta^\delta} = -\|[\xi^\alpha e_\alpha, \eta^\beta e_\beta]\|^2,$$

for bounded symmetric domains the degree of strong nondegeneracy of the bisectional curvature is the smallest positive integer s for which it is not possible to find two nonzero complex subspaces V and W of $\mathfrak{p}^{1,0}$ and $\mathfrak{p}^{0,1}$, respectively, such that $[v, w] = 0$ for $v \in V$, $w \in W$, and $\dim_C V + \dim_C W > s$. The computation of s is reduced to a computation of the largest dimension of certain kinds of abelian subalgebras.

For any irreducible bounded symmetric manifold D, Calabi and Vesentini [18] and Borel [12] introduced a number $\gamma(M)$ so that $H^q(M, T_M)$ vanishes for $q < \gamma(D) - 1$ for any compact quotient M of D. This number $\gamma(D)$ can be alternatively described as follows [12]. If X is the compact Hermitian symmetric manifold dual to D and g is a suitable generator of the infinite cyclic group $H^2(X, \mathbb{Z})$, then $\gamma(D)g$ is the first Chern class of X. The degree s of strong nondegeneracy of the bisectional curvature of D is equal to dim $D + 2 - \gamma(D)$.

3.2.5. The Strong Rigidity of Compact Quotient of Polydisks

In the above discussion on strong rigidity for compact quotients of bounded symmetric domains we have been assuming that the bounded symmetric domain is irreducible and of complex dimension at least 2. What happens when the bounded symmetric domain is reducible and some factor is of complex dimension 1? When the domain is reducible, we know from a theorem of Cartan that any biholomorphism from a product of bounded domains to another product of bounded domains can be broken up into biholomorphisms between the factors. Thus, by taking a subgroup of finite index and going to the factors, we can reduce the question on strong rigidity to the case of irreducible bounded symmetric domains if each domain has complex dimension at least 2. Concerning the question of dimension 1, we know that in the case of compact quotients of the unit disk there is no strong rigidity because the deformation space of a marked hyperbolic Riemannian surface is the Teichmüller space. The only situation involving complex dimension 1 that is interesting is the case of an irreducible compact quotient Δ^n/Γ of a polydisk. (The irreducibility of Δ^n/Γ means that no finite covering of Δ^n/Γ is the product of two compact quotients of polydisks of lower dimension.) The result in this case is the following.

If Δ^n/Γ is an irreducible compact quotient of the n-disk with $n \geq 2$ and f is a harmonic map from a compact Kähler manifold N to Δ^n/Γ, which is a homotopy equivalence, then each component f_i of the map $(f_1, \ldots, f_n)$ from the universal covering $\tilde{N}$ of N to Δ^n is either holomorphic or antiholomorphic.

The main idea of the proof is as follows. From the Bochner–Kodaira formula for harmonic maps we get $\bar{D}\partial f_i = 0$ and $\partial f_i \wedge \bar{\partial} f_i = 0$. The condition $\bar{D}\partial f_i = 0$ means that if the pull-back of the tangent bundle of Δ by f_i is given the complex structure so that a smooth section is holomorphic if and only if its $(0, 1)$ covariant derivative vanishes, then ∂f_i is a holomorphic section of its tensor product with the cotangent bundle. As a consequence the zero set of ∂f_i (in the tangent bundle of N) is a complex subvariety. The condition $\partial f_i \wedge \bar{\partial} f_i = 0$ means that at points of N where ∂f_i is not zero, the level set of f_i agrees with the integral submanifold for ∂f_i. Thus the level sets of f_i are complex subvarieties of N. From the irreducibility of Δ^n/Γ we know that at a generic point of Δ^n/Γ these level sets intersect normally.

The case of complex dimension 2 using the theory of complex surfaces was given by Jost and Yau [35, 36]. The general case, whose proof was outlined above, was due to Mok [45].

3.2.6. The Case of Riemannian Target Manifolds

Sampson [65] applied this Bochner-Kodaira argument to the case where only the domain manifold N is Kähler and the target manifold M is Riemannian. Let h_{ij} be the Riemannian metric of N. The Bochner-Kodaira-type formula becomes

$$\partial\bar\partial(h_{ij}\,\bar\partial f^i \wedge \partial f^j) = - h_{ij}D\,\bar\partial f^i \wedge \bar D\,\partial f^j - R_{ijkl}\,\partial f^k \wedge \bar\partial f^l \wedge \bar\partial f^i \wedge \partial f^j.$$

After integrating over N the exterior product of both sides with ω_N^{n-2}, we obtain from the second term the vanishing of $\sum \int_N R(X_p, \overline{X_p}, X_q, \overline{X_q})$, where $X_p = \partial f/\partial z^p$, and $\partial/\partial z^p$ $(1 \le p \le n)$ is an orthonormal basis, and the summation is over $1 \le p < q \le n$. As a consequence, the image of ∂f must be in an abelian subspace of $\mathfrak{p} \otimes \mathbb{C}$. Sampson used the divergence of $*\overline{(\partial f)}$ to derive the formula. By determining the maximum dimensions of abelian subalgebras, Sampson showed that a harmonic map from a compact Kähler manifold into a locally symmetric space $SO_0(p, q)/SO(p) \times SO(q)$ has rank ≤ 2 if $p = 1$, and $\le 2q$ if $p = 2$ and $q \ge 2$.

3.2.7. Maximum Abelian Subalgebras

After applying the Bochner-Kodaira technique to harmonic maps, the problem of drawing geometric consequences from it is reduced to determining the maximum dimensions of certain abelian subalgebras. Carlson and Toledo [20] proved the following result concerning such abelian subalgebras. Let G/K be a symmetric space of noncompact type, and let $\mathfrak{g} = \mathfrak{k} + \mathfrak{p}$ be the Cartan decomposition. Let $W \subset \mathfrak{p} \otimes \mathbb{C}$ be an abelian subalgebra. Then $\dim_{\mathbb{C}} W \le (1/2) \dim_{\mathbb{C}} \mathfrak{p} \otimes \mathbb{C}$. Moreover, in the case of equality the pair $(\mathfrak{g}, \mathfrak{k})$ must be Hermitian symmetric, and the following holds. Let $\mathfrak{g}_i, \mathfrak{k}_i, \mathfrak{p}_i$ be the irreducible components of the pair $(\mathfrak{g}, \mathfrak{k})$, and let $W_i = W \cap \mathfrak{p}_i$. Then $W = \oplus W_i$ and, for each i such that $\mathfrak{g}_i$ is not isomorphic to $\mathfrak{sl}(2, \mathbb{R})$, one has $W_i = \mathfrak{p}^{1,0}$ for one of the two invariant complex structures on G_i/K_i. The strong rigidity of compact quotients of irreducible bounded symmetric domains of complex dimension at least 2 then follows immediately from this result on maximum abelian subaglebras after the application of the Bochner-Kodaira-type formula.

An application of the result of Carlson and Toledo is the following. Let Γ be a torsion-free cocompact discrete subgroup of $SO(1, n)$ with $n > 2$.

Then Γ is not the fundamental group of a compact Kähler manifold. Using harmonic maps, one needs to verify that the rank of the harmonic map is at least 3. In case the rank is less than 3, the map factors through a circle or a compact Riemann surface (because of the complex analyticity of level sets as in the discussion of the strong rigidity of irreducible compact quotients of polydisks). Since the universal covering spaces of the circle and a compact Riemann surface are contractible, the cohomological dimension of Γ cannot be more than 2, contradicting the cocompactness of Γ in $SO(1, n)$.

The corresponding result for other general simple Lie groups for non-Hermitian symmetric spaces, which is formulated below as a problem, is still unknown.

PROBLEM 3.1. Let Γ be a torsion-free cocompact discrete subgroup of the isometry group of a non-Hermitian symmetric space. Then Γ cannot be the fundamental group of a compact Kähler manifold.

Carlson and Toledo also showed that for any torsion-free discrete subgroup Γ of $SO(2p, q)$ the space $\Gamma \backslash SO(2p, q)/SO(p) \times SO(q)$ $(p > 1, q \neq 2)$ is not of the homotopy type of a compact Kähler manifold.

3.2.8. Superrigidity

Margulis [42] proved the following superrigidity result. Let G and H be two simple algebraic groups over $\mathbb{R}$ so that G is noncompact and the $\mathbb{R}$-rank of G is at least 2. Suppose Γ is a lattice in G and $\pi : \Gamma \to H$ is a homomorphism with $\pi(\Gamma)$ Zariski dense in H. Then π can be extended to a homomorphism $G \to H$ defined over $\mathbb{R}$.

The same statement holds when G is semisimple without any compact factor and Γ is irreducible in the sense that the image of Γ onto the quotient of G by a normal subgroup of positive dimension is dense.

Superrigidity is more general than strong rigidity in that the image of the discrete subgroup is no longer assumed to be discrete and the homomorphism may not be injective. On the other hand, superrigidity requires that the domain group be of rank at least 2. Margulis's superrigidity theorem does not work in the rank 1 case. Mostow [54, pp. 274–275] gave two examples to illustrate two reasons why superrigidity does not work in the rank 1 case.

The first example uses the fact that automorphisms of $\mathbb{C}$, in general, are not continuous. The example is a nonarithmetic lattice Γ in $PU(2, 1)$ so that for some field automorphism σ of $\mathbb{C}$ the image $^\sigma\Gamma$ of Γ under σ is no longer a lattice. Then the monomorphism $\Gamma \to {}^\sigma\Gamma$ cannot be extended to a homomorphism of $PU(2, 1)$ to itself.

The second example uses the fact that a simple Lie group cannot admit an infinite normal subgroup. The example is a pair of cocompact discrete subgroups Γ_1 and Γ_2 with a homomorphism from Γ_1 to Γ_2 with infinite kernel.

The strong rigidity theorem of Mostow in the case of complex variables can be extended to the more general case where the domain manifold is merely compact Kähler instead of locally symmetric of noncompact type. It is natural to ask whether there is any similar extension of Margulis's superrigidity theorem to the general case of a compact Kähler manifold. Instead of a lattice in G one considers the fundamental group $\pi_1(M)$ of a compact Kähler manifold M. To stay in the category of complex Kähler manifolds, one considers only the case where G is the automorphism group of an irreducible bounded symmetric domain. Suppose one has a homomorphism $\pi_1(M) \to G$ whose image is Zariski dense when G is considered as an algebraic group over $\mathbb{R}$. We can formulate the notion of local superrigidity as follows.

Clearly we have to identify deformations generated by inner automorphisms of G. Let Γ be a finitely presented group. Choose elements $\gamma_1, \ldots, \gamma_k$ of Γ generating Γ so that a finite number of relations $R_i(\gamma_1, \ldots, \gamma_k) = 1$ $(1 \le i \le l)$ generate all relations. Consider the set Hom (Γ, G) of all homomorphisms from Γ to G. The space Hom (Γ, G) is simply the $\mathbb{R}$-subvariety in the product of k copies of G defined by the equations $R_i(g_1, \ldots, g_k) = 1$ $(1 \le i \le l)$ for $g_1, \ldots, g_k$ in G. Let G act on Hom (Γ, G) by conjugation on the target G. We are interested in the orbit space Hom $(\Gamma, G)/G$. In general, the orbit space Hom $(\Gamma, G)/G$ may not be Hausdorff. To overcome this difficulty, one considers only an open subset Hom^s (Γ, G) of Hom (Γ, G) invariant under G so that Hom^s $(\Gamma, G)/G$ is Hausdorff. That subset Hom^s (Γ, G) is the set of all *stable* homomorphisms from Γ to G. The local rigidity discovered by Weil can be reformulated as saying that Hom^s $(\Gamma, G)/G$ is isolated at points $\rho \in$ Hom (Γ, G) where the image of ρ is a cocompact discrete subgroup of G. For local superrigidity one removes the condition that the image of ρ is cocompact discrete and replaces it by the condition that ρ is stable. Since it is easier to handle complex analytic varieties than real analytic ones and deformation in the complex points $G_{\mathbb{C}}$ of the real algebraic group G is even more general, one considers the situation of homomorphisms from Γ to $G_{\mathbb{C}}$ (the set of all complex points of G) instead of from Γ to G. A homomorphism ρ from Γ to $G_{\mathbb{C}}$ is called *stable* if the orbit $G_{\mathbb{C}} \cdot \rho$ is closed in Hom $(\Gamma, G_{\mathbb{C}})$ and the stability subgroup $Z(\rho) \subset G_{\mathbb{C}}$ of ρ for the conjugation action of $G_{\mathbb{C}}$ in Hom $(\Gamma, G_{\mathbb{C}})$ is finite. Homomorphisms $\rho : \Gamma \to G_{\mathbb{C}}$ with Zariski dense image in $G_{\mathbb{C}}$ are stable. Let $X(\Gamma, G_{\mathbb{C}})$ be the quotient of the stable part Hom^s $(\Gamma, G_{\mathbb{C}})$ of Hom $(\Gamma, G_{\mathbb{C}})$ by $G_{\mathbb{C}}$. The investigation of deformations of stable homomorphisms $\rho : \Gamma \to G_{\mathbb{C}}$ up to conjugation by elements of $G_{\mathbb{C}}$ is equivalent to the study of the complex analytic variety $X(\Gamma, G_{\mathbb{C}})$. Such deformation spaces $X(\Gamma, G_{\mathbb{C}})$ were studied in Johnson and Millson [34]. Recently Corlette [22] obtained a local superrigidity result in this context. For his result he needs a condition on ρ stronger than stability and even stronger than the Zariski density of its image. Before stating his result we need some notation.

Let M be a compact Kähler manifold with fundamental group Γ, and let G be a simple real algebraic group whose quotient group G/K by a maximum compact subgroup K is a bounded symmetric domain of complex dimension at least 2. Assume that G/K is not the open unit ball and its compact dual is not the complex hyperquadric of odd complex dimension ≥ 3. In other words, G/K is neither of the form $U(n, 1)/U(n) \times U(1)$ nor of the form $SO(2n + 1, 2)/S(O(2n + 1) \times O(2))$. Let $\rho : \Gamma \to G$ be a homomorphism, and let P be the flat principal G-bundle over M corresponding to ρ. Since G/K is contractible, there is a smooth section f of the G/K-bundle over M associated with the principal G-bundle P. Another way to describe f is that f is a smooth map from the universal covering space $\tilde{M}$ of M to G/K that is equivariant with respect to the homomorphism $\rho : \Gamma \to G$. Let ω be the G-invariant volume form of G/K. Define $\mathrm{Vol}(\rho)$ to be the integral of $f^*\omega$ over M. (When $\rho(\Gamma)$ is a torsion-free discrete subgroup of G, $\mathrm{Vol}(\rho)$ is the integral over M of the pull-back of the volume form of $\rho(\Gamma)\backslash G/K$ by the map $M \to \rho(\Gamma)\backslash G/K$.)

Corlette's result is that if $\mathrm{Vol}(\rho)$ is nonzero, then the homomorphism $\rho : \Gamma \to G$ is locally rigid as a homomorphism from Γ to $G_{\mathbb{C}}$. In other words, the complex analytic variety $X(\Gamma, G_{\mathbb{C}})$ is isolated at the point defined by ρ. The idea of his proof is as follows. If $X(\Gamma, G_{\mathbb{C}})$ is not isolated at the point defined by ρ, then there exists a real analytic family of homomorphisms $\rho_t : \Gamma \to G_{\mathbb{C}}$ so that $\rho_0 = \rho$, and for $t \neq 0$ the image of ρ_t is not contained in the set $X(\Gamma, G)$ of real points of $X(\Gamma, G_{\mathbb{C}})$. He then sets out to get a contradiction by showing that for t sufficiently small one can find $g_t \in G_{\mathbb{C}}$ so that the image of the homomorphism $g_t\rho_t g_t^{-1}$ from Γ to $G_{\mathbb{C}}$ is contained in the set G of real points. We apply the Bochner technique to the section f, which we can assume without loss of generality to be harmonic. The nonvanishing of $\mathrm{Vol}(\rho)$ implies that f is a holomorphic (or antiholomorphic) section. The differential ∂f of type $(1, 0)$ has value in $\mathfrak{p}^{(1,0)}$, where $\mathfrak{g}_{\mathbb{C}} = \mathfrak{k}_{\mathbb{C}} + \mathfrak{p}$ and $\mathfrak{p} = \mathfrak{p}^{(1,0)} + \mathfrak{p}^{(0,1)}$. For every t, from the homomorphism ρ_t we obtain a flat bundle over M who fiber is $G_{\mathbb{C}}/H$, where H is the maximum compact subgroup of $G_{\mathbb{C}}$, and we have a smooth section f_t of this flat bundle. We can assume without loss of generality that f_t is smooth in t and is harmonic. From the Bochner formula we conclude that the image of ∂f_t is an abelian subalgebra of $\mathfrak{g}_{\mathbb{C}}$. So for every point in a Zariski open subset of M we obtain a deformation of $\mathfrak{p}^{(1,0)}$ as an abelian subalgebra of $\mathfrak{g}_{\mathbb{C}}$. One then verifies from the root system that all such abelian subalgebra deformations of $\mathbf{p}^{(1,0)}$ are conjugate to $\mathbf{p}^{(1,0)}$ by elements of $\mathfrak{g}_C$ and for the conjugation the same element g_t of $\mathfrak{g}_C$ can be chosen for every point in the Zariski open subset of M for t sufficiently small. This means that the image of the homomorphism $g_t \rho_t g_t^{-1}$ from Γ to G_C is contained in the set G of real points of G_C.

The problem remains of proving Corlette's result for the two cases G/K equal to $U(n, 1)/U(n) \times U(1)$ or $SO(2n + 1, 2)/S(O(2n + 1) \times O(2))$ not covered in Corlette's paper.

For global superrigidity, because of the two examples of Mostow one gives the following formulation.

PROBLEM 3.2. Let M be a compact Kähler manifold, and let G be a simple algebraic group such that its quotient G/K by a maximum compact subgroup K is an irreducible bounded symmetric domain of complex dimension at least 2. Suppose $\rho_\nu : \pi_1(M) \to G$ ($\nu = 1, 2$) are two *monomorphisms* with image Γ_ν so that Vol (ρ_ν) is nonzero (i.e., the pull-back of the invariant form of G/K by a smooth section f_ν of the flat bundle over M with fiber G/K corresponding to ρ_ν has nonzero integral over M). Then there exists an element g of G and a field automorphism σ of $\mathbb{C}$ such that $g\Gamma_1 g^{-1} = {}^\sigma\Gamma_2$, where ${}^\sigma\Gamma_2$ is the image of Γ_2 under σ.

Even the case when M is a compact quotient of the ball is unknown.

A stronger statement is that the assumption is weakened so that the monomorphism ρ_1 is from $\pi_1(M)$ to G but the monomorphism ρ_2 is from $\pi_1(M)$ to G_C, and the nonvanishing of Vol (ρ_2) is replaced by the corresponding statement that the rank of df is equal to the real dimension of G/K. For the conclusion the element g is in G_C instead of in G.

3.2.9. Problem of Strong Rigidity for Vector Bundles

A natural concern is the strong rigidity for vector bundles. We say that a holomorphic vector bundle V over a complex manifold M is *strongly rigid* if any other holomorphic vector bundle W over M that is topologically isomorphic to V is biholomorphically isomorphic to it (or in a weaker formulation biholomorphically isomorphic to it after tensoring with a holomorphic line bundle). Grauert [27] showed that all holomorphic vector bundles over Stein manifolds are strongly rigid. The problem of characterizing strongly rigid holomorphic vector bundles over compact complex manifolds is completely open. The main difficulty is that one cannot use any positive or negative curvature condition on V, because a correct curvature condition on V should first guarantee the vanishing of $H^1(M, V \otimes V^*)$ (where V^* is the dual of V) and the vector bundle $V \otimes V^*$ is self-dual.

As a first step one should look for curvature conditions on V for the vanishing of $H^1(M, V \otimes V^*)$.

3.3. Characterization of Compact Type by Curvature Conditions

3.3.1. The Original Frankel Conjecture

A characterization of the complex projective space is formulated in the Frankel conjecture [26], which states that a compact Kähler manifold with positive holomorphic bisectional curvature is biholomorphic to the complex projective space. The case of dimension 2 for the Frankel conjecture was verified by Andreotti and Frankel [26]. The three-dimensional case was verified by Mabuchi [41] and Mori and Sumihiro [52]. The general case was proved by Mori [51], using the method of algebraic geometry. He actually proved the more general statement that a compact complex manifold whose tangent bundle is ample is biholomorphic to the complex projective space. Later a differential geometric proof of the Frankel conjecture was given in Ref. 77.

The key point of the proof of the Frankel conjecture is to construct a projective line of minimal degree inside the given compact complex manifold M. Mori's method is to take first any complex curve and then try to deform it with one of its points fixed. After one gets such deformations parametrized by a compact complex curve C, one can easily get a rational curve by considering the minimal model of a surface of the union of all curves parametrized by C. Then one deforms the rational curve with two points fixed to get rational curves of low degree. Though the tangent bundle of M is positive, we have trouble deforming curves of high genus in M. Mori's idea is to apply the Frobenius transformation in characteristic p to the curve so that the Chern class of the tangent bundle of M pulled back to the Frobenius transform of the curve increases but the genus of the Frobenius transform of the curve remains unchanged. This method produces rational curves of low degree, and then one goes back to the case of the complex number field from the case of characteristic p.

The differential geometric method of producing the rational curve of minimal degree is to use energy-minimizing harmonic maps from the 2-sphere and apply the second variation to get holomorphicity of the energy-minimizing harmonic maps from the positivity of the holomorphic bisectional curvature.

After the existence of the rational curves of minimal degree is proved, one uses the positivity of the tangent bundle of M to move such rational curves around with one point fixed to conclude that M is biholomorphic to the complex projective space.

Mori's method of using characteristic p to produce a rational curve of low degree is more powerful than the differential geometric method. It can show that such a rational curve exists by assuming only that the anticanonical line bundle is positive. The differential geometric method is applicable only in the case of positive holomorphic bisectional curvature.

PROBLEM 3.3. Find a differential geometric proof of the statement that a rational curve exists in a compact Kähler manifold with positive anticanonical line bundle (which is the same as positive Ricci from the method of Yau's solution of the Calabi conjecture [88]).

3.3.2. The Generalized Frankel Conjecture

The original Frankel conjecture characterizes the complex projective space by curvature conditions. One can naturally formulate a generalized Frankel conjecture for a general compact Hermitian symmetric manifold by assuming the holomorphic bisectional curvature to be semipositive. A curvature characterization of the hyperquadric was given in Ref. 68. Bando [9] proved the generalized Frankel conjecture for complex dimension 3. Mok [48] proved the generalized Frankel conjecture for all dimensions. Mok's result is the following. Let (X, g) be an n-dimensional compact Kähler manifold of nonnegative holomorphic bisectional curvature. Let $(\tilde{X}, \tilde{g})$ be its universal covering space. Then there exist nonnegative integers $k, N_1, \ldots, N_l$ and irreducible compact Hermitian symmetric spaces $M_1, \ldots, M_p$ of rank ≥ 2 such that $(\tilde{X}, \tilde{g})$ is isometrically biholomorphic to

$$(\mathbb{C}^k, g_0) \times (\mathbb{P}^{N_1}, \theta_1) \times \cdots \times (\mathbb{P}^{N_l}, \theta_l) \times (M_1, g_1) \times \cdots \times (M_p, g_p),$$

where g_0 denotes the Euclidean metric on $\mathbb{C}^k$, $g_1, \ldots, g_p$ are canonical metrics on $M_1, \ldots, M_p$, and θ_i, $1 \leq i \leq l$, is a Kähler metric on $\mathbb{P}^{N_i}$ carrying nonnegative holomorphic bisectional curvature.

Earlier Mok and Zhong [50] proved the special case with the additional assumption that the scalar curvature (X, g) is constant.

The main ideas in Mok's proof of the generalized Frankel conjecture are as follows. One first does a preliminary reduction to split M into factors so that the second Betti number can be assumed to be 1 and the Ricci curvature is not identically zero. By deforming the Kähler metric by the heat equation in the direction opposite to that of the Ricci curvature, one can assume that the Ricci curvature is positive definite at every point. One then uses Mori's result of the Frobenius transform in characteristic p to produce a rational curve $f: \mathbb{P}_1 \to M$ with $f^*(K_M^{-1}) = \mathcal{O}(q)$ for some minimum $0 < q < n + 1$, where $\mathcal{O}(q)$ means the line bundle over $\mathbb{P}_1$ with Chern class q. Using the strict pointwise positivity of the Ricci curvature and the nonnegativity of the bisectional curvature, one can deform the rational curve $f: \mathbb{P}_1 \to M$ so that there are such rational curves through each point. Let $\mathfrak{M} \subset \mathbb{P}(T_M^{1,0})$ consist of all tangent directions X such that the rational curve $f: \mathbb{P}_1 \to M$ along that direction X does not satisfy $f^*(T_M) = \mathcal{O}(2) \oplus (\bigoplus_{i=1}^{n-1} \mathcal{O}(1))$. If there exists a point without such a direction ξ, then Mori's proof of the Frankel conjecture implies already that M is biholomorphic to $\mathbb{P}_n$. The final step is verify that $\mathfrak{M}$ is invariant under parallel transport, and then one invokes the theorem of Berger that if the holonomy group at a point of a complete simply connected irreducible Riemannian manifold does not act transitively on the unit tangent sphere, then the Riemannian manifold is a symmetric space of higher rank [11, 66].

In the case of compact Hermitian manifolds of rank ≥ 2 there is also a phenomenon of metric rigidity, which will be discussed in Section 3.7.

3.4. Characterization of Noncompact Type by Curvature Conditions

3.4.1. The Example of Nonlocally Symmetric Negatively Curved Compact Surface

Suppose M is a compact Kähler manifold with negative curvature in some suitable sense. Can we conclude that M is the compact quotient of a bounded symmetric domain? The answer is negative. An example of a compact Kähler surface was constructed in Ref. 55, whose universal covering is not biholomorphic to the ball and yet whose cotangent bundle is negative in the sense of Nakano (which, in particular, implies negative sectional curvature).

The example in Ref. 55 can be described as follows. In the complex 2-ball B take three complex lines L_i $(1 \leq i \leq 3)$ and a complex reflection R_i $(1 \leq i \leq 3)$ about each line. Here a complex reflection means rotation by a certain angle θ_i about the complex line. One considers the subgroup Γ in the automorphism group of B generated by the three complex reflections R_i $(1 \leq i \leq 3)$. A theory concerning the discreteness of such a subgroup Γ generated by three complex reflections was developed by Mostow [54]. With a suitable choice of L_i and θ_i the subgroup Γ in the automorphism group of B is almost discrete in the following sense. There exists a complex manifold B and a holomorphic map $f: \tilde{B} \to B$ with finite-order branching along an infinite number of complex lines of B such that there exists a discrete subgroup $\tilde{\Gamma}$ of the automorphism group of $\tilde{B}$ corresponding to Γ. The compact Kähler manifold M is the quotient of $\tilde{B}$ by $\tilde{\Gamma}$. The Kähler metric is obtained by patching together the invariant metric of B and the Bergman metric for the domain of the form

$$\{(z_1, z_2) \in \mathbb{C}^2 \,||z_1|^2 + |z_2|^{2m} < 1\}$$

with $m > 1$.

3.4.2. The Kähler–Einstein Metric with Curvature Pinching

One seeks some natural additional condition in order to conclude that M is the compact quotient of a bounded symmetric domain. So far, no natural additional condition has been found for which one has a proof for the conclusion. A good candidate for the natural additional condition is that the Kähler metric be Einstein. The question whether every Kähler–Einstein compact complex manifold of complex dimension 2 with negative sectional curvature is biholomorphic to a compact quotient of the complex 2-ball is still open. In Ref. 75 a positive answer to this question is obtained with some additional pinching assumption on the sectional curvature. More precisely, let $K_{av}(P)$ (respectively, $K_{max}(P)$, $K_{min}(P)$) be the average (respectively, maximum, minimum) sectional curvature at the point P. The additional pinching assumption is that $K_{av}(P) - K_{min}(P) \leq \chi(K_{max}(P) - K_{min}(P))$ is satisfied at every point P of M for some $\chi < 2/3(1 - \sqrt{6/11})$. The number $2/3(1 - \sqrt{6/11})$ is approximately 0.38346. In general, for any Kähler–Einstein surface without any curvature assumption one always has

$$\tfrac{1}{3}(K_{max}(P) - K_{min}(P)) \leq K_{av}(P) - K_{min}(P) \leq \tfrac{2}{3}(K_{max}(P) - K_{min}(P)).$$

The idea of the proof there is as follows. Let N be the set of balllike points (i.e., points where the holomorphic sectional curvature is constant). The set N is a real analytic subvariety. If N is not all of M, the real codimension of N is at least 2; otherwise from the Kähler–Einstein condition it follows that all covariant derivatives of the curvature tensor vanish at every point of N, which by the real analyticity of Kähler–Einstein metrics implies that M is locally symmetric. For every point P in $M - N$ choose a unitary frame e_1, e_2 so that the minimum sectional curvature is achieved at e_1. Let $R_{\alpha\bar{\beta}\gamma\bar{\delta}}$ be the components of the curvature tensor with respect to the basis e_1, e_2. Let

$$\phi(P) = 6|R_{1\bar{2}1\bar{2}}|^2 - (R_{1\bar{1}1\bar{1}} - 2R_{1\bar{1}2\bar{2}})^2.$$

If M is not locally symmetric, then one verifies by direct computation that the curvature pinching condition implies that ϕ^λ is strictly superharmonic on $M - N$ for some sufficiently small positive number λ. Since ϕ^λ is bounded and the real codimension of N in M is at least 2, ϕ^λ can be extended to a superharmonic function on M, contradicting the compactness of M and the strict superharmonicity of ϕ^λ on $M - N$.

Polombo improved on the pinching constant χ in Ref. 63. There, instead of the function ϕ given above, he chooses another function ϕ defined as follows. For a Riemannian manifold of dimension 4 the curvature tensor, when interpreted as an endomorphism of the space of 2-forms, can be broken into three parts. The first part is a constant times the identity endomorphism. The second part corresponds to the trace-free part of the Ricci tensor. The third part W that remains is known as the Weyl curvature tensor. The star operator breaks up the space of 2-forms into two eigenspaces of real dimension 3 each. One restricts the Weyl curvature tensor to one of the three-dimensional eigenspaces and obtains three eigenvalues $\lambda_1 \leq \lambda_2 \leq \lambda_3$. The function ϕ that Polombo uses is $-(8/7)(\lambda_1^2 + \lambda_2^2) + \lambda_3^2$. He shows that for $\chi = 0.48$ the function ϕ is strictly superharmonic on $M - N$.

3.4.3. The Case of Rank at Least 2

In the case when the rank of the manifold is at least 2, according to the result obtained by Ballmann, Brin, and Spatzier in a series of papers [5–7], the manifold must be locally symmetric if its volume is finite and its curvature is nonpositive and bounded from below [6, theorem]. The rank of a Riemannian manifold is defined as the largest number r so that along any geodesic the dimension of the space of parallel Jacobi fields is at least r. As in the semipositive case, one now uses the theorem of Berger that if the holonomy group at a point of a complete, simply connected, irreducible Riemannian manifold does not act transitively on the unit tangent sphere, then the Riemannian manifold is a symmetric space of higher rank [11, 66].

In the case of Hermitian locally symmetric manifolds of rank ≥ 2 of noncompact type (just as in the case of compact type), there is also a phenomenon of metric rigidity, which will be discussed in Section 3.7.

3.5. Characterization of Euclidean Space by Curvature Conditions

The starting point of this characterization is the following conjecture of Greene and Wu [28]. If M is a simply connected complete Kähler manifold with sectional curvature K satisfying $-C/r^{2+\varepsilon} \leq K \leq 0$ for some constant C, where r is the distance measured from a certain point of M, then M is biholomorphic to the Euclidean space $\mathbb{C}^n$. The conjecture was solved by Siu and Yau [76]. Their method was to use the L^2 estimates of $\bar{\partial}$ to produce n holomorphic functions of growth order $r^{1+\delta}$ for some sufficiently small positive δ so that they give local coordinates at one point. Then these n functions are used to give the biholomorphism between M and $\mathbb{C}^n$.

Later, Greene and Wu [29] proved that under the condition of nonpositive curvature of decay faster than quadratic the curvature is necessarily zero. In other words, there is a gap for the growth order of the curvature, which is at least quadratic order if not identically zero. They actually proved this gap phenomenon for curvature growth for Riemannian manifolds. The gap phenomenon holds not only for the case of nonpositive curvature but also for the case of nonnegative curvature. The precise statements of their results are as follows. Suppose M is a complete simply connected Riemannian manifold of either everywhere nonpositive or everywhere nonnegative sectional curvature. In the case of nonnegative curvature one assumes that the exponential map at some point P_0 of M is a global diffeomorphism. Let $k(s)$ be the supremum of the absolute value of all the sectional curvatures at points of distance s from P_0. Assume $\lim_{s \to \infty} s^2 k(s) = 0$ in the case of nonnegative curvature and $\int_{s=0}^{\infty} sk(s) \, ds =$ finite in the case of nonpositive curvature. If the dimension of M is not equal to 2 in the case of nonpositive curvature and is not equal to 2, 4, or 8 in the case of nonnegative curvature, then M is isometric to the Euclidean space. When the dimension n of M is equal to 4 or 8 and the curvature is nonnegative, M is isometric to the Euclidean space under the additional assumption that for some constant C the ball of radius r centered at P_0 has volume at least Cr^{n-1}.

Their proof uses the Gauss–Bonnet formula and comparison theorems. To illustrate their method, consider the case of odd-dimensional M and nonnegative curvature. Let S_r be the sphere of radius r centered at P_0, and let σ_r be the second fundamental form of S_r. Then by using the comparison theorem, one concludes that $\int_{S_r} \det \sigma_r$ is nonnegative and no greater than the volume E_0 of the Euclidean unit sphere in $\mathbb{R}^n$ and, if $\limsup \int_{S_r} \det \sigma_r = E_0$ as $r \to \infty$, then M is isometric to $\mathbb{R}^n$. The curvature Ω of the induced metric on S_r can be expressed in terms of its second fundamental form σ_r and the curvature Ω of M. By the Gauss–Bonnet formula an integral of a function of Ω over S_r gives the Euler characteristic of S_r. Thus one gets an integral over S_r involving $\tilde{\Omega}$ and $\det \sigma_r$ with value equal to the Euler characteristic of S_r. The decay condition on $\tilde{\Omega}$ and the above statement concerning $\limsup \int_{S_r} \det \sigma_r$ gives us the isometry between M and the Euclidean space.

Ballmann, Gromov, and Schroeder [8] give the above results on the curvature gap phenomenon in an exercise with the condition of the global diffeomorphism of the exponential map at some point P_0 of M removed in the case of nonnegative curvature.

In Mok, Siu and Yau [49] the following two results on the characterization of the Euclidean space were obtained, which properly belong to the category of Kähler manifolds instead of to the category of Riemannian manifolds.

1. Suppose M is a complete Kähler manifold of nonnegative holomorphic bisectional curvature of complex dimension $n \geq 2$. Let R be the scalar curvature and $B(x, r)$ the geodesic ball of radius r centered at x. If M is Stein, $R \leq 1/r^{2+\varepsilon}$, and the volume of $B(x, r)$ is $\geq Cr^{2n}$ (where $\varepsilon > 0$ and $C > 0$ are constants), the M is isometrically biholomorphic to $\mathbb{C}^n$.

2. Suppose M is a complete Kähler manifold of complex dimension $n \geq 2$ and the exponential map at some point P_0 is a global diffeomorphism. Assume that the absolute value of the sectional curvature is bounded by $A_\varepsilon/(1 + r^{2+\varepsilon})$, where A_ε is a positive constant depending on ε. Then M is biholomorphic to $\mathbb{C}^n$.

For the proof of these two results, besides refinements of the method introduced in Siu and Yau [76], the method of solving the Poincaré-Lelong equation $\partial\bar{\partial}u = \rho$ for a closed $(1, 1)$-form ρ is used. The Poincaré-Lelong equation was solved either by using the Bochner formula for the Laplacian or by using the L^2 estimates of $\bar{\partial}$.

3.6. Characterization of Noncompact Type by Compact Quotients

Because of the uniformization theorem for one complex dimension, one has the following conjecture for higher dimensions.

CONJECTURE 3.6. *If M is a compact Kähler manifold of negative sectional curvature, then there are enough bounded holomorphic functions on the universal covering $\tilde{M}$ of M that distinguish points in $\tilde{M}$ and give local coordinates of $\tilde{M}$.*

So far nobody knows how to produce any nonconstant bounded holomorphic function on $\tilde{M}$. There is a sufficient topological condition obtained in Ref. 73 that enables one to produce nonconstant bounded holomorphic functions on the universal cover of a compact Kähler manifold M. However, there is no good way of verifying such a topological condition. The topological condition is the following. There exists a continuous map from M to a compact hyperbolic Riemannian manifold that is nonzero on the second homology group.

The uniformization theory for one complex variable tells us not only that there are enough bounded holomorphic functions on the universal cover of a compact Riemann surface of genus at least 2, but it also tells us the universal cover is the open disk. Since we have trouble producing nonconstant bounded holomorphic functions on the universal cover of a compact Kähler manifold of negative sectional curvature, we ask the easier problem of assuming first that the universal is a domain and trying to show that the domain is bounded symmetric. The first result of this kind was obtained by Wong [86], who proved that if a smooth bounded domain covers a compact manifold, then the domain must be biholomorphic to the ball.

There are several proofs of B. Wong's result. One proof uses the fact that at a strongly pseudoconvex boundary point from the asymptotic expansion of the Bergman kernel one knows that asymptotically the holomorphic sectional curvature is constant. When the domain covers a compact manifold, every point can be mapped arbitrarily close to a strongly pseudoconvex boundary point, so the Bergman metric has constant holomorphic sectional curvature at every point and the domain is a ball. Rosay [64] made some improvements on Wong's result.

Recently, Frankel obtained the following general result concerning the characterization of bounded symmetric domains by the existence of compact quotients. A bounded Euclidean convex domain Ω in $\mathbb{C}^n$ with a cocompact fixed-point-free discrete subgroup Γ in Aut Ω must be a bounded symmetric domain. The main ideas of Frankel's proof are as follows.

First one uses the following high-dimensional distortion theorem to produce a continuous one-parameter subgroup of Aut Ω: the family of holomorphic maps from Ω to $\mathbb{C}^n$ with prescribed value and prescribed Jacobian matrix at one point is normal if the image of each map is a convex domain (which may depend on the map). (A distortion theorem in $\mathbb{C}^2$ with better estimates for normalized univalent mappings with convex image is given by R. W. Barnard, C. H. FitzGerald, and S. Gong [10]. Some other references on earlier work on the subject can be found in the references at the end [10].) The tangent space of a submanifold can be constructed by using dilations and taking limits. In the same way, at a point P_* on the boundary of Ω one uses a sequence of elements γ_ν in Γ with $\gamma_\nu P \to P_*$ (after normalization of value and Jacobian matrix at one point) to construct a limiting domain Ω_* from Ω. Because of the convexity of Ω, the limiting domain Ω_* is a holomorphic family of upper half-planes. The half-planes correspond to the complex normal direction. The limiting domain Ω_* is biholomorphic to Ω because of the high-dimensional distortion theorem and the convexity of Ω. The family of translations along the boundary of the upper half-planes is a one-parameter subgroup of Aut Ω_*.

Let G be the identity component of the group generated by Γ and a one-parameter subgroup in Aut Ω. We want to prove that G is semisimple. Suppose G is not semisimple. Then there exists a nontrivial abelian subgroup A of G normalized by Γ. The group A is obtained by taking the radical R

of G and the central descending series of the radical constructed by using commutators.

We claim that the group A has no fixed points. The set of all fixed points of A is invariant under Γ and consists of all P such that $f_a(P) - P = 0$ for all $a \in A$. Any bounded holomorphic function F on Ω that vanishes on one orbit of Γ must be identically zero, because we can cover Ω by ΓD and also by $\Gamma D'$ for some relatively compact open balls $D \Subset D'$ centered at a zero point P of F and, by applying the Schwarz lemma to each pair $\gamma D \Subset \gamma D'$, we can conclude that the supremum of $|F|$ must be zero.

The group A has no compact subgroups; otherwise by a modification of Cartan's argument of using the "center of mass," we have a fixed point for A. The orbits of A define in the tangent bundle T_Ω of Ω a subbundle E that is totally real because Ω is a bounded domain. Let $\tilde{E}$ be the complex subbundle of $T_{\Omega/\Gamma}$ defined by E. Since E is totally real, the first Chern class of $\tilde{E}$ must be zero. Moreover, $\tilde{E}$ is holomorphic. The existence of the Bergman kernel of Ω implies that the first Chern class of Ω/Γ is negative, and we have a Kähler-Einstein metric on Ω/Γ that implies the stability of $T_{\Omega/\Gamma}$. On the other hand, $c_1(T_{\Omega/\Gamma})$ is negative, and by the stability of $T_{\Omega/\Gamma}$ we know that $c_1(\tilde{E})$ cannot be zero, which is a contradiction. So G must be semisimple.

Without loss of generality we assume that the domain Ω cannot be decomposed as the product of two domains. Let K be a maximum compact subgroup of G. Since K has a fixed point by the Cartan argument of the "center of mass," we know that by its maximum property K is the isotropy subgroup of its fixed point. Let X be the fixed-point set of K. The set X is contractible, because when we join two points by a path the "centers of mass" of the orbits are also joined by a path consisting of the "centers of mass." The set GX is invariant under Γ, because two maximum compact subgroups are always conjugate to each other. Being diffeomorphic to $(G/K) \times X$, the set GX is contractible, because K is a maximum compact subgroup of the semisimple group G. It follows that the $2n$th cohomology of GX/Γ and Ω/Γ are isomorphic under the inclusion map and GX must equal Ω.

Take a point P in Ω. The isotropy group of P in G is a maximum compact subgroup of G that we can assume without loss of generality to be equal to K. Take a metric of G invariant under K. The map π from $T_{\Omega,P}$ to itself defined by averaging over K projects the tangent space $T_{\Omega,P}$ onto the tangent space $T_{X,P}$ of X at P. Then the tangent space $T_{GP,P}$ of GP at P belongs to the kernel of the projection; otherwise the action of K would leave invariant every point of the geodesic issuing from P and in the direction of a nonzero K-invariant vector and would have other fixed points on GP different from P. Since $T_{X,P}$ is invariant under the almost complex operator

J, it follows that its orthogonal complement $T_{GP,P}$ is also invariant under J, and we conclude that $X \times (G/K)$ is a holomorphic splitting of Ω contradicting the indecomposability of Ω unless X is a single point. So we know that G acts transitively on Ω, and Ω is homogeneous. By a well-known result (see Ref. 82) a bounded homogeneous domain covering a compact manifold must be symmetric.

A problem is how to extend the above result to the case of a quotient that is a Zariski open subset of a projective algebraic manifold (instead of a compact quotient). One of the reasons for such an extension is to prove the following conjecture concerning the Teichmüller space.

CONJECTURE 3.7. *The Teichmüller space is not biholomorphic to a bounded convex domain.*

The mapping class group Γ is a discrete subgroup of the automorphism group of the Teichmüller space Ω. The quotient Ω/Γ is not compact, but can be compactified to a compact projective algebraic manifold.

Recently Nadel obtained the following result, which is related to Frankel's result and gives more information after one obtains in Frankel's proof a one-parameter subgroup of automorphisms. If the universal cover $\tilde{M}$ of a compact complex surface M with ample canonical bundle is not Hermitian symmetric (i.e., if $\tilde{M}$ is neither the complex ball nor the bidisk), then the automorphism group of $\tilde{M}$ is discrete, acts properly discontinuously on $\tilde{M}$, and contains the group of covering transformations as a subgroup of finite index.

3.7. Metric Rigidity

In Sections 3.3 and 3.4 we referred to the phenomenon of metric rigidity for Hermitian symmetric manifolds of rank ≥ 2. For the case of noncompact type the phenomenon of metric rigidity is given by the following result due to Mok [46]. If (M, g) is a locally symmetric Hermitian manifold of finite volume uniformized by an irreducible bounded symmetric domain of rank ≥ 2 and h is a Hermitian metric on M such that the holomorphic bisectional curvature of (M, h) is seminegative and h is dominated by a constant multiple of g, then $h = cg$ for some constant $c > 0$.

To [79] removed the assumption that h is dominated by a constant multiple of g. As an immediate corollary one obtained the following result. If (M, g) is a locally symmetric Hermitian manifold of finite volume uniformized by an irreducible bounded symmetric domain of rank ≥ 2, (N, h) is a Hermitian manifold with seminegative holomorphic bisectional curvature, and if $f: M \to N$ is a nonconstant holomorphic mapping, then up to a normalizing constant f is a totally geodesic isometric immersion.

For the rank 1 case clearly there is no metric rigidity because the Kähler metric can be slightly perturbed without destroying the negative curvature condition. However, there is still metric rigidity when one considers only a special class of metrics. An example of this kind of metric rigidity is the following result of Cao and Mok [19]. Any holomorphic immersion from a compact quotient of the complex n-ball to a compact quotient of the complex m-ball must be a totally geodesic isometric immersion if $m \le 2n - 1$. This result gives metric rigidity among metrics obtained by pulling back the invariant metric of the complex ball by holomorphic immersion. It is the dual of Feder's result [24] that a holomorphic immersion from $\mathbb{P}_n$ to $\mathbb{P}_m$ must be projective linear if $m \le 2n - 1$.

The main arguments in the proof of metric rigidity for rank ≥ 2 are the following. The universal covering D of M is the quotient of a semisimple Lie group G by its maximum compact subgroup K. Let $\mathfrak{g} = \mathfrak{k} + \mathfrak{p}$ be the Cartan decomposition of the Lie algebra $\mathfrak{g}$ of G, where $\mathfrak{k}$ is the Lie Algebra of K. The complexification of $\mathfrak{p}$ can be decomposed into $\mathfrak{p}^{1,0}$ and its complex conjugate so that $\mathfrak{p}^{1,0}$ is identified with the space of tangent vectors of type $(1, 0)$ in D at a general point of D. For X in $\mathfrak{p}^{1,0}$ we consider the null space $N_X = \{Y \in \mathfrak{p}^{1,0} \,|\, [X, \bar{Y}] = 0\}$. Let d be the maximum dimension of all N_X for X in set of all $\mathfrak{p}^{1,0}$. Let $\mathfrak{M}$ be the set of all tangent vectors X of type $(1, 0)$ in D with $\dim N_X = d$. We regard the set $\mathfrak{M}$ as a subset of the projectivization $\mathbb{P}(T_D^{1,0})$ of the tangent bundle $T_D^{1,0}$ of D. In terms of curvature N_X is the set of all tangent vectors Y of type $(1, 0)$ such that the holomorphic bisectional curvature for the two tangent vectors X and Y vanish. So $\mathfrak{M}$ consists of all tangent vectors that form zero holomorphic bisectional curvature with the tangent subspaces of the largest dimension. There is an algebraic geometric description of the set $\mathfrak{M}$. Let D^* be the Hermitian symmetric manifold of compact type dual to D. The domain D is canonically embedded as an open subset of D^*. We call a projective line C in D^* representing a generator of $H_2(D^*, \mathbb{Z})$ a minimal projective line. Over any minimal projective line C the tangent bundle of D^* splits into a direct sum of line bundles in which there are precisely d trivial summands. The set of tangent vectors to C, as C ranges over all possible minimal projective lines, form a set $\mathfrak{M}^*$ in $\mathbb{P}(T_{D^*}^{1,0})$. The subset $\mathfrak{M}$ of $\mathbb{P}(T_D^{1,0})$ is the restriction of $\mathfrak{M}^*$ to $\mathbb{P}(T_D^{1,0})$.

Suppose we have two Hermitian metrics g and h. We consider their sum $g + h$. The holomorphic bisectional curvature $\text{Bisect}_{g+h}(X, Y)$ with respect to $g + h$ is equal to the sum of $\text{Bisect}_g(X, Y)$ and $\text{Bisect}_h(X, Y)$ if and only if $\nabla_X^g h(Y, Z)$ vanishes for all Z, where ∇^g means covariant differentiation with respect to the metric g. The computation is as follows:

$$R_{\alpha\bar{\beta}\gamma\bar{\delta}}^g = -\partial_\alpha \partial_{\bar{\beta}} g_{\gamma\bar{\delta}} + g^{\bar{\lambda}\mu} \partial_\alpha g_{\gamma\bar{\lambda}} \overline{\partial_\beta g_{\delta\bar{\mu}}}.$$

So using normal coordinates for g, we have

$$R^{g+h}_{\alpha\bar{\beta}\gamma\bar{\delta}} = -\partial_\alpha\partial_{\bar{\beta}}(g_{\gamma\bar{\delta}} + h_{\gamma\bar{\delta}}) + (g+h)^{\bar{\lambda}\mu}\,\partial_\alpha h_{\gamma\bar{\lambda}}\,\overline{\partial_\beta h_{\delta\bar{\mu}}}$$

$$= -R^{g}_{\alpha\bar{\beta}\gamma\bar{\delta}} - R^{h}_{\alpha\bar{\beta}\gamma\bar{\delta}} - (h^{\bar{\lambda}\mu} - (g+h)^{\bar{\lambda}\mu})\,\partial_\alpha h_{\gamma\bar{\lambda}}\,\overline{\partial_\beta h_{\delta\bar{\mu}}}.$$

We have $R^{g+h}_{\alpha\bar{\beta}\alpha\bar{\beta}} = R^{g}_{\alpha\bar{\beta}\alpha\bar{\beta}} + R^{h}_{\alpha\bar{\beta}\alpha\bar{\beta}}$ if and only if $\partial_\alpha h_{\beta\bar{\lambda}} = 0$ for every λ.

Let Γ be the discrete subgroup of G such that $M = D/\Gamma$. The integration of the $(d+1)$st power of the hyperplane section line bundle for the fiber of $\mathbb{P}(T^{1,0}_M)$ over $\mathfrak{M}/\Gamma$ (after exteriorly multiplied by an appropriate power of the Kähler form of M) must vanish. This forces the equality between $\text{Bisect}_{g+h}(X, Y)$ and the sum of $\text{Bisect}_g(X, Y)$ and $\text{Bisect}_h(X, Y)$ for X in $\mathfrak{M}/\Gamma$ and Y in N_X. Thus we have the vanishing of $\nabla^g_X h(Y, Z)$ for X in $\mathfrak{M}/\Gamma$, Y in N_X and for all Z. As a consequence, the two metrics g and h agree up to a constant factor.

Another immediate corollary of the metric rigidity for the noncompact type is another proof of the theorem of Matsushima [43] that the first Betti number of a compact Hermitian locally symmetric manifold of rank at least 2 must be zero. The reason is that, if ω is the invariant Kähler form of M and φ is a holomorphic 1-form on M, then $\omega + \sqrt{-1}\varphi \wedge \bar{\varphi}$ is a Kähler form of M whose holomorphic bisectional curvature is seminegative. Thus $\omega + \sqrt{-1}\varphi \wedge \bar{\varphi}$ is proportional to ω, from which one concludes that φ must be zero.

For the case of compact type Mok [47] proved the following corresponding result. If (M, g) is a Hermitian symmetric manifold of compact type of rank ≥ 2 and h is a Hermitian metric on M such that the holomorphic bisectional curvature of (M, h) is semipositive, then for some biholomorphism ϕ of M and some constant $c > 0$ one has $h \cdot \phi = cg$.

Metric rigidity for the compact type was generalized to the following result on holomorphic maps by Tsai [80]. Any nonconstant holomorphic map between two irreducible Hermitian symmetric manifolds of compact type must be a biholomorphism if the rank of the target manifold is at least 2.

References

1. Adler, The second fundamental forms of S^6 and $P^n(\mathbb{C})$, *Amer. J. Math* **91**, 657–670 (1969).
2. A. Andreotti, On the complex structures of a class of simply connected manifolds, in *Algebraic Geometry and Topology. A Symposium in Honor of S. Lefschetz*, R. H. Fox, D. C. Spencer, A. W. Tucker, eds. Princeton University Press, Princeton pp. 53–77 (1957).
3. T. Aubin, Métriques riemanniennes et courbure, *J. Diff. Geom.* **4**, 382–424 (1970).
4. T. Aubin, Equations du type de Monge-Ampère sur les variétés kählerienne compacts, *C. R. Acad. Sci. Paris* **283**, 119–121 (1976).

5. W. Ballmann, Manifolds of nonpositive curvature, in *Arbeitstagung Bonn 1984*, Springer, Lecture Notes in Math., Vol. 1111, pp. 261-267.

6. W. Ballmann, Nonpositively curved manifolds of higher rank, *Ann. of Math.* **122**, 597-609 (1985).

7. W. Ballmann, M. Brin, and R. Spatzier, Structure of manifolds of nonpositive curvature. I, II; *Ann. of Math.* **122**, 171-203 (1985), 205-235.

8. W. Ballmann, M. Gromov, and V. Schroeder, *Manifolds of Nonpositive Curvature*, Birkhauser, Boston (1985).

9. S. Bando, On three-dimensional compact Kähler manifolds of nonnegative bisectional curvature, *J. Diff. Geom.* **19**, 283-297 (1984).

10. R. W. Barnard, C. H. FitzGerald, and S. Gong, A distortion theorem for biholomorphic mappings in C^2, preprint (1988).

11. M. Berger, Sur les groupes d'holonomie homogène des variétés à connexion affine et des variétés Riemanniennes, *Bull. Soc. Math. France* **83**, 279-330 (1955).

12. A. Borel, On the curvature tensor of the Hermitian symmetric manifolds, *Ann. of Math.* **71**, 508-521 (1960).

13. E. Brieskorn, Ein Satz über Satz über die komplexen Quadriken, *Math. Ann.* **155**, 184-193 (1964).

14. E. Brieskorn, Über holomorphe P_n-Bündel über P_1, *Math. Ann.* **157**, 343-357 (1965).

15. E. Calabi, The variation of Kähler metrics. I: The structure of the space; II: A minimum problem, *Bull. Am. Math. Soc.* **60**, 168, Abstract Nos. 293, 394 (1954).

16. E. Calabi, The space of Kähler metrics, in *Proc. Int. Congress Math.*, Amsterdam, Vol. 2, pp. 206-207 (1954).

17. E. Calabi, On Kähler manifolds with vanishing canonical class, in *Algebraic Geometry and Topology, A Symposium in Honor of S. Lefschetz*, pp. 77-89 (R. H. Fox, D. C. Spencer, A. W. Tucker, eds.), Princeton University Press, Princeton (1957).

18. E. Calabi and E. Vesentini, On compact locally symmetric Kähler manifolds, *Ann. of Math.* **71**, 472-507 (1960).

19. H.-D. Cao and N. Mok, Holomorphic immersion between complex hyperbolic space forms. *Invent. Math.* **100**, 49-61 (1990).

20. J. Carlson and D. Toledo, Harmonic mappings of Kähler manifolds to locally symmetric spaces, preprint (1988).

21. B.-Y. Chen and K. Ogiue, Some characterizations of complex space forms in terms of Chern classes, *Quart. J. Math.* **26**, 456-464 (1975).

22. K. Corlette, Rigid representations of Kählerian fundamental groups, University of Chicago, preprint (1989).

23. J. Eells and J. H. Sampson, Harmonic of Riemannian manifolds, *Am. J. Math.* **86**, 109-160 (1964).

24. S. Feder, Immersions and embeddings in complex projective spaces, *Topology* **4**, 143-158 (1965).

25. S. Frankel, Complex geometry of convex domains that cover varieties, *Acta Math.* **163**, 109-149 (1989).

26. T. Frankel, Manifolds with positive curvature, *Pacific J. Math.* **11**, 165-174 (1961).

27. H. Grauert, Analytische Faserungen über holomorphvollständigen Räumen, *Math. Ann.* **68**, 263-273 (1958).

28. R. E. Greene and H. Wu, Analysis of noncompact Kähler manifolds, in *Proc. Symp. Pure Math*, Vol. 30, Part 1, pp. 69-100, American Mathematical Society, Providence, RI (1977).

29. R. E. Greene and H. Wu, Gap theorems for noncompact Riemannian manifolds, *Duke Math. J.* **49**, 731-756 (1982).

30. H. Guggenheimer, Über vierdimensionale Einsteinräume, *Experientia* **VIII/11**, 420-421 (1952).

31. F. Hirzebruch, Über eine Klasse von einfach-zusammenhängenden komplexen Mannifaltigkeiten, *Math. Ann.* **124**, 77-86 (1951/52).

32. F. Hirzebruch and K. Kodaira, On the complex projective spaces, *J. Math. Pures Appl.* **36**, 201-216 (1957).

33. C.-C. Hsiung, Nonexistence of a complex structure on the six-sphere, *Bull. Inst. Math., Academia Sinica, Taiwan* **14**, 231-247 (1986).

34. D. Johnson and J. Millson, Deformation spaces associated to compact hyperbolic manifolds, in *Discrete Groups in Geometry and Analysis: Papers in Honor of G. D. Mostow on His Sixtieth Birthday*, pp. 48-106. (R. Howe, ed.), *Progress in Math.*, vol. 67, Birkhäuser, Boston (1987).

35. J. Jost and S.-T. Yau, Harmonic mappings and Kähler manifolds, *Math. Ann.* **262**, 145-166 (1983).

36. J. Jost and S.-T. Yau, A strong rigidity theorem for a certain class of compact analytic surfaces. *Math. Ann.* **271**, 143-152 (1985).

37. Srinivasacharyulu Kilambi, Sur la déformation de certaines variétés complexes, *C. R. Acad. Sci.* (*Paris*) **252**, 3377-3378 (1961).

38. K. Kodaira, On stability of compact submanifolds of complex manifolds, *Am. J. Math.* **85**, 79-94 (1963).

39. K. Kodaira and D. C. Spencer, On deformation of complex analytic structures. II, *Ann. of Math.* **67**, 403-466 (1958).

40. N. Kuhlmann, On deformation of $\mathbb{P}_n$, Q_n, $G(1, 4)$, preprint.

41. T. Mabuchi, C^3-actions and algebraic threefolds with ample tangent bundle, *Nagoya Math. J.* **69**, 33-64 (1978).

42. G. A. Margulis, Discrete groups of motions of manifolds of nonpositive curvature, *Amer. Math. Soc. Trans.* **109**, 33-45 (1977).

43. Y. Matsushima, On the first Betti number of compact quotient spaces of higher-dimensional symmetric spaces, *Ann. of Math.* **75**, 312-330 (1962).

44. B. Moishezon, On n-dimensional compact varieties with n algebraically independent meromorphic functions, *Amer. Math. Soc. Trans.* **63**, 51-177 (1967).

45. N. Mok, The holomorphic or anti-holomorphic character of harmonic maps into irreducible compact quotients of polydiscs. *Math. Ann.* **272**, 197-216 (1985).

46. N. Mok, Uniqueness theorem of Hermitian metric of seminegative curvature on quotients of bounded symmetric domains, *Ann. of Math.* **125**, 105-152 (1987).

47. N. Mok, Uniqueness theorem of Kähler metrics of semipositive holomorphic bisectional curvature on compact Hermitian symmetric space, *Math. Ann.* **276**, 177-204 (1987).

48. N. Mok, The uniformization theorem for compact Kähler manifolds of nonnegative holomorphic bisectional curvature, *J. Diff. Geom.* **27**, 179-214 (1988).

49. N. Mok, Y. T. Siu, and S.-T. Yau, The Poincaré-Lelong equation on complete Kähler manifolds, *Compositio Math.* **44**, 183-218 (1981).

50. N. Mok and J.-Q. Zhong, Curvature characterization of compact Hermitian symmetric spaces. *J. Diff. Geom.* **23**, 15-67 (1986).

51. S. Mori, Projective manifolds with ample tangent bundles, *Ann. of Math.* **110**, 593-606 (1979).

52. S. Mori and H. Sumihiro, On Hartshorne's conjecture, *J. Math. Kyoto Univ.* **18**, 523-533 (1978).

53. G. D. Mostow, *Strong Rigidity of Locally Symmetric Spaces*, Ann. of Math. Studies, vol. 78, Princeton University Press, Princeton (1973).

54. G. D. Mostow, On a remarkable class of polyhedra in complex hyperbolic space, *Pacific J. Math.* **86**, 171-276 (1980).

55. G. D. Mostow and Y. T. Siu, A compact Kähler surface of negative curvature not covered by the ball, *Ann. of Math.* **112**, 321-360 (1980).

56. A. Nadel, Semisimplicity of the group of biholomorphisms of the universal cover of a compact complex manifold with ample canonical bundle, preprint (1988).
57. I. Nakamura, Moishezon threefolds homeomorphic to $\mathbb{P}^3$, *J. Math. Soc. Japan.* **39**, 522-535 (1987).
58. I. Nakamura, Characterizations of $\mathbb{P}^3$ and hyperquadrics Q^3 in $\mathbb{P}^4$, *Proc. Japan Acad. Ser. A*, **62**, 230-233 (1986).
59. I. Nakamura, Threefolds homeomorphic to a hyperquadric in $\mathbb{P}^4$, Hokkaido University preprint series (1987).
60. S. Nakano, On complex analytic vector bundles, *J. Math. Soc. Japan* **7**, 1-12 (1955).
61. T. Peternell, A rigidity theorem for $\mathbb{P}_3(\mathbb{C})$, *Manuscripta Math.* **50**, 397-428 (1985).
62. T. Peternell, Algebraic structures on certain 3-folds, *Math. Ann.* **274**, 133-428 (1986).
63. A. Polombo, Condition d'Einstein et courbure négative en dimension 4, *C. R. Acad. Sci. Paris Sér. I*, **307**, 667-670 (1988).
64. J. P. Rosay, Characterisation de la boule parmi son groupe d'automorphismes, *Ann. Inst. Fourier* **29**, 91-97 (1979).
65. J. H. Sampson, Applications of harmonic maps to Kähler geometry, *Contemp. Math.* **49**, 125-133 (1986).
66. J. Simons, On the transitivity of holonomy systems, *Ann. of Math.* **76**, 213-234 (1962).
67. Y.-T. Siu, The complex-analyticity of harmonic maps and strong rigidity of compact Kähler manifolds, *Ann. of Math.* **112**, 73-111 (1980).
68. Y.-T. Siu, Curvature characterization of hyperquadrics, *Duke Math. J.* **47**, 641-654 (1980).
69. Y.-T. Siu, Some remarks on the complex-analyticity of harmonic maps, *Southeast Asian Bull. Math.* **3**, 240-253 (1979).
70. Y.-T. Siu, Strong rigidity of compact quotients of exceptional bounded symmetric domains, *Duke Math. J.* **48**, 857-871 (1981).
71. Y.-T. Siu, Complex analyticity of harmonic maps, vanishing and Lefschetz theorems, *J. Diff. Geom.* **17**, 55-138 (1982).
72. Y.-T. Siu, Every $K3$ surface is Kähler, *Invent. Math.* **73**, 139-150 (1983).
73. Y.-T. Siu, Strong rigidity for Kähler manifolds and the construction of bounded holomorphic functions, in *Discrete Groups in Geometry and Analysis* (*Proc. Conf. in honor of G. D. Mostow*), pp. 124-151 (Roger Howe, ed.), Birkhäuser-Verlag (1987).
74. Y.-T. Siu, Nondeformability of the complex projective space, *J. Reine Angew. Math.* **399**, 208-219 (1989).
75. Y.-T. Siu and P. Yang, Compact Kähler-Einstein surfaces of nonpositive bisectional curvature, *Invent. Math.* **64**, 471-487 (1981).
76. Y.-T. Siu and S.-T. Yau, Complete Kähler manifolds with nonpositive curvature of faster than quadratic decay, *Ann. Math.* **105**, 225-264 (1977).
77. Y.-T. Siu and S.-T. Yau, Compact Kähler manifolds of positive bisectional curvature, *Invent. Math.* **59**, 189-204 (1980).
78. W. Thimm, *Über algebraische Relation zwischen meromorphen Funktionen abgeschlossenen Räumen.* Thesis, Königsber (1939).
79. W.-K. To, Hermitian metrics of seminegative curvature on quotients of bounded symmetric domains, *Invent, Math.* **95**, 559-578 (1989).
80. I.-H. Tsai, Rigidity of holomorphic maps between Hermitian symmetric spaces of compact type, preprint.
81. H. Tsuji, Deformation of complex projective space, preprint.
82. J. Vey, Sur la division des domaines de Siegel, *Ann. Sci. École Norm. Sup.* **3**, 479-506 (1970).
83. A. Weil, On discrete subgroups of Lie groups, *Ann. of Math.* **72**, 369-384 (1960).
84. A. Weil, On discrete subgroups of Lie groups, II, *Ann. of Math.* **75**, 578-602 (1962).
85. A. Weil, Remarks on cohomology of groups, *Ann. of Math.* **80**, 149-157 (1964).

86. B. Wong, Characterization of the ball by its automorphism group, *Invent. Math.* **41**, 253–257 (1977).
87. S.-T. Yau, Calabi's conjecture and some new results in algebraic geometry, *Proc. Nat. Acad. Sci. U.S.A.* **74**, 1789–1799 (1977).
88. S.-T. Yau, On the Ricci curvature of a compact Kähler manifold and the complex Monge–Ampère equation. I, *Comm. Pure Appl. Math.* **31**, 339–411 (1978).
89. J.-Q. Zhong, The degree of strong nondegeneracy of the bisectional curvature of exceptional bounded symmetric domains, in *Several Complex Variables* (Proc. 1981 Hangzhou Conf.), pp. 127–139 (J. J. Kohn, Q.-K. Lu, R. Remmert, Y.-T. Siu, eds.), Birkhäuser (1984).

Part II

Selected Papers of Zhong Jia-Qing

4

Harmonic Analysis on Rotation Groups: Abel Summability

The fundamental theorem of harmonic analysis on compact groups is just the Peter-Weyl theorem. It claims that every continuous function on a compact group can be approximated as closely as desired by linear combinations of irreducible representations of the group. Professor Hua modified this theorem in Ref. 1. He defined Fourier series for continuous functions on the unitary group $U(n)$ and proved that any continuous function on the unitary group can be obtained through Fourier series via Abel summability.

If we consider $U(n)$ as the characteristic manifold of the first classical domain $R_1(n)$: $I - Z\bar{Z}' > 0$, from the theory of harmonic analysis on classical domains we know that any continuous function $u(U)$ on $U(n)$ determines uniquely a harmonic function $u(Z)$ on $R_1(n)$. By the Poisson integral,

$$u(Z) = \frac{1}{c} \int_{U(n)} P(Z, U) u(U) \dot{U}, \tag{4.1}$$

with $u(Z) \to u(U)$, $Z \in R_1(n)$, $Z \to U$. Here $P(Z, U)$ is the so-called Poisson kernel given by

$$P(Z, U) = \frac{\det (I - Z\bar{Z}')^n}{|\det (I - Z\bar{U}')|^{2n}}. \tag{4.2}$$

Choosing $Z = rV$, $V \in U(n)$ in (4.1), we obtain

$$u(rV) = \frac{1}{c} \int_{U(n)} P(rV, U)u(U)\dot{U}$$

$$= \frac{1}{c} \int_{U(n)} P(rI, U\bar{V}')u(U)\dot{U} \rightarrow u(V) \qquad (r \rightarrow 1). \qquad (4.3)$$

Hua [1] proved that $P(rI, U)$ can be expanded according to the irreducible representation

$$P(rI, U) = \sum \rho_f(r)\sigma_f(U). \qquad (4.4)$$

Here $\sigma_f(U)$ represents the character of the irreducible representation of $U(n)$ with signature $f = (f_1, \ldots, f_n)$ and $\rho_f(r)$ are constants depending on f.

Substitution of (4) into (3) gives

$$u(rV) = \frac{1}{c} \int_{U(n)} P(rI, U\bar{V}')u(U)\dot{U}$$

$$= \sum_{f \geq 0} \rho_f(r) \cdot \frac{1}{c} \int_{U(n)} \bar{\sigma}_f(U\bar{V}')u(U)\dot{U}$$

$$= \sum_{f \geq 0} \rho_f(r) \cdot \frac{1}{c} \int_{U(n)} \sum_{i,j} \varphi_{i,j}^f(V)\bar{\varphi}_{i,j}^f(U)u(U)\dot{U}$$

$$= \sum_{f \geq 0} \rho_f(r) \cdot \frac{1}{c} \sum_{i,j} \varphi_{ij}^f(V)a_{ij}^f \rightarrow u(V) \qquad (r \rightarrow 1). \qquad (4.5)$$

$\sum_{f \geq 0} \sum_{i,j} a_{ij}^f \varphi_{ij}^f(U)$ is defined as the Fourier series of $U(n)$ in Ref. 1, a_{ij}^f denotes its Fourier coefficients, and (4.5) then means that this series may be Abel summable to $U(n)$.

On the basis of the above work, Professor Kung [5] gave a formula for $\rho_f(r)$, thus completing the study of Abel summability on unitary groups.

This Chapter will study the same problem on rotation groups. The main idea is the same as that in Ref. 1, but the methods used are quite different. In proving that the Poisson kernel on the rotation group can be expanded into a series of its irreducible representations [see (4.4)], we find at the same time the representation corresponding to $\rho_f(r)$. Considering the method of Ref. 5 in calculating $\rho_f(r)$ to be too complicated to apply to the rotation group, we employ a method provided by Murnaghan [7] so that we solve this problem simply and directly.

As a direct application of the method under consideration, a problem of group representation is discussed. It includes two special cases in a general problem proposed in Ref. 2, and the results we obtain are parallel to those in Ref. 2.

This work was finished and reported at the seminar held by Professor Hua in 1965. The author is very grateful to him for his guidance and encouragement.

4.1. Abel Summability on Rotation Groups

By rotation group, we mean $O^+(n) = \{\Gamma \in GL(n, \mathbb{R}) \mid \Gamma\Gamma' = I, \det \Gamma = 1\}$. From the theory of harmonic analysis on real classical domains, we know the following.

THEOREM 4.1 (*K. H. Look* [4]). *Let $u(\Gamma)$ be a continuous function on $O^+(n)$. Then its Poisson integral*

$$u(x) = \frac{1}{c} \int_{O^+(n)} P(x, \Gamma) u(\Gamma) \dot\Gamma, \qquad \frac{1}{c} \int_{O^+(n)} \dot\Gamma = 1,$$

satisfies $u(x) \to u(\Gamma)$ when $I - xx' > 0$, $x \to \Gamma$. Here

$$P(x, \Gamma) = \frac{\det (I - xx')^{(n-1)/2}}{\det (I - x\Gamma')^{n-1}}. \tag{4.6}$$

In particular, pick $\Gamma_0 \in O^+(n)$ and $x = r\Gamma_0$, $0 \le r \le 1$. Then

$$u(r\Gamma_0) = \frac{1}{c} \int_{O^+(n)} P(r\Gamma_0, \Gamma) u(\Gamma) \dot\Gamma \to u(\Gamma_0), \qquad r \to 1. \tag{4.7}$$

By (4.7) our task is to expand $P(r\Gamma_0, \Gamma)$ in a series of irreducible representations of $O^+(n)$. Notice that

$$P(r\Gamma_0, \Gamma) = \frac{(1 - r^2)^{n(n-1)/2}}{\det (I - r\Gamma_0\Gamma')^{n-1}} = \frac{(1 - r^2)^{n(n-1)/2}}{\det (I - r(\Gamma\Gamma_0')')^{n-1}} = P(rI, \Gamma\Gamma_0').$$

It is then sufficient to carry out the expansion of $P(rI, \Gamma)$. Recall that the representations of the conjugate classes of the elements of $O^+(n)$ can be divided into two cases (n odd or even).

Case 1. $n = 2k$.

$$O^+ \sim \begin{pmatrix} c(\theta_1) & & \\ & \ddots & \\ & & c(\theta_k) \end{pmatrix}, \qquad c(\theta_i) = \begin{pmatrix} \cos\theta_i & \sin\theta_i \\ -\sin\theta_i & \cos\theta_i \end{pmatrix}. \qquad (4.8)$$

Case 2. $n = 2k + 1$.

$$O^+ \sim \begin{pmatrix} c(\theta_1) & & & \\ & \ddots & & \\ & & c(\theta_k) & \\ & & & 1 \end{pmatrix}, \qquad c(\theta_i) = \begin{pmatrix} \cos\theta_i & \sin\theta_i \\ -\sin\theta_i & \cos\theta_i \end{pmatrix}. \qquad (4.9)$$

Here "$\sim$" stands for conjugation.

As is well known, irreducible representations of $O^+(n)$ are characterized by $k = [n/2]$ nonnegative integers $m = (m_1, \ldots, m_k)$, $m_1 \geq \cdots \geq m_k \geq 0$; m is called the signature of the representation. Let $\sigma_m(\Gamma)$ denote the character of the irreducible representation of $O^+(n)$ with signature m. Then we have the following fundamental identity. Let $t_1, \ldots, t_k$ be k independent variables, and $f_i = f_i(t_i, \Gamma) = \det(I - t_i\Gamma)$. Then

$$\frac{\prod_{i \leq j}^{k} (1 - t_it_j)}{f_1 \cdots f_k} = \sum_{m \geq 0} \sigma_m(\Gamma)\chi_m(t). \qquad (4.10)$$

Here

$$\chi_m(t) = \frac{A_{l_1 \cdots l_k}(t_1, \ldots, t_k)}{\Delta_k(t_1, \ldots, t_k)},$$

$$A_{l_1 \cdots l_k}(t_1, \ldots, t_k) = \begin{vmatrix} t_1^{l_1} & \cdots & t_k^{l_1} \\ \vdots & & \vdots \\ t_1^{l_k} & \cdots & t_k^{l_k} \end{vmatrix}, \qquad l_i = m_i + k - i, \quad i = 1, 2, \ldots, k,$$

$$\Delta_k(t_1, \ldots, t_k) = \begin{vmatrix} t_1^{k-1} & \cdots & t_k^{k-1} \\ t_1^{k-2} & & t_k^{k-2} \\ \vdots & & \vdots \\ 1 & & 1 \end{vmatrix} = \prod_{i < j}^{k} (t_i - t_j).$$

When $\max |t_i| < r < 1$, (4.10) converges uniformly for Γ (see Ref. 7, p. 255).

Replacing the left side of (4.10) by $\prod_{i \leq j}^{n-1} (1 - t_it_j)/f_1 \cdots f_{n-1}$ and putting $t_i = r$, we have

$$\left. \frac{\prod_{i \leq j}^{n-1} (1 - t_it_j)}{f_1 \cdots f_{n-1}} \right|_{t_1 = \cdots = t_{n-1} = r} = \frac{(1 - r^2)^{n(n-1)/2}}{\det(I - r\Gamma)^{n-1}} = P(rI, \Gamma). \qquad (4.11)$$

Thus along with the formula (4.10) extended from k to $n-1$, the problem as to the expansion of $P(rI, \Gamma)$ is solved.

Throughout, we adopt the following conventions:

$$T_i = 1 + t_i^2, \quad \Delta_s(t) = \prod_{i<j}^s (t_i - t_j), \quad L_s(t) = \prod_{i<j}^s (1 - t_i t_j);$$

$$\bar{L}_s(t) = \prod_{i \leq j}^s (1 - t_i t_j) = L_s(t) \prod_{i=1}^s (1 - t_i^2);$$

$$\langle g_1(t), \ldots, g_s(t) \rangle = \begin{vmatrix} g_1(t_1) & \cdots & g_s(t_1) \\ \vdots & & \vdots \\ g_1(t_s) & \cdots & g_s(t_s) \end{vmatrix},$$

where $g_i(t)$ are functions of t.

LEMMA 4.2. *For $s > 0$, we have the following identity:*

$$\Delta_{s+1}(t) L_{s+1}(t) = \langle t^s, t^{s-1} T, \ldots, tT^{s-1}, T^s \rangle$$

$$= \langle t^s, t^{s-1} + t^{s+1}, \ldots, t + t^{2s-1}, 1 + t^{2s} \rangle. \qquad (4.12)$$

PROOF.

$$\langle t^s, t^{s-1} T, \ldots, T^s \rangle = (T_1 \cdots T_{s+1})^s \left\langle \left(\frac{t}{T}\right)^s, \left(\frac{t}{T}\right)^{s-1}, \ldots, \frac{t}{T}, 1 \right\rangle$$

$$= (T_1 \cdots T_{s+1})^s \prod_{i<j}^{s+1} \left(\frac{t_i}{T_i} - \frac{t_j}{T_j}\right) = \prod_{i<j}^{s+1} (t_i T_j - t_j T_i)$$

$$= \prod_{i<j}^{s+1} (t_i + t_i t_j^2 - t_j - t_j t_i^2) = \prod_{i<j}^{s+1} (t_i - t_j)(1 - t_i t_j)$$

$$= \Delta_{s+1}(t) L_{s+1}(t).$$

As for $\langle t^s, t^{s-1} T, \ldots, T^s \rangle = \langle t^s, t^{s-1} + t^{s+1}, \ldots, 1 + t^{2s} \rangle$, its right-hand side is achieved by subtracting an appropriate linear combination of the first $i-1$ rows from the ith row, in the determinant $\langle t^s, t^{s-1} T, \ldots, T^s \rangle$. The lemma is trivial. $\qquad \square$

LEMMA 4.3. *Assume that $s \geq k = [n/2]$. Given*

$$\frac{\Delta_s(t) \bar{L}_s(t)}{f_1 \cdots f_s} = \sum_{m \geq 0} \sigma_m(\Gamma) \langle g_1(t), \ldots, g_s(t) \rangle, \qquad (4.13)$$

where $f_i = \det(I - t_i \Gamma)$ and $\sigma_m(\Gamma)$ indicates the character of the irreducible *representation of $O^+(n)$ with signature $m = (m_1, \ldots, m_k)$, we have*

$$\frac{\Delta_{s+1}(k)\bar{L}_{s+1}(k)}{f_1 \cdots f_{s+1}} = \sum_{m \geq 0} \sigma_m(\Gamma)\langle g_1(t)t, \ldots, g_s(t)t, d(t)(1 + t^2)^{s-k}\rangle, \quad (4.14)$$

with

$$d(t) = \begin{cases} 1 - t^2 & \text{for } \Gamma \in O^+(2k), \\ 1 + t & \text{for } \Gamma \in O^+(2k + 1). \end{cases}$$

PROOF. *Step 1.* Assume that $\Gamma \in O^+(2k)$. By (4.8)

$$f = \det(I - r\Gamma) = \prod_{i=1}^{k} (1 - 2t \cos \theta_i + t^2)$$

$$= \prod_{i=1}^{k} (T - 2t \cos \theta_i) = T^k + \cdots,$$

the unwritten term "$\cdots$" being a linear combination of $t^i T^{k-i}$, $i = 1, 2, \ldots, k$. Thus,

$$T^s = T^k T^{s-k} = T^{s-k}(f + \cdots) = T^{s-k}f + \cdots, \quad (4.15)$$

the unwritten term "$\cdots$" being a linear combination of $t^i T^{k-i}$, $i = 1, 2, \ldots, k$. By Lemma 4.2 and (4.15) we obtain

$$\Delta_{s+1}(t)L_{s+1}(t) = \langle t^s, t^{s-1}T, \cdots, T^s \rangle = \langle t^s, t^{s-1}T, \cdots, tT^{s-1}, T^{s-k}f \rangle$$

$$= \sum_{i=1}^{s+1} (-1)^{s+1-i} T_i^{s-k} f_i t_1 \cdots \hat{t}_i \cdots t_{s+1}$$

$$\times \begin{vmatrix} t_1^{s-1} & t_1^{s-2}T_1 & \cdots & t_1 T_1^{s-1} & T_1^{s-1} \\ \vdots & \vdots & & \vdots & \vdots \\ \hat{t}_i^{s-1} & \hat{t}_i^{s-2}\hat{T}_i & \cdots & \hat{t}_i\hat{T}_i^{s-1} & \hat{T}_i^{s-1} \\ \vdots & \vdots & & \vdots & \vdots \\ t_{s+1}^{s-1} & t_{s+1}^{s-2}T_{s+1} & \cdots & t_{s+1}T_{s+1}^{s-1} & T_{s+1}^{s-1} \end{vmatrix}$$

$$= \sum_{i=1}^{s+1} (-1)^{s+1-i} T_i^{s-k} f_i t_1 \cdots \hat{t}_i \cdots t_{s+1} \cdot \Delta_s^{(i)}(t)L_s^{(i)}(t), \quad (4.16)$$

where $\hat{t}_i$ means the absence of t_i, $\Delta_s^{(i)}(t) = \Delta_s(t_1, \ldots, \hat{t}_i, \ldots, t_{s+1})$, $L_s^{(i)}(t) = L_s(t_1, \ldots, \hat{t}_i, \ldots, t_{s+1})$. Now multiply $\prod_{i=1}^{s+1} (1 - t_i^2)/f_1 \cdots f_{s+1}$ on both sides of (4.16) and notice the relation between $L_s(t)$ and $\bar{L}_s(t)$. Then

$$\frac{\Delta_{s+1}(t)\bar{L}_{s+1}(t)}{f_1 \cdots f_{s+1}} = \sum_{i=1}^{s+1} (-1)^{s+1-i} T_i^{s-k} f_i t_1 \cdots \hat{t}_i \cdots t_{s+1}(1 - t_i^2)\frac{\Delta_s^{(i)}(t)\bar{L}_s^{(i)}(t)}{f_1 \cdots f_{s+1}}$$

$$= \sum_{i=1}^{s+1} (-1)^{s+1-i} T_i^{s-k}(1 - t_i^2)t_1 \cdots \hat{t}_i \cdots t_{s+1} \frac{\Delta_s^{(i)}(t)\bar{L}_s^{(i)}(t)}{f_1 \cdots \hat{f}_i \cdots f_{s+1}}.$$

By the assumption, we get

$$
\frac{\Delta_s^{(i)}(t)\bar{L}_s^{(i)}(t)}{f_1 \cdots \hat{f}_i \cdots f_{s+1}} = \sum_{m \geq 0} \sigma_m(\Gamma) \begin{vmatrix} g_1(t_1) & \cdots & g_s(t_1) \\ \vdots & & \vdots \\ \hat{g}_1(t_i) & & \hat{g}_s(t_i) \\ \vdots & & \vdots \\ g_1(t_{s+1}) & \cdots & g_s(t_{s+1}) \end{vmatrix}.
$$

Hence,

$$
\frac{\Delta_{s+1}(t)\bar{L}_{s+1}(t)}{f_1 \cdots f_{s+1}} = \sum_{m \geq 0} \sigma_m(\Gamma) \sum_{i=1}^{s+1} (-1)^{s+1-i} T_i^{s-1}(1 - t_i^2)t_1 \cdots \hat{t}_i \cdots t_{s+1}
$$

$$
\times \begin{vmatrix} g_1(t_1) & \cdots & g_s(t_1) \\ \vdots & & \vdots \\ g_1(t_{i-1}) & & g_s(t_{i-1}) \\ g_1(t_{i+1}) & & g_s(t_{i+1}) \\ \vdots & & \vdots \\ g_1(t_{s+1}) & \cdots & g_s(t_{s+1}) \end{vmatrix}
$$

$$
= \sum_{m \geq 0} \sigma_m(\Gamma) \langle f_1(t)t, g_2(t)t, \cdots, g_s(t)t, (1 - t^2)T^{s-k} \rangle.
$$

The lemma is true for $\Gamma \in O^+(2k)$.

Step 2. Assume that $\Gamma \in O^+(2k + 1)$. Equation (4.9) implies

$$
f = \det(I - t\Gamma) = (1 - t)\prod_{i=1}^{k}(1 - 2t\cos\theta_i + t^2) = (1 - t)h,
$$

$$
h = \prod_{i=1}^{k}(T - 2t\cos\theta_i) = T^k + \cdots.
$$

Thus $T^s = T^{s-k}T^k = T^{s-k}h + \cdots$. The remainder of the proof is identical to step 1; i.e., replacing T^s by $T^{s-k}h$ in $\Delta_{s+1}(t)L_{s+1}(t) = \langle t^s, t^{s-1}T, \ldots, tT^{s-1}, T^s \rangle$, we arrive at

$$
\frac{\Delta_{s+1}(t)\bar{L}_{s+1}(t)}{f_1 \cdots f_{s+1}} = \sum_{i=1}^{s+1} (-1)^{s+1-i} T_i^{s-k}(1 - t_i^2)h_i t_1 \cdots \hat{t}_i \cdots t_{s+1} \frac{\Delta_s^{(i)}(t)\bar{L}_s^{(i)}(t)}{f_1 \cdots f_{s+1}}.
$$

Then using $(1 - t_i^2)h_i = (1 + t_i)(1 - t_i)h_i = (1 + t_i)f_i$, we have

$$
\frac{\Delta_{s+1}(t)\bar{L}_{s+1}(t)}{f_1 \cdots f_{s+1}} = \sum_{i=1}^{s+1} (-1)^{s+1-i}(1 + t_i)T_i^{s-k}t_1 \cdots \hat{t}_i \cdots t_{s+1} \frac{\Delta_s^{(i)}(t)\bar{L}_s^{(i)}(t)}{f_1 \cdots \hat{f}_i \cdots f_{s+1}}.
$$

Resorting to (4.13), we get

$$\frac{\Delta_{s+1}(t)\bar{L}_{s+1}(t)}{f_1\cdots f_{s+1}} = \sum_{m\geq 0} \sigma_m(\Gamma)\langle g_1(t)t,\ldots,g_s(t)k,(1+t)T^{s-k}\rangle.$$

This completes the proof of the lemma.   □

Now, let us start from the fundamental equality (4.10) and rewrite it as follows:

$$\frac{\Delta_k(t)\bar{L}_k(t)}{f_1\cdots f_k} = \sum_{m\geq 0} \sigma_m(\Gamma)\langle t_1^{l_1},\ldots,t_k^{l_k}\rangle. \qquad (4.17)$$

Using Lemma 4.3 repeatedly, we get the following sequence for the coefficients of $\sigma_m(\Gamma)$:

$$\langle t_1^{l_1},\ldots,t_k^{l_k}\rangle \xrightarrow{k+1} \langle t^{l_1+1},\ldots,t^{l_k+1},d(t)\rangle \xrightarrow{k+2} \langle t^{l_1+2},\ldots,t^{l_k+2},$$

$$d(t)t, d(t)(1+t^2)\rangle$$

$$\xrightarrow{k+3}\cdots\xrightarrow{n-1} \langle t^{l_1+n-1-k},\ldots,t^{l_k+n-1-k},d(t)t^{n-k-2},$$

$$d(t)t^{n-k-3}(1+t^2),\ldots,d(t)(1+t^2)^{n-k-2}\rangle.$$

$$(4.18)$$

Denoting the last term in (4.18) by $C_{n-1}(t_1,\ldots,t_{n-1})$, we can formulate the following theorem.

THEOREM 4.4. *Adopt the same notations as used in Lemma 4.3. The following identity is valid:*

$$\frac{\prod_{i<j}^{n-1}(t_i-t_j)\prod_{i\leq j}^{n-1}(1-t_it_j)}{f_1\cdots f_{n-1}} = \sum_{m\geq 0}\sigma_m(\Gamma)C_{m-1}(t_1,\ldots,t_{n-1}), \quad (4.19)$$

where $f_i = \det(I - t_i\Gamma)$, $\Gamma \in O^+(n)$, $C_{m-1}(t_1,\ldots,t_{n-1})$ is defined as in (4.18), and $\sigma_m(\Gamma)$ represents the character of the irreducible representation of $O^+(n)$ with signature $m = (m_1,\ldots,m_k)$.

From (4.19) we get

$$\frac{\prod_{i\leq j}^{n-1}(1 - t_i t_j)}{f_1 \cdots f_{n-1}} = \sum_{m\geq 0} \sigma_m(\Gamma) \frac{C_{n-1}(t_1, \ldots, t_{n-1})}{\Delta_{n-1}(t_1, \ldots, t_{n-1})}. \tag{4.20}$$

Let $t_1 = t_2 = \cdots = t_{n-1} = r$. From (4.11) the following corollary arises.

Corollary 4.5. The Poisson kernel $P(rI, \Gamma)$ of the rotation group $O^+(n)$ allows the following expansion:

$$P(rI, \Gamma) = \frac{(1 - r^2)^{n(n-1)/2}}{\det (I - r\Gamma)^{n-1}} = \sum_{m\geq 0} \rho_m(r)\sigma_m(\Gamma), \tag{4.21}$$

$$\rho_m(r) = \frac{C_{n-1}(t_1, \ldots, t_{n-1})}{\Delta_{n-1}(t_1, \ldots, t_{n-1})}\bigg|_{t_1 = \cdots = t_{n-1} = r}.$$

Now we will give a more precise formula for $\rho_m(r)$. By the assertion of Lemma 4.2, $C_{n-1}(t_1, \ldots, t_{n-1})$ can also be given as

$$C_{n-1}(t_1, \ldots, t_{n-1}) = \langle t^{l_1+n-1-k}, \ldots, t^{l_k+n-1-k}, d(t)t^{n-k-2},$$

$$d(t)(t^{n-k-3} + t^{n-k-1}),$$

$$d(t)(t^{n-k-4} + t^{n-k}), \ldots, d(t)(1 + t^{2n-2k-4})\rangle,$$

where

$$d(t) = \begin{cases} 1 + t & \text{if } n = 2k + 1, \\ 1 - t & \text{if } n = 2k. \end{cases}$$

Now consider the following two cases:

Case 1. $n = 2k + 1$. Since

$$d(t)t^{n-k-2} = (1 + t)t^{n-k-2} = t^{n-k-2} + t^{n-k-1},$$

$$d(t)(t^{n-k-3} + t^{n-k-1}) = (1 + t)(t^{n-k-3} + t^{n-k-1})$$

$$= t^{n-k-3} + t^{n-k} + (t^{n-k-2} + t^{n-k-1}),$$

$$d(t)(t^{n-k-4} + t^{n-k}) = (1 + t)(t^{n-k-4} + t^{n-k})$$

$$= t^{n-k-4} + t^{n-k+1} + (t^{n-k-3} + t^{n-k}),$$

$$\vdots$$

$$d(t)(1 + t^{2n-2k-4}) = 1 + t^{2n-2k-3} + (t + t^{2n-2k-4}),$$

by the properties of determinants, we have

$$
\begin{aligned}
C_{n-1}(t_1, \ldots, t_{n-1}) &= \langle t^{l_1+n-1-k}, \ldots, t^{l_k+n-1-k}, d(t)t^{n-k-2}, \\
&\qquad d(t)(t^{n-k-3} + t^{n-k-1}), \ldots, d(t)(1 + t^{2n-2k-1}) \rangle \\
&= \langle t^{l_1+n-k-1}, \ldots, t^{l_k+n-k-1}, (t^{n-k-2} + t^{n-k-1}), \\
&\qquad (t^{n-k-3} + t^{n-k}), \ldots, (1 + t^{2n-2k-3}) \rangle \\
&= \sum_{(p_1,\ldots,p_{n-k-1}) \in D} \varepsilon(p_1, \ldots, p_{n-k-1}) \\
&\qquad \times \langle t^{l_1+n-1-k}, \ldots, t^{l_k+n-1-k}, t^{p_1}, \ldots, t^{p_n-k-1} \rangle, \qquad (4.22)
\end{aligned}
$$

where the index set D is defined by $(p_1, \ldots, p_{n-k-1}) \in D$ and subject to the conditions: (i) $2n - 2k - 3 \geq p_1 > \cdots > p_{n-k-1} \geq 0$; (ii) For $i \neq j$, there exists $p_i + p_j \neq 2n - 2k - 3$ and the sign function $\varepsilon(p_1, \ldots, p_{n-k-1})$ in (4.22) can be determined:

$$
\varepsilon(p_1, \ldots, p_{n-k-1}) =
\begin{cases}
+1 & \text{if } p_1 + \cdots + p_{n-k-1} \\
& \qquad - \dfrac{(n-k-2)(n-k-1)}{2} = 0, \\[2ex]
+1 & \text{if } p_1 + \cdots + p_{n-k-1} \\
& \qquad - \dfrac{(n-k-2)(n-k-1)}{2} \equiv 1, 2 \pmod 4, \\[2ex]
-1 & \text{if } p_1 + \cdots + p_{n-k-1} \\
& \qquad - \dfrac{(n-k-2)(n-k-1)}{2} \equiv 3, 4 \pmod 4.
\end{cases}
$$

So, (4.21) becomes

$$
\begin{aligned}
\rho_m(r) &= \frac{C_{n-1}(t_1, \ldots, t_{n-1})}{\Delta_{n-1}(t_1, \ldots, t_{n-1})} \Bigg|_{t_1 = \cdots = t_{n-1} = r} \\
&= \sum_{(p_1,\ldots,p_{n-k-1}) \in D} \varepsilon(p_1, \ldots, p_{n-k-1}) \\
&\qquad \times \frac{\langle t^{l_1+n-k-1}, \ldots, t^{l_k+n-k-1}, t^{p_1}, \ldots, t^{p_{n-k-1}} \rangle}{\Delta_{n-1}(t_1, \ldots, t_{n-1})} \Bigg|_{t_1 = \cdots = t_{n-1} = r}. \qquad (4.23)
\end{aligned}
$$

Setting $q_1 = p_1 - n + 2 + k$, $q_2 = p_2 - n + 3 + k$, ..., $q_{n-k-1} = p_{n-k-1}$, from $l_1 = m_1 + k - 1$, $l_2 = m_2 + k - 2$, ..., $l_k = m_k$, we get

$$\langle t^{l_1 + n - k - 1}, \ldots, t^{l_k + n - 1 - k}, t^{p_1}, \ldots, t^{p_{n-k-1}} \rangle$$

$$= \langle t^{m_1 + n - 2}, t^{m_2 + n - 1}, \ldots, t^{m_k + n - k - 1}, t^{q_1 + n - k - 2}, \ldots, t^{q_n - k - 1} \rangle.$$

Thus, (4.23) leads to

$$\rho_m(r) = \sum_{(q_1, \ldots, q_{n-k-1}) \in \overset{\circ}{D}} \varepsilon(q_1, \ldots, q_{n-k-1})$$

$$\times \frac{\langle t^{m_1 + n - 2}, \ldots, t^{m_k + n - k - 1}, t^{q_1 + n - k - 2}, \ldots, t^{q_n - k - 1} \rangle}{\Delta_{n-1}(t_1, \ldots, t_{n-1})} \Bigg|_{t_1 = \cdots = t_{n-1} = r}$$

$$= \sum_{(q_1, \ldots, q_{n-k-1}) \in \overset{\circ}{D}} \varepsilon(q_1, \ldots, q_{n-k-1})$$

$$\times N(m_1, \ldots, m_k, q_1, \ldots, q_{n-k-1}) r^{(\sum_i^k m_i + \sum_1^{n-k-1} q_j)}. \tag{4.24}$$

Here the new index set $\overset{\circ}{D}$ comes from D:

$$(q_1, \ldots, q_{n-k-1}) \in \overset{\circ}{D} \quad \Leftrightarrow \quad (p_1, \ldots, p_{n-k-1}) \in D.$$

It is easy to check that $\overset{\circ}{D}$ as defined by $(q_1, \ldots, q_{n-k-1}) \in \overset{\circ}{D}$ is subject to the conditions: (i) $n - k - 1 \geq q_1 \geq \cdots \geq q_{n-k-1} \geq 0$; (ii) for $i \neq j$, there exists $q_i + q_j \neq i + j - 1$. The sign function $\varepsilon(q_1, \ldots, q_n)$ is written as

$$\varepsilon(q_1, \ldots, q_n) = \begin{cases} +1 & \text{if } \sum_{i=1}^{n-k-1} q_i = 0, \\ +1 & \text{if } \sum q_i \equiv 1, 2 \pmod 4, \\ -1 & \text{if } \sum q_i \equiv 3, 4 \pmod 4, \end{cases} \tag{4.25}$$

while $N(m_1, \ldots, m_k, q, \ldots, q_{n-k-1})$ signifies the dimension of the irreducible representation of the $(n - 1)$-dimensional unitary group $O(n - 1)$ with signature $(m_1, \ldots, m_k, q_1, \ldots, q_{n-k-1})$. Equation (4.24) gives the formula of $\rho_m(r)$ for $n = 2k + 1$.

Case 2. $n = 2k$. Since

$$d(t)t^{n-k-2}$$

$$= (1 - t^2)t^{n-k-2} = t^{n-k-2} - t^{n-k},$$

$$d(t)(t^{n-k-3} + t^{n-k-1})$$

$$= (1 - t^2)(t^{n-k-3} + t^{n-k-1}) = t^{n-k-3} - t^{n-k-1},$$

$$d(t)(t^{n-k-1} + t^{n-k})$$

$$= (1 - t^2)(t^{n-k-4} + t^{n-k}) = t^{n-k-4} - t^{n-k+2} - (t^{n-k-2} - t^{n-k}),$$

$$\vdots$$

$$d(t)(1 + t^{2n-2k-4})$$

$$= (1 - t^2)(1 + t^{2n-2k-4}) = (1 - t^{2n-2k-2}) - (t^2 - t^{2n-2k-4}),$$

from the properties of determinants, we have

$$C_{n-1}(t_1, \ldots, t_{n-1}) = \langle t^{l_1 + n - 1 - k}, \ldots, t^{l_k + n - 1 - k},$$

$$\times d(t)t^{n-k-2}, \ldots, d(t)(1 + t^{2n-2k-4}) \rangle$$

$$= \langle t^{l_1 + n - 1 - k}, \ldots, t^{l_k + n - 1 - k}, (t^{n-k-2} - t^{n-2}),$$

$$\times (t^{n-k-3} - t^{n-k-1}), \ldots (1 - t^{2n-2k-2}) \rangle$$

$$= \sum_{(p_1, \ldots, p_{n-1}) \in E} \varepsilon(p_1, \ldots, p_{n-1})$$

$$\times \langle t^{l_1 + n - 1 - k}, \ldots, t^{l_k + n - 1 - k}, t^{p_1}, \ldots, t^{p_{n-k-1}} \rangle.$$

The index set E is determined by the conditions: (i) $2n - 2k - 2 \geq p_1 \geq \cdots \geq p_{n-k-1} \geq 0$; (ii) for any i, there exists $p_i \neq n - k - 1$; (iii) for $i \neq j$, there exists $p_i + p_j \neq 2n - 2k - 2$. The sign function can be proved to be (see also next section)

$$\varepsilon(p_1, \ldots, p_n) = (-1)^{[(p_1 + \cdots + p_{n-k-1})/2] - [(n-k-2)(n-k-1)/4]}. \tag{4.26}$$

Thus, $\rho_m(r)$ becomes

$$\rho_m(r) = \left. \frac{C_{n-1}(t_1, \ldots, t_{n-1})}{\Delta_{n-1}(t_1, \ldots, t_{n-1})} \right|_{t_1 = \cdots = t_{n-1} = r}$$

$$= \sum_{(p_1,\dots,p_{n-k-1})\in E} \varepsilon(p_1,\dots,p_{n-k-1})$$

$$\times \frac{\langle t^{l_1+n-1-k},\dots,t^{l_k+n-1-k},t^{p_1},\dots,t^{p_{n-k-1}}\rangle}{\Delta_{n-1}(t_1,\dots,t_{n-1})}\Bigg|_{t_1=\cdots=t_{n-1}=r}.$$

Similarly, set $q_1 = p_1 - n + 2 + k$, $q_2 = p_2 - n + 3 + k, \dots, q_{n-k-1} = p_{n-k-1}$. Then from the known $l_1 = m_1 + k - 1, \dots, l_k = m_k$, we get

$$\langle t^{l_1+n-1-k},\dots,t^{l_k+n-1-k},t^{p_1},\dots,t^{p_{n-k-1}}\rangle$$

$$= \langle t^{m_1+n-2}, t^{m_2+n-3},\dots,t^{m_k+n-k-1}, t^{q_2+n-k-2},\dots,t^{q_{n-k-1}}\rangle.$$

Substituting this into the above equality, we deduce

$$\rho_m(r) = \sum_{(q_1,\dots,q_{n-k-1})\in \mathring{E}} \varepsilon(q_1,\dots,q_{n-k-1})$$

$$\cdot \frac{\langle t^{m_1+n-2},\dots,t^{m_k+n-k-1}, t^{q_1+n-k-2},\dots,t^{q_{n-k-1}}\rangle}{\Delta_{n-1}(t_1,\dots,t_{n-1})}\Bigg|_{t_1=\cdots=t_{n-1}=r}$$

$$= \sum_{(q_1,\dots,q_{n-k-1})\in \mathring{E}} (-1)^{(q_1+\cdots+q_{n-1})/2} N(m_1,\dots,m_k,$$

$$q_1,\dots,q_{n-k-1})r^{\sum m_i + \sum q_j}. \quad (4.27)$$

The new index set $\mathring{E}$ is derived from E as follows:

$$(q_1,\dots,q_{n-k-1}) \in \mathring{E} \iff (p_1,\dots,p_{n-k-1}) \in E.$$

It is easy to verify that the condition of $\mathring{E}$ demands that (i) $n - k \geq q_1 \geq \cdots \geq q_{n-k-1} \geq 0$; (ii) for any i there exists $q_i \neq i$; (iii) for any $i \neq j$ there exists $q_i + q_j \neq i + j$. The sign in (4.27) comes from (4.26). Equation (4.27) is just the formula of $\rho_m(r)$ for $n = 2k$.

Summing up the discussions above, we obtain the next result.

THEOREM 4.6. *The Poisson kernel $P(rI, \Gamma)$ of the rotation group $O^+(n)$ allows the expansion*

$$P(rI, \Gamma) = \sum_{m\geq 0} \rho_m(r)\sigma_m(\Gamma),$$

where $m = (m_1,\dots,m_k)$ refers to the signature of the irreducible representations of $O^+(n)$, $\sigma_m(\Gamma)$ is the character of the irreducible representation, and

$$\rho_m(r) = \sum_{(q_1,\dots,q_{n-k-1})\in \mathring{D}=\mathring{E}} \varepsilon(q_1,\dots,q_{n-k-1})$$

$$\times N(m_1,\dots,m_k,q_1,\dots,q_{n-k-1})r^{\sum m_i + \sum q_j}.$$

Here, for odd n, we use index set $\mathring{D}$, and the corresponding sign $\varepsilon(q_1, \ldots, q_{n-k-1})$ is determined by (4.25); for even n, we use the index set $\mathring{E}$, and the corresponding sign $\varepsilon(q_1, \ldots, q_{n-k-1})$ is equal to $(-1)^{(q_1 + \cdots + q_{n-k-1})/2}$. $N(m_1, \ldots, m_k, q_1, \ldots, q_{n-k-1})$ denotes the dimension of the irreducible representation of the $(n-1)$-dimensional unitary group $U(n-1)$ with signature $(m_1, \ldots, m_k, q_1, \ldots, q_{n-k-1})$.

In addition to the above expressions we will show that $\rho_m(r)$ can also be expressed by determinants. For this, let us turn back to (4.20):

$$\rho_m(r) = \frac{C_{n-1}(t_1, \ldots, t_{n-1})}{\Delta_{n-1}(t_1, \ldots, t_{n-1})} \bigg|_{t_1 = \cdots = t_{n-1} = r}.$$

With the aid of the well-known identity ([2], p. 24, Theorem 1.2.4)

$$\lim_{\substack{x_1 \to x \\ \vdots \\ x_n \to x}} \frac{\begin{vmatrix} f_1(x_1) & \cdots & f_n(x_1) \\ \vdots & & \vdots \\ f_1(x_n) & \cdots & f_n(x_n) \end{vmatrix}}{\Delta_n(x_1, \ldots, x_n)} = \frac{(-1)^{n(n-1)/2}}{1! \, 2! \cdots n!} \begin{vmatrix} f_1(x) & \cdots & f_n(x) \\ f_1'(x) & \cdots & f_n'(x) \\ \vdots & & \vdots \\ f_1^{(n-1)}(x) & \cdots & f_n^{(n-1)}(x) \end{vmatrix},$$

the expression of $\rho_m(r)$ may be written as

$$\rho_m(r) = \frac{(-1)^{(n-1)(n-2)/2}}{1! \, 2! \cdots n!} \begin{vmatrix} \xi_1(r) & \cdots & \xi_{n-1}(r) \\ \xi_1'(r) & & \xi_{n-1}'(r) \\ \vdots & & \vdots \\ \xi_1^{(n-2)}(r) & \cdots & \xi_{n-1}^{(n-2)}(r) \end{vmatrix}, \qquad (4.28)$$

where

$$\xi_1(r) = r^{m_1 + n - 2}, \; \xi_2(r) = r^{m_2 + n - 3}, \ldots, \xi_k(r) = \gamma^{m_n + n - k - 1},$$

$$\xi_{k+1}(r) = \begin{cases} r^{n-k-2} + r^{n-k-1}, & n = 2k+1, \\ r^{n-k-2} - r^{n-k}, & n = 2k, \end{cases}$$

$$\xi_{k+2}(r) = \begin{cases} r^{n-k-3} + r^{n-k}, & n = 2k+1, \\ r^{n-k-3} - r^{n-k-1}, & n = 2k, \end{cases}$$

$$\vdots$$

$$\xi_{n-1}(r) = \begin{cases} 1 + r^{2n-2k-3}, & n = 2k+1, \\ 1 - r^{2n-2k-4}, & n = 2k. \end{cases}$$

EXAMPLE 4.1. Now we give the expression of $\rho_m(r)$ for the simplest case $n = 3$. In this case $k = 1$, an irreducible representation of $O^+(3)$ can be characterized by a nonnegative integer $m \geq 0$. We have

$$\xi_1(r) = r^{m+1}, \qquad \xi_2(r) = 1 + r,$$

$$\rho_m(r) = (-1) \begin{vmatrix} r^{m+1} & 1 + r \\ (m+1)r^m & 1 \end{vmatrix} = (m+1)r^m + mr^{m-1}.$$

In the light of the expansion of the Poisson kernel $P(rI, \Gamma)$, let us discuss the Abel summability on $O^+(n)$. Now pass to (1.7). For a given continuous function $u(\Gamma)$ on $O^+(n)$, use its Poisson integral

$$u(r\Gamma_0) = \frac{1}{c} \int_{O^+(n)} P(r\Gamma_0, \Gamma) u(\Gamma) \dot{\Gamma} = \frac{1}{c} \int_{O^+(n)} P(rI, \Gamma\Gamma_0') u(\Gamma) \dot{\Gamma}.$$

Referring to Theorem 4.6, we have

$$P(rI, \Gamma\Gamma_0') = \sum_{m \geq 0} \rho_m(r) \sigma_m(\Gamma\Gamma_0') = \sum_{m \geq 0} \rho_m(r) \sum_{i,j}^{N(m)} \varphi_{ij}^m(\Gamma) \varphi_{ij}^m(\Gamma_0), \quad (4.29)$$

where $\Gamma \to (\varphi_{ij}^m(\Gamma))$ is the irreducible representation of $O^+(n)$ with signature $m = (m_1, \ldots, m_k)$, and $N(m)$ is the dimension of this representation. Write

$$a_{ij}^m = \frac{N(m)}{c} \int_{O^+(n)} \varphi_{ij}^m(\Gamma) u(\Gamma) \dot{\Gamma}. \quad (4.30)$$

Then we obtain

$$u(r\Gamma) = \sum_{m \geq 0} \rho_m(r) \sum_{i,j}^{N(m)} \varphi_{ij}^m(\Gamma_0) \frac{1}{c} \int_{O^+(n)} \varphi_{ij}^m(\Gamma) u(\Gamma) \dot{\Gamma}$$

$$= \sum_{m \geq 0} \rho_m(r) \sum_{i,j}^{N(m)} \frac{a_{ij}^m}{N(m)} \varphi_{ij}^m(\Gamma_0)$$

$$= \sum_{m \geq 0} \frac{\rho_m(r)}{N(m)} \sum_{i,j}^{N(m)} a_{ij}^m \varphi_{ij}^m(\Gamma_0) \to u(\Gamma_0), \qquad r \to 1. \quad (4.31)$$

Accordingly, we can define $\sum_{m \geq 0} \sum_{i,j}^{N(m)} a_{ij}^m \varphi_{ij}^m(\Gamma)$ as the Fourier series of $u(\Gamma)$, and a_{ij}^m as the Fourier coefficients. Then (4.31) turns out to be the Abel summation of Fourier series on the rotation groups $O^+(n)$. Here $\rho_m(r)$ is determined by (4.28) or Theorem 4.6.

Fix $u(\Gamma) = \sigma_{m'}(\Gamma)$ and $\Gamma_0 = I$ in (4.31). Then we infer

$$u(r\Gamma) = \frac{1}{c} \int_{O^+(n)} P(rI, \Gamma) \sigma_{m'}(\Gamma) \dot{\Gamma} = \sum_{m \geq 1} \rho_m(r) \cdot \frac{1}{c} \int_{O^+(n)} \sigma_m(\Gamma) \sigma_{m'}(\Gamma) \dot{\Gamma}$$

$$= \rho_{m'}(r) \to \sigma_{m'}(I) = N(m'), \qquad \text{as } r \to 1,$$

That is, for any signature $m = (m_1, \ldots, m_k)$, there exists $\rho_m(r) \to N(m)$ $(r \to 1)$. The orthogonality of the characters has been used in the proof. From the viewpoint of Abel summability, the conclusion is apparent.

4.2. A Problem in Group Representations

As a direct application of Lemma 4.2, we consider a general problem about the decomposition of combinations of group representations proposed by Hua [2, p. 10].

Let $X \to A_{f_1, \ldots, f_n}(X)$ be the irreducible representation of GL(n) with signature $(f_1, \ldots, f_n)$, and let $N = N(f_1, \ldots, f_n)$ be its dimension. Again let $Y \to B_{g_1, \ldots, g_N}(Y)$ be the irreducible representation of GL(N) with signature $(g_1, \ldots, g_N)$. Then

$$X \to B_{g_1, \ldots, g_N}(A_{f_1, \ldots, f_N}(X))$$

is the representation of GL(n). We ask: What direct sum of irreducible representations may be decomposed by this representation? Two special cases of this problem have been solved in [2] as follows:

Let

$$A_{f,0,\ldots,0}(X) = X^{[f]}, \qquad A_{1,\ldots,1,0,\ldots,0}(X) = (X)^{(g)}.$$
$$\underset{g}{}$$

Then the irreducible decompositions of $X \to (X^{(2)})^{[f]}$ and $X \to (X^{[2]})^{[f]}$ are expressed as

$$\sigma((X^{[2]})^{[f]}) = \sum_{\substack{f_1 + \cdots + f_n = f \\ f_1 \geq \cdots \geq f_n \geq 0}} \chi_{2f_1, \ldots, 2f_n}(X), \qquad (4.33)$$

$$\sigma((X^{(2)})^{[f]}) = \sum_{\substack{f_1+\cdots+f_n=f \\ f_1\geq\cdots\geq f_n\geq 0}} \chi_{f_1,f_1,f_2,f_2,\dots}(X), \qquad (4.34)$$

where σ denotes the trace of the matrix, and $\chi_{2f_1,\dots,2f_n}$ means the character of the irreducible representation of $GL(N)$ with signature $(2f_1,\dots,2f_n)$. It is known that the decompositions (4.33) and (4.34) are fundamental results in the harmonic analysis on the second and the third classical domains of several complex variables. In this section, we use Lemma 4.2 to study $X \to (X^{(2)})^{(f)}$ and $X \to (X^{[2]})^{(f)}$. Our results are parallel to (4.33) and (4.34).

Recall that if $X = [x_1,\dots,x_n]$ (diagonal matrix with $x_1,\dots,x_n$ as diagonal elements), then we may put

$$X^{(2)} = [x_1x_2, x_1x_3,\dots,x_1x_n, x_2x_3,\dots,x_2x_n,\dots,x_{n-1}x_n],$$

$$X^{[2]} = [x_1^2, x_1x_2,\dots,x_1x_n, x_2^2,\dots,x_2x_n,\dots,x_{n-1}^2, x_{n-1}x_n, x_n^2]$$

the former being a $n(n-1)/2$ diagonal matrix with diagonal elements x_ix_j $(i<j)$, the latter a $n(n+1)/2$ diagonal matrix with diagonal elements x_ix_j $(i\leq j)$. Let t be an independent variable. Then

$$\prod_{i<j} (1 - tx_ix_j) = \det(I - tX^{(2)}), \qquad (4.35)$$

$$\prod_{i\leq j} (1 - tx_ix_j) = \det(I - tX^{[2]}), \qquad (4.36)$$

but it is easy to see that

$$\det(I - tX^{(2)}) = \sum_{q=0}^{n(n-1)/2} (-1)^q t^q \sigma^{\overset{q}{1,\dots,1,0,\dots,0}}(X^{(2)})$$

$$= \sum_{q=0}^{n(n-1)/2} (-1)^q t^q \sigma((X^{(2)})^{(q)}), \qquad (4.37)$$

$$\det(I - tX^{[2]}) = \sum_{q=0}^{n(n+1)/2} (-1)^q t^q \sigma^{\overset{q}{1,\dots,1,0,\dots,0}}(X^{[2]})$$

$$= \sum_{q=0}^{n(n+1)/2} (-1)^q t^q \sigma((X^{[2]})^{(q)}). \qquad (4.38)$$

On the other hand, Lemma 4.2 implies

$$\prod_{i<j} (x_i - x_j) \prod_{i<j} (1 - x_i x_j)$$

$$= \begin{vmatrix} x_1^{n-1} & x_1^{n-2}(1+x_1^2) & \cdots & (1+x_1^2)^{n-1} \\ \vdots & \vdots & & \vdots \\ x_n^{n-1} & x_n^{n-2}(1+x_n^2) & \cdots & (1+x_n^2)^{n-1} \end{vmatrix}$$

$$= \begin{vmatrix} x_1^{n-1} & x_1^{n-2}+x_1^n & x_1^{n-3}+x_1^{n+1} & \cdots & 1+x_1^{2n-2} \\ \vdots & \vdots & \vdots & & \vdots \\ x_n^{n-1} & x_n^{n-2}+x_n^n & x_n^{n-3}+x_n^{n+1} & \cdots & 1+x_n^{2n-2} \end{vmatrix}$$

$$= \sum_{(n_1,\ldots,n_n)\in D_n} \pm \begin{vmatrix} x_1^{n_1} & x_1^{n_2} & \cdots & x_1^{n_n} \\ \vdots & & & \vdots \\ x_n^{n_1} & x_n^{n_2} & \cdots & x_n^{n_n} \end{vmatrix}, \tag{4.39}$$

where the condition of D_n is subject to (a) $2n - 2 \geq n_1 \geq \cdots \geq n_n \geq 0$; (b) there exists an i, $1 \leq i \leq n$, $n_i = n - 1$; (c) for any $i \neq j$, $n_i + n_j \neq 2n - 2$, and clearly there are 2^{n-1} elements in the set D_n. Therefore, we can state

$$\prod_{i<j} (1 - x_i x_j) = \sum_{(n_1,\ldots,n_n)\in D_n} \pm \begin{vmatrix} x_1^{n_1} & x_1^{n_2} & \cdots & x_1^{n_n} \\ \vdots & & & \vdots \\ x_n^{n_1} & x_n^{n_2} & \cdots & x_n^{n_n} \end{vmatrix} \bigg/ \prod_{i<j} (x_i - x_j)$$

$$= \sum_{(n_1,\ldots,n_n)\in D_n} \pm \chi_{(n_1-n+1,\, n_2-n+2,\ldots,n_n)}(X), \tag{4.40}$$

where $X = [x_1, \ldots, x_n]$. Replace x_i by $\sqrt{t}\, x_i$ in (4.36). It yields

$$\prod_{i<j} (1 - t x_i x_j) = \sum_{(n_1,\ldots,n_n)\in D_n} \pm t^{[\sum n_i - (n(n-1)/2)]} \chi_{(n_1-n+1,\ldots,n_n)}(X).$$

Set $f_1 = n_1 - n + 1, f_2 = n_2 - n + 2, \ldots, f_n = n_n$; the above equality turns out to be

$$\prod_{i<j} (1 - t x_i x_j) = \sum_{(f_1,\ldots,f_n)\in \mathring{D}_n} \pm t^{\sum f_i/2} \chi_{(f_1,\ldots,f_n)}(X), \tag{4.41}$$

where $\mathring{D}_n$ comes from D_n by $(f_1, \ldots, f_n) \in \mathring{D}_n \Leftrightarrow (n_1, \ldots, n_n) \in D_n$. It is not hard to check that conditions of $\mathring{D}_n$ demand (a) $n - 1 \geq f_1 \geq \cdots \geq f_n \geq 0$; (b) there exists $-f_k = k - 1$ in $f_1, \ldots, f_n$; (c) for any $i \neq j$ there exists $f_i + f_j \neq i + j - 2$. Comparing (4.41) with (4.37) and observing that no negative coefficient term appears in the representation of irreducible decomposition, we see that

$$\sigma((X^{[2]})^{(q)}) = \sum_{\substack{f_1+\cdots+f_n=2q \\ (f_1,\ldots,f_n)\in\check{D}_n}} \chi_{(f_1,\ldots,f_n)}(X). \tag{4.42}$$

Now consider the decomposition of $(X^{[2]})^{(q)}$. From Lemma 4.2, it follows that

$$\prod_{i<j} (x_i - x_j) \prod_{i\leq j} (1 - x_i x_j)$$

$$= \begin{vmatrix} x_1^{n-1} & x_1^{n-2} + x_1^n & \cdots & x_1^{2n-2} + 1 \\ \vdots & & & \vdots \\ x_n^{n-1} & x_n^{n-2} + x_n^n & \cdots & x_n^{2n-2} + 1 \end{vmatrix} \prod_i (1 - x_i^2)$$

$$= \begin{vmatrix} (1 - x_1^2)x_1^{n-1} & (1 - x_1^2)(x_1^{n-2} + x_1^n) & \cdots & (1 - x_1^2)(1 + x_1^{2n-2}) \\ \vdots & & & \vdots \\ (1 - x_n^2)x_n^{n-1} & (1 - x_n^2)(x_n^{n-2} + x_n^n) & \cdots & (1 - x_n^2)(x_n^{2n-2} + 1) \end{vmatrix}.$$

Notice the linear combinations of

$$(1 - x_1^2)x_1^{n-1} = x_1^{n-1} - x_1^{n+1},$$

$$(1 - x_1^2)(x_1^{n-2} + x_1^n) = x_1^{n-2} + x_1^{n+2},$$

$$(1 - x_1^2)(x_1^{n-3} + x_1^{n+1}) = x_1^{n-3} - x_1^{n+3} + (x_1^{n-1} - x_1^{n+1}),$$

$$\vdots$$

With the same argument we can prove the validity of other, similar, formulas, and the above determinant can be written as

$$\prod_{i<j} (x_i - x_j \prod_{i\leq j} (1 - x_i x_j)) = \begin{vmatrix} x_1^{n-1} - x_1^{n+1} & x_1^{n-2} - x_1^{n+2} & \cdots & 1 - x_1^{2n} \\ \vdots & & & \vdots \\ x_n^{n-1} - x_n^{n+1} & x_n^{n-2} - x_n^{n+2} & \cdots & 1 - x_n^{2n} \end{vmatrix}$$

$$= \sum_{(n_1,\ldots,n_n)\in S_n} \pm \begin{vmatrix} x_1^{n_1} & x_1^{n_2} & \cdots & x_1^{n_n} \\ \vdots & & & \vdots \\ x_n^{n_1} & x_n^{n_2} & \cdots & x_n^{n_n} \end{vmatrix}.$$

Here the conditions of index set S_n are (a) $2n \geq n_1 \geq \cdots \geq n_n \geq 0$; (b) for any i there exists $n_i \neq n$; (c) if $i \neq j$, then $n_i + n_j \neq 2n$. In the following the discussion is parallel to the above. It suffices to write down the final result.

We have

$$\sigma((X^{[2]})^{(q)}) = \sum_{\substack{f_1+\cdots+f_n=2q \\ (f_1,\ldots,f_n)\in \mathring{S}_n}} \chi_{f_1\cdots f_n}(X). \tag{4.13}$$

Here the set $\mathring{S}_n$ is subject to (a) $n+1 \geq f_1 \geq \cdots \geq f_n \geq 0$; (b) for any i there exists $f_i \neq i$; (c) if $i \neq j$, then $f_i + f_j \neq i + j$.

Decompositions (4.42) and (4.43) are exactly what we wanted.

References

1. L. K. Hua, A convergent theorem of the space of continuous functions on compact groups, *Sci. Record* (*N.S*) No. 9 (1958).
2. L. K. Hua, *Harmonic Analysis of Functions of Several Complex Variables in the Classical Domain*, Science Press, Beijing (1958) (in Chinese); American Mathematical Society, Providence, RI (1968).
3. L. K. Hua and K. H. Look, Theory of harmonic functions in the classical domain. I, *Acta Math. Sinica* **8**, 531–547 (1958); *Sci. Sinica* **9**, 1031–1094 (1958) (in English).
4. K. H. Look, *Classical Manifolds and Classical Domains*, Shanghai Science and Technical, Shanghai (1963).
5. S. Kung, Fourier analysis on unitary groups. I, *Acta Math. Sinica* **10**, 239–261 (1960); *Chinese Math. Acta* **3**, 249–272 (1962).
6. S. Kung, Fourier analysis on unitary groups. II, *Acta Math. Sinica* **12**, 17–32 (1962); *Chinese Math. Acta* **3**, 19–33 (1963).
7. E. D. Murnaghan, *The Theory of Group Representations*, Johns Hopkins University Press, Baltimore (1938).

5

Coxeter–Killing Transformations of Simple Lie Algebras

To determine the Betti numbers of complex simple Lie groups (algebras) is a classical problem. It is known that these Betti numbers are simply the coefficients of the Poincaré polynominal

$$p(t) = (1 + t^{2m_1+1}) \cdots (1 + t^{2m_l+l})$$

of the corresponding Lie group, where $m_1, \ldots, m_l$ are called the Poincaré indices of the Lie algebras. Brauer and Pontryagin (see Ref. 1) already determined these numbers for four classes of classical simple Lie algebras, and in 1950 C. Chevalley [1] announced the result for several exceptional Lie algebras.

Afterwards, A. Coleman [2], H. M. Coxeter [3], and B. Kostant [4] further showed that these indices can be obtained by investigating a special class of elements of the Weyl group of the simple Lie algebras. These elements are called Coxeter elements. Suppose that g is a complex simple algebra, and $\mathfrak{h}$ is its Cartan subalgebra with rank l and root system Δ. Let $\Pi = \{\alpha_1, \ldots, \alpha_l\}$ be a basic root system of Δ, and let W be its Weyl group. It is well known that W is a finite group generated by R_{α_i} $(i = 1, \ldots, l)$, where R_{α_i} is the reflection determined by α_i. Let

$$\gamma = R_{\alpha_1} R_{\alpha_2} \cdots R_{\alpha_l}. \tag{5.1}$$

We call γ defined by (5.1) and its equivalent elements the Coxeter–Killing transformation of g. Denote the positive root set by Δ^+. For each $\alpha \in \Delta^+$, we can write $\alpha = \sum n_i \alpha_i$, where n_i are nonnegative integers. The symbol $o(\alpha) = \sum n_i$ is called the order of the root. A partial order is determined by means of Π in Δ. The highest root under this partial order is written as ψ (i.e., the first member of the adjoint representation ad g of g that is irreducible). Let $S_0 = 1 + o(\psi)$. Then the main results in Refs. 2-4 can be summed up as follows.

THEOREM 5.1. (*Coxeter-Coleman-Konstant*). *If* $\gamma = R_{\alpha_1} \cdots R_{\alpha_l}$, *its eigenvalues are* w^{m_j}, $j = 1, \ldots, l$, $w = e^{2\pi i / s_0}$, $s_0 = 1 + o(\psi)$, *and* $m_1, \ldots, m_l$ *are the Poincaré indices of* g.

In view of the importance of the Coxeter elements in investigating the indices in simple algebras, the main purpose of the chapter is to describe its characters. We develop a method by which we prove:

THEOREM 5.2. *For* $\sigma \in W$ *suppose we have*

1. 1 *is not an eigenvalue of* σ.
2. *The number of permutations contained in* σ *is* $\leq l$.

Then σ *is a Coxeter element.*

Notice that $\gamma = R_{\alpha_1} \cdots R_{\alpha_l}$ satisfies conditions 1 and 2 [4]. Thus in order that σ be a Coxeter-Killing transformation, it is sufficient and necessary that conditions 1 and 2 hold.

Using the same method we also proved the following theorems easily.

THEOREM 5.3. *Suppose that* ψ *is the highest root of a simple algebra* g *with rank* l, $s_0 = 1 + o(\psi)$, *and* t *is the number of positive roots in* Δ. *Then* $s_0 l = 2t$.

THEOREM 5.4. *Suppose that* m_i, $i = 1, \ldots, l$, *are the Poincaré indices of* g. p_k *denotes the number of* m_i $(i = 1, \ldots, l)$ *that are equal to* k, *and* q_k *denotes the number of positive roots that are equal to* k *in the root system. Let* $s_0 = 1 + o(\psi)$. *Then*

$$p_k = q_k + q_{s_0-k} - l.$$

These two theorems are the basic results of Kostant [4]. The first theorem is the key to proving the Coxeter-Coleman-Kostant theorem on the basis of Ref. 4. The second theorem gives the formula determining the Poincaré indices from the radical system of g. The proofs of these two theorems in Ref. 4 are based on the systematic discussion of the basic three-dimensional subalgebra. It is easy to see that our method is entirely different and much easier than his.

The meanings of g, $\mathfrak{h}$, Δ, Π, W are as above. The real linear space generated by $\alpha_1, \ldots, \alpha_l$ is written as $\mathfrak{h}^{\#}$. For $\alpha \in \Delta$, denote by e_α the root vector corresponding to α. It is well known that we can assign a permutation of Δ to every $\sigma \in W$. So σ can be represented as the product of the different permutation of the roots. Every permutation corresponds to an orbit of σ:

$$\varphi \in \Delta, \qquad \varphi \to \sigma\varphi \to \sigma^2\varphi \to \cdots \to \sigma^{m-1}\varphi, \qquad \sigma^m\varphi = \varphi,$$

where m is called the length of the permutation.

The inner automorphism group of g is written $A(g)$. It is well known that each $\sigma \in W$ can be extended to an inner automorphism A_σ of g:

$$\sigma \to A_\sigma, \qquad A_\sigma(e_\alpha) = (e_{\sigma_\alpha}),$$

where (e_α) represents the one-dimensional subspace generated by e_α.

LEMMA 5.5. *Suppose that $\sigma \in W$ does not have 1 as its eigenvalue, the order of σ is h, the number of the permutation contained in σ is L, and A_σ is an automorphism of g extended by σ. Then under some basis of g, A_σ^n can be represented as*

$$\begin{pmatrix} I^{(l)} & & & & \\ & \lambda_1 I^{(n_1)} & & & \\ & & \ddots & & \\ & & & \lambda_L I^{(n_L)} \end{pmatrix}, \qquad \lambda_i^{n_j} = 1, \quad i = 1, \ldots, L, \qquad (5.2)$$

where n_i, $i = 1, \ldots, l$, is the length of each permutation.

PROOF. Evidently, for $x \in \mathfrak{h}$, we have $A_\sigma^h x = x$. Therefore,

$$A_\sigma^h[x, e_\varphi] = [x, A_\sigma^h e_\varphi] = \varphi(x) A_\sigma^h e_\varphi,$$

therefore,

$$A_\sigma^h e_\varphi = \lambda_\varphi e_\varphi.$$

Operating A_σ on the two sides, we have

$$A_\sigma^{h+1} e_\varphi = A_\varphi A_\sigma e_\varphi = \lambda_\varphi(e_{\sigma\varphi}),$$

$$A_\sigma^{h+1} e_\varphi = A_\sigma^h A_\sigma e_\varphi = A_\sigma^h(e_{\sigma\varphi}) = \lambda_{\sigma\varphi}(e_{\sigma\varphi}), \qquad (5.3)$$

$$\Rightarrow \qquad \lambda_\varphi = \lambda_{\sigma\varphi}.$$

So if Γ_i $(i = 1, \ldots, L)$ represents the different orbit of σ in Δ, then $\Delta = \sum \Gamma_i$. Let $E_i = \sum_{\varphi \in \Gamma_i} (e_\varphi)$. Obviously we have $g = \mathfrak{h} + \sum_{i=1}^{L} E_i$. From (5.3) it follows that A_σ^h is a scalar operator $\lambda_i I$ on E_i. Since A_σ^h is an identity operator on $\mathfrak{h}$, there is $x \in \mathfrak{h}$ such that $A_\sigma^h = e^{\operatorname{ad} x}$. If $\varphi \in \Gamma_i$, then $\lambda_i = e^{(x,\varphi)}$. Then we have

$$\lambda_i^{n_i} = \lambda_\varphi \cdot \lambda_{\sigma\varphi} \cdots \lambda_{\sigma\varphi}^{n_i-1} = e^{(x, \varphi + \sigma\varphi + \cdots + \sigma^{n_i-1}\varphi)} = 1.$$

The last equation holds because n_i is the length of Γ_i, $(\sigma^{n_i} - 1)\varphi = 0$ but 1 is not an eigenvalue of σ and hence $\sigma - 1$ is nonsingular. Therefore $(1 + \sigma + \sigma^2 + \cdots + \sigma^{n_i-1})\varphi = 0$. The lemma is proved.   □

LEMMA 5.6. σ *is as in Lemma 5.5. Then under some basis of g, A_σ can be represented as*

$$
\begin{pmatrix}
\sigma & & & & \\
& \zeta_1 \begin{pmatrix} 1 & & & \\ & w_1 & & \\ & & \ddots & \\ & & & w_1^{n_1-1} \end{pmatrix} & & & \\
& & \ddots & & \\
& & & \zeta_L \begin{pmatrix} 1 & & & \\ & w_L & & \\ & & \ddots & \\ & & & w_L^{n_L-1} \end{pmatrix} &
\end{pmatrix},
\tag{5.4}
$$

where $w_j = e^{2\pi i/n_j}$, $|\zeta_j| = 1$, and $j = 1, \ldots, L$. In particular, A_σ is semisimple.

PROOF. By the proof of Lemma 5.5 we can see that

$$g = \mathfrak{h} + \sum_{j=1}^{L} E_j, \qquad E_j = \sum_{\varphi \in \Gamma_j} (e_\varphi),$$

where A_σ restricting on $\mathfrak{h}$ is σ and we can choose a basis of e_φ, $A_\sigma e_\varphi, \ldots, A_\sigma^{n_j-1} e_\varphi$, $\varphi \in \Gamma_j$, under which A_σ can be represented as

$$
\begin{pmatrix}
0 & 1 & 0 & \cdots & 0 \\
0 & 0 & 1 & & \\
\vdots & & \ddots & \ddots & \\
\vdots & & & \ddots & 1 \\
\xi_j & & & 0 & 0
\end{pmatrix}, \qquad A_\sigma^{n_j} e_\varphi = \xi_j e_\varphi,
$$

whose eigenequation is $\lambda^{n_j} - \xi_j = 0$ and its eigenvalues are $\xi_j^{1/n_j} e^{2\pi i p/n_j} = \zeta_j \omega_j^p$, $\zeta_j = \xi_j^{1/n_j}$, $p = 0, 1, \ldots, n_j - 1$. By Lemma 5.5,

$$A_\sigma^h e_\varphi = \lambda_j e_\varphi,$$

$$A_\sigma^h e\varphi = (A_\sigma^{n_j})^{h/n_j} e_\varphi = \xi_j^{(h/n_j)} e_\varphi$$

so

$$\xi_j^{h/n_j} = \lambda_j$$

Since $\lambda_j^{n_j} = 1$, $|\zeta_j| = 1$. The lemma is proved. $\square$

COROLLARY 5.7. *If 1 is not an eigenvalue of σ, then $L \geq l$ and there are at least l values of $\zeta_1, \ldots, \zeta_L$ equal to 1.*

PROOF. Note that A_σ is semisimple. Let g^{A_σ} be the subspace formed by fixed points under A_σ. It is well known from Ref. 6 that dim $g^{A_\sigma} \geq l$. Since 1 is not an eigenvalue of σ, $g^{A_\sigma} \cap \mathfrak{h} = 0$. Therefore there are at least l eigenvalues of A_σ on $\sum_{j-1}^{L} E_j$ equal to 1. $\square$

COROLLARY 5.8. *Assume 1 is not an eigenvalue of σ. Let $N(\sigma)$ denote the number of positive roots that change sign under σ. Then $N(\sigma) \geq l$.*

PROOF. For any Γ_j, $j = 1, \ldots, L$, in the different orbits $\Gamma_1, \ldots, \Gamma_L$ of σ, if $\varphi \in \Gamma_j$, by the proof of Lemma 5.5, we have $\varphi + \sigma\varphi + \cdots + \sigma^{n_j-1}\varphi = 0$, and hence not all $\varphi, \sigma\varphi, \ldots, \sigma^{n_j-1}\varphi$ are positive or negative. Therefore there exists $\varphi \in \Gamma_j$ such that φ and $\sigma\varphi$ have opposite signs. In other words, $N(\sigma) \geq L \geq l$. $\square$

LEMMA 5.9. *If $\sigma \in W$ and the number of permutations of σ is L, whose lengths are $n_1, \ldots, n_L$, respectively, then every eigenvalue of σ can be represented as $e^{2\pi i p_j/n_j}$, $0 \leq p_j < n_j$ for some j.*

PROOF. Let Γ_j $(j - 1, \ldots, L)$ be the different orbits of σ, and let K_i be the subspace spanned by the elements in Γ_i. Evidently K_i is a σ-invariant subspace, and for every $x \in K_i$, $\sigma^{n_i} x = x$. Suppose that ξ is an eigenvalue of σ on K_i with corresponding eigenvector x. Then $\sigma x = \xi x$, $\sigma^{n_i} x = \xi^{n_i} x = x$, and so $\xi^{n_i} = 1$; i.e., ξ is an n_i-th root of 1. Arrange the eigenvalues and the corresponding eigenvectors of K_i as follows:

	K_1	K_2	$\cdots$
Eigenvalues	$\xi_1, \ldots, \xi_{s_1}$	$\eta_1, \ldots, \eta_{s_2}$	$\cdots$
Eigenvectors	$x_1, \ldots, x_{s_1}$	$y_1, \ldots, y_{s_2}$	$\cdots$

Then examine whether $y_1 \in K_1$. If so, then discard it; otherwise let $K_1' = K_1 \cup \{y_1\}$. Then examine whether $y_2 \in K_1'$. Continue this process until we finally obtain l ($l = \dim \mathfrak{h}$) eigenvectors whose eigenvalues are some n_i-th root of 1. The lemma is proved.   □

DEFINITION 5.10. The set $\{x \mid (x, \alpha_i) \geq 0, (x, \psi) \leq 1, \psi$ is the highest root in $\Delta\}$ if $\mathfrak{h}^*$ is called the elementary simplex of $\mathfrak{h}$ and written as T.

LEMMA 5.11. *If the automorphism* $A \in A(g)$ *of* g *satisfies* $A\mathfrak{h} = \mathfrak{h}$ *and the norms of its eigenvalues all equal 1, then* A *can be written as* $e^{2\pi i \, \mathrm{ad} \, x}$, $x \in T$, *in the sense of conjugation.*

PROOF. See Ref. 4, Section 8.6.   □

COROLLARY 5.12. *If 1 is not an eigenvalue of* $\sigma \in W$ *and* A_σ *is an automorphism induced by* σ, *then the eigenvalues can be represented as*

$$\{1, 1, \ldots, 1, e^{2\pi i (x, \varphi)}, \varphi \in \Delta, x \in T\}. \tag{5.5}$$

PROOF. By Lemma 5.6, A_σ is semisimple and the norms of its eigenvalues all equal 1. But it is well known from Ref. 6 that g^{A_σ} contains a Cartan subalgebra $\tilde{\mathfrak{h}}$ on which A_σ is an identity. Then by Lemma 5.11, in the conjugation sense, A_σ can be written as $e^{2\pi i \, \mathrm{ad} \, \tilde{x}}$, $\tilde{x} \in \tilde{T}$ (elementary simplex of $\tilde{\mathfrak{h}}$). But Cartan subalgebras are conjugate to each other; i.e., there is $B \in A(g)$ such that $B\mathfrak{h} = \tilde{\mathfrak{h}}$. The root system and the root vectors of g with respect to $\mathfrak{h}$, $(\alpha, e_\alpha) \to^B (B\alpha, Be_\alpha) = (\tilde{\alpha}, e_{\tilde{\alpha}})$, are those with respect to $\tilde{\mathfrak{h}}$. It is easy to verify that $B^{-1} e^{2\pi i \, \mathrm{ad} \, \tilde{x}} = e^{2\pi i \, \mathrm{ad} \, x}$, where $\tilde{x} \in \tilde{T}$, $x \in T$, and $(x, \alpha) = (\tilde{x}, \tilde{\alpha})$; i.e., $\tilde{x}$ and x have the same coefficients with respect to the corresponding root system of $\tilde{\mathfrak{h}}$ and $\mathfrak{h}$, respectively. Therefore in the conjugation sense, A_σ can be written as $e^{2\pi i \, \mathrm{ad} \, x}$, $x \in T$. So all the eigenvalues of A_σ can be written as in (5.5).

We now need to use two properties of Coxeter elements.

LEMMA 5.13. *Let* γ *be a Coxeter element with order* h. *Then* γ *satisfies the following*:

1. 1 *is not an eigenvalue of* γ.
2. r *is the product of* l *permutations with length* h.
3. *There exists a regular eigenvector* z_0 *where the eigenvalue is* $e^{2\pi i/h}$.

PROOF. See Refs. 2 and 4 and Section 8.   □

We prove the following.

THEOREM 5.14. *Suppose that* ψ *is the highest root of the simple algebra* g. *Let* $s_0 = 1 + o(\psi)$. *Then* $s_0 l = 2t$, *where* t *is the number of all the positive roots.*

PROOF. Choose a Coxeter element γ. By Lemma 5.13, γ is the product of l permutations with length h, and 1 is not the eigenvalue of γ. By Corollary 5.12, the positive eigenvalues of A_γ are

$$\{1, 1, \ldots, 1, e^{2\pi i(x,\varphi)}, \varphi \in \Delta, x \in T\}. \tag{5.6}$$

On the other hand, by Lemma 5.6 and Corollary 5.7 A_γ is conjugate to

$$\left[\begin{pmatrix} \delta_1 & & \\ & \ddots & \\ & & \delta_l \end{pmatrix} \begin{pmatrix} 1 & & & \\ & e^{2\pi i/h} & & \\ & & \ddots & \\ & & & e^{2\pi i(h-1)/h} \end{pmatrix} \ddots \begin{pmatrix} 1 & & & \\ & e^{2\pi j/h} & & \\ & & \ddots & \\ & & & e^{2\pi i(h-1)/h} \end{pmatrix} \right], \tag{5.7}$$

where $\delta_1, \ldots, \delta_l$ are eigenvalues of γ. By Lemma 5.13(3), without loss of generality we may write $\delta_1 = e^{2\pi i/h}$. All the δ_i's are not 1, since 1 is not an eigenvalue of γ. By Lemma 5.9, they are h-th roots of 1.

We write $x = \sum x_i \varepsilon_i / h$ in (5.6), where $\{\varepsilon_i\}$ are the dual basis of $\{\alpha_i\}$. Since the eigenvalues of A_γ are h-th roots of 1 and 1 appears l times among them, $(x, \varphi) \neq 0$ if $\varphi \in \Delta$. Noticing again that $x \in T$, we know all $x_i \geq 1$ and are positive integers and $(x, \psi) < 1$.

Comparing (5.6) with (5.7), for the highest root ψ, we have

$$(x, \psi) = \frac{h-1}{h}. \tag{5.8}$$

Therefore $e^{2\pi i(x,-\psi)} = e^{2\pi i(1-h)/h} = e^{2\pi i/h}$. For the other positive roots $\beta < \psi$, if $(x, \beta) \neq 1/h$, then $1/h < (x, \beta) < (h-1)/h$. Naturally we have $e^{(x,-\beta)} \neq e^{2\pi i/h}$. But from (5.7), we know that $e^{2\pi i/h}$ appears at least $l+1$ times (including $\delta_1 = e^{2\pi i/h}$). Then by the above discussion, there are, besides $-\psi$, at least l positive roots $\beta_1, \ldots, \beta_l$ such that

$$(x, \beta_i) = \frac{1}{h} \qquad (i = 1, \ldots, l). \tag{5.9}$$

These l positive roots are necessarily elementary roots, because for any nonsimple positive roots $\alpha = \sum k_i \alpha_i$, we have $\sum k_i > 1$. Since $(x, \alpha_i) \geq 1/h$,

$$(x, \alpha) = \sum k_i(x, \alpha_i) > \frac{1}{h}.$$

Therefore we can assume that $\beta_i = \alpha_i$ $(i = 1, \ldots, l)$. Then (5.9) becomes

$$(x, \beta_i) = (x, \alpha_i) = \frac{1}{h} \left(\sum x_i \varepsilon_i, \alpha_i \right) = \frac{x_i}{h} = \frac{1}{h}.$$

Therefore

$$x_i = 1, \qquad x = \frac{\sum \varepsilon_j}{h}.$$

Substituting this expression in (5.8), we obtain

$$\frac{h-1}{h} = (x, \psi) = \frac{1}{h} \left(\sum \varepsilon_i, \psi \right) = \frac{1}{h} o(\psi),$$

so

$$h = 1 + o(\psi) = s_0.$$

Because of (5.7) we have $hl = 2t$, so $s_0 l = 2t$, where t is the number of all positive roots. The proof of the theorem is completed. $\square$

In the proof of the theorem, we know the following.

COROLLARY 5.15. *If γ is a Coxeter element, then A_γ conjugates to* $e^{2\pi i \, \mathrm{ad}(\sum \varepsilon_j/s_0)}$.

Because of the Coxeter-Coleman-Kostant theorem, the eigenvalues of the Coxeter element γ are w^{m_j} $(j = 1, \ldots, l)$, $w = e^{2\pi i/s_0}$, $s_0 = 1 + o(\psi)$, where m_j are the Poincaré indexes of g. How do we determine these m_j concretely? If p_k represents the number of m_i, $i = 1, \ldots, l$, being equal to k and q_k represents the number of positive roots whose orders equal k, then

$$p_k = \sum_{m_j = k} 1, \qquad q_k = \sum_{o(\varphi) = k, \varphi \in \Delta^+} 1. \tag{5.10}$$

THEOREM 5.16. *For the Poincaré indexes of g, we have the equalities*

$$p_k = q_k + q_{s_0-k} - l, \qquad k = 1, \ldots, s_0 - 1. \tag{5.11}$$

PROOF. By Corollary 5.15 the extended automorphism A_γ of the Coxeter element γ conjugates to $e^{2\pi i \, \mathrm{ad} \, \Sigma \, \varepsilon_i / s_0}$ and so the eigenvalues of A_γ are

$$\{1, 1, \ldots, 1, \, e^{(2\pi i/s_0)(\Sigma \, \varepsilon_i, \varphi)} = e^{(2\pi i/s_0)o(\varphi)} = \omega^{o(\varphi)}, \, \varphi \in \Delta, \, \omega = e^{2\pi i/s_0}\}. \quad (5.12)$$
$$\underbrace{\qquad\qquad}_{l}$$

On the other hand, (5.7) becomes

$$\begin{pmatrix} w^{m_1} & & & & \\ & \ddots & & & \\ & & w^{m_1} & & \\ & & & \begin{pmatrix} 1 & & & \\ & w & & \\ & & \ddots & \\ & & & w^{s_0-1} \end{pmatrix} & \\ & & & & \ddots \\ & & & & & \begin{pmatrix} 1 & & & \\ & w & & \\ & & \ddots & \\ & & & w^{s_0-1} \end{pmatrix} \end{pmatrix}, \quad w = e^{2\pi i/s_0}. \quad (5.13)$$

For the sake of easily comparing (5.12) with (5.13), we make the change in (5.12), when $-\varphi \in \Delta^-$,

$$w^{o(-\varphi)} = w^{-o(\varphi)} = w^{s_0-o(\varphi)}.$$

Then according to (5.12) and (5.13), all the eigenvalues of A_γ can be written as in the following table:

Eigenvalue	Order	Eigenvalue	Order
1	l	1	l
w	$q_1 + q_{s_0-1}$	w	$l + p_1$
w^2	$q_2 + q_{s_0-2}$	w^2	$l + p_2$
$\vdots$	$\vdots$	$\vdots$	$\vdots$
w^{s_0-1}	$q_{s_0-1} + q_1$	w^{s_0-1}	$l + p_{s_0-1}$

Comparing both sides of the table, we obtain $p_k + l = q_k + q_{s_0-k}$ ($k = 1, \ldots, s_0 - 1$). Then (5.11) follows, since there are l simple roots and 1 is the highest root; i.e., $q_1 = l$, $q_{s_0-1} = 1$. Therefore $p_1 = p_{s_0-1} = 1$, so $(m_1, \ldots, m_l) = (1, m_2, \ldots, m_{l-1}, s_0 - 1)$ and $m_2 > 1$, $m_{l-1} < s_0 - 1$.

Using Theorem 5.16, we can compute the Poincaré index table of each simple algebra. We now list the following index table, which could be looked up in Refs. 1 and 2, since it will be useful in the remainder of the chapter.

$$
\begin{array}{ll}
A_n & 1, 2, 3, \ldots, n \\
B_n, C_n & 1, 3, 5, \ldots, 2n - 1 \\
D_n & 1, 3, 5, \ldots, 2n - 3, 2n - 1 \\
G_2 & 1, 5 \\
F_4 & 1, 5, 7, 11 \\
E_6 & 1, 4, 5, 7, 8, 11 \\
E_7 & 1, 5, 7, 9, 11, 13, 17 \\
E_8 & 1, 7, 11, 13, 14, 19, 23, 29.
\end{array}
\tag{5.14}
$$

The following lemma is fundamental in the proof of the main result (Theorem 5.18).

LEMMA 5.17. *Suppose that Δ is the root system of the simple Lie algebra g with rank l, $\Pi = \{\alpha_1, \ldots, \alpha_l\}$ is the simple root system of Δ, and ψ is the highest root of Δ. Choosing arbitrarily $\alpha_{i_1}, \ldots, \alpha_{i_s}$ $(s < l)$ of Π, we obtain the inequalities*

the number of $\{\alpha_{i_j} + \alpha_{i_k} \in \Delta^+\}$ + the number of $\{\psi - \alpha_{i_j} \in \Delta^+\} \leq s$.   (5.15)

PROOF. It is well known that the Dynkin diagram of simple Lie algebras are as follows

$$
\begin{array}{ll}
\Gamma(A_n) & \text{·—·—·—· } \cdots \text{ ·—·} \\
\Gamma(B_n) & \text{·=o—o—o } \cdots \text{ o—o} \\
\Gamma(C_n) & \text{o=·—·—· } \cdots \text{ ·—·} \\
\Gamma(D_n) & \text{·>·—·—· } \cdots \text{ ·—·} \\
\Gamma(E_n) & \text{·—·—· } \cdots \text{ ·—·} \qquad (n = 6, 7, 8) \\
\\
\Gamma(F_4) & \text{·—·=o—o} \\
\Gamma(G_2) & \text{·≡o}
\end{array}
$$

where $\cdot$ represents the shorter vectors, and $\circ$ the longer vectors. The one, two, and three lines between the two roots indicate that the angles between the two roots are 120°, 135°, and 150°, respectively.

It is well known that if $\overset{\alpha}{\cdot} - \overset{\beta}{\cdot}$ or $\overset{\alpha}{\circ} - \overset{\beta}{\circ}$, then α, β, $\alpha + \beta \in \Delta^+$.

If $\overset{\alpha}{\cdot} = \overset{\beta}{\circ}$, then α, β, $\alpha + \beta$, $\beta + 2\alpha \in \Delta^+$.

If $\cdot \equiv \circ$, then α, β, $\beta + \alpha$, $\beta + 2\alpha$, $\beta + 3\alpha \in \Delta^+$.

If α, β do not connect directly, then $\alpha + \beta \notin \Delta^+$.

Therefore the necessary and sufficient condition for $\alpha_i + \alpha_j \in \Delta^+$ is that α_i, α_j connect directly in the Dynkin diagram. Since there are no cocycles in the Dynkin diagrams, arbitrarily choosing $\alpha_{i_1}, \ldots, \alpha_{i_s} \in \Pi$, the number of $\alpha_i + \alpha_j \in \Delta^+ \le s - 1$, $1 \le i, j \le l$. In particular, when $s = l$, the number of $\alpha_i + \alpha_j \in \Delta^+ = l - 1$, $1 \le i, j \le l$. Using the notation of Theorem 5.16, $q_2 = l - 1$.

On the other hand, because of (11), $p_k = q_k + q_{s_0-k} - l$. Taking $k = 2$, then

$$p_2 = q_2 + q_{s_0-2} - l$$

$$= l - 1 + q_{s_0-2} - l$$

$$= q_{s_0-2} - 1. \tag{5.16}$$

But from the Poincaré index table (5.14) of simple algebras, one can see that the indices of simple algebras are not equal to 2 except for A_n (i.e., $p_2 = 0$). For A_n, the index table contains 2 one time (i.e., $p_2 = 1$). Therefore from (5.16) we have

$$g \ne A_n, \quad \text{then } q_{s_0-2} = 1,$$

$$g = A_n, \quad \text{then } q_{s_0-2} = 2. \tag{5.17}$$

We now prove the lemma by separating the following two cases.

Case 1. When $g \ne A_n$, we know the number of $\alpha_{i_j} + \alpha_{i_k} \in \Delta^+ \le s - 1$, $(1 \le j, k \le s)$, and if $\psi - \alpha_{i_j} \in \Delta^+$, then $o(\psi - a_{i_j}) = o(\psi) - 1 = s_0 - 2$. Because of (5.17), $q_{s_0-2} = 1$; i.e., the number of $\psi - \alpha_{i_j} \in \Delta^+ \le 1$ for $1 \le j \le s$. Therefore number of $\alpha_{i_j} + \alpha_{i_k} \in \Delta^+$ $(1 \le j, k \le s)$ + the number of $\psi - \alpha_{i_j}$ $(1 \le j \le s) \in \Delta^+ \le s$.

Case 2. When $g = A_n$, by (5.17), there are possibly two of $\{\alpha_i\}$ such that $\psi - \alpha_i \in \Delta^+$; i.e., $\psi - \alpha_1 = \alpha_2 + \cdots + \alpha_n$ and $\psi - \alpha_n = \alpha_1 + \cdots + \alpha_{n-1}$. So if $\{\alpha_{i_1}, \ldots, \alpha_{i_s}\}$ does not contain α_1 and α_n, then the number of $\psi - \alpha_{i_j} \in \Delta^+ \le 1$ for $1 \le j \le s$. Then the lemma would be proved.

If $\{\alpha_{i_1}, \ldots, \alpha_{i_s}\}$ contains α_1 and α_n, then since $s < l$, there exists j, $1 < j < l$, such that $\alpha_j \notin \{\alpha_{i_1}, \ldots, \alpha_{i_s}\}$, and $\alpha_{i_1}, \ldots, \alpha_{i_s}$ belong to one of the following two cases:

$$\overset{\alpha_1}{\bullet}\!\!-\!\!\overset{\alpha_2}{\bullet}\!\!-\!\!\cdots\!\!-\!\!\overset{\alpha_{j-1}}{\bullet} \quad \text{and} \quad \overset{\alpha_{j+1}}{\bullet}\!\!-\!\!\overset{\alpha_{j+2}}{\bullet}\!\!-\!\!\cdots\!\!-\!\!\overset{\alpha_n}{\bullet}.$$

In this case, the number of $\alpha_{i_j} + \alpha_{i_k} \in \Delta^+ \leq s - 2$, $(1 \leq j, k \leq s)$, and the number of $\psi - \alpha_{i_j} \in \Delta^+ = 2$ for $1 \leq j \leq k$. This completes the proof of the lemma. $\qquad\square$

THEOREM 5.18. *If 1 is not an eigenvalue of $\sigma \in W$, which contains at most l permutations, then we have the following:*

1. *The lengths of the permutations are the same and equal to $s_0 = 1 + o(\psi)$.*
2. *σ is a Coxeter element.*

PROOF. Since 1 is not an eigenvalue of σ, by Corollary 5.7 the number L of the permutations contained in $\sigma \geq l$, and hence $L = l$. Extending σ to an automorphism A_σ of g, by Corollary 5.7 one can represent A_σ with respect to a basis of g as

$$
\begin{pmatrix}
e^{2\pi i \rho_1} & & & & & & \\
& \ddots & & & & & \\
& & e^{2\pi i \rho_l} & & & & \\
& & & \begin{pmatrix} 1 & & & \\ & e^{2\pi i/n_1} & & \\ & & \ddots & \\ & & & e^{(2\pi i/n_1)(n_1-1)} \end{pmatrix} & & & \\
& & & & \ddots & & \\
& & & & & \begin{pmatrix} 1 & & & \\ & e^{2\pi i/n_l} & & \\ & & \ddots & \\ & & & e^{2\pi i(n_l-1)/n_l} \end{pmatrix}
\end{pmatrix}
$$

$$(5.18)$$

where $n_1, \ldots, n_l$ are the lengths of the permutations of σ and $e^{2\pi i \rho_1}, \ldots, e^{2\pi i \rho_l}$ are the eigenvalues of σ. By Lemma 5.9, we have

$$\min_i \frac{1}{n_i} \leq \rho_j \leq \max_i \frac{n_i - 1}{n_i}, \qquad j = 1, \ldots, l. \qquad (5.19)$$

On the other hand, by (5.5) of Lemma 5.11, the eigenvalues of A_σ are

$$\{1, 1, \ldots, 1, e^{2\pi i(x, \varphi)}, \varphi \in \Delta, x \in T \text{ (elementary simplex of } g)\}. \tag{5.20}$$

We can write $x = \sum x_i \varepsilon_i \in T$, where (ε_i) is the dual basis of $\{\alpha_i\}$. There are only l eigenvalues among all the eigenvalues of A_σ that are equal to 1, so that $x_i = (x, \delta_i) > 0$ and for the highest root ψ we must have $(x, \psi) < 1$. The proof can be completed by the following steps.

Step 1. There is no harm in assuming

$$A = n_1 = n_2 = \cdots = n_\gamma > n_{r+1} > \cdots > n_l > 1. \tag{5.21}$$

We write the exponents of the entries of (5.18) (ignoring the coefficients $2\pi i$ and eigenvalue 1) as follows:

$$\rho_1, \ldots, \rho_l, \frac{1}{A}, \ldots, \frac{A-1}{A}; \frac{1}{A}, \ldots, \frac{A-1}{A}, \ldots, \frac{1}{A}, \ldots, \frac{A-1}{A},$$

$$r$$

$$\frac{1}{n_{r+1}}, \ldots, \frac{n_{r+1}}{n_{r+1}}; \ldots, \frac{1}{n_l}, \ldots, \frac{n_{l-1}}{n_l}. \tag{5.22}$$

where by (5.19) and (5.21), the smallest of them is $1/A$ and it appears at least r times; the largest of them is $(A-1)/A$ and it also appears r times. By (5.20), for the highest root ψ we have

$$(x, \psi) = \frac{A-1}{A},$$

so $(x, -\psi) = (1 - A)/A$ and $e^{2\pi i(x, -\psi)} = e^{2\pi i/A}$.

For another positive root β, $\beta < \psi$, if $(x, \beta) \neq 1/A$, then

$$\frac{1}{A} < (x, \beta) < \frac{A-1}{A},$$

so $e^{2\pi i(x, -\beta)} \neq e^{2\pi i/A}$; if $(x, \beta) = 1/A$, then $e^{2\pi i(x, -\beta)} \neq e^{2\pi i/A}$. Thus, besides $-\psi$ there are at least $r - 1$ positive roots $\beta_1, \ldots, \beta_{r-1}$ such that

$$(x, \beta_i) = \frac{1}{A} \qquad (i = 1, 2, \ldots, r - 1),$$

and these $r - 1$ roots are simple, since for any nonsimple positive root $\alpha = \sum k_i \alpha_i$, $\sum k_i > 1$. In view of $(x, \alpha_i) \geq 1/A_1$ we have

$$(x, \alpha) = \sum k_i(x, \alpha_i) > \frac{1}{A}.$$

Then we can assume $\beta_i = \alpha_i$ $(i = 1, \ldots, r - 1)$ as well, so x can be written as

$$x = \frac{1}{A}\,\varepsilon_1 + \cdots + \frac{1}{A}\,\varepsilon_{r-1} + x_r\varepsilon_r + \cdots + x_l\varepsilon_l \tag{5.23}$$

and $x_i > 1/A$ $(i = r, r + 1, \ldots, l)$. Without loss of generality we may assume $x_r \leq x_{r+1} \leq \cdots \leq x_l$.

Step 2. Furthermore, we can prove $x_r = 1/A$. If the claim is false, we may assume $x_r > 1/A$. By (5.22), there are at least r eigenvalues of A_σ that equal $e^{(2\pi i/A)\cdot 2}$, but from (5.20) and (5.23) we have

$$x = \varepsilon_1/A + \cdots + \varepsilon_{r-1}/A + x_r\varepsilon_r + \cdots + x_l\varepsilon_l.$$

For each $\alpha = \alpha_i + \alpha_j \in \Delta^+$, $i, j \leq r - 1$, we have

$$(x, \alpha) = \frac{1}{A} + \frac{1}{A} = \frac{2}{A},$$

and if $\alpha \neq \alpha_i + \alpha_j (i, j \leq r - 1)$, then by the assumption of $x_l \geq \cdots \geq x_r > 1/A$, we have $(x, \alpha) > 1/A + 1/A = 2/A$.

On the other hand, it is easy to prove that for each $-\beta \in \Delta^-$, $e^{2\pi i(x, -\beta)} = e^{(2\pi i)2/A}$ if and only if $(x, \beta) = (A - 2)/A$. Next we prove that the positive roots β satisfying $(x, \beta) = (A - 2)/A$ must have the form of $\psi - \alpha_i$ $(i = 1, \ldots, r - 1)$. This is because if $\alpha \in \Delta^+$, $\alpha \neq \alpha_i$ $(i = 1, \ldots, r - 1)$, then by the assumption of $x_l \geq \cdots \geq x_r \geq 1/A$, we know $(x, \alpha) > 1/A$, so $(x, \beta + \alpha) > (A - 2)/A + 1/A = (A - 1)/A$ and $\beta + \alpha \notin \Delta^+$. Meanwhile, since β is not the highest root, there exists a positive root α such that $\beta + \alpha \in \Delta^+$. Therefore, we have $\alpha = \alpha_i$ for some i. If $\beta + \alpha_i \in \Delta^+$, then $(x, \beta + \alpha_i) = (A - 2)/A + 1/A = (A - 1)/A = (x, \psi)$. As ψ is the highest root, $\beta + \alpha_i = \psi$, $\beta = \psi - \alpha_i$. Then it follows that if $e^{2\pi i(x,\beta)} = e^{2\pi i(2/A)}$, then $\beta - \alpha_i + \alpha_j \in \Delta^+$ $(i, j \leq r - 1)$ or $\beta = \alpha_i - \psi \in \Delta^-$ $(i = 1, \ldots, r - 1)$. By Lemma 5.17, the number of β satisfying the above equalities $\leq r - 1$.

This agrees with the fact that there are at least r eigenvalues of A_σ equal to $e^{2\pi i(2/A)}$, so $x_r = 1/A$. Furthermore, x can be written as

$$x = \frac{1}{A}\,\varepsilon_1 + \cdots + \frac{1}{A}\,\varepsilon_r + x_{r+1}\varepsilon_{r+1} + \cdots + x_l\varepsilon_l \tag{5.24}$$

Step 3. By (5.24) the number of roots equal to $e^{2\pi i/A}$ is at least $r + 1$. From (5.22), under the assumption $A = n_1 = \cdots = n_r > n_{r+1} \geq n_{r+2} \geq \cdots \geq n_l$, every group of $(1/A, \ldots, (A-1)/A)$ only contains one $1/A$, and there are r groups. Then there is at least one eigenvalue of σ equal to $e^{2\pi i/A}$.

Step 4. Use the Molien equality of Weyl groups (see Ref. 2)

$$g \prod_{i=1}^{l} (1 - t^{k_i})^{-1} = \sum_{\sigma \in W} \prod_{i=1}^{l} (1 - w_i^\sigma t)^{-1}, \tag{5.25}$$

where g is the order of the Weyl group W, $k_i = m_i + 1$, m_i are the Poincaré indices of g, and w_i^σ $(i = 1, \ldots, l)$ are the eigenvalues of $\sigma \in W$.

By comparing the poles on the two sides of (5.25), we find that for any $\sigma \in W$, its eigenvalues can be written in the form $e^{2\pi i(q/p)}$ and $(p, q) = 1$. Then $p | k_i$, i.e., $p | m_i + 1$ for some i. And so if $e^{2\pi i/A}$ is a root of σ, then $A | m_i + 1$. Therefore for some i, we have

$$A \leq m_l + 1 = s_0.$$

In this case (5.21) becomes

$$s_0 \geq n_1 = n_2 = \cdots = n_r > n_{r+1} \geq \cdots \geq n_l. \tag{5.26}$$

But since $\sum_{i=1}^{l} n_i = 2t$, where t is the number of all positive roots, and by Theorem 5.14, $s_0 l = 2t$, we have $\sum_{i=1}^{l} n_i = s_0 l$ and from (5.26) we obtain the only possible case:

$$n_1 = n_2 = \cdots = n_l = s_0,$$

and the first part of the theorem has been proved.

Step 5. By the above proof we find without difficulty that $x = \sum \varepsilon_i / s_0$ (i.e., A_σ conjugates to $e^{2\pi i(\sum \varepsilon_j / s_0)}$), and by Corollary 5.15 the extension A_r of the Coxeter element $\gamma = R_{\alpha_1} \cdots R_{\alpha_l}$ also conjugates to $e^{2\pi i\,\mathrm{ad}(\sum \varepsilon_j / s_0)}$. So there exists $B \in A(g)$ such that

$$A_\sigma = BA_r B^{-1}. \tag{5.27}$$

By Lemma 5.13(3), γ has a regular eigenvector z_0 whose eigenvalue is $e^{2\pi i/s_0}$. Suppose that $\alpha_1, \ldots, \alpha_l$ is the basis of $\mathfrak{h}$. If $z_0 = \sum_{i=1}^{l} c_i \alpha_i$, then define $\bar{z}_0 = \sum_{i=1}^{l} \bar{c}_i \alpha_i$. Since $\gamma \in W$, all elements of its representation matrix are rational numbers. So, if $rz_0 = e^{2\pi i/s_0} z_0$, we have

$$r\bar{z}_0 = e^{-2\pi i/s_0} z_0 = e^{2\pi i(s_0-1)/s_0}\bar{z}_0. \tag{5.28}$$

In other words, $\bar{z}_0$ is an eigenvector corresponding to $e^{2\pi i(s_0-1)/s_0}$. As z_0 is regular (i.e., for each $\alpha \in \Delta$, we have $(z_0, \alpha) \neq 0$, hence $(\bar{z}_0, \alpha) = (\bar{z}_0, \bar{\alpha}) = \overline{(z_0, \alpha)} \neq 0$), $\bar{z}_0$ is also regular.

Choose $y \in \mathfrak{h}$ such that $\nabla^{-1}y = \lambda y$, $\lambda \neq e^{2\pi i/s_0}$. By (5.27), we have

$$[A_\sigma Bz_0, y] = [BA_r z_0, y] = e^{2\pi i/s_0}[Bz_0, y],$$

and

$$[A_\sigma Bz_0, y] = [Bz_0, A_\sigma^{-1}y] = [Bz_0, \sigma^{-1}y] = [Bz_0, \lambda y].$$

Therefore,

$$[Bz_0, (\lambda - e^{2\pi i/s_0})y] = 0, \qquad (\lambda - e^{2\pi i/s_0})[z_0, B^{-1}y] = 0.$$

Since $\lambda \neq e^{2\pi i/s_0}$ and z_0 is regular, $B^{-1}y \in \mathfrak{h}$. In other words, if the element y of $\mathfrak{h}$ is not the corresponding eigenvector of $e^{2\pi i/s_0}$, we have $B^{-1}y \in \mathfrak{h}$. As in the above proof, replacing z_0 by $\bar{z}_0$, one can see that if y is not an eigenvector corresponding to $e^{-2\pi i/s_0}$, then $B^{-1}y \in \mathfrak{h}$. Therefore for any $y \in \mathfrak{h}$, we always have $B^{-1}y \in \mathfrak{h}$. But B is an automorphism, so $B^{-1}\mathfrak{h} = \mathfrak{h}$; i.e., $B\mathfrak{h} = \mathfrak{h}$. It is well known [6] that every inner automorphism that keeps the Cartan subalgebra invariant restricting to $\mathfrak{h}$ is an element of the Weyl group; i.e., there is $\tau \in W$ such that $B|_\mathfrak{h} = \tau$. So restricting (5.27) to $\mathfrak{h}$ yields

$$\sigma = \tau\gamma\tau^{-1}, \qquad \tau \in W; \tag{5.29}$$

i.e., σ is a Coxeter element and the theorem is completely proved. $\square$

References

1. C. Chevalley, The Betti numbers of the exceptional simple Lie groups, in *Proc. Int. Congress of Mathematicians*, II, pp. 21–24 (1951).
2. A. J. Coleman, The Betti numbers of the simpler Lie groups, *Canad. J. Math.* **10**, (1958).
3. H. S. M. Coxeter, The product of the generators of a finite group generated by reflections, *Duke. Math. J.* **18**, (1951).
4. B. Kostant, *Am. J. Math.* **81**, 973–1032 (1959).
5. Л. И. Мальцев, О лолулростых лодгруллах груллЛИ. АН СССР. **8**, (1944).
6. Wan Ze-xian, *Lie algebras*, Science Press, Beijing (1964) (in Chinese).
7. H. Weyl, *The Classical Groups*, Princeton University Press, Princeton (1946).

6

Dimensions of the Rings of Invariant Differential Operators on Bounded Homogeneous Domains

Let D be a bounded homogeneous domain in $\mathbb{C}^n$, and suppose $G(D)$ is the identity component of the analytic automorphism group of D. By a famous theorem of H. Cartan, $G(D)$ is a real Lie group. We might as well assume that $0 \in D$. From a well-known result in function theory of several complex variables the isotropy group $G_0(D)$ of $G(D)$ at $0 \in D$ has the following representation:

$$\omega = zA + o(z), \tag{6.1}$$

where $z = (z_1, \ldots, z_r)$, $\omega = (\omega_1, \ldots, \omega_n)$, and A is an $n \times n$ complex matrix constituting the linear parts $\{A\}$ of $G_0(D)$, which form a compact subgroup of the unitary group $U(n)$ denoted by K.

Let

$$T = \sum a_{i_1 \cdots i_n j_1 \cdots j_n}(z) \frac{\partial^{i_1 + \cdots + i_n + j_1 + \cdots + j_n}}{\partial z_1^{i_1} \cdots \partial z_n^{i_n} \partial \bar{z}_1^{j_1} \cdots \partial \bar{z}_n^{j_n}} \tag{6.2}$$

$$= \sum a_{(i,j)}(z) \frac{\partial^{(i+j)}}{\partial z_1^{i_1} \cdots \partial \bar{z}_n^{j_n}}$$

be a finite-order differential operator on D, $i = (i_1, \ldots, i_n)$. T is said to be invariant if any differentiable function $f(z)$ on D and $g \in G(D)$ satisfy

$$(Tf(z))_{zg} = (Tf(zg))_z, \qquad (6.3)$$

and the ring generated by all invariant differential operators on D is written $I(D)$.

Let $T \in I(D)$. As shown by (6.2), for any $z \in D$ and from the transitivity of D, there exists a $g \in G(D)$ such that $g: 0 \to z$. Then by (6.3),

$$(Tf(z))_z = (Tf(zg))_0.$$

As a consequence, T is completely determined by the group $G(D)$ and by its value at zero,

$$\sum a_{(i,j)}(0) \frac{\partial^{(i+j)}}{\partial z_1^{i_1} \cdots \partial \bar{z}_n^{j_n}}.$$

A differential operator is said to be of order (k, l) if its highest orders for $\partial/\partial z$ and $\partial/\partial \bar{z}$ are k and l, respectively; i.e., it can be expressed as

$$T = \sum_{\substack{|i|=k \\ |j|=l}} a_{(i,j)}(z) \frac{\partial^{(i+j)}}{\partial z_1^{i_1} \cdots \partial \bar{z}_n^{j_n}} + \text{terms with lower order}; \qquad (6.4)$$

here $|i| = i_1 + \cdots + i_n$ and $|j| = j_1 + \cdots + j_n$. We use $I_{k,l}(D)$ to denote the linear space generated by all invariant differential operators of order (k, l).

DEFINITION 6.1. A polynomial $p(z, \bar{z})$, where $z = (z_1, \ldots, z_n)$, $\bar{z} = (\bar{z}_1, \ldots, \bar{z}_n)$, is said to be homogeneous of order (k, l) if it is a homogeneous polynomial of Z of degree k and of $\bar{Z}$ of degree l. $p(z, \bar{z})$ is said to be K-invariant if, for any $A \in K$, $p(zA, \overline{zA}) = p(z, \bar{z})$. $P_{k,l}$ is used to denote the space generated by all K-invariant homogeneous polynomials of order (k, l).

Concerning the ring $I(D)$ of invariant differential operators, Selberg [5] proved that it has a finite basis. The outline of his proof is as follows. Choose arbitrarily

$$T = \sum_{\substack{|i| \le k \\ |j| \le l}} a_{(i,j)}(z) \frac{\partial^{(i+j)}}{\partial z_1^{i_1} \cdots \partial \bar{z}_n^{j_n}} \in I_{k,l}(D)$$

and fix its value at zero, expressed as

$$T_0 = \sum_{\substack{|i| \le k \\ |j| \le l}} a_{(i,j)}(0) \frac{\partial^{(i+j)}}{\partial z_1^{i_1} \cdots \partial \bar{z}_n^{j_n}}$$

and fix further the highest term in T_0 to be

$$\hat{T}_0 = \sum_{\substack{|i| = k \\ |j| = l}} a_{(i,j)}(0) \frac{\partial^{(i+j)}}{\partial z_1^{i_1} \cdots \partial \bar{z}_n^{j_n}}.$$

Replacing $\partial/\partial z_i$ by z_i and $\partial/\partial \bar{z}_j$ by $\bar{z}_j$ (throughout we denote this process by $\partial/\partial z \to z$, and vice versa), we obtain a corresponding homogeneous polynomial of order (k, l) written as

$$t(z, \bar{z}) = \sum_{\substack{|i| = k \\ |j| = l}} a_{(i,j)}(0) z_1^{i_1} \cdots z_n^{i_n} \bar{z}_1^{j_1} \cdots \bar{z}_n^{j_n}.$$

It is easy to show that $t(z, \bar{z}) \in P_{k,l}$. The above process can be illustrated as

$$T \in I_{k,l}(D) \xrightarrow{\text{evaluated at 0}} T_0 \xrightarrow{\text{choose the highest term}} \hat{T}_0 \xrightarrow{\partial/\partial z \to z} t(z, \bar{z}) \in P_{k,l}.$$

$$(6.5)$$

Conversely, for any polynomial $p \in P_{k,l}$, using the transformations $z_i \to \partial/\partial z_i$ and $\bar{z}_j \to \partial/\partial \bar{z}_j$, we get a homogeneous differential operator $p(\partial/\partial z, \partial/\partial \bar{z})$ that is K invariant. Selberg [5] has verified that any differential operator of constant coefficients can go through the process of "averaging" to obtain an element of the ring $I(D)$,

$$p \in P_{k,l} \xrightarrow{z \to \partial/\partial z} p\left(\frac{\partial}{\partial z}, \frac{\partial}{\partial \bar{z}}\right) \xrightarrow{\text{aver.}} \tilde{p} \in I_{k,l}(D), \qquad (6.6)$$

and is subject to the following conditions:

1. The homogeneous part of the highest-order terms of $\tilde{p}$ is p; i.e.,

$$(\hat{\tilde{p}})_0 = p. \qquad (6.7)$$

2. If $p \xrightarrow{\text{aver.}} \tilde{p}$ and $q \xrightarrow{\text{aver.}} \tilde{q}$, then the homogeneous part of the highest-order terms of $\tilde{p} \cdot \tilde{q}$ at 0 is $p \cdot q$; i.e.,

$$(\widehat{\tilde{p} \cdot \tilde{q}})_0 = p \cdot q \text{ (this process will be denoted by h-t at 0).} \quad (6.8)$$

Thus, starting from any $T \in I_{k,l}(D)$, with the aid of (6.5) and (6.6), we find

$$I_{k,l}(D) \ni T \xrightarrow{\text{h-t at 0}} \hat{T}_0 \xrightarrow{\text{aver.}} (\tilde{\hat{T}}_0) \in I_{k,l}(D).$$

It follows from (6.7) that the highest-order terms of T and $(\hat{T}_0)$ turn out to be $\hat{T}_0$. We may write

$$T - (\tilde{\hat{T}}_0) = \text{invariant differential operator with lower order.} \quad (6.9)$$

We can now state the following conclusion: for any element in $I_{k,l}(D)$, its homogeneous parts of highest-order terms at 0 correspond to an element in $P_{k,l}$; each element in $P_{k,l}$ is able to generate an element in $I_{k,l}(D)$, the leading term of which is the very same homogeneous form. Using this fact together with (6.9) we can sum up the foregoing in the following theorem.

THEOREM 6.2.

$$I_{k,l}(D) \Big/ \sum_{(k',l')<(k,l)} I_{k',l'}(D) \cong P_{k,l}. \quad (6.10)$$

Here $(k', l') < (k, l) \Leftrightarrow k - k' \geq 0, l - l' \geq 0, (k - k')^2 + (l - l')^2 \neq 0$.

According to function theory of several complex variables, if the domains D and D' are analytically equivalent, then $G(D) \cong G(D')$; hence, $I_{k,l}(D) \cong I_{k,l}(D')$. From (6.10), dim $P_{k,l}$ is an analytic invariant of D.

DEFINITION 6.3. $T_1, T_2 \in I_{k,l}(D)$ are said to be equivalent if $T_1 - T_2 =$ invariant differential operator with lower order.

From Theorem 6.2, dim $P_{k,l}$ gives the number of linearly independent and inequivalent invariant differential operators of order (k, l) on a complex domain. The main purpose of this chapter is to give a formula for calculating $C_{k,l} = \dim P_{k,l}$.

Let K be the group formed by the linear parts of $G_0(D)$ as given in (6.1), and K is a compact subgroup of $U(n)$. For $z = (z_1, \ldots, z_n)$, write

$$z^{[k]} = \left(\sqrt{\frac{k!}{k_1! \cdots k_n!}} \, z_1^{k_1} \cdots z_n^{k_n} \right), \qquad k = k_1 + k_2 + \cdots + k_n, \quad k_i \geq 0,$$

$z^{[k]}$ being a C^k_{n+k+1}-dimensional vector. For $A \in U(n)$, the transformation $z \to zA$ induces a transformation $z^{[k]} \to z^{[k]}A^{[k]}$, where $A \to A^{[k]}$ represents the irreducible representation of $U(n)$ with signature $(k, 0, \ldots, 0)$ (see Ref. 1). But as soon as this representation is restricted to $K \subseteq U(n)$, in general, it is no longer irreducible on K. Now let $\chi_{(k, 0, \ldots, 0)}(A)$ be the character of the representation $A \to A^{[k]}$ on A. Then the following result is valid.

THEOREM 6.4.

$$C_{k,l} = \int_K \chi_{(k, 0, \ldots, 0)}(A)\bar{\chi}_{(l, 0, \ldots, 0)}(A) \, dA, \qquad (6.11)$$

where $\int_K dA = 1$.

PROOF. $C_{k,l} = \dim P_{k,l}$. Choose arbitrarily $F(z, \bar{z}) \in P_{k,l}$. Then $F(zA, \overline{zA}) = F(z, \bar{z})$, $\forall A \in K$.

First, we express $F(z, \bar{z})$ in the matrix form $F(z, \bar{z}) = z^{[k]}D\bar{z}^{[l]'}$, D referring to a $C^k_{n+k-1} \times C^l_{n+l-1}$ matrix, and prime standing for matrix transpose. From invariance, we must have

$$z^{[k]}A^{[k]}D\bar{A}^{[l]'}\bar{z}^{[l]'} = z^{[k]}D\bar{z}^{[l]'};$$

$$A^{[k]}D = DA^{[l]}. \qquad (6.12)$$

As stated before, $A^{[k]}$ and $A^{[l]}$ may not be irreducible for K. Now we proceed to the decomposition of the irreducible representation of K with $A^{[k]}$ and $A^{[l]}$:

$$A^{[k]} = n_1 D_1 + n_2 D_2 + \cdots + n_R D_r + m_1 E_1 + \cdots + m_s E_s, \qquad (6.13)$$

$$A^{[l]} = \xi_1 D_1 + \xi_2 D_2 + \cdots + \xi_r D_r + \zeta_1 E'_1 + \cdots + \zeta_t E'_t, \qquad (6.14)$$

where $+$ stands for direct sum, D_i, $i = 1, \ldots, r$, signify the terms of irreducible representations that appear commonly in decompositions, E_i and E'_i are the terms of irreducible representations that appear individually in decompositions and are equivalent neither to each other nor to D_i, n_i, m_i, ξ_i, ζ_i denote the multiplicities taken place by each irreducible representation. Since the calculations of (6.13) and (6.14) run along the same line to general decomposition, for brevity we may fix $r = 1$, $s = 1$, $t = 1$, namely,

$A^{[k]} = n_1 D_1 + m_1 E_1$, $A^{[l]} = \xi_1 D_1 + \zeta_1 E_1'$. Then there exists a $C_{n+k-1}^k \times C_{n+k-1}^k$ nonsingular matrix P and a $C_{n+l-1}^l \times C_{n+l-1}^l$ nonsingular matrix Q such that

$$A^{[k]} = P \left(\begin{pmatrix} D_1 & & \\ & \ddots & \\ & & D_1 \end{pmatrix}_{n_1} \quad \begin{pmatrix} E_1 & & \\ & \ddots & \\ & & E_1 \end{pmatrix}_{m_1} \right) P^{-1}$$

$$A^{[l]} = Q \left(\begin{pmatrix} D_1 & & \\ & \ddots & \\ & & D_1 \end{pmatrix}_{\xi_1} \quad \begin{pmatrix} E_1 & & \\ & \ddots & \\ & & E_1 \end{pmatrix}_{\zeta_1} \right) Q^{-1}$$

From (6.12), it follows that

$$\left(\begin{pmatrix} D_1 & & \\ & \ddots & \\ & & D_1 \end{pmatrix}_{n_1} \quad \begin{pmatrix} E_1 & & \\ & \ddots & \\ & & E_1 \end{pmatrix}_{m_1} \right) P^{-1} DQ$$

$$= P^{-1} DQ \left(\begin{pmatrix} D_1 & & \\ & \ddots & \\ & & D_1 \end{pmatrix}_{\xi_1} \quad \begin{pmatrix} E_1' & & \\ & \ddots & \\ & & E_1' \end{pmatrix}_{\zeta_1} \right)$$

Therefore in the light of the Schur lemma, we arrive at

$$P^{-1} DQ = \left(\begin{pmatrix} A_{11} & \cdots & A_{1\xi_1} \\ \vdots & & \vdots \\ A_{n_1 1} & \cdots & A_{n_1 \xi_1} \end{pmatrix} \quad 0 \\ 0 \qquad\qquad 0 \right)$$

and

$$\begin{pmatrix} D_1 & & \\ & \ddots & \\ & & D_1 \end{pmatrix}_{n_1} \begin{pmatrix} A_{11} & \cdots & A_{1\xi_1} \\ \vdots & & \vdots \\ A_{n_1 1} & \cdots & A_{n_1 \xi_1} \end{pmatrix} = \begin{pmatrix} A_{11} & \cdots & A_{1\xi_1} \\ \vdots & & \vdots \\ A_{n_1 1} & \cdots & A_{n_1 \xi_1} \end{pmatrix} \begin{pmatrix} D_1 & & \\ & \ddots & \\ & & D_1 \end{pmatrix}_{\xi_1},$$

which means $D_1 A_{ij} = A_{ij} D_1$. The same Schur lemma gives rise to $A_{ij} = \lambda_{ij} I$, $\lambda_{ij} \in C$; hence,

$$D = P \begin{pmatrix} \lambda_{11} I & \cdots & \lambda_{1\xi_1} I & \\ \vdots & & \vdots & 0 \\ \lambda_{n_1 1} I & \cdots & \lambda_{n_1 \xi_1} I & \\ & 0 & & 0 \end{pmatrix} Q^{-1}.$$

Thus the degree of freedom of D amounts to $n_1 \cdot \xi_1$.

By the same argument, if the irreducible decompositions of $A^{[k]}$ and $A^{[l]}$ are given as (6.13), (6.14), then the degree of freedom of D turns out to be

$$\dim P_{k,l} = C_{k,l} = \sum_{i=1}^{r} n_i \xi_i.$$

If we consider the set of all irreducible representations of K as a basis and regard the multiplicity taken place in each irreducible representation at $A^{[k]}$ as its components, then (6.13), (6.14) can be rewritten as

$$D_1 D_2 \cdots D_r E_1 \cdots E_s E_1' \cdots E_t'$$

$$A^{[k]} \sim (n_1, n_2, \ldots, n_r, m_1, \ldots, m_s, 0, \ldots, 0),$$

$$A^{[l]} \sim (\xi_1, \xi_2, \ldots, \xi_r, 0, \ldots, 0, \zeta_1, \ldots, \zeta_t).$$

As a result $\sum n_i \xi_i$ is the inner product of these two vectors. Provided we put all irreducible representations of K into a sequence (the irreducible representation of the compact Lie group is of finite dimension, and the set of all irreducible representations is countable) $(D_1, D_2, \ldots, D_i, \ldots)$, then we have $A^{[k]} = \sum n_i D_i$, $A^{[l]} = \sum \xi_i D_i$ (since K is compact, only a finite number of n_i and ξ_i is nonzero) and

$$\chi_{(k, 0, \ldots, 0)}(A) = \sum_{i=1}^{\infty} n_i \chi_i(A),$$

$$\chi_{(l, 0, \ldots, 0)}(A) = \sum_{i=1}^{\infty} \xi_i \chi_i(A).$$

Here $\chi_i(A)$ is the character of the irreducible representation D_i of K on A. By the orthonormality of characters, this becomes

$$C_{k,l} = \sum_{i=1}^{\infty} n_i \xi_i = \int_K \chi_{(k,0,\dots,0)}(A) \bar{\chi}_{(l,0,\dots,0)}(A) \, dA.$$

This completes the proof.   □

COROLLARY 6.5. $C_{k,l} = C_{l,k}$.

PROOF. From (6.11), the proof is trivial.   □

In some concrete cases, such as the classical symmetric domains and Siegel domains, D is always given by matrices and the linear subgroup K of the isotropy group $G_0(D)$ is usually of the following forms:

$$Z \to AZB, \tag{6.15}$$

$$Z \to AZA', \tag{6.16}$$

where Z indicates an $m \times n$ matrix in (6.15) and Z denotes a symmetric matrix in (6.16). For these cases, we propose the next theorem.

THEOREM 6.6. *If the linear subgroup K of the isotropy group $G_0(D)$ is given by $Z \to AZB$, where Z is an $m \times n$ ($m \le n$) matrix, $\{A\}$ forms a compact subgroup Γ_1 of $U(m)$, and $\{B\}$ forms a compact subgroup Γ_2 of $U(n)$. Let $a_{(f_1,\dots,f_m)}(\Gamma_1)$ be the number of inequivalent irreducible representations of Γ_1 in the decomposition of $\pi(f_1, f_2, \dots, f_m)|_{\Gamma_1}$, where $\pi(f_1, \dots, f_m)$ is the irreducible representation of $U(m)$ with signature $(f_1, \dots, f_m)$, $f_1 \ge f_2 \ge \cdots \ge f_m$, and let $a_{(f_1,\dots,f_n)}(\Gamma_2)$ be similarly defined. Then we find*

$$C_{k,k} = \sum_{\substack{f_1 \ge \cdots \ge f_m \\ \sum f_i = k}} a_{(f_1 \cdots f_m)}(\Gamma_1) a_{(f_1,\dots,f_m,0,\dots,0)}(\Gamma_2). \tag{6.17}$$

PROOF. Arrange the elements of Z as

$$z = (z_{11}, \dots, z_{1n}, z_{21}, \dots, z_{2n}, \dots, z_{m1}, \dots, z_{mn}).$$

Then $Z \to AZB$ can be written as $z \to z(A \times B)$, where $A \times B$ refers to the Kronecker product of A and B. As in Theorem 6.4, let $F(z, \bar{z}) \in P_{k,l}$ be given in matrix forms [i.e., $F(z, \bar{z}) = Z^{[k]} D \bar{z}^{[k]'}$] with D a $C_{mn+k-1}^k \times C_{mn+k-1}^k$ matrix. By invariance, we then get

$$(A \times B)^{[k]} D (\overline{A \times B})^{[k]'} = D, \quad \text{for arbitrary } A \in \Gamma_1, B \in \Gamma_2.$$

Since $\Gamma_1 \subset U(m)$, $\Gamma_2 \subset U(n)$, we have $(A \times B)^{[k]} \in U(C^k_{mn+k-1})$ and

$$(A \times B)^{[k]}D = D(A \times B)^{[k]}. \tag{6.18}$$

In general, as a representation of the mn-dimensional unitary subgroup $\Gamma_1 \times \Gamma_2$, the $A \times B \to (A \times B)^{[k]}$ may not be irreducible. Suppose $n_1 D_1 + \cdots + n_r D_r$ is its irreducible decomposition, with D_i $(i = 1, \ldots, r)$ the irreducible representation of $\Gamma_1 \times \Gamma_2$. By the Schur lemma, there exists a matrix P with det $P \neq 0$ such that $D = P(n_1\lambda_1 I + \cdots + n_r\lambda_r I)P^{-1}$, $\lambda_i \in C$; therefore dim $P_{k,k} = r$. The completion of the proof therefore depends on the calculation of r.

Applying a formula of Hua [1, p. 31, (1.4.3)], we obtain

$$\text{tr}\,[(X \times Y)^{[k]}] = \sum_{\substack{f_1+\cdots+f_m=k \\ f_1 \geq \cdots \geq f_m \geq 0}} \chi_{(f_1,\ldots,f_m)}(X)\chi_{(f_1,\ldots,f_m,0,\ldots,0)}(Y). \tag{6.19}$$

Here $\chi_{(f_1,\ldots,f_m)}(X)$ is the character of the irreducible representation of GL $(m, \mathbb{C})$ with signature $(f_1,\ldots,f_m)$ on X. We know that the representation $X \times Y \to (X \times Y)^{[k]}$ of GL $(m, \mathbb{C}) \times$ GL $(n, \mathbb{C})$ can be decomposed as a direct sum of the tensor product of $\pi(f_1,\ldots,f_m)$ and $\pi(f_1,\ldots,f_m,0,\ldots,0)$, in which each tensor product has multiplicity 1, and $\pi(f_1,\ldots,f_m)$ is used to denote the irreducible representation of GL $(m, \mathbb{C})$ with signature $(f_1,\ldots,f_m)$.

Let $A = X$, $B = Y$. However, in general, for $A \in \Gamma_1$, the representation $\pi(f_1,\ldots,f_m)(A)|_{\Gamma_1}$ is still not irreducible and can be again decomposed into a direct sum of some irreducible representations of Γ_1. Suppose $a_{(f_1,\ldots,f_m)}(\Gamma_1)$ is the number of inequivalent irreducible representations (multiplicity is not counted) in this decomposition. A similar reasoning applies to $a_{(f_1,\ldots,f_m,0,\ldots,0)}(\Gamma_2)$. Then in the product $\pi(f_1,\ldots,f_m)(A)|_{\Gamma_1} \cdot \pi(f_1,\ldots,f_m,0,\ldots,0)(B)|_{\Gamma_2}$, the total number of inequivalent representations is $a_{(f_1,\ldots,f_m)}(\Gamma_1)a_{(f_1,\ldots,f_m,0,\ldots,0)}(\Gamma_2)$. Thus (6.19) implies (6.17). $\square$

THEOREM 6.7. *Suppose the linear parts of $G_0(D)$ are given as $Z \to AZA'$, where Z is an $n \times n$ symmetric matrix, and $\{A\}$ forms a compact subgroup Γ of $U(n)$. Provided that $\pi(2f_1,\ldots,2f_n)$ signifies the irreducible representation of $U(n)$ with signature $(2f_1,\ldots,2f_n)$, then $C_{k,k}$ is the number of inequivalent irreducible representations (multiplicity is not counted) of Γ in the decompositions of $\pi(2f_1,\ldots,2f_n)|_\Gamma$. For all $f_1 \geq \cdots \geq f_n$, the relation $\sum_{i=1}^{n} f_i = k$ holds.*

PROOF. Arrange the elements of Z as

$$z = (z_{11},\ldots,z_{1n},z_{21},\ldots,z_{2n},\ldots,z_{n1},\ldots,z_{nn}).$$

Then $Z \to AZA'$ can be written as $z \to z(A \times A)_s$, where $(A \times A)_s$ is the symmetric Kronecker product (see Ref. 3). Applying the preceding method for $F(z, \bar{z}) \in P_{k,k}$, we write $F(z, \bar{z}) = z^{[k]} D \bar{z}^{[k]'}$, where D is a $C^k_{(n(n+1)/2)+k-1} \times C^k_{(n(n+1)/2)+k-1}$ matrix. By invariance, we then get

$$(A \times A)^{[k]}_s D = D(A \times A)^{[k]}_s.$$

In general, as the representation of the unitary subgroup Γ, $A \to (A \times A)^{[k]}_s$ may be reducible. Assume that $n_1 D_1 + \cdots + n_r D_r$ stands for its irreducible decomposition, where D_i are the irreducible representations of Γ. Then on the basis of Schur's lemma, there exists a matrix P with $\det(P) \neq 0$, such that $D = P(n_1 \lambda_1 I + \cdots + n_r \lambda_r I) P^{-1}$, $\lambda_i \in \mathbb{C}$. Hence $C_{k,k} = r$ is the number of all inequivalent irreducible representations in the decomposition of the representation $\Gamma \ni A \to (A \times A)^{[k]}_s$.

Recall that $A \to (A \times A)_s$ is the irreducible representation of $GL(n, \mathbb{C})$ with signature $(2, 0, \ldots, 0)$—that is, $(A \times A)_s = A^{[2]}$ with the preceding notation. Thus, $A \to (A \times A)^{[k]}_s = (A^{[2]})^{[k]}$.

By a formula of Hua [1, p. 32, (1.4.6)]

$$\text{tr}\,[(A^{[2]})^{[k]}] = \sum_{\substack{f_1+\cdots+f_n=k \\ f_1 \geq \cdots \geq f_n \geq 0}} \chi_{(2f_1,\ldots,2f_n)}(A), \qquad (6.20)$$

we know that, as a representation of $U(n)$, $(A^{[2]})^{[k]}$ can be decomposed as a direct sum of $\pi(2f_1, \ldots, 2f_n)$ in which each $\pi(2f_1, \ldots, 2f_n)$ has multiplicity 1. Now since $A \in \Gamma$, when $\pi(2f_1, \ldots, 2f_n)$ are restricted to the subgroup Γ, it can also be decomposed into a direct sum of inequivalent irreducible representations of Γ. Notice that for different $(2f_1, \ldots, 2f_n)$, $\pi(2f_1, \ldots, 2f_n)|_\Gamma$ may have the same irreducible components; therefore, for all $f_1 \geq \cdots \geq f_n \geq 0$, there exists $\sum f_i = k$. The number of all inequivalent irreducible representations in the decompositions of $\pi(2f_1, \ldots, 2f_n)|_\Gamma$ for all $(2f_1, \ldots, 2f_n)$ is r.   □

THEOREM 6.7'. *Suppose the linear part K of $G_0(D)$ is given as $Z \to AZA'$, where Z is an $n \times n$ skew-symmetric matrix and $\{A\}$ forms a compact subgroup Γ of $U(n)$. Let $\pi(f_1, f_1, f_2, f_2, \ldots, f_\nu)$ be the irreducible representation of $U(n)$ with signature $(f_1, f_1, f_2, f_2, \ldots, f_\nu, f_\nu)$, $\nu = [n/2]$. Then $C_{k,k}$ is the number of all inequivalent irreducible representations (multiplicity is not counted) of $\pi(f_1, f_1, f_2, f_2, \ldots, f_\nu, f_\nu)$. For all $f_1 \geq f_2 \geq \cdots \geq f_\nu \geq 0$, there exist $\sum f_i = k$, $\nu = [n/2]$.*

PROOF. The proof is similar to that of Theorem 6.7, with (6.20) replaced by [1, p. 31, Theorem 1.4.3]

$$\text{tr}\,[(A^{[2]})^{[k]}] = \sum_{\substack{\sum f_i = k \\ f_1 \geq \cdots \geq f_\nu \geq 0}} \chi_{(f_1, f_1, f_2, f_2, \ldots, f_\nu, f_\nu)}(A). \qquad \square \quad (6.21)$$

We now study two examples by applying the result of the preceding section. The second example concerns nonsymmetric domains. As far as the author knows there are only a few works about invariant differential operators on a nonsymmetric domain. Calculation shows that in comparison with symmetric domains there are many more invariant differential operators available on the nonsymmetric domains. ☐

EXAMPLE 6.8. The four classes of classical domains.
The classical domains and their isotropy group are as follows:

Domain	Isotropy group
$R_I(I - Z\bar{Z}' > 0)$	$Z \to UZV,$
$R_{II}(I - Z\bar{Z} > 0, Z = Z')$	$Z \to UZU',$
$R_{III}(I - Z\bar{Z}' > 0, Z = -Z')$	$Z \to UZU',$
$R_{IV}(1 + |zz'|^2 - 2z\bar{z}' > 0, 1 - |zz'|^2 > 0)$	$z \to e^{i\theta}z\Gamma, \Gamma\Gamma' = I.$

where U, V are unitary matrices. For each of R_I, R_{II}, R_{III}, and R_{IV} introduced by Theorems 6.6, 6.7, 6.7', and 6.4, respectively, show that $C_{1,1} = 1$. More precisely, for each of the above classical domains, there exists a unique invariant differential operator of order $(1, 1)$, that is, the Laplace–Beltrami operator:

$$\Delta_I = \text{tr}\left\{[(I - \bar{Z}Z') \times (I - \bar{Z}'Z)]\frac{\partial^2}{\partial z'\partial\bar{z}}\right\},$$

$$\Delta_{II} = \text{tr}\left\{[(I - \bar{Z}Z) \times (I - \bar{Z}Z)]_s\frac{\partial^2}{\partial z'\partial\bar{z}}\right\},$$

$$\Delta_{III} = \text{tr}\left\{[(I + \bar{Z}Z) \times (I + \bar{Z}Z)]_{sk}\frac{\partial^2}{\partial z'\partial\bar{z}}\right\},$$

$$\Delta_{IV} = (1 + |zz'|^2 - 2z\bar{z}')\sum_{\alpha,\beta=1}^{N}(\delta_{\alpha\beta} - zz_\alpha\bar{z}_\beta)\frac{\partial^2}{\partial z_\alpha\,\partial\bar{z}_\beta}$$

$$+ 2\sum_{\alpha,\beta=1}^{N}(\bar{z}_\alpha - \overline{zz'}z_\alpha)(z_\beta - zz'\bar{z}_\beta)\frac{\partial^2}{\partial z_\alpha\,\partial\bar{z}_\beta}, \qquad z = (z_1, \ldots, z_N).$$

EXAMPLE 6.9. Consider the domain

$$D: \frac{1}{2i}(Z - \bar{Z}) > 0, \qquad Z = \begin{pmatrix} z_{11} & z_{12} & z_{13} \\ z_{12} & z_{22} & 0 \\ z_{13} & 0 & z_{33} \end{pmatrix}. \tag{6.22}$$

It is not hard to check that this is a Siegel domain of the first kind, and that it is biholomorphic to a bounded domain. The analytic automorphism group $G(D)$ of D contains the following transitive affine subgroup $A(D)$:

$$A(D): Z \to B(Z - X_0)B', \qquad Z_0 = X_0 + iY_0 \in D, \qquad (6.23)$$

$$B = \begin{pmatrix} b_{11} & b_{12} & b_{13} \\ 0 & b_{22} & 0 \\ 0 & 0 & b_{33} \end{pmatrix} \text{ satisfies } BY_0B' = I.$$

Evidently, we have $Z_0 \to iI$ in (6.23); therefore $A(D)$ acts transitively on D.

Resorting to the theory of analytic automorphism groups of Siegel's domain (see Ref. 4), one can show that the isotropy group $G_{iI}(D)$ of D at iI is given by (6.24), the proof of which is omitted because it is not closely related to our topic.

$$Z \to (A + ZC)^{-1}(-C + ZA), \qquad (6.24)$$

where

$$A = \begin{pmatrix} 1 & & \\ & a_2 & \\ & & a_3 \end{pmatrix}, \quad C = \begin{pmatrix} 0 & & \\ & c_2 & \\ & & c_3 \end{pmatrix}, \quad a_2^2 + c_2^2 = a_3^2 + c_3^2 = 1,$$

and $a_2, c_2, a_3, c_3 \in \mathbb{R}$.

In order to apply Theorem 6.4, we first transform D into a bounded domain. Let

$$\psi: Z \to W = (Z - iI)(Z + iI)^{-1}, \qquad Z \in D, \qquad (6.25)$$

and $D^* = \psi(D)$. Then it is easy to verify that D^* can be given as

$$D^*: I - W\bar{W} > 0, \qquad (6.26)$$

$$W = \begin{pmatrix} w_{11} & w_{12} & w_{13} \\ w_{12} & w_{22} & \varphi(w) \\ w_{13} & \varphi(w) & w_{33} \end{pmatrix}, \qquad \varphi(w) = \frac{w_{12}w_{13}}{w_{11} - 1},$$

and $\psi(iI) = 0 \in D^*$. The diagram

$$
\begin{array}{ccc}
D & \xrightarrow{\ G_{iI}(D)\ } & D \\
\downarrow{\scriptstyle\psi} & & \downarrow{\scriptstyle\psi} \\
D^* & \xrightarrow{\ G_0(D^*)\ } & D^*
\end{array}
\tag{6.27}
$$

shows $G_0(D^*) = \psi_0 G_{iI}(D) \cdot \psi^{-1}$, with $\psi^{-1}: Z = i(I - W)^{-1}(I + W)$. A direct calculation shows that $G_0(D^*)$ is given by

$$W \to \begin{pmatrix} 1 & & \\ & e^{i\theta} & \\ & & e^{i\varphi} \end{pmatrix} W \begin{pmatrix} 1 & & \\ & e^{i\theta} & \\ & & e^{i\varphi} \end{pmatrix}. \tag{6.28}$$

PROPOSITION 6.10. *For the domain D^*, we have*

$$c_{1,1}(D^*) = 5, \qquad c_{2,0}(D^*) = c_{0,2}(D^*) = 1.$$

PROOF. Rewrite (6.28) as

$$
\begin{pmatrix} w_{11} \\ w_{12} \\ w_{13} \\ w_{22} \\ w_{23} \end{pmatrix} \to
\begin{pmatrix} 1 & & & & \\ & e^{i\theta} & & & \\ & & e^{i\varphi} & & \\ & & & e^{2i\theta} & \\ & & & & e^{2i\varphi} \end{pmatrix}
\begin{pmatrix} w_{11} \\ w_{12} \\ w_{13} \\ w_{22} \\ w_{23} \end{pmatrix},
$$

$$
\left\{ \begin{pmatrix} 1 & & & & \\ & e^{i\theta} & & & \\ & & e^{i\varphi} & & \\ & & & e^{2i\theta} & \\ & & & & e^{2i\varphi} \end{pmatrix} \right\} = \Gamma.
\tag{6.29}
$$

From a well-known fact of the theory of group representations, we have

$$\chi_{(1,0,0,0,0)}(\Gamma) = 1 + e^{i\theta} + e^{i\varphi} + e^{2i\theta} + e^{2i\varphi},$$

$$\chi_{(0,0,0,0,0)}(\Gamma) = 1,$$

$$\chi_{(2,0,0,0,0)}(\Gamma) = \sum_{1\le i\le j\le 5} \lambda_i\lambda_j, \qquad \Gamma = \begin{pmatrix} \lambda_1 & & & & \\ & \lambda_2 & & & \\ & & \lambda_3 & & \\ & & & \lambda_4 & \\ & & & & \lambda_5 \end{pmatrix}$$

Hence by (6.11),

$$C_{1,1}(D^*) = \left(\frac{1}{2\pi}\right)^2 \int_{\substack{0\le\theta\le2\pi \\ 0\le\varphi\le2\pi}} (1 + e^{i\theta} + e^{i\varphi} + e^{2i\theta} + e^{2i\varphi})$$

$$\times (1 + e^{i\theta} + e^{i\varphi} + e^{2i\theta} + e^{2i\varphi})\, d\theta\, d\varphi = 5,$$

$$C_{2,0}(D^*) = C_{0,2}(D^*) = \left(\frac{1}{2\pi}\right)^2 \int_{\substack{0\le\theta\le2\pi \\ 0\le\varphi\le2\pi}} \chi_{(2,0,0,0,0)}(\Gamma)\cdot 1\, d\theta\, d\varphi$$

$$= \left(\frac{1}{2\pi}\right)^2 \int_{\substack{0\le\theta\le2\pi \\ 0\le\varphi\le2\pi}} (1 + e^{i\theta} + \cdots)\cdot 1\, d\theta\, d\varphi = 1.$$

This completes the proof.   □

As has been mentioned, $C_{k,l}$ are biholomorphic invariants; therefore we also have $C_{1,1}(D) = 5$, $C_{2,0}(D) = C_{0,2}(D) = 1$.

In the following, we will concretely determine the seven bases of invariant differential operators of order 2 on D. By (6.28) we know that

$$dw_{11}d\bar{w}_{11},\ dw_{12}d\bar{w}_{12},\ dw_{13}d\bar{w}_{13},\ dw_{22}d\bar{w}_{22},\ dw_{33}d\bar{w}_{33}\ \text{(type of (1, 1))},$$

$$(dw_{11})^2\ \text{(type of (2, 0))},\ (d\bar{w}_{11})^2\ \text{(type of (0, 2))} \qquad (6.30)$$

form a basis of invariant differentials on D^* with respect to $G_0(D^*)$.

Now we pass to D. From formula (6.25),

$$dW = dZ(Z + iI)^{-1} - (Z - iI)(Z + iI)^{-1} dZ (Z + iI)^{-1}$$

$$= [I - (Z - iI)(Z + iI)^{-1}] dZ (Z + iI)^{-1}$$

$$= 2i(Z + iI)^{-1} dZ (Z + iI)^{-1}.$$

If $Z = iI$, $W = 0$, then $dW = (1/2i) dZ$, and (6.30) becomes a linear basis of invariant differentials of order 2 with respect to $G_{iI}(D)$:

$$dz_{11}d\bar{z}_{11}, dz_{12}d\bar{z}_{12}, dz_{13}d\bar{z}_{13}, dz_{22}d\bar{z}_{22}, dz_{33}d\bar{z}_{33}, dz_{11}^2, d\bar{z}_{11}^2. \quad (6.31)$$

Their duals that form a basis of differential operators of order 2, which are invariant with respect to $G_{iI}(D)$, correspond to

$$\frac{\partial^2}{\partial z_{11} \partial \bar{z}_{11}}, \frac{\partial^2}{\partial z_{12} \partial \bar{z}_{12}}, \frac{\partial^2}{\partial z_{13} \partial \bar{z}_{13}}, \frac{\partial^2}{\partial z_{22} \partial \bar{z}_{22}}, \frac{\partial^2}{\partial z_{33} \partial \bar{z}_{33}}, \frac{\partial^2}{\partial z_{11}^2}, \frac{\partial^2}{\partial \bar{z}_{11}^2}. \quad (6.32)$$

From this basis, we resort to the transitive affine group $A(D)$ to get a basis of invariant differential operators of order 2 on $G(D)$.

Assume that T is an invariant differential operator on D, and let T_0 be the value of T at $iI \in D$. Then T_0 is a differential operator with constant coefficients and is invariant with respect to $G_{iI}(D)$. Choose $g \in A(D) \subset G(D)$. The invariance requires

$$T \circ g = g \circ T_0.$$

Therefore

$$T = g^{-1} \circ T_0 \circ g. \quad (6.33)$$

Now we may take T_0 to be one of the operators in (6.32), and fix $g \in A(D)$ such that $g(iI) = Z_0 = X_0 + iY_0$:

$$g: Z \to W = BZB' + X_0, \quad BB' = Y, \quad B = \begin{pmatrix} * & * & * \\ 0 & * & 0 \\ 0 & 0 & * \end{pmatrix}. \quad (6.34)$$

Through calculation (details are omitted) from (6.32) to (6.34), we obtain the bases of the following seven invariant differential operators of order 2 with respect to D:

(i) $\quad a_0^2(z) = \dfrac{\partial^2}{\partial z_{11} \, \partial \bar{z}_{11}};$

(ii) $\quad a_0(z)\left[\dfrac{2(z_{12} - \bar{z}_{12})^2}{i(z_{22} - \bar{z}_{22})} \dfrac{\partial^2}{\partial z_{11} \, \partial \bar{z}_{11}} + \dfrac{z_{22} - \bar{z}_{22}}{2i} \dfrac{\partial^2}{\partial z_{12} \, \partial \bar{z}_{12}} \right.$

$$\left. + \dfrac{z_{12} - \bar{z}_{12}}{i}\left(\dfrac{\partial^2}{\partial z_{11} \, \partial \bar{z}_{12}} + \dfrac{\partial^2}{\partial z_{12} \, \partial \bar{z}_{11}} \right) \right];$$

(iii) $\quad a_0(z)\left[\dfrac{2(z_{13} - \bar{z}_{13})^2}{i(z_{33} - \bar{z}_{33})} \dfrac{\partial^2}{\partial z_{11} \, \partial \bar{z}_{11}} + \dfrac{z_{33} - \bar{z}_{33}}{2i} \dfrac{\partial^2}{\partial z_{13} \, \partial \bar{z}_{13}} \right.$

$$\left. + \dfrac{z_{13} - \bar{z}_{13}}{i}\left(\dfrac{\partial^2}{\partial z_{11} \, \partial \bar{z}_{13}} + \dfrac{\partial^2}{\partial z_{13} \, \partial \bar{z}_{11}} \right) \right];$$

(iv) $\quad -\dfrac{(z_{12} - \bar{z}_{12})^4}{4(z_{22} - \bar{z}_{22})^2} \dfrac{\partial^2}{\partial z_{11} \, \partial \bar{z}_{11}} - \dfrac{(z_{12} - \bar{z}_{12})^2}{4} \dfrac{\partial^2}{\partial z_{12} \, \partial \bar{z}_{12}}$

$$-\dfrac{(z_{22} - \bar{z}_{22})^2}{4} \dfrac{\partial^2}{\partial z_{22} \, \partial \bar{z}_{22}} - \dfrac{(z_{12} - \bar{z}_{12})^2}{4(z_{22} - \bar{z}_{22})}\left(\dfrac{\partial^2}{\partial z_{11} \, \partial \bar{z}_{12}} + \dfrac{\partial^2}{\partial \bar{z}_{11} \, \partial z_{12}} \right)$$

$$-\dfrac{(z_{12} - \bar{z}_{12})^2}{4}\left(\dfrac{\partial^2}{\partial z_{11} \, \partial \bar{z}_{22}} + \dfrac{\partial^2}{\partial \bar{z}_{11} \, \partial z_{22}} \right)$$

$$-\dfrac{(z_{22} - \bar{z}_{22})(z_{12} - \bar{z}_{12})}{4}\left(\dfrac{\partial^2}{\partial z_{12} \, \partial \bar{z}_{22}} + \dfrac{\partial^2}{\partial \bar{z}_{12} \, \partial z_{22}} \right);$$

(v) $\quad -\dfrac{(z_{13} - \bar{z}_{13})^4}{4(z_{33} - \bar{z}_{33})^2} \dfrac{\partial^2}{\partial z_{11} \, \partial \bar{z}_{11}} - \dfrac{(z_{13} - \bar{z}_{13})^2}{4} \dfrac{\partial^2}{\partial z_{13} \, \partial \bar{z}_{13}}$

$$-\dfrac{(z_{33} - \bar{z}_{33})^2}{4} \dfrac{\partial^2}{\partial z_{33} \, \partial \bar{z}_{33}} - \dfrac{(z_{13} - \bar{z}_{13})^3}{4(z_{33} - \bar{z}_{33})}\left(\dfrac{\partial^2}{\partial z_{11} \, \partial \bar{z}_{13}} + \dfrac{\partial^2}{\partial \bar{z}_{11} \, \partial z_{13}} \right)$$

$$-\dfrac{(z_{13} - \bar{z}_{13})^2}{4}\left(\dfrac{\partial^2}{\partial z_{11} \, \partial \bar{z}_{33}} + \dfrac{\partial^2}{\partial z_{33} \, \partial \bar{z}_{11}} \right)$$

$$-\dfrac{(z_{33} - \bar{z}_{33})(z_{13} - \bar{z}_{13})}{4}\left(\dfrac{\partial^2}{\partial z_{13} \, \partial \bar{z}_{33}} + \dfrac{\partial^2}{\partial \bar{z}_{13} \, \partial z_{33}} \right);$$

(vi) $a^2(z)\dfrac{\partial^2}{\partial z_{11}^2}$;

(vii) $a^2(z)\dfrac{\partial^2}{\partial \bar{z}_{11}^2}$,

where

$$a(z) = \frac{z_{11} - \bar{z}_{11}}{2i} - \frac{(z_{12} - \bar{z}_{12})^2}{2i(z_{22} - \bar{z}_{22})} - \frac{(z_{13} - \bar{z}_{13})^2}{2i(z_{33} - \bar{z}_{33})}.$$

In these bases, (i)–(v) are of type $(1, 1)$, (vi) is of type $(2, 0)$, and (vii) is of type $(0, 2)$.

References

1. L. K. Hua, *Harmonic Analysis of Functions of Several Complex Variables in the Classical Domains*, American Mathematical Society, Providence, RI (1963).
2. Q. K. Lu (K. H. Look), *Introduction to Function Theory of Several Complex Variables*, China Academic, Beijing (1961) (in Chinese).
3. Q. K. Lu, *Classical Manifolds and Classical Domains*, Shanghai Science and Technical publishers, China (1963) (in Chinese).
4. Shingo Murakami, On automorphisms of Siegel domains, in Lecture Notes in Math., vol. 286, Springer-Verlag, New York (1972).
5. A. Selberg, Harmonic analysis and discontinuous group in weakly symmetric Riemannian space with applications to Dirichlet series, *J. Ind. Math. Soc.* **20** (1965).

On Prime Ideals of the Ring of Differential Operators

As everybody knows, the fundamental facts about the theory of harmonic functions in the classical domain of several complex variables consist of the following. Let $\mathscr{R}$ be a classical domain, Γ its characteristic manifold, and Δ the Laplace-Beltrami operator of $\mathscr{R}$. If $\Delta f = 0, f \in C^\infty(\mathscr{R})$ is said to be harmonic. Suppose f stands for the harmonic function whose continuous boundary value is φ on Γ. Then in $\mathscr{R}, f$ is given by the following Poisson integral (see Ref. 3):

$$f(Z) = \int_\Gamma P(Z, U)\varphi(U)\dot{U}, \qquad \forall Z \in \mathscr{R}. \tag{7.1}$$

The Poisson kernels of all classical domains have been given precisely in Ref. 2. Take $\mathscr{R}_\mathrm{I}(n): I - Z\bar{Z}' > 0$ as an example, where Z denotes all $n \times n$ matrix, the characteristic manifold of $\mathscr{R}_\mathrm{I}(n)$ is $\Gamma = U(n) = \{U \,|\, U\bar{U}' = I\}$, the Laplacian is

$$\Delta = \mathrm{tr}\,[(I - Z\bar{Z}')\partial_{\bar{z}}(I - \bar{Z}'Z)\partial'_z], \tag{7.2}$$

and the corresponding Poisson kernel is

$$P(Z, U) = \frac{\det\,(I - Z\bar{Z}')^n}{|\det\,(I - Z\bar{U}')|^{2n}}, \qquad U \in \Gamma = U(n). \tag{7.3}$$

As is well known, $\Delta P(Z, U) = 0$ is valid for any $U \in U(n)$. Hua pointed out [2] that Δ annihilates $P(Z, U)$ and so do the following differential operators:

$$\{\partial_{\bar{Z}'}(I - \bar{Z}'Z)\partial'_Z\}_{ij} \quad \text{i.e.,} \quad \sum_{\alpha,\beta=1}^n \left(\delta_{\alpha\beta} - \sum_{k=1}^n \bar{z}_{k\alpha}z_{k\beta} \right) \frac{\partial^2}{\partial\bar{z}_{i\alpha}\,\partial\bar{z}_{j\beta}}. \quad (7.4)$$

In connection with this kind of differential operator equipped with a "common Poisson kernel," Hua proposed to investigate the algebraic structure of all differential operators which annihilate the same Poisson kernel (in the ring of all differential operators these constitute a left ideal, obviously).

We were inspired by a comparison of Hua's problem with Hilbert's Nullstellensatz, which claims the following: suppose $g_1, \ldots, g_m \in \mathbb{R}[x_1, \ldots, x_n]$, where $\mathbb{R}$ is the field of real numbers, and let $\Omega = \{z \in \mathbb{C}^n \mid g_1(z) = \cdots = g_m(z) = 0\}$. If $f \in \mathbb{R}[x_1, \ldots, x_n]$ and $f|_\Omega \equiv 0$, then there exists a positive integer σ such that $f^\sigma \equiv 0 \,(\text{mod}(g_1, \ldots, g_m))$. Furthermore, if the ideal $(g_1, \ldots, g_m)$ is in fact prime, then $\sigma = 1$; i.e., $f \equiv 0 \,(\text{mod} \,(g_1, \ldots, g_m))$.

Now let $T_1, \ldots, T_m$ be m differential operators defined on a domain $\mathcal{D}$, and let Ω be a set of common solutions. Then from $T_1, \ldots, T_m$ we have two left ideals in the ring of differential operators.

1. $(T_1, \ldots, T_m) = \sum_{i=1}^m A_i\left(\frac{\partial}{\partial x_1}, \ldots, \frac{\partial}{\partial x_n}\right) T_i.$

2. $H(T_1, \ldots, T_m) = \{T \mid T \text{ a differential operator on } \mathcal{D}, Tf = 0 \forall f \in \Omega\}.$

Obviously, $(T_1, \ldots, T_m) \subseteq H(T_1, \ldots, T_m)$. We introduce the following definition.

DEFINITION 7.1. The ideal $(T_1, \ldots, T_m)$ is prime if $(T_1, \ldots, T_m) = H(T_1, \ldots, T_m)$.

Now go back to the classical domain. If Δ is a Laplacian, our question becomes: When will (Δ) be prime? If $H(\Delta) \neq (\Delta)$, then what is a basis for $H(\Delta)$?

The aim of this chapter is to deal first with the above-mentioned questions. Some conditions are given for $(T_1, \ldots, T_m)$ to be a prime ideal; they seem to summarize the cases of classical domains. Section 7.1 proves that if Δ is the Laplacian of $\mathbb{R}^n$ or of the unit ball $'B_n: Z\bar{Z}' < 1$, $Z = (z_1, \ldots, z_n) \in \mathbb{C}^n$, then $H(\Delta) = (\Delta)$; that means (Δ) is prime. Section 7.2 gives a sufficient condition for $(T_1, \ldots, T_m)$ to be prime in the case that $T_1, \ldots, T_m$ have constant coefficients. Section 7.3 studies the case when

$T_1, \ldots, T_m$ have variable coefficients. By requiring that they have a common "Poisson kernel" and be invariant under some kind of group, we give a sufficient condition for $(T_1, \ldots, T_m)$ to be prime. Section 7.4, by taking $\mathcal{R}_1(2)$ as an example, presents a basis for $Z(P(Z, U)) = \{T \mid T$ a differential operator of $\mathcal{R}_1(2), T(P(Z, U)) = 0, \forall U \in U(2)\}$.

7.1. Euclidean Space and Its Unit Ball

Recall that the fundamental solutions of the Laplacian

$$\Delta = \frac{\partial^2}{\partial x_1^2} + \cdots + \frac{\partial^2}{\partial x_n^2} \tag{7.5}$$

of $\mathbb{R}^n$ are $1/r^{n-2}$, $r = (\sum_{i=1}^{n} (x_i - a_i)^2)^{1/2}$, $\forall (a_1, \ldots, a_n) \in \mathbb{R}^n$. What we attempt to show is that $Z(1/r^{n-2}) = \{T \mid T(1/r^{n-2}) = 0, \forall (a_1, \ldots, {}_n) \in \mathbb{R}^n\}$ is a principal ideal generated by Δ.

We use P_m to denote the space of all homogeneous polynomials of degree m, and we let $H_m = \{f \in P_m \mid \Delta f = 0\}$ be the space of all harmonic homogeneous polynomials of the same degree.

LEMMA 7.2. *Every* $p_m(x_1, \ldots, x_n) \in P_m$ *can be expressed uniquely as*

$$p_m(x_1, \ldots, x_n) = p_{m-2}(x_1, \ldots, x_n)r^2 + h_m(x_1, \ldots, x_n), \tag{7.6}$$

with $p_{m-2} \in P_{m-2}$, $h_m \in H_m$, *and* $r^2 = (x_1^2 + \cdots + x_n^2)$.

PROOF. Consider the homomorphism $\Delta: P_m \to P_{m-2}$, $p_m \to \Delta p_m$. Then evidently $H_m = \ker \Delta$. If $r^2 P_{m-2}$ is a subspace of P_m, since $\dim r^2 P_{m-2} = \dim P_{m-2}$ and $\Delta(r^2 p_{m-2}) = 0 \Rightarrow p_{m-2} = 0$, we get $\Delta(r^2 P_{m-2}) = P_{m-2}$. So for any $p_m \in P_m$, $\Delta p_m \in P_{m-2}$, there exists a $p_{m-2} \in P_{m-2}$ such that $\Delta(r^2 p_{m-2}) = \Delta p_m$. Set $h = p_m - r^2 p_{m-2}$. It is clear that $h_m \in H_m$; i.e., $p_m = r^2 p_{m-2} + h_m$. Since the uniqueness for decomposition is obvious, the lemma follows. $\square$

LEMMA 7.3. *Let T be a differential operator. If* $T(\sum (x_i - a_i)^2)^{-(n-2)/2} = 0$, $\forall (a_1, \ldots, a_n) \in \mathbb{R}^n$, *then* $Th = 0$, $\forall h \in H_m$, $m = 1, 2, \ldots$.

PROOF. Write $x = (x_1, \ldots, x_n)$, $a = (a_1, \ldots, a_n)$, $(x, x) = \sum x_i^2$, $r = \sqrt{(x, x)}$, and $\xi = x/r \in \mathbb{R}^n$. Considering $\sum (a_i - tx_i)^2$ with $(a, a) = 1$, we have

$$\sum (a_i - tx_i)^2 = t^2 \sum \left(x_i - \frac{a_i}{t} \right)^2 = t^2 r^2 + \sum a_i^2 - 2t(a, x)$$

$$= 1 + t^2 r^2 - 2tr(a, \xi). \tag{7.7}$$

By a well-known equality [1] we have

$$(1 + t^2 r^2 - 2tr(a, \xi))^{-(n-2)/2} = \sum_{m=0}^{\infty} h_m(x_1, \ldots, x_n) t^m, \qquad (7.8)$$

with $h_m(x_1, \ldots, x_n) \in H_m$, and the set $\{h_m(x_1, \ldots, x_n)\}$ contains a linear basis of H_m when a ranges over the unit sphere $(a, a) = 1$.

Thus if $T(\sum (x_i - a_i)^2)^{-(n-2)/2} = 0$, then from (7.7) we have

$$T(1 + t^2 r^2 - 2tr(a, \xi))^{-(n-2)/2} = 0$$

as well for all a and t. By (7.8), $Th_m = 0$ is valid for all a. As a consequence, $Th = 0$, $\forall h \in H_m$.

THEOREM 7.4. For $\Delta = \partial^2/\partial x_1^2 + \cdots + \partial^2/\partial x_n^2$, the ideal (Δ) is prime.

PROOF Since $1/r^{n-2}$, $r = (\sum (x_i - a_i^2))^{1/2}$, $\forall a_i \in \mathbb{R}$ are the fundamental solutions of Δ, it suffices to show that if $T(1/r^{n-2}) = 0$, then $T \in (\Delta)$.

Suppose

$$T = \sum a_{i_1 \cdots i_n}(x) \frac{\partial^{i_1 + \cdots + i_n}}{\partial x_1^{i_1} \cdots \partial x_n^{i_n}} \in H(\Delta).$$

(For $H(\Delta)$ and (Δ), see the introduction.) Thus we have

$$T(\sum (x_i - a_i)^2)^{-(n-2)/2} = 0, \forall a_i \in \mathbb{R}.$$

Arbitrarily fix a point x^0 and write $y = x - x^0$. Then

$$0 = T(\sum (x_i - a_i)^2)^{-(n-2)/2}\big|_x$$

$$= \sum a_{i_1 \cdots i_n}(y + x^0) \frac{\partial^{i_1 + \cdots + i_n}}{\partial y_1^{i_1} \cdots \partial y_n^{i_n}} (\sum (y_i + x_i^0 - a_i)^2)^{-(n-2)/2}\big|_y.$$

Defining

$$\tilde{T}_y = \sum a_{i_1 \cdots i_n}(y + x^0) \frac{\partial^{i_1 + \cdots + i_n}}{\partial y_1^{i_1} \cdots \partial y_n^{i_n}},$$

we have $\tilde{T}_y(\sum(y_i + x_i^0 - a_i)^2)^{-(n-2)/2} = 0$, $\forall a_i \in \mathbb{R}$. Since a_i can be prescribed arbitrarily, we arrive at

$$\tilde{T}_y(\sum (y_i - a_i)^2)^{-(n-2)/2} = 0, \qquad \forall a_i \in \mathbb{R}. \qquad (7.9)$$

Denote the value of $\tilde{T}_y$ at 0 by $\tilde{T}_0$; that is,

$$\tilde{T}_0 = \tilde{T}_y(0) = \sum a_{i_1 \cdots i_n}(x^0) \frac{\partial^{i_1 + \cdots + i_n}}{\partial y_1^{i_1} \cdots \partial y_n^{i_n}}, \qquad (7.10)$$

which becomes a differential operator of $\partial/\partial y_i$, $i = 1, 2, \ldots, n$, with constant coefficients. From (7.9) and the arbitrariness of a_i, we have $\tilde{T}_0(\sum (y_i - a_i)^2)^{-(n-2)/2} = 0$. Now denote $\partial/\partial y_i$ by T_i; then $\tilde{T}_0$ can be written as

$$\tilde{T}_0 = \sum b_{i_1 \cdots i_n} T_1^{i_1} \cdots T_n^{i_n}, \qquad b_{i_1 \cdots i_n} = a_{i_1 \cdots i_n}(x^0).$$

Thanks to Lemma 7.2, there exist differential operators $q(T_1, \ldots, T_n)$ and $h(T_1, \ldots, T_n)$ with constant coefficients such that

$$\tilde{T}_0 = q(T_1, \ldots, T_n)(T_1^2 + \cdots + T_n^2) + h(T_1, \ldots, T_n), \qquad (7.11)$$

where if T_i is replaced by y_i in $h(T_1, \ldots, T_n)$, it turns out to be a harmonic polynomial $h(y_1, \ldots, y_n)$. Since $\tilde{T}_0(\sum (y_i - a_i)^2)^{-(n-2)/2} = 0$, we infer from Lemma 7.3 that $\tilde{T}_0$ annihilates all harmonic polynomials. Now by the condition $h(T_1, \ldots, T_n) = \sum c_{i_1 \cdots i_n} T_1^{i_1} \cdots T_n^{i_n}$, we take the corresponding polynomial $h(y_1, \ldots, y_n) = \sum c_{i_1 \cdots i_n} y_1^{i_1} \cdots y_n^{i_n}$ and obtain

$$h(T_1, \ldots, T_n)h(y_1, \ldots, y_n)\big|_{y=0} = \sum c_{i_1 \cdots i_n} \frac{\partial^{i_1 + \cdots + i_n}}{\partial y_1^{i_1} \cdots \partial y_n^{i_n}} (\sum c_{i_1 \cdots i_n} y_1^{i_1} \cdots y_n^{i_n})$$

$$= \sum i_1! \cdots i_n! \, c_{i_1 \cdots i_n}^2.$$

Making (7.11) to act on $h(y_1, \ldots, y_n)$, performing the calculations at $y = 0$, and observing that $\tilde{T}_0 h(y_1, \ldots, y_n) = 0$, we claim

$$0 = \tilde{T}_0 h(y_1, \ldots, y_n)\big|_{y=0}$$

$$= [q(T_1, \ldots, T_n)(T_1^2 + \cdots + T_n^2)$$

$$+ h(T_1, \ldots, T_n)]h(y_1, \ldots, y_n)\big|_{y=0}$$

$$= h(T_1, \ldots, T_n)h(y_1, \ldots, y_n)\big|_{y=0}$$

$$= \sum i_1!, \ldots, i_n! \, c_{i_1 \cdots i_n}^2.$$

This means $c_{i_1 \cdots i_n} = 0$, and then $h(T_1, \ldots, T_n) = 0$. Hence,

$$\tilde{T}_0 = q(T_1, \ldots, T_n)(T_1^2 + \cdots + T_n^2); \qquad (7.12)$$

that is,

$$\sum a_{i_1 \cdots i_n}(x^0) \frac{\partial^{i_1 + \cdots + i_n}}{\partial y_1^{i_1} \cdots \partial y_n^{i_n}} = q\left(\frac{\partial}{\partial y_1}, \ldots, \frac{\partial}{\partial y_n}\right)\left(\frac{\partial^2}{\partial x_1^2} + \cdots + \frac{\partial^2}{\partial x_n^2}\right).$$

Now we turn to $x = y + x^0$. Notice that $\partial/\partial x_i = \partial/\partial y_i$ and that the coefficients of $q(\partial/\partial y_1, \ldots, \partial/\partial y_n)$ depend only on x^0. Since x^0 can be chosen arbitrarily, we have

$$T = \sum a_{i_1 \cdots i_n}(x) \frac{\partial^{i_1 + \cdots + i_n}}{\partial x_1^{i_1} \cdots \partial x_n^{i_n}}$$

$$= q\left(\frac{\partial}{\partial x_1}, \ldots, \frac{\partial}{\partial x_n}\right)\left(\frac{\partial^2}{\partial x_1^2} + \cdots + \frac{\partial^2}{\partial x_n^2}\right)$$

$$= q \cdot \Delta \subset (\Delta).$$

The theorem is proved.   □

COROLLARY 7.5. *If* $Z((\sum x_i^2)^{-(n-2)/2}) = \{T \mid T$ *a differential operator with constant coefficient,* $T((x_1^2 + \cdots + x_n^2)^{-(n-2)/2}) = 0\}$, *then* $Z((\sum x_i^2)^{-(n-2)/2})$ *is a principal ideal generated by* Δ.

PROOF. Because $P(\partial/\partial x_1, \ldots, \partial/\partial x_n)$ is an operator with constant coefficients and $P((x_1^2 + \cdots + x_n^2)^{-(n-2)/2}) = 0$, this ensures that

$$P((\sum (x_i - a_i)^2)^{-(n-2)/2}) = 0, \quad \forall a_i \in \mathbb{R}.$$

By Theorem 7.4, Corollary 7.5 immediately follows.   □

In the following, let $z = (z_1, \ldots, z_n)$, $u = (u_1, \ldots, u_n)$, $z_i, u_i \in \mathbb{C}$ throughout this paper. Define $\Delta_0 = \sum \partial^2/\partial z_i \, \partial \bar{z}_i$. A polynomial f is called a harmonic homogeneous polynomial of degree m if f is homogeneous of the same degree satisfying $\Delta_0 f = 0$. We use H_m to denote the space of all harmonic homogeneous polynomials of degree m.

Let $B: z\bar{z}' < 1$, $z \in \mathbb{C}^n$, be the unit ball in $\mathbb{C}^n$. From Ref. 3 the Laplacian on B is given by

$$\Delta = (1 - z\bar{z}') \sum (\delta_{ij} - z_i \bar{z}_j) \frac{\partial^2}{\partial z_i \, \partial \bar{z}_j}. \qquad (7.13)$$

and according to the definition of the unit ball of the theory of harmonic functions, the Poisson kernel of B is

$$P(z, u) = \frac{(1 - z\bar{z}')^n}{|1 - z\bar{u}'|^{2n}}, \qquad z \in B, \quad u \in \partial B = \{u \mid u\bar{u}' = 1\}.$$

First we shall study the relationship between the power series of $P(z, u)$ and harmonic polynomials. The identity

$$(1 - x)^{-n} = \sum_{k=0}^{\infty} a_k x^k, \qquad a_k = \frac{\Gamma(n + k)}{\Gamma(n)\Gamma(k + 1)}$$

implies

$$P(z, u) = (1 - z\bar{z}')^n \left(\sum_{k=0}^{\infty} a_k (z\bar{u}')^k \right) \left(\sum_{k=0}^{\infty} a_k (\bar{z}u')^k \right) \tag{7.14}$$

$$= \sum_{m=0}^{\infty} h_m(z, u),$$

$$h_m(z, u) = \sum_{k+l=m} a_k a_l (z\bar{u}')^k (\bar{z}u')^l$$

$$- c_n^1 (z\bar{z}') \sum_{k+l=m-2} a_k a_l (z\bar{u}')^k (\bar{z}u')^l$$

$$+ \cdots + (-1)^r c_n^r (z\bar{z}')^r$$

$$\times \sum_{k+l=m-2r} a_k a_l (z\bar{u}')^k (\bar{z}u')^l. \tag{7.15}$$

Here $r = \min\{n, [m/2]\}$. Equation (7.15) shows that $h_m(z, u)$ is homogeneous for $z, \bar{z}$ and for $u, \bar{u}$ as well. However, $u_i, \bar{u}_i$, $i = 1. 2, \ldots, n$, are not all independent and so a "reduction" must be carried out under the condition

$$u_1\bar{u}_1 + u_2\bar{u}_2 + \cdots + u_n\bar{u}_n = 1. \tag{7.16}$$

The reduction is achieved by replacing $u_1\bar{u}_1$ by $1 - u_2\bar{u}_2 - \cdots - u_n\bar{u}_n$ without exception. A monomial is called irreducible if the above procedure is impossible. The monomial $u_1^{i_1} \cdots u_n^{i_n} \bar{u}_1^{j_1} \cdots \bar{u}_n^{j_n}$ is irreducible if and only if $i_1 j_1 = 0$.

LEMMA 7.6. *Let $f(z, u)$ be a polynomial of degree m for z, $\bar{z}$ and u, $\bar{u}$, satisfying the following:*

1. *$f(z, u) = F(z\bar{u}', \bar{z}u')$, $f(z, u)$ being a homogeneous polynomial of degree m with real coefficients for $z\bar{u}'$ and $\bar{z}u'$.*
2. *$f(z, u)$ contains all monomials $z_1^{i_1} \cdots z_n^{i_n} \bar{z}_1^{j_1} \cdots z_n^{j_n}$, $\forall i, j$, satisfying $\sum i_l + \sum j_l = m$.*

Then replacing $u_1\bar{u}_1$ by $1 - u_2\bar{u}_2 - u_3\bar{u}_3 - \cdots - u_n\bar{u}_n$, $f(z, u)$ can be reduced to

$$f(z, u) = \sum a_i(z) b_i(u)$$

$$+ \text{ terms in which degrees of } u, \bar{u}$$
$$\text{are less than } m - 2, \tag{7.17}$$

where $\{b_i(u)\}$ are irreducible independent monomials of degree m for u, $\bar{u}$, and $\{a_i(z)\}$ satisfy the following conditions:

All $a_i(z)$ are harmonic homogeneous polynomials of degree m with real coefficients for z, $\bar{z}$.

$\{a_i(z)\}$ are linearly independent.

$\{a_i(z)\}$ forms a linear basis for the space H_m of all harmonic homogeneous polynomials of degree m.

PROOF. *Step 1.* For a fixed u, $u\bar{u}' = 1$, choose U, $U\bar{U}' = 1$ such that

$$U = \begin{pmatrix} \bar{u}_1 & * & \cdots & * \\ \vdots & \vdots & & \vdots \\ \bar{u}_n & * & \cdots & * \end{pmatrix}. \tag{7.18}$$

Consider the transformation $w = zU$. It is easy to verify $\Delta_0 = \sum \partial^2/\partial z_i \, \partial\bar{z}_i = \sum \partial^2/\partial w_i \, \partial\bar{w}_i$, and then

$$\Delta_0 f(z, u) = \sum_{i=1}^{n} \frac{\partial^2}{\partial z_i \, \partial\bar{z}_i} f(z, u) = \sum_{i=1}^{n} \frac{\partial^2}{\partial z_i \, \bar{z}_i} F(z\bar{u}', \bar{z}u')$$

$$= \sum_{i=1}^{n} \frac{\partial^2}{\partial w_i \, \partial\bar{w}_i} F(w_1, \bar{w}_1) \qquad \text{(from (7.18), } w_1 = z\bar{u}')$$

$$= \sum_{i+j=m-2} a_{ij} w_1^i \bar{w}_1^j = \sum_{i+j=m-2} a_{ij}(z\bar{u}')^i (\bar{z}u')^j. \tag{7.19}$$

On the other hand, under the assumption of (7.17) we get

$$\Delta_0 f(z, u) = \sum_i (\Delta_0 a_i(z)) b_i(u)$$

$$+ \text{ terms in which degrees of } U, \bar{U}$$
$$\text{are less than } m. \tag{7.20}$$

By reduction, (7.19) can be rewritten as

$$\sum_{i+j=m-2} a_{ij}(Z\bar{U}')^2(\bar{Z}U')^j = \text{terms in which degrees of } U, \bar{U} \text{ are}$$
$$\text{less than } m. \tag{7.21}$$

Comparing (7.19)–(7.21) and noticing that $b_i(u)$ are independent of each other, we obtain $\Delta_0 a_i(z) = 0$, $a_i(z) \in H_m$. The $a_i(z)$ have real coefficients because $f(z, u) = F(z\bar{u}', \bar{z}u')$ has real coefficients. The reduction will not change the reality of the coefficients.

Step 2. Now we prove that $\{a_i(z)\}$ are linearly independent. By assumption, $f(z, u) = F(z\bar{u}', \bar{z}u')$, and so in each monomial of $f(z, u)$ the degrees of x_i, u_i and $\bar{z}_j$, $\bar{u}_j$ will be the same. Since $f(z, u)$ contains all monomials $z_1^{i_1} \cdots z_n^{i_n} \bar{z}_1^{j_1} \cdots \bar{z}_n^{j_n}$, we see that

$$f(z, u) = \sum a_{ij} z_1^{i_1} \cdots z_n^{i_n} \bar{z}_1^{j_1} \cdots \bar{z}_n^{j_n} \bar{u}_1^{i_1} \cdots \bar{u}_n^{i_n} u_1^{j_1} \cdots u_n^{j_n},$$
$$a_{ij} \neq 0, \quad \forall i = (i_1, \ldots, i_n), \quad j = (j_1, \ldots, j_n), \quad \sum_l i_l + \sum_l j_l = m. \tag{7.22}$$

With the aid of the reduction, each monomial $u_1^{i_1} \cdots u_n^{i_n} \bar{u}_1^{j_1} \cdots \bar{u}_n^{j_n}$ will be reduced to an irreducible monomial. Let

$$b_{i_0}(u) = u_1^{i_1^0} \cdots u_n^{i_n^0} \bar{u}_1^{j_1^0} \cdots \bar{u}_n^{j_n^0}, \ i_1^0 j_1^0 = 0, \ \sum i_l^0 + \sum_l {}_l^0 = m$$

be an irreducible term in (7.17), with $a_{i_0}(z)$ its coefficient. By (7.22), $a_{i_0}(z)$ must contain a monomial of the form $z_1^{i_1^0} \cdots z_n^{j_n^0} \bar{z}_1^{i_1^0} \cdots \bar{z}_n^{j_n^0}$, and this monomial is not contained in any other $a_i(z)$, $i \neq i_0$ (because in each $a_i(z)$ all monomials but one come from reducible terms of the form $z_1^{i_1} \cdots z_n^{i_n} \bar{z}_1^{j_1} \cdots \bar{z}_n^{j_n} u_1^{i_1} \cdots u_n^{i_n} \bar{u}_1^{i_1} \cdots \bar{u}_n^{i_n}$ with $i_1 \geq 0, j_1 \geq 1$). Since each $a_i(z)$ contains a monomial which does not contain any other $a_j(z)$, $j \neq i$, we conclude that $\{a_i(z)\}$ are linearly independent.

Step 3. $\{a_i(z)\}$ forms a linear basis for the space H_m. The method of proof depends on the calculation of dimension. According to the assumption $f(z, u)$ contains all irreducible monomials, and so that number of irreducible monomials is the same as that of $\{a_i(z)\}$. Denote by $d_{m,k}$ the dimension of the space formed by all homogeneous polynomials of degree m in $\mathbb{R}[t_1, \ldots, t_k]$. It is well known that

$$d_{m,k} = \frac{\Gamma(m+k)}{\Gamma(k)\Gamma(m+1)} = C_{m+k-1}^{k-1}. \tag{7.23}$$

Furthermore, the irreducible momomials can be divided into three types, namely $i_1 = 0, j_1 \neq 0; i_1 \neq 0, j_1 = 0; i = 0, j = 0$. For $i_1 = 0$, the total number of irreducible monomials is $d_{m,2n-1}$ and is the same for $j_1 = 0$ and $i_1 = j_1 = 0$. Therefore the number of all irreducible monomials is

$$2(d_{m,2n-1} - d_{m,2n-2}) + d_{m,2n-2} = 2d_{m,2n-1} - d_{m,2n-2}.$$

Concerning the dimension of the harmonic polynomial of degree m for z and $\bar{z}$, from Lemma 7.2 we know that $\dim H_m = d_{m,2n} - d_{m-2,2n}$. Consequently it suffices to verify

$$d_{m,2n} - d_{m-2,2n} = 2d_{m,2n-1} - d_{m,2n-2},$$

this is easy to check by (7.23). The proof is finished. $\qquad\square$

REMARK 7.7. Lemma 7.6 is of independent interest. It gives a way to find a linear basis of H_m. We may start with any homogeneous polynomial satisfying the lemma—for example, $\sum_{k+i=m} (z\bar{u}')^k(\bar{z}u')^i$—to get a basis of H_m via reduction.

Now we embark on studying the Poisson kernel of the unit ball B. By (7.14), the Poisson kernel has the expansion

$$P(z, u) = \sum_{m=0}^{\infty} h_m(z, u),$$

and by reducing u, $\bar{u}$ with $h_m(z, u)$, (7.15) implies

$$h_m(z, u) = \sum_{k+l=m} a_k a_l (z\bar{u}')^k(\bar{z}u')^l + \text{terms in which degrees}$$
$$\text{of } u, \bar{u} \text{ are less than } m$$

$$= \sum_{i=1}^{d_m} a_m^{(i)}(z) b_m^{(i)}(z) + \text{terms in which degrees}$$
$$\text{of } u, \bar{u} \text{ are less than } m. \tag{7.24}$$

Since $\sum_{k+l} a_k a_l (z\bar{u}')^k (\bar{z}u')^l$ satisfies Lemma 7.6, $\{a_m^{(i)}(z)\}$ forms a linear basis of H_m and $d_m = \dim H_m$.

LEMMA 7.8. *Let T be a differentiable operator with constant coefficients of the highest order l. Suppose T satisfies*

$$TP(z, u)\big|_{z=0} = 0. \tag{7.25}$$

Expanding T to $T = T^{(l)} + T^{(l-1)} + \cdots + T^{(0)}$, $T^{(i)}$ *being the homogeneous differential operator of order i, we find*

$$T^{(l)}H_l = 0. \tag{7.26}$$

PROOF. Equations (7.14), (7.25), and (7.24) imply

$$0 = TP(z, u)\big|_{z=0} = \sum_{m=0}^{\infty} Th_m(z, u) = \sum_{m=0}^{l} T^{(m)}h_m(z, u)$$

$$= \sum_{m=0}^{l} T^{(m)} \sum_{i=0}^{d_m} a_m^{(i)}(z) b_m^{(i)}(u)$$
$$+ \text{ terms in which degrees}$$
$$\text{of } u, \bar{u} \text{ are less than } m$$

$$= \sum_{i=0}^{d_l} T^{(l)} a_l^{(i)}(z) b_l^{(i)}(u)$$
$$+ \text{ terms in which degrees}$$
$$\text{of } u, \bar{u} \text{ are less than } l.$$

Since $\{b_l^{(i)}(u)\}$ are linearly independent, there exists

$$T^{(l)}a_l^{(i)}(z) = 0, \qquad 1 \le i \le d_l.$$

But $\{a_l^{(i)}(z)\}$ forms a linear basis of H_l. Hence $T^{(l)}H_l = 0$. □

THEOREM 7.4. *Let*

$$\Delta = (1 - z\bar{z}') \sum_{i,j=1}^{n} (\delta_{ij} - z_i \bar{z}_j) \frac{\partial^2}{\partial z_i \, \partial \bar{z}_j}$$

be the Laplacian on the unit ball B: $z\bar{z}' < 1$, $z \in \mathbb{C}^n$. *Then* $H(\Delta) = (\Delta)$; *i.e.,* (Δ) *is prime.*

PROOF. Since Δ has a Poisson kernel $P(z, u)$, i.e., $\Delta P(z, u) = 0$, $\forall u, u\bar{u}' = 1$, we set

$$T\left(z, \frac{\partial}{\partial z}\right) = \sum_{i+j \le l} a_{i_1 \cdots i_n j_1 \cdots j_n}(z, \bar{z}) \frac{\partial^{i_1 + \cdots + i_n + j_1 + \cdots + j_n}}{\partial z_1^{i_1} \cdots \partial z_n^{i_n} \partial \bar{z}_1^{j_1} \cdots \partial \bar{z}_n^{j_n}} \in H(0),$$

which results in $TP(z, u) = 0$, $\forall u$, $u\bar{u}' = 1$. Put $z = 0$ and set

$$T\left(0, \frac{\partial}{\partial z}\right) = \sum_{i+j \le l} a_{i_1 \cdots i_n j_1 \cdots j_n}(0) \frac{\partial^{i_1 + \cdots + i_n + j_1 + \cdots + j_n}}{\partial z_1^{i_1} \cdots \partial z_n^{i_n} \partial \bar{z}_1^{j_1} \cdots \partial \bar{z}_n^{j_n}}.$$

$$= T_0 = T_0^{(l)} + T_0^{(l-1)} + \cdots + T_0^{(0)}. \qquad (7.27)$$

Here $T_0^{(i)}$ is a homogeneous differential operator of order i. On the basis of $TP(z, u) = 0$, we can state that $T_0 P(z, u)|_{z=0} = 0$. From Lemma 7.8,

$$T_0^{(l)} H_l = 0, \qquad (7.28)$$

where H_l indicates the space spanned by all harmonic polynomials of $z, \bar{z}$ with degree l.

By Lemma 7.2, there exist differential operators with constant coefficients $q(\partial/\partial z, \partial/\partial \bar{z})$ and $h(\partial/\partial z, \partial/\partial \bar{z})$ of orders $l - 2$ and l, respectively, such that

$$T_0^{(l)} = q\left(\frac{\partial}{\partial z}, \frac{\partial}{\partial \bar{z}}\right)\Delta_0 + h\left(\frac{\partial}{\partial z}, \frac{\partial}{\partial \bar{z}}\right), \qquad (7.29)$$

and a harmonic homogeneous polynomial $h(z, \bar{z}) \in H_l$ is obtained when $\partial/\partial z$ is replaced by z in h. By Lemma 7.6, $\{a_l^{(i)}(z)\}$, $1 \le i \le d_l$, in (7.24) forms a linear basis of H_l. Thus,

$$h(z, \bar{z}) = \sum_{i=1}^{d_l} b_i a_l^{(i)}(z) = \sum b_{i_1 \cdots i_n j_1 \cdots j_n} z_1^{i_1} \cdots z_n^{i_n} \bar{z}_1^{j_1} \cdots \bar{z}_n^{j_n}$$

for some $b_i \in \mathbb{C}$. Since the coefficients of $a_l^{(i)}(z)$ are real (Lemma 7.6), the equality

$$\hat{h}(z, \bar{z}) = \sum_{i=1}^{d_l} \bar{b}_i a_l^{(i)}(z) = \sum \bar{b}_{i_1 \cdots i_n j_1 \cdots j_n} z_1^{i_1} \cdots z_n^{i_n} \bar{z}_1^{j_1} \cdots \bar{z}_n^{j_n}$$

remains in H_l. By making (7.29) act on $\hat{h}(z, \bar{z})$ and noticing that $T_0^{(l)} H_l = 0$, $\Delta_0 H_l = 0$, we find

$$0 = h\left(\frac{\partial}{\partial z}, \frac{\partial}{\partial \bar{z}}\right)\hat{h}(z, \bar{z})$$

$$= \sum b_{i_1 \cdots i_n j_1 \cdots j_n} \frac{\partial^{i_1 + \cdots + j_n}}{\partial z_1^{i_1} \cdots \partial z_n^{i_n} \partial \bar{z}_1^{j_1} \cdots \partial \bar{z}_n^{j_n}} (\sum \bar{b}_{i_1 \cdots i_n j_1 \cdots j_n} z_1^{i_1} \cdots z_n^{i_n} \bar{z}_1^{j_1} \cdots \bar{z}_n^{j_n}$$

$$= \sum i_1! \cdots i_n! j_1! \cdots j_n! |b_{i_1 \cdots i_n j_1 \cdots j_n}|^2.$$

Thus $b_{i_1\cdots i_n j_1\cdots j_n} = 0$, and (7.29) becomes

$$T_0^{(l)} = q\left(\frac{\partial}{\partial z}, \frac{\partial}{\partial z}\right)\Delta_0. \tag{7.30}$$

From Ref. 2 we know that for the unit ball B, any point $(z, u) \in B \times \partial B$ can be transformed to (w, v) by an automorphism $\gamma : (z, u) \to (w, v)$ of B, with w being specified beforehand. Furthermore, the Poisson kernel satisfies

$$P(w, v) = \frac{P(z, u)}{P(z_0, u)}, \tag{7.31}$$

where $z_0 = \gamma^{-1}(0)$.

Now let $z_0 \in B$ be arbitrarily prescribed, and consider $\gamma : w = f(z, z_0)$ to be an automorphism of B such that

$$\gamma : z_0 \to 0, \qquad (z, u) \to (w, v). \tag{7.32}$$

If $w = f(z, z_0)$ is allowed to be an expression of γ, then by (7.31) we deduce

$$TP(z, u)\big|_{z_0} = [T_w P(w, v)]_{w=0} P(z_0, u) = 0, \tag{7.33}$$

where the operator T_w comes from the operator $T(z, \partial/\partial z)$ transformed by γ; that is,

$$T_w\left(0, \frac{\partial}{\partial w}\right) = T\left(z_0, \frac{\partial}{\partial z}\right). \tag{7.34}$$

Because of $P(z_0, u) \neq 0$, (7.33) yields $T_w(0, \partial/\partial w)P(w, v)\big|_{w=0} = 0, \forall v \in \partial B$. Summing up the discussion above, we obtain

$$T_w\left(0, \frac{\partial}{\partial w}\right) = Q\left(\frac{\partial}{\partial w}, \frac{\partial}{\partial \bar{w}}\right)\left(\frac{\partial^2}{\partial w_1 \, \partial \bar{w}_1} + \cdots + \frac{\partial^2}{\partial w_n \, \partial \bar{w}_n}\right)$$

$$+ \text{ terms with lower order.} \tag{7.35}$$

Now we consider the Laplacian on B:

$$\Delta\left(z, \frac{\partial}{\partial z}\right) = (1 - z\bar{z}')\sum_{i,j=1}^{n} (\delta_{ij} - z_i\bar{z}_j)\frac{\partial^2}{\partial z_i \, \partial \bar{z}_j}.$$

Recall that Δ is an invariant differential operator on B; that is, if $\gamma: z \to w$ is an automorphism of B, then according to Ref. 2 we have

$$\Delta\left(z, \frac{\partial}{\partial z}\right) = \Delta\left(w, \frac{\partial}{\partial w}\right). \tag{7.36}$$

Choose γ to be the automorphism of (7.32). Then

$$\Delta\left(z_0, \frac{\partial}{\partial z}\right) = \Delta\left(0, \frac{\partial}{\partial w}\right) = \sum_{i=1}^{n} \frac{\partial^2}{\partial w_i \, \partial \bar{w}_i};$$

thus

$$T\left(z_0, \frac{\partial}{\partial z}\right) = Q^*\left(\frac{\partial}{\partial z}, \frac{\partial}{\partial \bar{z}}\right) \Delta\left(z_0, \frac{\partial}{\partial z}\right)$$
$$+ \text{terms with lower order}, \tag{7.37}$$

by (7.34) and (7.35). Because z_0 is an arbitrary fixed point in B, the above equality is exactly the same as

$$T = Q^*\Delta + T_1; \text{ order of } T_1 < \text{order of } T. \tag{7.38}$$

However, $T_1 = T - Q^*\Delta \in H(\Delta)$. By induction on the order of T, we get a differential operator $\tilde{T}$ of order 1 such that $\tilde{T}P(z, u) = 0$. In addition, the differential operator with order 1, which annihilates the Poisson kernel, should be zero, because from Lemma 7.8 such a differential operator with order 1 annihilates all z_i and $\bar{z}_j$. Finally, we obtain

$$T = Q\Delta \in (\Delta).$$

The theorem is proved. $\square$

7.2. A Criterion for Primality

This section is devoted to the sufficient condition for $(T_1, \ldots, T_m)$ to be prime in the case that T_i are differential operators with constant coefficients. Our discussion is based on a result of H. Maass [4].

THEOREM 7.10 (*H. Maass*). *Let* $f_1(x), \ldots, f_q(x) \in \mathbb{R}[x_1, \ldots, x_n]$ *be homogeneous polynomials, and let* $\Omega = \{x \in \mathbb{C}^n | f_1(x) = \cdots = f_q(x) = 0\}$. *If the ideal* $(f_1, \ldots, f_q)$ *is prime, then any homogeneous polynomial* $u(x)$ *of degree* k *can be expressed as*

$$u(x) = \sum_{i=1}^{q} p_i(x)f_i(x) + \sum (xa_i')^k, \quad a_i = (a_i^1, \ldots, a_i^n) \in \Omega, \tag{7.39}$$

where xa_i' *refers to the inner product of* a_i *and* x.

Notice that if $f(x) = \sum_{\sum i_t = q} a_{i_1 \cdots i_n} x_1^{i_1} \cdots x_n^{i_n}$ signifies a homogeneous polynomial of degree q, then it is easy to verify

$$f\left(\frac{\partial}{\partial x}\right)(xa')^k = \sum_{\sum i_t = q} a_{i_1 \cdots i_n} \frac{\partial^{i_1 + \cdots + i_n}}{\partial x_1^{i_1} \cdots \partial x_n^{i_n}} (xa')^k$$

$$= \begin{cases} q! \, c_k^q f(a)(xa')^{k-q}, & k \geq q \\ 0, & k < q \end{cases}. \qquad (7.40)$$

In particular, if $f(a) = 0$, then

$$f\left(\frac{\partial}{\partial x}\right)(xa')^k = 0, \qquad \forall k \in \mathbb{Z}^+. \qquad (7.41)$$

THEOREM 7.11. *Suppose*

$$T_1(\partial/\partial x_1, \ldots, \partial/\partial x_n), \ldots, T_m(\partial/\partial x_1, \ldots, \partial/\partial x_n)$$

are homogeneous differential operators with constant coefficients in the usual sense. If the ideal $(T_1, \ldots, T_m)$ generated by the corresponding homogeneous polynomials

$$T_1(x_1, \ldots, x_n), \ldots, T_m(x_1, \ldots, x_n)$$

is prime, so is

$$(T_1(\partial/\partial z_1, \ldots, \partial/\partial x_n), \ldots, T_m(\partial/\partial x_1, \ldots, \partial/\partial x_n))$$

in the ring of differential operators with constant coefficients.

PROOF. Let the differential operator with constant coefficients $G(\partial/\partial x_1, \ldots, \partial/\partial x_n) \in H(T_1, \ldots, T_m)$, and let $H(T_1, \ldots, T_m)$ be defined as in the introduction. We want to show that $G(\partial/\partial x_1, \ldots, \partial/\partial x_n) \in (T_1, \ldots, T_m)$. Replacing $\partial/\partial x$ by x, we get a polynomial $G = G(x_1, \ldots, x_n)$. Let $G = G_1 + \cdots + G_l$ be its homogeneous decomposition. Application of (7.39) in Maass's theorem to each G_i gives

$$G(x_1, \ldots, x_n) = \sum_{i=1}^{m} Q_i(x_1, \ldots, x_n) T_i(x_1, \ldots, x_n) + \sum (xa_i')^l + \sum (xb_i')^{l-1}$$

$$+ \cdots + \sum (xc'^i), \qquad a_i, b_i, \ldots, c_i \in \Omega, \qquad (7.42)$$

where $\Omega = \{z \in \mathbb{C}^n \mid T_1(z) = \cdots = T_m(z) = 0\}$. So, by (7.41) we infer

$$T_j\left(\frac{\partial}{\partial x_1}, \ldots, \frac{\partial}{\partial x_n}\right)(xa_i')^l = 0, \ldots, T_j\left(\frac{\partial}{\partial x_1}, \ldots, \frac{\partial}{\partial x_n}\right)(xc_i') = 0,$$

$$1 \le j \le m. \quad (7.43)$$

From the assumption of the theorem all $T_j(\partial/\partial x_1, \ldots, \partial/\partial x_n)$ possess real coefficients. Thus $(x\bar{a}_i')^l$, $(x\bar{b}_i')^{l-1}, \ldots, (x\bar{c}_i')$ also satisfy

$$T_j\left(\frac{\partial}{\partial x_1}, \ldots, \frac{\partial}{\partial x_n}\right)(x\bar{a}_i')^l = 0, \ldots, T_j\left(\frac{\partial}{\partial x_1}, \ldots, \frac{\partial}{\partial x_n}\right)(x\bar{c}_i') = 0,$$

$$1 \le j \le m. \quad (7.44)$$

If we set

$$\sum (xa_i')^l = \sum_{i_1 + \cdots + i_n = l} a_{i_1 \cdots i_n} x_1^{i_1} \cdots x_n^{i_n}, \ldots, \sum (xc_i') = \sum_{i_1 + \cdots + i_n = 1} c_{i_1 \cdots i_n} x_1^{i_1} \cdots x_n^{i_n},$$

then (7.42) becomes

$$G\left(\frac{\partial}{\partial x_1}, \ldots, \frac{\partial}{\partial x_n}\right) = \sum_{i=1}^{m} Q_i\left(\frac{\partial}{\partial x_1}, \ldots, \frac{\partial}{\partial x_n}\right) T_i\left(\frac{\partial}{\partial x_1}, \ldots, \frac{\partial}{\partial x_n}\right)$$

$$+ \sum_{i_1 + \cdots + i_n = l} a_{i_1 \cdots i_n} \frac{\partial^l}{\partial x_1^{i_1} \cdots \partial x_n^{i_n}} + \sum_{i_1 + \cdots + i_n = l-1} b_{i_1 \cdots i_n}$$

$$\times \frac{\partial^{l-1}}{\partial x_1^{i_1} \cdots \partial x_n^{i_n}} + \cdots + \sum_{i_1 + \cdots + i_n = 1} c_{i_1 \cdots i_n} \frac{\partial}{\partial x_1^{i_1} \cdots \partial x_n^{i_n}}.$$

$$(7.45)$$

From (7.44), $\sum (xa_i')^l$ is the common solution for all operators $T_1(\partial/\partial x_1, \ldots, \partial/\partial x_n), \ldots, T_m(\partial/\partial x_1, \ldots, \partial/\partial x_n)$; therefore it is also the solution for $G(\partial/\partial x_1, \ldots, \partial/\partial x_n)$ (since $G(\partial/\partial x_1, \ldots, \partial/\partial x_n) \in H(T_1, \ldots, T_m)$). Now applying (7.45) to $\sum (x\bar{a}_i')^l = \sum \bar{a}_{i_1 \cdots i_n} x_1^{i_1} \ldots x_n^{i_n}$ yields

$$0 = \sum |a_{i_1 \cdots i_n}|^2 i_1! \ldots i_n! + \text{terms of } x \text{ with degree higher than 1}.$$

Thus $a_{i_1 \cdots i_n} = 0$ is obtained; that is, $\sum (xa_i')^l = \sum (x\bar{a}_i')^l = 0$. On the basis of this, again applying (7.45) to $\sum (xb_i')^{l-1}$, we get $\sum (xb_i')^{l-1} = \sum (x\bar{b}_i')^{l-1} = 0$. Repeat this procedure until the following formula is obtained:

$$G\left(\frac{\partial}{\partial x_1}, \ldots, \frac{\partial}{\partial x_n}\right) = \sum Q_i\left(\frac{\partial}{\partial x_1}, \ldots, \frac{\partial}{\partial x_n}\right) T_i\left(\frac{\partial}{\partial x_1}, \ldots, \frac{\partial}{\partial x_n}\right) \in (T_1, \ldots, T_m).$$

The theorem can now be readily proved.   □

COROLLARY 7.12. *If* $F(x_1, \ldots, x_n)$ *is an irreducible homogeneous polynomial with real coefficients, then* $(F(\partial/\partial x_1, \ldots, \partial/\partial x_n))$ *is prime in the ring of differential operators with constant coefficients.*

PROOF. It follows from the Nullstellensatz that the ideal $(F(x_1, \ldots, x_n))$ is prime in the ring of polynomials. Then by Theorem 7.11, Corollary 7.12 is easy to prove.   □

7.3. Extension to Variable Coefficients

In the following, we consider differential operators with variable coefficients. Let M be a domain in $\mathbb{R}^n$. We might as well assume that $0 \in M$ and that G is a transitive automorphism group of M.

DEFINITION 7.13. A set of linear differential operators

$$T_1(x, \partial/\partial x), \ldots, T_m(x, \partial/\partial x)$$

on M is called G-invariant if for any $g \in G$ and $x, y \in M$, $y = xg$, there exist $a_{ij}(y) \in C^\infty(M)$ such that

$$T_i\left(x, \frac{\partial}{\partial x}\right) = \sum a_{ij}(y) T_j\left(y, \frac{\partial}{\partial y}\right), \qquad i = 1, 2, \ldots, m. \qquad (7.46)$$

DEFINITION 7.14. A set of linear differential operators

$$T_1(x, \partial/\partial x), \ldots, T_m(x, \partial/\partial x)$$

is said to have a Poisson kernel $P(x, t)$, $x \in M$, $t \in N$, where N is a manifold and G acts transitively on N, if the following two conditions are satisfied:

1. For every i, $1 \le i \le m$, the differential operator $T_i(x, \partial/\partial x)$ annihilates the Poisson kernel; i.e.,

$$T_0\left(x, \frac{\partial}{\partial x}\right) P(x, t) = 0, \quad \forall t \in N, \qquad 1 \le i \le m. \qquad (7.47)$$

2. For any $g \in G$, $g: x_0 \to 0$, $(x, t) \to (y, t')$, we have

$$P(x, t) = P(y, t')f(x_0, t, t'), \qquad f(x_0, t, t') \ne 0. \qquad (7.48)$$

We also need the following definition.

DEFINITION 7.15. Assume that the system of linear differential operators $\{T_1(x, \partial/\partial x), \ldots, T_m(x, \partial/\partial x)\}$ is equipped with a Poisson kernel $P(x, t)$ satisfying (7.47) and (7.48). The system of linear differential operators is said to be complete with respect to the Poisson kernel $P(x, t)$ is any differential operator $T(\partial/\partial x)$ with constant coefficients satisfies

$$T\left(\frac{\partial}{\partial x}\right)P(x, t)|_{x=0} = 0, \qquad \forall t \in N. \tag{7.49}$$

This leads to

$$T \in \left(T_1\left(0, \frac{\partial}{\partial x}\right), \ldots, T_m\left(0, \frac{\partial}{\partial x}\right)\right). \tag{7.50}$$

THEOREM 7.16. *Suppose M is transitive under group G.*

$$T_1(x, \partial/\partial x), \ldots, T_m(x, \partial/\partial x)$$

is a system of linear differential operators on M satisfying the conditions:

1. *The system is G-invariant.*
2. *The system is endowed with a Poisson kernel $P(x, t)$ and is complete with respect to $P(x, t)$.*

Then the ideal $(T_1(x, \partial/\partial x), \ldots, T_m(x_1, \partial/\partial x)$ is prime.

PROOF. Given $T(x, \partial/\partial x) \in H(T_1, \ldots, T_m)$. Then we have

$$T(x, \partial/\partial x)P(x, t) = 0, \forall x \in M, t \in N.$$

Arbitrarily prescribe $x \in M$, and let G act transitively on M. There exists $g_0 \in G$ such that $g_0: x_0 \to 0$, $(x, t) \to (x, t')$. Consequently, from (7.48) we have

$$T\left(x, \frac{\partial}{\partial x}\right)P(x, t)|_{x=x_0} = T^*\left(y, \frac{\partial}{\partial y}\right)P(y, t')f(x_0, t, t')|_{y=0} = 0,$$

where $T^*(y, \partial/\partial y)$ comes from $T(x, \partial/\partial x)$ via the translation $y = xg_0$. Hence $T(x_0, \partial/\partial x) = T^*(0, \partial/\partial y)$. From the assumption in (7.48), $f(x_0, t, t') \neq 0$, we get

$$T^*\left(0, \frac{\partial}{\partial y}\right)P(y, t')|_{y=0} = 0, \tag{7.51}$$

where t' ranges over N because G also acts on N transitively. Now by the completeness of the system $\{T_i\}$ with respect to $P(x, t)$, we get

$$T^*\left(0, \frac{\partial}{\partial y}\right) = \sum_{i=1}^{n} Q^*\left(\frac{\partial}{\partial y}\right) T_i\left(0, \frac{\partial}{\partial y}\right). \tag{7.52}$$

Resorting to the G-variant of operators $T_1, \ldots, T_m$, choosing $g = g_0$ in (3.46), and then calculating at $y = 0$, we infer

$$T_i\left(0, \frac{\partial}{\partial y}\right) = \sum_{i-1}^{m} b_{ij}(x_0) T_j\left(x_0, \frac{\partial}{\partial x}\right), \qquad 1 \leq i \leq m. \tag{7.53}$$

By (7.52), (7.53), and $T^*(0, \partial/\partial y) = T(x_0, \partial/\partial x)$, this leads to

$$T\left(x_0, \frac{\partial}{\partial x}\right) = \sum Q_i\left(x_0, \frac{\partial}{\partial y}\right) T_i\left(x_0, \frac{\partial}{\partial x}\right).$$

Since x_0 can be any point on M, the above equation means

$$T\left(x, \frac{\partial}{\partial x}\right) \equiv 0 \qquad \left(T_1\left(x, \frac{\partial}{\partial x}\right), \ldots, T_m\left(x, \frac{\partial}{\partial x}\right)\right).$$

The proof is concluded. $\qquad\qquad\qquad\qquad\qquad\qquad\qquad\qquad\qquad$ □

In applications of Theorem 7.15, the main difficulty lies in checking the completeness of the system $\{T_i\}$ with respect to the Poisson kernel. Theorem 7.17 will provide a method for it. For this we introduce another definition.

DEFINITION 7.16. Let $T_1(\partial/\partial x), \ldots, T_m(\partial/\partial x)$ be differential operators with constant coefficients. A polynomial $f(x_1, \ldots, x_n)$ is said to be $\{T_1, \ldots, T_m\}$-harmonic if

$$T_i\left(\frac{\partial}{\partial x}\right) f(x_1, \ldots, x_n) = 0, \qquad 1 \leq i \leq m. \tag{7.54}$$

THEOREM 7.17. Assume that $T_1(x, \partial/\partial x), \ldots, T_m(x, \partial/\partial x)$ are endowed with a Poisson kernel and satisfy the conditions:

1. In the differential operators $T_1(0, \partial/\partial x), \ldots, T_m(0, \partial/\partial x)$ with real coefficients, if we replace $\partial/\partial x_i$ by x_i to obtain the polynomials $T_j(x_1, \ldots, x_n)$, $1 \leq j \leq m$, then the ideal $(T_1(x_1 \ldots, x_n), \ldots, T_m(x, \ldots, x_n))$ generated by the polynomials $T_1(x_1, \ldots, x_n), \ldots, T_m(x_1, \ldots, x_n)$ is prime;
2. In a neighborhood of $x = 0$, $P(x, t)$ has an expansion

$$P(x, t) = P_0(t) + P_1(x, t) + P_2(x, t) + \cdots, \tag{7.55}$$

in which the term with degree l can be expressed as

$$P_l(x, t) = \sum_i h_{ll}^{(i)}(x)\varphi_{ll}^{(i)}(t) + \sum_{k<l} \sum_j h_{lk}^{(j)}(x)\varphi_{lk}^{(j)}(t). \tag{7.56}$$

Here $h_{ll}^{(i)}(x)$, $h_{lk}^{(i)}(x)$ denote homogeneous polynomials of degree l, and $\{h_{ll}^{(i)}(x)\}$ forms a linear basis of the space of all $\{T_1(0, \partial/\partial x), \ldots, T_m(0, \partial/\partial x)\}$-harmonic homogeneous polynomials with degree l. Moreover, $\{\varphi_{ll}^{(i)}(t), \varphi_{lk}^{(i)}(t))\}$ are linearly independent over N. Then $\{T_1(x, \partial/\partial x), \ldots, T_m(x, \partial/\partial x)\}$ is complete with respect to $P(x, t)$.

PROOF. By (7.49) and (7.50), it is sufficient to prove that T belongs to the ideal $(T_i(0, \partial/\partial x), \ldots, T_m(0, \partial/\partial x))$ for any differential operator T with constant coefficients satisfying $TP(x, t)|_{x=0} = 0$, $\forall t \in N$.

Write $T = T^{(1)} + T^{(2)} + \cdots + T^{(l)}$ for the homogeneous decomposition of T. By (7.55) it follows that

$$0 = TP(x, t)|_{x=0} = T(P_0 + P_1(x, t) + P_2(x, t) + \cdots +)|_{x=0}$$

$$= \sum_{i=1}^{l} T^{(i)} P_i(x, t)|_{x=0} = \sum_i (T^{(l)} h_{ll}^{(i)}(x))\varphi_{ll}^{(i)}(t) + \sum_{\substack{j,k,s \\ k \le l, s \le l}} (*)\varphi_{ks}^{(j)}(t).$$

The linear independence of $\varphi_{ll}^{(i)}(t)$, $\varphi_{lk}^{(j)}(t)$ gives

$$T^{(l)} h_{ll}^{(i)}(x) = 0, \qquad 1 \le i \le l.$$

But under the assumption, $\{h_{ll}^{(i)}(x)\}$ means a linear basis of the space of $\{T_1(0, \partial/\partial x), \ldots, T_m(0, \partial/\partial x)\}$-harmonic polynomials with degree l. Accordingly, $T^{(l)}$ may annihilate any $\{T_1(0, \partial/\partial x), \ldots, T_m(0, \partial/\partial x)\}$-harmonic polynomial of degree l.

From Condition 1 the ideal $(T_1(x_1, \ldots, x_n), \ldots, T_m(x_1, \ldots, x_n))$ is prime. In the light of Maass's theorem there exist polynomials $Q_i(x)$, $1 \le i \le m$, and a $\{T_1(0, \partial/\partial x), \ldots, T_m(0, \partial/\partial x)\}$-harmonic polynomial $h_l(x)$ (notice that in (7.39) and (7.41), $(xa_i')^l$ is harmonic) such that

$$T^{(l)}x = \sum_{i=1}^{m} Q_i(x) T_i(x_1, \ldots, x_n) + h_l(x), \tag{7.57}$$

where $T^{(l)}(x)$ denotes the corresponding homogeneous polynomial of the differential operator $T^{(l)}$. According to the above discussion, we have $T^{(l)}(\partial/\partial x)h_l(x) = 0$.

Now rewrite (7.57) as an equation of differential operators:

$$T^{(l)}\left(\frac{\partial}{\partial x}\right) = \sum_{i=1}^{m} Q_i\left(\frac{\partial}{\partial x}\right) T_i\left(0, \frac{\partial}{\partial x}\right) + h_i\left(\frac{\partial}{\partial x}\right). \tag{7.58}$$

Since $T_j(0, \partial/\partial x)$ have real coefficients, $\bar{h}_l(x)$ is also $\{T_1(0, \partial/\partial x), \ldots, T_m(0, \partial/\partial x)\}$-harmonic. Operating on $\bar{h}(x)$ with (7.58) gives rise to $h_l(\partial/\partial x)\bar{h}_l(x) = 0$. This means $h_l(x) = 0$. Hence,

$$T^{(l)}\left(\frac{\partial}{\partial x}\right) \equiv 0 \quad \left(T_1\left(0, \frac{\partial}{\partial x}\right), \ldots, T_m\left(0, \frac{\partial}{\partial x}\right)\right),$$

and similarly for $T - T^{(l)} = T^{(l-1)} + \cdots + T^{(1)}$, etc. Finally, we arrive at $T \equiv 0 \ (T_1(0, \partial/\partial x), \ldots, T_m(0, \partial/\partial x))$. This completes the proof. $\quad\square$

7.4. An Example

In this section, we shall take $R_1(2)$ as an example to discuss some applications of the theorems studied in Section 7.3. Since we only deal with a special example, some details will be omitted. As pointed out in the introduction, from Ref. 2 for $R_1(n)$: $I - Z\bar{Z}' > 0$, the differential operators

$$\partial_{\bar{z}}(I - \bar{Z}'Z)\partial_z', \quad \partial_z = \left(\frac{\partial}{\partial z_{ij}}\right)_{1 \le i,j \le n}, \quad \partial_{\bar{z}} = \left(\frac{\partial}{\partial \bar{z}_{ij}}\right)_{1 = i,j \le n} \tag{7.59}$$

annihilate the Poisson kernel of $R_1(n)$. In fact, another system of differential operators

$$\partial_z'(I - Z\bar{Z}')\partial_{\bar{z}} \tag{7.60}$$

annihilates the Poisson kernel. We can see this from the following: recall that the transformation group G,

$$G: W = (AZ + B)(CZ + D)^{-1},$$

$$\overline{\begin{pmatrix} A & B \\ C & D \end{pmatrix}}' \begin{pmatrix} I & 0 \\ 0 & -I \end{pmatrix} \begin{pmatrix} A & B \\ C & D \end{pmatrix} = \begin{pmatrix} I & 0 \\ 0 & -I \end{pmatrix},$$

acts transitively on $R_1(n)$ and there holds

$$(I - W\bar{W}')\partial_{\bar{w}}(I - \bar{W}'W)\partial_w'$$

$$= (Z\bar{B}' + \bar{A}')^{-1}(I - Z\bar{Z}')\partial_{\bar{z}}(I - \bar{Z}'Z)\partial_z'(Z\bar{B}' + \bar{A}'),$$

$$(I - \bar{W}'W)\partial_w'(I - W\bar{W}')\partial_{\bar{w}}$$

$$= \overline{(ZC + D)}'^{-1}(I - \bar{Z}'Z)\partial_z'(I - Z\bar{Z}')\partial_{\bar{z}}\overline{(ZC + D)}'.$$

Thus both (7.59) and (7.60) are G-invariant. In order to prove that system (7.60) annihilates the Poisson kernel, we need only verify it at $Z = 0$. That is,

$$\partial'_Z \partial_{\bar{Z}} P(Z, U)\Big|_{Z=0} = \partial'_Z \partial_{\bar{Z}} \frac{\det(I - Z\bar{Z}')^n}{|\det(I - Z\bar{U}')|^{2n}}\Big|_{Z=0} = n^2 \bar{U}'U - n^2 I = 0.$$

Now we are going to show that all the operators in systems (7.59) and (7.60) generate a prime ideal, and they are exactly the ring of differential operators which annihilate the Poisson kernel.

Consider $\partial_{\bar{Z}}\partial'_Z, \partial'_Z \partial_{\bar{Z}} \to \bar{Z}Z', Z'\bar{Z}$. In the light of Theorems 2.15 and 7.17, the final step is to verify the following:

1. Elements of matrices $\bar{Z}Z', Z'\bar{Z}$ generate a prime ideal in the ring $\mathbb{C}[z_1, z_2, \ldots, z_N]$.
2. The system $\{\partial_{\bar{Z}}\partial'_Z, \partial'_Z \partial_{\bar{Z}}\}$ is complete with respect to the Poisson kernel $P(Z, U)$.

For proving Condition 2 of the theorem we have to discuss the so-called U-harmonic polynomials.

DEFINITION 7.18. Let $Z = (z_{ij})$ be an $n \times n$ matrix. A polynomial $f(Z)$ of $z_{ij}, \bar{z}_{ij}$ is said to be U-harmonic if

$$\partial_{\bar{Z}}\partial'_Z f(Z) = \partial'_Z \partial_{\bar{Z}} f(Z) = 0. \tag{7.61}$$

In the Poisson kernel $P(Z, U)$, we have $U \in U(n)$, and $U(n)$ can be considered as a manifold N. Since $U = (u_{ij})$ and not all monomials of $U, \bar{U}$ are independent, a reduction must be performed under $U\bar{U}' = \bar{U}'U = I$ [see (7.16)]. For this we propose a lemma.

LEMMA 7.19. *Let $f(Z, \bar{Z})$ be a homogeneous polynomial of degree m for $Z, \bar{Z}$, satisfying*

$$f(Z\bar{U}, \bar{Z}U) = f(\bar{U}Z, \bar{Z}U), \qquad \forall U \in U(n).$$

Then under the condition $U\bar{U}' = \bar{U}'U = I, f(Z\bar{U}, \bar{Z}U)$ can be reduced to

$$f(Z\bar{U}, \bar{Z}U) = \sum_{i=1}^{N_m} \varphi_i(Z, \bar{Z})\xi_i(U) + \text{terms with lower degree of } U, \bar{U}, \tag{7.62}$$

where $\{\xi_i(U)\}$ are linearly independent irreducible monomials of U, $\bar{U}$ with degree m, and $\varphi_i(Z, \bar{Z})$ are U-harmonic homogeneous polynomials with the same degree.

PROOF. Since

$$\frac{\partial}{\partial \bar{Z}} f(Z\bar{U}, \bar{Z}U) = \frac{\partial}{\partial Y} f(Z\bar{U}, Y)|_{Y = \bar{Z}U} \cdot U',$$

$$\frac{\partial'}{\partial Z} f(\bar{U}Z, \bar{Z}U) = \frac{\partial'}{\partial X} f(X, \bar{Z}U)|_{X = \bar{U}Z} \cdot \bar{U},$$

we have

$$\partial_{\bar{Z}} \partial'_Z f(Z\bar{U}, \bar{Z}U) = \partial_{\bar{Z}} \partial'_Z f(\bar{U}Z, \bar{Z}U) = \partial_{\bar{Z}} \{\partial'_X f(X, \bar{Z}U)\}_{X = \bar{U}Z} \cdot \bar{U}$$

$$= \partial_{\bar{X}} \{\partial'_X f(X, \bar{X})\}_{\substack{X = \bar{U}Z \\ \bar{X} = \bar{Z}U}} \cdot U'\bar{U}' = \partial_{\bar{X}} \partial'_X f(X, \bar{X})|_{\substack{X = \bar{U}Z \\ \bar{X} = \bar{Z}U}}$$

By (7.62) it follows that

$$\sum_i \partial_{\bar{Z}} \partial'_Z \varphi_i(Z, \bar{Z}) \xi_i(U) + \text{terms with lower degree of } U, \bar{U}$$

$$= \text{terms with lower degree of } U, \bar{U}.$$

Hence

$$\sum_i \partial_{\bar{Z}} \partial'_Z \varphi_i(Z, \bar{Z}) \xi_i(U) = 0.$$

Resorting to the linear independence of $\xi_i(U)$, we have $\partial_{\bar{Z}} \partial'_Z \varphi_i(Z, \bar{Z}) = 0$. Applying the same argument, we have $\partial'_Z \partial_{\bar{Z}} \varphi_i(Z, \bar{Z}) = 0$. This completes the proof. $\square$

Now we are going to expand the power series of $P(Z, U)$ at $Z = 0$. Assume that

$$\det (I - X)^{-n} = \sum_{i=0}^{\infty} A_j(x),$$

$A_j(x)$ being a homogeneous polynomial of X with degree j, and then we have

$$\det (I - Z\bar{U}')^{-n} = \sum_{j=0}^{\infty} A_j(Z\bar{U}'); \quad \det (I - \bar{Z}U')^{-n} = \sum_{j=0}^{\infty} A_j(\bar{Z}U').$$

Therefore

$$P(Z, U) = \det(I - Z\bar{Z}')^n \det(I - Z\bar{U}')^{-n} \det(I - \bar{Z}U')^{-n}$$

$$= (1 - n\,\mathrm{tr}\,(Z\bar{Z}') + \cdots) \sum_{j=0}^{\infty} A_j(Z\bar{U}') \cdot \sum_{l=0}^{\infty} A_l(\bar{Z}U')$$

$$= \sum_{m=0}^{\infty} p_m(Z, U)$$

and

$$p_m(Z, U) = \sum_{i+j=m} A_i(Z\bar{U}')A_j(\bar{Z}U')$$

$$- n\,\mathrm{tr}\,(Z\bar{Z}') \sum_{i+j=m-2} A_i(Z\bar{U}')A_j(\bar{Z}U') + \cdots$$

$$= \sum_{i+j=m} A_i(Z\bar{U}')A_j(\bar{Z}U')$$

$$+ \text{ terms with lower degree of } U, \bar{U}. \tag{7.63}$$

In (7.63), we might as well regard U' as U, and put

$$f(Z, \bar{Z}) = \sum_{i+j=m} A_i(Z)A_j(\bar{Z}).$$

By the relations

$$\det(I - Z\bar{U})^{-n} = \det(U'(I - \bar{U}Z)\bar{U})^{-n} = \det(I - \bar{U}Z)^{-n},$$

we get $A_i(Z\bar{U}) = A_i(\bar{U}Z)$, so $f(Z\bar{U}, \bar{Z}U) = f(\bar{U}Z, \bar{Z}U)$. In view of Lemma 7.19 and for $\sum_{i+j=m} A_i(Z\bar{U})A_j(\bar{Z}U)$, reduction of $U\bar{U} = \bar{U}U = I$ yields

$$p_m(Z, U) = \sum_{i+j=m} A_i(Z\bar{U})A_j(\bar{Z}U) + \text{ terms with lower degree of } U, \bar{U}$$

$$= \sum_i \varphi_i(Z, \bar{Z})\xi_i(U) + \text{ terms with lower degree of } U, \bar{U}, \tag{7.64}$$

where $\varphi_i(Z, \bar{Z})$ are U-harmonic and $\xi_i(U)$ are irreducible monomials of $U, \bar{U}$ with degree m.

If Z is a 2×2 matrix, we can prove furthermore the validity of the following conclusions.

PROPOSITION 7.20. $\{\varphi_i(Z, \bar{Z})\}$ in (7.64) forms a linear basis of the space of all $\{\partial_{\bar{Z}}\partial'_Z, \partial'_Z\partial_{\bar{Z}}\}$-harmonic homogeneous polynomials with degree m.

PROPOSITION 7.21. On the assumption that

$$X = \begin{pmatrix} x_{11} & x_{12} \\ x_{21} & x_{22} \end{pmatrix}, \qquad Y = \begin{pmatrix} y_{11} & y_{12} \\ y_{21} & y_{22} \end{pmatrix},$$

the elements of XY, YX generate a prime ideal. Thus if we set $\bar{Z} = X$, $Y = Z'$, then the elements of $\bar{Z}Z'$, $Z'\bar{Z}$ generate a prime ideal.

For all 2×2 matrices, these propositions can be established with direct, although involved, calculations. Owing to the limitation of space, we omit the proof in detail.

Now when $n = 2$, by Theorem 7.17 we have that $\{\partial_{\bar{Z}}\partial'_Z, \partial'_Z\partial_{\bar{Z}}\}$ is complete with respect to the Poisson kernel $P(Z, U)$, and by Theorem 7.15 we deduce the following theorem.

THEOREM 7.22. For $n = 2$, the ideal generated by the system of differential operators on $R_I(2)$,

$$(\partial_{\bar{Z}}(I - \bar{Z}'Z)\partial'_Z, \partial'_Z(I - Z\bar{Z}')\partial_{\bar{Z}}), \tag{7.65}$$

is prime.

REMARK 7.23. Since $\operatorname{tr}(\partial'_Z(I - Z\bar{Z}')\partial_Z) = \operatorname{tr}(\partial_{\bar{Z}}(I - \bar{Z}'Z)\partial'_Z)$, there exists a linear relation between the elements of (7.65).

On $R_I(2)$, the solutions of the Laplacian

$$\Delta = \operatorname{tr}[(I - Z\bar{Z}')\partial_{\bar{Z}}(I - \bar{Z}'Z)\partial'_Z]$$

are given by the Poisson integral (see the introduction in Ref. 2). For a $T \in H(\Delta)$, of necessity $TP(Z, U) = 0$. Our theorem thus gives

$$T \in (\partial_{\bar{Z}}(I - \bar{Z}'Z)\partial_Z, \partial'_Z(I - Z\bar{Z}')\partial_{\bar{Z}}).$$

Conversely, if $T \in (\partial_Z(I - \bar{Z}'Z)\partial'_Z, \partial'_Z(I - Z\bar{Z}')\partial_{\bar{Z}})$, then $TP(Z, U) = 0$, so $Th = 0$ for all harmonic functions h. Therefore $T \in H(\Delta)$. Thus

$$H(\Delta) = (\partial_{\bar{Z}}(I - \bar{Z}'Z)\partial'_Z, \partial'_Z(I - Z\bar{Z}')\partial_{\bar{Z}}).$$

Obviously, $(\Delta) \neq (\partial_{\bar{Z}}(I - \bar{Z}'Z)\partial'_Z, \partial'_Z(I - Z\bar{Z}')\partial_{\bar{Z}})$; therefore

$$(\Delta) \neq H(\Delta) = (\partial_{\bar{Z}}(I - \bar{Z}'Z)\partial'_Z, \partial'_Z(I - Z\bar{Z}')\partial_{\bar{Z}}).$$

References

1. A. Erdelyi, *et al.*, *Higher Transcendental Functions*, New York (1953).
2. L.-K. Hua, *Harmonic Analysis of Functions of Several Complex Variables in the Classical Domain*, Science Press, Beijing (1958) (in Chinese); American Mathematical Society, Providence, RI (1968).
3. L.-K. Hua and L.-H. Look, Theory of harmonic function in the classical domain. I, *Acta Math. Sinica* **8**, 531-547 (1958); *Sci. Sinica* **9**, 1031-1094 (1958) (in English).
4. H. Maass, Zur Theorie der Harmonischen Formen, *Math. Ann.*, **137**, 142-149 (1959).

8

On the Sum of Class Functions
of Weyl Groups

It is well known that the Weyl groups play a key role in the theory of the structure of Lie algebras and their linear representations, as well as in the theory of Chevalley groups. Suppose g is a complex semisimple Lie algebra, h its Cartan subalgebra, Σ and Π denote its root system and the basis of the root system, respectively, and w_{α_i} indicates the reflection determined by $\alpha_i \in \Pi$. The so-called Weyl group, denoted by W, refers to the finite group generated by all w_{α_i} ($\alpha_i \in \Pi$). There have been many works published on the structure and properties of W [6, 7]. The following Molien identity related to the global structure of W is well known [3]:

$$\sum_{w \in W} \frac{1}{\det(1 - tw)} = \frac{1}{\prod(1 - t^{d_i})},\tag{8.1}$$

where d_i signify the Poincaré indices of g. In 1956, by topological methods based on Morse's theory, Bott proved another identity [2]:

$$\sum_{w \in W} t^{l(w)} = \prod \frac{1 - t^{d_i}}{1 - t},\tag{8.2}$$

where $l(w)$ refers to the Π-length of w; i.e., w may be expressed by the number of minimal products of w_{α_i} ($\alpha_i \in \Pi$). This important formula may shed new light on our understanding the structure of W. In 1966, Solomon offered a purely algebraic proof of (8.2), thus solving the order problem of the Chevalley simple groups [5, 3].

The left-hand sides of (8.1) and (8.2) represent the form $\sum_{w \in W} f(w)$, $f(w)$ being a function on W. This chapter investigates the summation of $\sum f(w)$ for the class functions (they are considered to be the most important functions) on the four classical simple algebras. The so-called class functions are those of the conjugacy classes—i.e., all $w, w_1 \in W$, satisfying $f(w) = f(w_1 w w_1^{-1})$.

Our results are as follows. Section 8.1 proposes some preliminary lemmas (which are also true for the exceptional simple algebras). Section 8.2 presents some recursion formulas on $\Sigma f(w)$. Section 8.3 contains the following formulas: when $f(w)$ belongs to some concrete functions, they lead to some new significant formulas, which include the formula on $\Sigma t^{\bar{l}(w)}$, with $\bar{l}(w)$ as the Σ-length of w, the formula on $\sum_{w^2=1} t^{\bar{l}(w)}$, giving the total numbers of involutions of various classes, and the formula on $\Sigma t^{\text{tr} \, w}$, giving the distribution of the traces of elements in W.

The author would like to express his thanks to Professor Wan Ze-xian for going over this manuscript and for his helpful comments.

8.1. Preliminaries

It is well known that the Weyl group W of a semisimple Lie algebra g can be regarded as a finite group generated by reflections of a σ root system in a finite-dimensional Euclidean space [7]. Let V be an l-dimensional Euclidean space, and let the reflection w_r determined by $r \in V$ be

$$x \in V, \qquad x \to w_r(x) = x - \frac{2(r, x)}{(r, r)} r.$$

By Σ, Σ^+, $\Pi = \{\alpha_1, \ldots, \alpha_l\}$ we denote the root system, positive root system, and basis of the root system, respectively. Any element of W can be expressed in the form $w = w_{r_1} \cdots w_{r_k}$, $r_i \in \Sigma^+$; the minimum of k in such expressions is denoted by $\bar{l}(w)$, called the Σ-length of w, so as to distinguish from the Π-length $l(w)$ of w. Put $V_1(w)$ for the eigenspace corresponding to the eigenvalue 1 of w, and $V_1^{\perp}(w)$ to be the orthogonal complement of $V_1(w)$. It is obvious that $V = V_1(w) \oplus V_1^{\perp}(w)$.

LEMMA 8.1.

$$\bar{l}(w) = \dim V_1^{\perp}(w). \tag{8.3}$$

PROOF [4]. Suppose $\bar{l}(w) = k$, $w = w_{r_1} \cdots w_{r_k}$, $r_i \in \Sigma$. Provided $H_{r_i}(i = 1, \ldots, k)$ is the hyperplane orthogonal to r_i, then, evidently, there exist $H_{r_1} \cap \cdots \cap H_{r_k} \subseteq V_1(w)$. Thus

$$\dim V_1^{\perp}(w) = l - \dim V_1(w) \leq l - \dim (H_{r_1} \cap \cdots \cap H_{r_k}).$$

However, since $\dim (H_{r_1} \cap \cdots \cap H_{r_k}) \geq l - k$, we admit $\dim V_1^{\perp}(w) \leq l - (l - k) = k = \bar{l}(w)$.

Next, let us prove $\bar{l}(w) \leq \dim V_1^{\perp}(w)$. For the dimension of V we use the method of induction. If $V_1(w) \neq 0$, then by a well-known result on finite reflection groups [6], w can be expressed as such a product of reflections w_r that each successive point in $V_1(w)$ is fixed under each w_r, and so $r \in V_1^{\perp}(w)$. All roots in $V_1^{\perp}(w)$ form a root system, and $\dim V_1^{\perp}(w) < l$. By the inductive hypothesis, $\bar{l}(w) \leq \dim V_1^{\perp}(w)$.

When $V_1(w) = 0$, we have $\dim V_1^{\perp}(w) = l$. For any $r \in \Sigma$, since $w - 1$ is nonsingular, there exists $v \in V$ satisfying $(w - 1)v = r$; i.e., $w(v) = v + r$. By $(w(v), w(v)) = (v, v) = (v + r, v + r)$, we have

$$\frac{2(v, r)}{(r, r)} = -1;$$

thus $w_r(v) = w(v)$, which means $w_r w(v) = v$. Thereby we find $V_1(w_r w) \neq 0$. Referring to the previous argument, we have $\bar{l}(w_r w) \leq \dim V_1^{\perp}(w_r w) \leq l - 1$; thus we conclude

$$\bar{l}(w) \leq \dim V_1^{\perp}(w) = l.$$

The lemma is readily proved. $\qquad\qquad\qquad\qquad\qquad\qquad\qquad\qquad\qquad\square$

LEMMA 8.2. *Let $w = w_{r_1} \cdots w_{r_k}$. Then $\bar{l}(w) = k$ if and only if $r_1, \ldots, r_k$ are linearly independent.*

PROOF. Considering the previous lemma, it suffices to show that $\dim V_1^{\perp}(w) = k \Leftrightarrow r_1, \ldots, r_k$ are linearly independent.

Necessity. If $\dim V_1^{\perp}(w) = k$, then $\dim V_1(w) = l - k$. Since $H_{r_1} \cap \cdots \cap H_{r_k} \subseteq V_1(w)$, it follows that $\dim (H_{r_1} \cap \cdots \cap H_{r_k}) \leq l - k$. On the other hand, obviously $\dim (H_{r_1} \cap \cdots \cap H_{r_k}) \geq l - k$, and so we arrive at $\dim (H_{r_1} \cap \cdots \cap H_{r_k}) = l - k$. Thus $r_1, \ldots, r_k$ are linearly independent.

Sufficiency. Under the constraint that $r_1, \ldots, r_k$ are linearly independent, we infer $\dim (H_{r_1} \cap \cdots \cap H_{r_k}) = l - k$. To show $\dim V_1^{\perp}(w) = k$, we need only show $V_1(w) \subseteq H_{r_1} \cap \cdots \cap H_{r_k}$. Thus it suffices to prove the validity of $V_1(w) \subseteq H_{r_1} \cap \cdots \cap H_{r_k}$. We can see easily that

$$w_{r_1} \cdots w_{r_{i-1}}(r_i) = r_i + (r_1, \ldots, r_{i-1}),$$

where $(r_1, \ldots, r_{i-1})$ means a linear combination of $r_1, \ldots, r_{i-1}$. However, $r_1, \ldots, r_k$ are linearly independent, and so the same holds for $r_1, w_{r_1}(r_2)$, $w_{r_1} w_{r_2}(r_3), \ldots, w_{r_1} \cdots w_{r_{k-1}}(r_k)$.

Choose arbitrarily $x \in V_1(w)$; that is, $w(x) = x$. What we want to show is $x \in H_{r_1} \cap \cdots \cap H_{r_k}$. Because

$$w(x) = w_{r_1} \cdots w_{r_k}(x) = w_{r_1} \cdots w_{r_{k-1}}\left(x - \frac{2(r_k, x)}{(r_k, r_k)} r_k\right)$$

$$= w_{r_1} \cdots w_{r_{k-1}}(x) - \frac{2(r_k, x)}{(r_k, r_k)} w_{r_1} \cdots w_{r_{k-1}}(r_k)$$

$$= \cdots$$

$$= x - \frac{2(r_1, x)}{(r_1, r_1)} r_1 - \frac{2(r_2, x)}{(r_2, r_2)} w_{r_1}(r_2) - \cdots$$

$$- \frac{2(r_k, r_k)}{(r_k, r_k)} w_{r_1} \cdots w_{r_{k-1}}(r_k) = x,$$

we must have $(r_i, x) = 0$ $(i = 1, \ldots, k)$; i.e., $x \in H_{r_1} \cap \cdots \cap H_{r_k}$. The lemma is proved.   □

Let J be a subset of Π, let W_J represent the subgroup of W generated by $\{w_r | r \in J\}$, and let Σ_J denote the root system spanned by J.

LEMMA 8.3. *If $w = w_r w'$, $w' \in W_J$, $r \notin \Sigma_J$, then we infer*

$$\bar{l}(w_r w') = 1 + \bar{l}(w'). \tag{8.4}$$

PROOF. Since $w' \in W_J$, w' can be written as a product of reflections determined by the roots in Σ_J. On account of $r \notin \Sigma_J$, it follows that r and Σ_J are linearly independent. From Lemma 8.2, the conclusion is apparent.

$w \in W$ is called an involution if $w^2 = I$. We have the following lemma related to involutions.

LEMMA 8.4. *Let $w = w_{r_1} \cdots w_{r_k}$, $\bar{l}(w) = k$. Then w is an involution if and only if any two of $r_1, \ldots, r_k$ are orthogonal.*

PROOF. The sufficiency is trivial since $(r_i, r_j) = 0 \Leftrightarrow w_{r_i}$ and w_{r_j} commute.

Necessity. Since $\bar{l}(w) = k$, considering Lemma 8.2, $r_1, \ldots, r_k$ are linearly independent. From the proof of the same lemma, we have $V_1(w) = H_{r_1} \cap \cdots \cap H_{r_k}$, $\dim V_1(w) = l - k$. Hence $V_1^{\perp}(w)$ is equal to the linear subspace $V(r_1, \ldots, r_k)$ generated by $r_1, \ldots, r_k$. But $w^2 = I$, the eigenvalues of w are either 1 or -1, and so $V = V_1(w) \oplus V_{-1}(w)$. Consequently, $V_{-1}(w) = V_1^{\perp}(w)$. Therefore r_i $(i = 1, \ldots, k)$ give rise to $w(r_i) = -r_i$; especially the equality $w(r_k) = w_{r_1} \cdots w_{r_k}(r_k) = -r_k$ leads to $w_{r_1} \cdots w_{r_{k-1}}(r_k) = r_k$.

However, there holds

$$w_{r_1} \cdots w_{r_{k-1}}(r_k) = r_k - \frac{2(r_1, r_k)}{(r_1, r_1)} w_{r_1}(r_2) - \cdots - \frac{2(r_{k-1}, r_k)}{(r_{k-1}, r_{k-1})} w_{r_1} \cdots w_{r_{k-1}}(r_{k-1}).$$

Modeling the proof of Lemma 8.2, we admit $w_{r_1}(r_2), \ldots, w_{r_1} \cdots w_{r_{k-2}}(r_{k-1})$ are linearly independent, which implies $(r_i, r_k) = 0$ $(i = 1, \ldots, k-1)$. Thus w_{r_k} commutes with $w_{r_1} \cdots w_{r_{k-1}}$. This means $w = w_{r_k} w'$, $w' = w_{r_1} \cdots w_{r_{k-1}}$ is also an involution. By induction, we come to the conclusion immediately.   □

8.2. Some Recursion Formulas

Let g be a semisimple Lie algebra of rank l, Σ its root system, and $\Pi = \{\alpha_1, \ldots, \alpha_l\}$ the basis of Σ. The Weyl group of g is denoted by $W(g)$ and abbreviated to W if no confusion rises. The Poincaré indices are written $d_1, \ldots, d_l$. The following lemmas are well known.

LEMMA 8.5 [*Molien (see in Ref.* 3)].

$$\sum_{w \in W} \frac{1}{\det (1 - tw)} = 1 \bigg/ \prod_{i=1}^{l} (1 - t^{d_i}). \tag{8.5}$$

LEMMA 8.6 (*Todd-Shepard* [3]).

$$\sum_{i=1}^{l} (d_i - 1) = N, \qquad \prod_{i=1}^{l} d_i = |W|, \tag{8.6}$$

where N means the number of the positive roots of g all together and $|W|$ means the order of w.

Suppose J is a subset of Π. The subgroup of W generated by w_α, $\alpha \in J$, is denoted by W_J, which can be regarded as a finite reflection group in the linear subspace V_J spanned by J. Define $D_J = \{w \in W | w(J) > 0\}$, such that the set of elements in W maps J into positive roots. Then the following decomposition theorem on W arises (for the proof see Ref. 3).

LEMMA 8.7. If J is a subset of Π, then every $w \in W$ can be written uniquely as $w = d_J w_J$, with $d_J \in D_J$, $w_J \in W_J$, and

$$l(w) = l(d_J) + l(w_J). \tag{8.7}$$

Here $l(w)$ denotes the minimal number of factors in the expressions w, which are written as a product of w_{α_i} ($\alpha_i \in \Pi$). It is well known that this number also equals that of positive roots, their sign being changed under W. Hence the Bott-Solomon formula (8.2) described in the introduction can be rewritten as

$$\sum_w t^{l(w)} = \sum_{d_J \in D_J} t^{l(d_J)} \sum_{w_J \in W_J} t^{l(w_J)} = \prod_{i=1}^l \frac{1-t^{d_i}}{1-t}.$$

$$\therefore \sum_{d_J \in D_J} t^{l(d_J)} = \prod_{i=1}^l \frac{1-t^{d_i}}{1-t} \bigg/ \sum_{w_J \in W_J} t^{l(w_J)} = \prod_{i=1}^l \frac{1-t^{d_i}}{1-t} \bigg/ \prod_{i=1}^{|J|} \frac{1-t^{d_i}}{1-t}.$$

Let $t = 1$,

$$\prod_{i=1}^l \frac{1-t^{d_i}}{1-t} = \prod_{i=1}^l d_i.$$

By (8.6),

$$|D_J| = \frac{|W|}{|W_J|} = \frac{d_1 \cdots d_l}{d_1' \cdots d_{|J|}'} \tag{8.8}$$

$d_1', \ldots, d_{|J|}'$ indicating the Poincaré indices of W_J and $|J|$ the number of elements in J.

The following discussion is restricted to the four classical simple algebras: A_l, B_l, C_l, D_l. Since $W(B_l) \simeq W(C_l)$, it is sufficient to consider

A_l, C_l, D_l only. Their Dynkin diagrams are as follows:

$$\Gamma(A_l): \qquad \overset{\alpha_1}{\bullet}\!\!-\!\!-\!\!\overset{\alpha_2}{\bullet}\!\!-\!\!-\!\!\overset{\alpha_3}{\bullet}\cdots\cdots\overset{\alpha_{l-1}}{\bullet}\!\!-\!\!-\!\!\overset{\alpha_l}{\bullet} \qquad l \geq 1,$$

$$\Gamma(B_l): \qquad \overset{\alpha_1}{\bullet}\!\!-\!\!-\!\!\overset{\alpha_2}{\bullet}\!\!-\!\!-\!\!\overset{\alpha_3}{\bullet}\cdots\cdots\overset{\alpha_{l-1}}{\bullet}\!\!=\!\!\overset{\alpha_l}{\bullet} \qquad l \geq 2,$$

$$\Gamma(C_l): \qquad \overset{\alpha_1}{\bullet}\!\!-\!\!-\!\!\overset{\alpha_2}{\bullet}\!\!-\!\!-\!\!\overset{\alpha_3}{\bullet}\cdots\cdots\overset{\alpha_{l-1}}{\bullet}\!\!=\!\!\overset{\alpha_l}{\bullet} \qquad l \geq 2, \qquad (8.9)$$

$$\Gamma(D_l): \qquad \overset{\alpha_1}{\bullet}\!\!-\!\!-\!\!\overset{\alpha_2}{\bullet}\!\!-\!\!-\!\!\overset{\alpha_3}{\bullet}\cdots\cdots\overset{\alpha}{\bullet}\!\!\!<\!\!\!\overset{\overset{\textstyle\alpha_{l-1}}{\bullet}}{\underset{\textstyle\alpha_l}{\bullet}} \qquad l \geq 4,$$

where $\bullet$ stands for a short vector, and $\circ$ a long one. If $\circ\!\!-\!\!-\!\!\bullet$ or $\bullet\!\!-\!\!-\!\!\circ$, then it follows that

$$\frac{2(\alpha, \beta)}{(\alpha, \alpha)} = \frac{2(\alpha, \beta)}{(\beta, \beta)} = -1, \qquad \alpha, \beta, \alpha + \beta \in \Sigma;$$

if $\circ\!\!=\!\!\bullet$, then it follows that

$$\frac{2(\alpha, \beta)}{(\alpha, \alpha)} = \frac{4(\alpha, \beta)}{(\beta, \beta)} = 2, \qquad \alpha, \beta, \beta + \alpha, \beta + 2\alpha \in \Sigma.$$

Moreover, recall that for A_l, B_l, C_l, D_l, the table of Poincaré indices is

$$d_1, d_2, d_3, \ldots, d_{l-1}, d_l$$

$$A_l \qquad 2, 3, 4, \ldots, l, l + 1$$

$$(8.10)$$

$$B_l, C_l \quad 2, 4, 6, \ldots, 2(l - 1), 2l$$

$$D_l \qquad 2, 4, 6, \ldots, 2(l - 1), l.$$

It is worth noting that a convention is made to fix the position of $\alpha_1, \ldots, \alpha_l$ in the Dynkin diagram (8.9) and the order of $d_1, \ldots, d_l$ in the table (8.10). For example, d_l is just l for D_l.

LEMMA 8.8. *If g is prescribed to be one of A_l, B_l, C_l, D_l and $J = \{\alpha_2, \alpha_3, \ldots, \alpha_l\}$, then*

$$|D_J| = \begin{cases} l + 1, & \text{if } g = A_l, \\ 2l, & \text{if } g = B_l, C_l, D_l. \end{cases} \qquad (8.11)$$

PROOF. By (8.8), we arrive at $|D_J| = |W|/|W_J|$ and W_J is clearly isomorphic to the Weyl group of a simple algebra of the same class with rank

$l - 1$. Resorting to the formulas (8.6) and (8.10), we come to the conclusion of the lemma.   □

For further study of the structure of D_J, we will first try to prove the next lemma.

LEMMA 8.9. *Let g be one of A_l, B_l, C_l, D_l, $J = \{\alpha_1, \ldots, \alpha_l\}$ and $D_J = \{w \in W \mid w(J) > 0\}$. Furthermore let $D_{J'}$ be the elements of the Weyl subgroup W_J which change $\alpha_3, \alpha_4, \ldots, \alpha_l$ to positive roots. Then we have $D_{J'} w_{\alpha_1} \in D_J$.*

PROOF. Take arbitrarily $w' \in D_{J'}$; that is, $w' \in W_J$. Simultaneously, there exist $w'(\alpha_i) > 0$, $i = 2, 3, \ldots, l$. We want to verify the correctness of $w' w_{\alpha_1} \in D_J$.

If $w' = 1$, then $w' w_{\alpha_1} = w_{\alpha_1} \in D_J$ is self-evident.

For $w' \neq 1$, necessarily $w'(\alpha_2) < 0$, or else $w'(\alpha_i) > 0$ for $i = 2, 3, \ldots, l$. At the same time, since $w' \in W_J$, there is no doubt that $w'(\alpha_1) > 0$. In this way we have $w'(\alpha_i) > 0$ for all $\alpha_i \in \Pi$. This means $w' = 1$, which induces a contradiction. Hence $w'(\alpha_2) < 0$. Put $w'' = w' w_{\alpha_2}$. Then it follows that $w'' \in W_J$; thus

$$w' w_{\alpha_1}(\alpha_2) = w'(\alpha_1 + \alpha_2) = w'' w_{\alpha_2}(\alpha_1 + \alpha_2) = w''(\alpha_1 + \alpha_2 - \alpha_2) = w''(\alpha_1)$$

$$= \alpha_1 + \text{a linear combination of } J > 0, \text{ if } g \neq C_2.$$

$$w' w_{\alpha_1}(\alpha_2) = 2\alpha_1 + \text{a linear combination of } J > 0, \text{ if } g = C_2.$$

On the other hand, we have $w' w_{\alpha_1}(\alpha_i) = w'(\alpha_i) > 0$, for $i = 3, 4, \ldots, l$. This results in $w' w_{\alpha_1} \in D_J$ and concludes the proof of the lemma.   □

From this lemma, we obtain the following result.

THEOREM 8.10. *Let g be one of A_l, B_l, C_l, D_l. The meaning of J, D_J, and $D_{J'}$ is assumed to be the same as before. Then*

$$D_J = \begin{cases} 1 \cup D_{J'} w_{\alpha_1}, & g = A_l, \\ 1 \cup D_{J'} w_{\alpha_1} \cup w_\tau, & g = B_l, C_l, \\ 1 \cup D_{J'} w_{\alpha_1} \cup w_{\tau_1} w_{\tau_2}, & g = D_l, \end{cases} \tag{8.12}$$

with

$$\tau = \alpha_1 + \alpha_2 + \cdots + \alpha_l \quad (g = B_l),$$

$$\tau = 2\alpha_1 + 2\alpha_2 + \cdots + 2\alpha_{l-1} + \alpha_l \quad (g = C_l),$$

$$\tau_1 = \alpha_1 + \cdots + \alpha_{l-2} + \alpha_{l-1},$$

$$\tau_2 = \alpha_1 + \cdots + \alpha_{l-2} + \alpha_l \quad (g = D_l)$$

PROOF. On the basis of Lemma 8.8 if $g = A_l$, then $|D_J| = l + 1$; if $g = B_l$, C_l, D_l, then $|D_J| = 2l$. Moreover, if the rank of g is l and $J = \{\alpha_2, \ldots, \alpha_l\}$, evidently W_J is isomorphic to the Weyl group of a simple algebra of the same type with rank $l - 1$; therefore the relation between $D_{J'}$ and W_J in Lemma 8.9 is the same as that between D_J and W. In this case, it turns out that

$$|D_{J'}| = \begin{cases} l, & g = A_l, \\ 2(l-1), & g = B_l, C_l, D_l. \end{cases}$$

By Lemma 8.9, we obtain $D_{J'} w_{\alpha_1} \in D_J$. Under this circumstance, for $g = A_l$, in addition to $D_{J'} w_{\alpha_1}$, there is another element in D_J, which is the identity $w = 1$, while for $g = B_l$, C_l, in addition to $D_{J'} w_{\alpha_1}$ ($2l - 2$ elements in total) two more elements exist, namely, 1 and w_τ, where $\tau = \alpha_1 + \cdots + \alpha_l$ ($g = B_l$) or $\tau = 2\alpha_1 + \cdots + 2\alpha_{l-1} + \alpha_l$ ($g = C_l$), which can be checked directly. Finally, for $g = D_l$, in addition to $D_{J'} w_{\alpha_1}$, there are 1 and $w_{\tau_1} w_{\tau_2}$, where $\tau_1 = \alpha_1 + \cdots + \alpha_{l-2} + \alpha_{l-1}$ and $\tau_2 = \alpha_1 + \cdots + \alpha_{l-2} + \alpha_l$, which can also be checked directly. Thus the theorem is proved. $\Box$

Now, let us investigate the sum $\sum_{w \in W} f(w)$, where $f(w)$ is a class function on W. Put $J = \{\alpha_2, \ldots, \alpha_l\}$. Then W_J is isomorphic to the Weyl group of a simple algebra of the same type with rank $l - 1$. Lemma 8.7 implies

$$\sum_{w \in W} f(w) = \sum_{d_J \in D_J, \, w_J \in W_J} f(d_J w_J) = \sum_{d_J \in D_J} \sum_{w \in W_J} f(d_J w).$$

From Theorem 8.10, the main part of D_J is just the set $D_{J'} w_{\alpha_1}$. If $d_J = d_{J'} w_{\alpha_1} \in D_{J'} w_{\alpha_1}$, then

$$\sum_{w \in W_J} f(d_J w) = \sum_{w \in W_J} f(d_{J'} w_{\alpha_1} w) = \sum_{w \in W_J} f(w_{\alpha_1} w d_{J'}) = \sum_{w \in W_J} f(w_{\alpha_1} w).$$

The last step of the proof comes from the fact that $d_{J'} \in D_{J'} \subset W_J$.

By the preceding analysis together with Theorem 8.10, we note that $|D_{J'}| = l$ ($g = A_l$) and $|D_{J'}| = 2(l - 1)$ ($g = B_l$, C_l, D_l), which lead to the main result of this section.

THEOREM 8.11. *With regard to a class function $f(w)$, the following formulas are valid:*

$$\sum_{w \in W(A_l)} f(w) = \sum_{w \in W_J(A_l)} f(w) + l \sum_{w \in W_J(A_l)} f(w_{\alpha_1} w), \tag{8.13}$$

$$\sum_{w \in W(B_l)} f(w) = \sum_{w \in W_J(B_l)} f(w) + (2l - 2) \sum_{w \in W_J(B_l)} f(w_{\alpha_1} w)$$

$$+ \sum_{w \in W_J(B_l)} f(w_\tau w), \quad (8.14)$$

$$\sum_{w \in W(C_l)} f(w) = \sum_{w \in W_J(C_l)} f(w) + (2l - 2) \sum_{w \in W_J(C_l)} f(w_{\alpha_1} w)$$

$$+ \sum_{w \in W_J(C_l)} f(w_\tau w), \quad (8.15)$$

$$\sum_{w \in W(D_l)} f(w) = \sum_{w \in W_J(D_l)} f(w) + (2l - 2) \sum_{w \in W_J(D_l)} f(w_{\alpha_1} w)$$

$$+ \sum_{w \in W_J(D_l)} f(w_{\tau_1} w_{\tau_2} w), \quad (8.16)$$

where w_τ, w_{τ_1}, w_{τ_2} are as in Theorem 8.10.

Since W_J is isomorphic to the Weyl group of a Lie algebra of the same type with rank $l - 1$—i.e., $W_J(A_l) \simeq W(A_{l-1})$, etc., then (8.15)-(8.16) allow us to find $\sum f(w)$ by recursion. In these formulas, the only difficulty lies in finding the term $\sum_{w \in W_J(D_l)} f(w_{\tau_1} w_{\tau_2} w)$ in (8.16). To do this, we embed $W(D_l)$ into $W(C_l)$ and reduce the problem to $\sum_{w \in W_J(C_l)} f(w)$.

Recall that [7], $W(C_l)$ is isomorphic to the following signed permutation group:

$$(\lambda_1, \ldots, \lambda_l) \to (\varepsilon_1 \lambda_{i_1}, \ldots, \varepsilon_l \lambda_{i_l}), \qquad \varepsilon_i = \pm 1,$$

where $(i_1, \ldots, i_l)$ is a permutation of $(1, \ldots, l)$. $W(D_l)$ is isomorphic to

$$(\lambda_1, \ldots, \lambda_l) \to (\varepsilon_1 \lambda_{i_1}, \ldots, \varepsilon_l \lambda_{i_l}), \qquad \varepsilon_i = \pm 1, \quad \prod_{i=1}^{l} \varepsilon_i = 1.$$

If we call an element $(\lambda_1, \ldots, \lambda_l) \to (\varepsilon_1 \lambda_{i_1}, \ldots, \varepsilon_l \lambda_{i_l})$ in $W(C_l)$ a positive (or negative) permutation according to $\prod_{i=1}^{l} \varepsilon_i = +1$ (or -1) and the set of all positive (or negative) permutations is denoted by $W^+(C_l)$ (or $W^-(C_l)$), then evidently $W^+(C_l)$ indicates a subgroup of $W(C_l)$ with index 2 and $W(D_l) = W^+(C_l)$.

In the classification of simple algebras, we must have $l \geq 2$ for C_l and $l \geq 4$ for D_l. However, when $W(C_l)$ and $W(D_l)$ are realized to be signed permutation groups in the form mentioned above, these restrictions can be removed provided $l \geq 1$. With this idea in mind we will carry on the subsequent discussions.

Consider $\sigma \in W(C_l)$ and assume that

$$\sigma: (\lambda_1, \ldots, \lambda_l) \to (\varepsilon_1 *, \ldots, \varepsilon_j \lambda_1, \ldots, \varepsilon_l *).$$

Case 1. We have $j = 1$, $\varepsilon_1 = 1$, namely,

$$\sigma: (\lambda_1, \ldots, \lambda_l) \to (\lambda_1, \varepsilon_2 *, \ldots, \varepsilon_l *), \quad \varepsilon_2 \cdots \varepsilon_l = -1.$$

Evidently, $\sigma \in W^-(C_{l-1})$.

Case 2. We have $j = 1$, $\varepsilon_1 = -1$, namely,

$$\sigma: (\lambda_1, \ldots, \lambda_l) \to (-\lambda_1, \varepsilon_2 *, \ldots, \varepsilon_l *), \quad \varepsilon_2 \cdots \varepsilon_l = -1.$$

Let $\hat{\zeta}_1: (\lambda_1, \ldots, \lambda_l) \to (-\lambda_1, \lambda_2, \ldots, \lambda_l)$. It is easy to check that it corresponds to the reflection w_τ in $W(C_l)$, $\tau = 2\alpha_1 + \cdots + 2\alpha_{l-1} + \alpha_l$. Clearly,

$$\hat{\zeta}_1 \sigma: (\lambda_1, \ldots, \lambda_l) \to (\lambda_1, \varepsilon_2 *, \ldots, \varepsilon_l *), \qquad \varepsilon_2 \cdots \varepsilon_l = 1.$$

Thus $\hat{\zeta}_1 \sigma \in W^+(C_{l-1})$; i.e., $\sigma \in \hat{\zeta}_1 W^+(C_{l-1})$.

Case 3. We have $j > 1$, $\varepsilon_j = 1$. By ζ_{1j} we denote the permutation

$$(\lambda_1, \ldots, \lambda_j, \ldots, \lambda_l) \to (\lambda_j, \lambda_2, \ldots, \overset{j}{\lambda_1}, \ldots, \lambda_l).$$

Then clearly, $\zeta_{1j} \sigma \in W^-(C_{l-1})$ and $\sigma \in \zeta_{1j} W^-(C_{l-1})$.

Case 4. We have $j > 1$, $\varepsilon_j = -1$. By ζ_j we denote the permutation

$$(\lambda_1, \ldots, \lambda_j, \ldots, \lambda_l) \to (\lambda_1, \ldots, \overset{j}{-\lambda_j}, \ldots, \lambda_l).$$

Then we can check that $\zeta_{1j} \zeta_j \sigma \in W^+(C_{l-1})$ and $\sigma \in \hat{\zeta}_j \zeta_{1j} W^+(C_{l-1})$.

Now let us discuss $\sum_{w \in W^-(C_l)} f(w)$, where $f(w)$ is a class function and w is prescribed to be one of the previous four cases. Then

$$\sum_{w \in W^-(C_l)} f(w) = \sum_{w \in W^-(C_{l-1})} f(w) + \sum_{w \in W^+(C_{l-1})} f(\hat{\zeta}_1 w)$$

$$+ \sum_{j=2}^{l} \left[\sum_{w \in W^-(C_{l-1})} f(\zeta_{1j} w) + \sum_{w \in W^+(C_{l-1})} f(\hat{\zeta}_j \zeta_{1j} w) \right]. \quad (8.17)$$

When $j \geq 2$, we infer $\hat{\zeta}_j \in W^-(C_{l-1})$, which gives rise to

$$\sum_{w \in W^+(C_{l-1})} f(\hat{\zeta}_j \zeta_{1j} w) = \sum_{w \in W^+(C_{l-1})} f(\zeta_{1j} w \hat{\zeta}_j) = \sum_{w \in W^-(C_{l-1})} f(\zeta_{1j} w), \quad (8.18)$$

where $\zeta_{1j} = \zeta_{j-1,j}\zeta_{j-2,j-1} \cdots \zeta_{12}\zeta_{23} \cdots \zeta_{j-1,j}$ with the understanding that ζ_{ij} denotes a permutation interchange of λ_i and λ_j and the other elements are fixed. Thus we find

$$\sum_{w \in W^-(C_{l-1})} f(\zeta_{1j}w) = \sum_{w \in W^-(C_{l-1})} f(\zeta_{j-1,j}\zeta_{j-2,j-1} \cdots \zeta_{12}\zeta_{23} \cdots \zeta_{j-1,j}w)$$

$$= \sum_{w \in W^-(C_{l-1})} f(\zeta_{j-2,j-1} \cdots \zeta_{12}\zeta_{23} \cdots \zeta_{j-1,j}w\zeta_{j-1,j})$$

$$= \sum_{w \in W^-(C_{l-1})} f(\zeta_{j-2,j-1} \cdots \zeta_{23}\zeta_{12}\zeta_{23} \cdots \zeta_{j-2,j-1}w)$$

$$= \cdots = \sum_{w \in W^-(C_{l-1})} f(\zeta_{12}w) \tag{8.19}$$

because for $j \geq 2$, $\zeta_{j-1,j}W^-(C_{l-1})\zeta_{j-1,j} = W^-(C_{l-1})$, etc.

Since in $W(C_l)$, ζ_{12} corresponds to w_{α_1} and ζ_1 to w_τ, equalities (8.17)–(8.19) give rise to

$$\sum_{w \in W^-(C_l)} f(w) = \sum_{w \in W^-(C_{l-1})} f(w) + \sum_{w \in W^+(C_{l-1})} f(w_\tau w)$$

$$+ (2l - 2) \sum_{w \in W^-(C_{l-1})} f(w_{\alpha_1}w). \tag{8.20}$$

Noting that $W^+(C_{l-1}) = W(D_{l-1})$ and $W(C_l) = W^+(C_{l-1}) \cup W^-(C_{l-1})$, we are in a position to say

$$\sum_{w \in W(C_l)} f(w) = \sum_{w \in W(D_l)} f(w) + \sum_{w \in W^-(C_l)} f(w)$$

$$= \sum_{w \in W(D_l)} f(w) + \sum_{w \in W^-(C_{l-1})} f(w) + \sum_{w \in W(D_{l-1})} f(w_\tau w)$$

$$+ (2l - 2) \sum_{w \in W^-(C_{l-1})} f(w_{\alpha_1}w)$$

$$= \sum_{w \in W(D_l)} f(w) + \sum_{w \in W(C_{l-1})} f(w) - \sum_{w \in W(D_{l-1})} f(w)$$

$$+ \sum_{w \in W(D_{l-1})} f(w_\tau w) + (2l - 2) \sum_{w \in W(C_{l-1})} f(w_{\alpha_1}w)$$

$$- (2l - 2) \sum_{w \in W(D_{l-1})} f(w_{\alpha_1}w).$$

Referring to (8.15), we can prove

$$\sum_{w \in W(C_{l-1})} f(w_\tau w) = \sum_{w \in W(C_l)} f(w) - \sum_{w \in W(C_{l-1})} f(w)$$

$$- (2l - 2) \sum_{w \in W(C_{l-1})} f(w_{\alpha_1} w).$$

Thus the preceding expression becomes

$$\sum_{w \in W(D_l)} f(w) = \sum_{w \in W(D_{l-1})} f(w) + (2l - 2) \sum_{w \in W(D_{l-1})} f(w_{\alpha_1} w)$$

$$+ \sum_{w \in W(C_{l-1})} f(w_\tau w) - \sum_{w \in W(D_{l-1})} f(w_\tau w). \qquad (8.21)$$

Equation (8.21) turns out to be the recursion formula needed to replace (8.16). In this formula $W(C_l)$ and $W(D_l)$ are regarded as signed permutation groups without the requirement of $l \geq 4$, $W(C_{l-1})$ is regarded as the subgroup of $W(C_l)$ keeping λ_1 fixed, and $W(D_{l-1})$ is regarded as the subgroup of $W^+(C_l)$ keeping λ_1 fixed.

8.3. New Identities

As in the preceding discussion, we prescribe g to be one of A_l, B_l, C_l, D_l. In this section, we use (8.13)-(8.15) and (8.21) to prove some identities related to the global structure of $W(g)$.

8.3.1. On $\Sigma\, t^{\bar{l}(w)}$

$\bar{l}(w)$ stands for the Σ-length of w mentioned in Section 8.1. We have the following result.

THEOREM 8.12. Prescribe g to be one of A_l, B_l, C_l, D_l and $d_1, \ldots, d_l$ to be its indices, respectively. Put $m_i = d_i - 1$, $i = 1, \ldots, l$. The following equality holds:

$$\sum_{w \in W(g)} t^{\bar{l}(w)} = \prod_{i=1}^{l} (1 + m_i t). \qquad (8.22)$$

PROOF. Set $J = \{\alpha_2, \ldots, \alpha_l\}$. By Lemma 8.3, when $r \notin \Sigma_J$, $w \in W_J$, we have $\bar{l}(w_r w) = 1 + \bar{l}(w)$. In (8.13)-(8.15), put $f(w) = t^{\bar{l}(w)}$, and note that the following formula holds

$$\sum_{w \in W_J(g)} f(w) = \sum_{w \in W(g_{l-1})} f(w),$$

$W(g_{l-1})$ being the Weyl group of a simple algebra of the same type with rank $l-1$. Thus for $g = A_l$, B_l, C_l, equalities (8.13)-(8.15) give directly

$$\sum_{w \in W(g)} t^{T(w)} = (1 + (d_l - 1)t) \sum_{w \in W(g_{l-1})} t^{T(w)} = (1 + m_l t) \sum_{w \in W(g_{l-1})} t^{T(w)}.$$

Equation (8.22) holds with the help of induction.

If $g = D_l$, by (8.21) we infer

$$\sum_{w \in W(D_l)} t^{T(w)} = (1 + (2l - 3)t) \sum_{w \in W(D_{l-1})} t^{T(w)} + \sum_{w \in W(C_{l-1})} t^{T(w_r w)}$$

$$= (1 + (2l - 3)t) \sum_{w \in W(D_{l-1})} t^{T(w)} + t \sum_{w \in W(C_{l-1})} t^{T(w)}.$$

By the formula (8.22) for $W(C_l)$, which has been proved, it follows that

$$\sum_{w \in W(C_{l-1})} t^{T(w)} = (1 + t)(1 + 3t) \cdots (1 + (2l - 3)t).$$

Let

$$F_k(t) = \sum_{w \in W(D_k)} t^{T(w)}.$$

Then the above equality becomes

$$F_l(t) = (1 + (2l - 3)t)F_{l-1}(t) + t(1 + t)(1 + 3t) \cdots (1 + (2l - 3)t).$$

By induction, we directly obtain

$$F_l(t) = (1 + t)(1 + 3t) \cdots (1 + (2l - 3)t)(1 + (l - 1)t) = \prod_{i=1}^{l} (1 + m_i t).$$

The theorem is proved.

8.3.2. On Involutions

If $w^2 = I$, then $w \in W$ is called an involution. The eigenvalues of involutions are either 1 or -1. Hence the Euclidean space V is decomposed into $V = V_1(w) \oplus V_{-1}(w)$, where $V_{\pm 1}(w)$ denote the eigenspace of w corresponding to $\pm l$. Obviously, for $w^2 = I$, the equality $\dim V_{-1}(w) = \bar{I}(w)$ holds. Thus an involution w satisfying $\bar{I}(w) = k$ is similar to $\binom{I^{(l-k)}}{\quad -I^{(k)}}$. If $k = 1$, it is called a reflection. Recall that the total number of reflections equals the number of all positive roots. Now a problem arises: How many involutions are there for other k's?

If we set

$$f(w) = \begin{cases} 0, & w^2 \neq I, \\ t^{\bar{I}(w)}, & w^2 = I, \end{cases}$$

then $\sum_{w \in W} f(w)$ is a polynomial of t. The coefficient of t^k in this polynomial turns out to be the number of involutions w analogous to $\binom{I^{(l-k)}}{\quad -I^{(k)}}$. To find this polynomial, we begin with the proof of the following lemma.

LEMMA 8.13. *Let g be one of A_l, B_l, C_l, D_l, $J = \{\alpha_2, \ldots, \alpha_l\}$, $J' = \{\alpha_3, \ldots, \alpha_l\}$ and $w = w_\alpha w'$ be an involution, having $w' \in W_J$. Then w' is just an involution and $w' \in W_{J'}$.*

PROOF. Let the irreducible representation of w' be written as $w' = w_{r_1} \cdots w_{r_k}$, $k \leq l - 1$, $r_i \in \Sigma_J$. According to Lemma 8.2, r_i's are linearly independent and so are $\alpha_1, r_1, \ldots, r_k$ because $\alpha_1 \in \Sigma_J$. Considering Lemma 8.4, $\alpha_1, r_1, \ldots, r_k$ are mutually orthogonal; thus w' is an involution as claimed.

Choose any $r_i = m_2 \alpha_2 + \cdots + m_l \alpha_l$, m_i's being nonnegative integers. By orthogonality,

$$0 = (\alpha_1, r_i) = m_2(\alpha_1, \alpha_2) + m_3(\alpha_1, \alpha_3) + \cdots + m_l(\alpha_1, \alpha_l).$$

However, the Dynkin diagram lets us easily ensure $(\alpha_1, \alpha_j) = 0$ $(j \geq 3)$. Hence $m_2 = 0$ and $r_i \in \Sigma_{J'}$, i.e., $w' \in W_{J'}$. The lemma is proved.

LEMMA 8.14. *Write $W = W(C_l)$, $J = \{\alpha_2, \ldots, \alpha_l\}$, $\tau = 2\alpha_1 + 2\alpha_2 + \cdots + 2\alpha_{l-1} + \alpha_l$, $w' \in W_J$. Then $w = w_\tau w'$ is an involution if and only if w' is.*

PROOF. It is easy to check directly that $(\tau, \alpha_i) = 0$, $i = 2, 3, \ldots, l$. Hence, w_τ and w_{α_i} $(i = 2, \ldots, l)$ are commutative. Since $w' \in W_J$, we may assert w_τ and w' are commutative, but $w_\tau w'$ amounts to an involution; thus we conclude that w' is also an involution.

The above two lemmas shed light on the fact that if $w_{\alpha_1} w'$ is an involution and $w' \in W_J$, then

$$\bar{l}(w_{\alpha_1} w') = 1 + \bar{l}(w'), \qquad w' \in W_J, \text{ and } w' \text{ is an involution.} \quad (8.23)$$

As for $W = W(C_l)$, provided $w_r w'$ is an involution, $w' \in W_J$, we find

$$\bar{l}(w_r w') = 1 + \bar{l}(w'), \qquad w' \in W_J \text{ and } w' \text{ is an involution.} \quad (8.24)$$

THEOREM 8.15. *Let g be one of A_l, B_l, C_l, D_l and let $d_1, \ldots, d_l$ be its indices. On the supposition of $m_i = d_i - 1$, $i = 1, \ldots, l$ and $F_l(t) = \sum_{w^2 = I, w \in W(g)} t^{\bar{l}(w)}$, we claim*

$$F_l(t) = 1 + a_1 t + a_2 t^2 + \cdots + a_{[(l+1)/2]} t^{[(l+1)/2]}, \qquad \text{if } g = A_l, \quad (8.25)$$

$$a_k = \sum_{i_1 - i_2 \geq 2, \, i_2 - i_3 \geq 2, \ldots, i_{k-1} - i_k \geq 2}^{l} m_{i_1} m_{i_2} \cdots m_{i_k}, \qquad 1 \leq k \leq \left[\frac{l+1}{2}\right],$$

$$F_l(t) = (1 + t)^l + b_1 t (1 + t)^{l-2} + \cdots + b_{[l/2]} t^{[l/2]} (1 + t)^{l-2[l/2]},$$

$$\text{if } g = B_l, C_l, \quad (8.26)$$

$$F_l(t) = \tfrac{1}{2}[(1 + t)^l + (1 - t)^l] + \tfrac{1}{2} b_1 t [(1 + t)^{l-2} + (1 - t)^{l-2}]$$

$$+ \cdots + \tfrac{1}{2} b_{[l/2]} t^{[l/2]} [(1 + t)^{l-2[l/2]} + (1 - t)^{l-2[l/2]}],$$

$$\text{if } g = D_l, \quad (8.27)$$

$$b_k = \sum_{i_1 - i_2 \geq 2, \, i_2 - i_3 \geq 2, \ldots, i_{k-1} - i_k \geq 2}^{l-1} d_{i_1} d_{i_2} \cdots d_{i_k}, \qquad 1 \leq k \leq \left[\frac{l}{2}\right].$$

PROOF. *Step 1.* $g = A_l$
By (8.13) and (8.23) we obtain

$$F_l(t) = F_{l-1}(t) + lt F_{l-2}(t). \quad (8.28)$$

Now we proceed to prove (8.25) by induction. Since $W(A_l) \simeq S_{l+1}$, the symmetric group of degree $l + 1$, it is easy to check that $F_1(t) = 1 + t$, $F_2(t) = 1 + 3t$; i.e., (8.25) holds for $l = 1, 2$. Assume that (8.25) is valid for

F_{l-2}, F_{l-1}. Then by (8.28),

$$F_l(t) = 1 + \left(\sum_{i=1}^{l-1} m_i \right) t + \left(\sum_{i, -i_2 \geq 2}^{l-1} m_{i_1} m_{i_2} \right) t^2$$

$$+ \cdots + \left(\sum_{i_1 - i_2 \geq 2, \ldots, i_{[l/2]-1} - i_{[l/2]} \geq 2}^{l-1} m_{i_1} \cdots m_{i_{[l/2]}} \right) t^{[l/2]}$$

$$+ m_1 t \left[1 + \left(\sum^{l-2} m_i \right) t + \cdots + \left(\sum^{l-2} m_{i_1} \cdots m_{i_{[(l-1)/2]}} \right) t^{[(l-1)/2]} \right],$$

where the coefficient of t^k on the right-hand side reads

$$a_k = \sum_{\substack{i_1 - i_2 \geq 2 \\ \cdots \\ i_{k-1} - i_k \geq 2}}^{l-1} m_{i_1} \cdots m_{i_k} + m_l \sum_{\substack{i_1 - i_2 \geq 2 \\ \cdots \\ i_{k-2} - i_{k-1} \geq 2}}^{l-2} m_{i_1} \cdots m_{i_{k-1}} = \sum_{\substack{i_1 - i_2 \geq 2 \\ \cdots \\ i_{k-1} - i_k \geq 2}}^{l} m_{i_1} \cdots m_{i_k}.$$

Thus (8.25) is proved.

Step 2. $g = C_l$ (since $W(B_l) \simeq W(C_l)$, the two results are alike).
By (8.74) and (8.23), (8.24), we obtain

$$F_l(t) = (1 + t)F_{l-1}(t) + 2(l - 1)tF_{l-2}(t)$$

$$= (1 + t)F_{l-1}(t) + d_{l-1}tF_{l-2}(t). \tag{8.29}$$

When $W(C_l)$ is regarded as a signed permutation group of degree l, then
it is easy to check that $F_1(t) = 1 + t$ and $F_2(t) = 1 + 4t + t^2 = (1 + t)^2 + 2t$.
In this case, (8.26) is true for $l = 1, 2$. Applying induction and the recursion
formula (8.29), we can prove (details are omitted)

$$F_l(t) = (1 + t)^l + b_1 t(1 + t)^{l-2} + \cdots + b_{[l/2]} t^{[l/2]}(1 + t)^{l-2[l/2]},$$

$$b_k = \sum_{i_1 - i_2 \geq 2, \, i_2 - i_3 \geq 2, \ldots, i_{k-1} - i_k \geq 2}^{l-1} d_{i_1} d_{i_2} \cdots d_{i_k}, \qquad 1 \leq k \leq \left[\frac{l}{2} \right].$$

Step 3. $g = D_l$.
Write

$$F_l(t) = \sum_{w^2 = I, \, w \in W(D_l)} t^{T(w)}, \qquad C_l(t) = \sum_{w^2 = I, \, w \in W(C_l)} t^{T(w)}.$$

Note that $C_l(t)$ is given by (8.26). Then (8.21) together with (8.23) and (8.24) yields

$$F_l(t) = F_{l-1}(t) + (2l-2)tF_{l-2}(t) + tC_{l-1}(t) - tF_{l-1}(t)$$

$$= (1-t)F_{l-1}(t) + d_{l-1}tF_{l-2}(t) + tC_{l-1}(t). \qquad (8.30)$$

By (8.29), we derive

$$tC_{l-1}(t) = C_l(t) - C_{l-1}(t) - d_{l-1}tC_{l-2}(t).$$

Substitution of the above equality into (3.9) leads to

$$C_l(t) - F_l(t) = (1-t)(C_{l-1}(t) - F_{l-1}(t))$$

$$+ d_{l-1}t(C_{l-2}(t) - F_{l-2}(t)) + tC_{l-1}(t).$$

Provided $S_l(t) = C_l(t) - F_l(t)$, the above equality becomes

$$S_l(t) = (1-t)S_{l-1}(t) + d_{l-1}tS_{l-2}(t) + tC_{l-1}(t). \qquad (8.31)$$

Subtracting (8.30) from (8.31) and putting $x_l(t) = S_l(t) - F_l(t) = C_l(t) - 2F_l(t)$, we arrive at

$$x_l(t) = (1-t)x_{l-1}(t) + d_{l-1}tx_{l-2}(t). \qquad (8.32)$$

Since the recursion formula (8.32) is parallel to (8.29), this reasoning then gives

$$x_l(t) = -\{(1-t)^l + b_1t(1-t)^{l-2} + \cdots + b_{[l/2]}t^{[l/2]}(1-t)^{l-2[l/2]}\}, \quad (8.33)$$

where $b_k = \sum_{i_1-i_2 \geq 2,\ldots,i_{k-1}-i_k \geq 2}^{l-1} d_{i_1} \cdots d_{i_k}$. To do this, we only need to regard $W(D_l)$ as a signed permutation group: $(\lambda_1, \ldots, \lambda_l) \to (\varepsilon_1\lambda_{i_1}, \ldots, \varepsilon_l\lambda_{i_l})$, $\prod_{i=1}^{l} \varepsilon_i = 1$. It is easy to verify $F_1(t) = 1$, $F_2(t) = (1+t)^2$, thus achieving

$$x_1(t) = C_1(t) - 2F_1(t) = -(1-t), \qquad x_2(t) = -[(1-t)^2 + 2t].$$

So (8.33) is true for $l = 1, 2$, and the other cases can be justified by induction. Thus by (8.26) and (8.23), we are in a position to assert

$$F_l(t) = \tfrac{1}{2}(C_l(t) - X_l(t))$$

$$= \tfrac{1}{2}[(1+t)^l + (1-t)^l] + \tfrac{1}{2}b_1t[(1+t)^{l-2} + (1-t)^{l-2}]$$

$$+ \cdots + \tfrac{1}{2}b_{[l/2]}t^{[l/2]}[(1+t)^{l-2[l/2]} + (1-t)^{l-2[l/2]}],$$

with b_k $(1 \leq k \leq [l/2])$ as in (8.27). The proof of the theorem is complete.

$\square$

8.3.3. On $\Sigma\, t^{\mathrm{tr}\, w}$.

THEOREM 8.16. *Let g be one of* A_l, B_l, C_l, D_l, *and suppose* $F_l(t) = \sum_{w \in W(g)} t^{\mathrm{tr}\, w}$. *The following is valid:*

$$F_l(t) = \frac{1}{t}[(t-1)^{l+1} + d_l(t-1)^l + d_l d_{l-1}(t-1)^{l-1}$$

$$+ \cdots + d_l d_{l-1} \cdots d_2(t-1)^2] + d_l \cdots d_1,$$

$$\text{if } g = A_l. \quad (8.34)$$

$$F_l(t) = \xi^l + d_l \xi^{l-1} + d_l d_{l-1} \xi^{l-2} + \cdots + d_l d_{l-1} \cdots d_1,$$

$$\xi = t^{-1}(1-t)^2, \quad \text{if } g = B_l, C_l. \quad (8.35)$$

$$F_l(t) = \tfrac{1}{2}[(t - t^{-1})^l + \xi^l] + d_l \xi^{l-1} + d_l d_{l-1} \xi^{l-2} + \cdots + d_l \cdots d_1,$$

$$\xi = t^{-1}(1-t)^2, \quad \text{if } g = D_l. \quad (8.36)$$

PROOF. For convenience, the subgroup W_J of W, $J = \{\alpha_2, \ldots, \alpha_l\}$, will be denoted by $W(\alpha_2, \ldots, \alpha_l)$, and the Weyl group of rank k (when it is not regarded as a subgroup of W) will be denoted by $W(A_k)$, etc.

Case 1. $g = A_l$. Put $f(w) = t^{\mathrm{tr}\, w}$ in (8.13). There holds

$$\sum_{w \in W(A_l)} f(w) = \sum_{w' \in W(\alpha_2, \ldots, \alpha_l)} f(w') + l \sum_{w' \in W(\alpha_2, \ldots, \alpha_l)} f(w_{\alpha_1} w'). \quad (8.37)$$

By Theorem 8.10, the coset $W(\alpha_2, \ldots, \alpha_l)/W(\alpha_3, \ldots, \alpha_l)$ contains l elements, namely, 1, w_{α_2}, $w_{\alpha_3} w_{\alpha_2}, \ldots, w_{\alpha_l} \cdots w_{\alpha_2}$. They are all of the form $*w_{\alpha_2}$ except the unit element 1, where $* \in W(\alpha_3, \ldots, \alpha_l)$. As a result, we have

$$\sum_{w' \in W(\alpha_2, \ldots, \alpha_l)} f(w_{\alpha_1} w')$$

$$= \sum_{w'' \in W(\alpha_3, \ldots, \alpha_l)} f(w_{\alpha_1} w'') + (l-1) \sum_{w'' \in W(\alpha_3, \ldots, \alpha_l)} f(w_{\alpha_1} * w_{\alpha_2} w'')$$

$$= \sum_{w'' \in W(\alpha_3, \ldots, \alpha_l)} f(w_{\alpha_1} w'') + (l-1) \sum_{w'' \in W(\alpha_3, \ldots, \alpha_l)} f(* w_{\alpha_1} w_{\alpha_2} w'')$$

$$= \sum_{w'' \in W(\alpha_3, \ldots, \alpha_l)} f(w_{\alpha_1} w'') + (l-1) \sum_{w'' \in W(\alpha_3, \ldots, \alpha_l)} f(w_{\alpha_1} w_{\alpha_2} w'' *)$$

$$= \sum_{w'' \in W(\alpha_3, \ldots, \alpha_l)} f(w_{\alpha_1} w'') + (l-1) \sum_{w'' \in W(\alpha_3, \ldots, \alpha_l)} f(w_{\alpha_1} w_{\alpha_2} w'').$$

The derivation above depended on $* \in W(\alpha_3, \ldots, \alpha_l)$, so that $*$ and W_α may be commuted. Hence (8.37) becomes

$$\sum_{w \in W(A_l)} f(w) = \sum_{w' \in W(\alpha_2, \ldots, \alpha_l)} f(w') + l \sum_{w'' \in W(\alpha_3, \ldots, \alpha_l)} f(w_{\alpha_1} w'')$$

$$+ l(l-1) \sum_{w'' \in W(\alpha_3, \ldots, \alpha_l)} f(w_{\alpha_1} w_{\alpha_2} w'').$$

Repeating this procedure yields

$$\sum_{w \in W(A_l)} f(w)$$

$$= \sum_{w' \in W(\alpha_2, \ldots, \alpha_l)} f(w') + l \sum_{w'' \in W(\alpha_3, \ldots, \alpha_l)} f(w_{\alpha_1} w'')$$

$$+ l(l-1) \sum_{w''' \in W(\alpha_4, \ldots, \alpha_l)} f(w_{\alpha_1} w_{\alpha_2} w''')$$

$$+ \cdots + l(l-1) \cdots (l-k+1) \sum_{w^{(k+1)} \in W(\alpha_{k+2}, \ldots, \alpha_l)} f(w_{\alpha_1} \cdots w_{\alpha_k} w^{(k+1)})$$

$$+ \cdots + l(l-1) \cdots 2 f(w_{\alpha_1} \cdots w_{\alpha_{l-1}}) + l! \, f(w_{\alpha_1} \cdots w_{\alpha_l}). \qquad (8.38)$$

However, it is well known that $w_{\alpha_1} \cdots w_{\alpha_k}$, restricted to $V(\alpha_1, \ldots, \alpha_k)$, belongs to a Coxeter element, the characteristic polynomial of which is expressed as $1 + t + \cdots + t^k$. Hence, its trace, restricted to the subspace $V(\alpha_1, \ldots, \alpha_k)$ gives

$$\mathrm{tr}_{V(\alpha_1, \ldots, \alpha_k)} w_{\alpha_1} \cdots w_{\alpha_k} = -1.$$

Thus, when $w^{(k+1)} \in w(\alpha_{k+2}, \ldots, \alpha_l)$, it implies

$$\mathrm{tr}_V w_{\alpha_1} \cdots w_{\alpha_k} w^{(k+1)} = 1 + \mathrm{tr}_{V(\alpha_1, \ldots, \alpha_k)} w_{\alpha_1} \cdots w_{\alpha_k} + \mathrm{tr}_{V(\alpha_{k+2}, \ldots, \alpha_l)} W^{(k+1)}$$

$$= \mathrm{tr}_{V(\alpha_{k+2}, \ldots, \alpha_l)} w^{(k+1)}.$$

$$\therefore \sum_{w^{(k+1)} \in W(\alpha_{k+2}, \ldots, \alpha_l)} f(w_{\alpha_1} \cdots w_{\alpha_k} w^{(k+1)}) = \sum_{w \in W(A_{l-k-1})} f(w) = F_{l-k-1}(t).$$

Hence (8.38) is reduced to

$$F_l(t) = t F_{l-1}(t) + l F_{l-2}(t) + \cdots + l(l-1) \cdots (l-k+1) F_{l-k-1}(t)$$

$$+ \cdots + l(l-1) \cdots 2 F_0 + l! \, t^{-1},$$

where $F_0 = 1$. Note that we have a similar recursion formula for $F_{l-1}(t)$ and easily arrive at

$$F_l(t) = (l + t)F_{l-1} + l(1 - t)F_{l-2}, \tag{8.39}$$

which can be rewritten as $F_l - (l + 1)F_{l-1} = (t - 1)(F_{l-1} - lF_{l-2})$. By recursion we have

$$F_l - (l + 1)F_{l-1} = (t - 1)(F_{l-1} - lF_{l-2}) = (t - 1)^2(F_{l-2} - (l - 1)F_{l-3})$$

$$= \cdots = (t - 1)^{l-1}(F_1 - 2F_0).$$

The easy verification of $F_1 = t + t^{-1}$ gives rise to

$$F_l(t) = (l + 1)F_{l-1}(t) + \frac{1}{t}(t - 1)^{l+1}$$

$$= \frac{1}{t}(t - 1)^{l+1} + \frac{1}{t}(l + 1)(t - 1)^l + (l + 1)lF_{l-2} = \cdots$$

$$= \frac{1}{t}(l + 1)(t - 1)^l + \frac{1}{t}(t - 1)^{l+1} + \frac{1}{t}(l + 1)l(t - 1)^{l-1}$$

$$+ \cdots + \frac{1}{t}(l + 1)\cdots 3(t - 1)^2 + (l + 1)\cdots 3 \cdot 2$$

$$= \frac{1}{t}[(t - 1)^{l+1} + d_l(t - 1)^l + d_l d_{l-1}(t - 1)^{l-1} + \cdots + d_l \cdots d_2(t - 1)^2]$$

$$+ d_l \cdots d_1,$$

The proof is concluded.

Case 2. $W(C_l)$. With the help of (8.14), it follows that

$$\sum_{w \in W(C_l)} f(w) = \sum_{w' \in W(\alpha_2, \ldots, \alpha_l)} f(w') + d_{l-1} \sum_{w' \in W(\alpha_2, \ldots, \alpha_l)} f(w_{\alpha_1} w')$$

$$+ \sum_{w' \in W(\alpha_2, \ldots, \alpha_l)} f(w_\tau w'). \tag{8.40}$$

Evidently, when $w' \in W(\alpha_2, \ldots, \alpha_l)$,

$$\text{tr}_V w' = 1 + \text{tr}_{V(\alpha_2,\ldots,\alpha_l)} w'.$$

Since $\tau = 2\alpha_1 + \cdots + 2\alpha_{l-1} + \alpha_l$ is orthogonal to $\alpha_2, \ldots, \alpha_l$ (Lemma 8.14), we confirm that

$$\text{tr}_V w_\tau w' = -1 + \text{tr}_{V(\alpha_2,\ldots,\alpha_l)} w'.$$

Choose $F_k(t) = \sum_{w \in W(C_k)} f(w)$. Then (8.40) becomes

$$F_l(t) = t F_{l-1}(t) + t^{-1} F_{l-1}(t) + d_{l-1} S_{l-1}(t),$$

with $S_{l-1}(t) = \sum_{w' \in W(\alpha_2,\ldots,\alpha_l)} f(w_{\alpha_1} w')$. Subsequently, we will compute this term.

By Theorem 8.10, there are $2l - 2$ elements in the coset $W(\alpha_2, \ldots, \alpha_l)/W(\alpha_3, \ldots, \alpha_l)$, except 1 and $w_{\tau'}(\tau' = 2\alpha_2 + \cdots + 2\alpha_{l-1} + \alpha_l)$, and the other d_{l-2} $(=2l - 4)$ elements are all of the form $*w_{\alpha_2}$, $* \in W(\alpha_3, \ldots, \alpha_l)$.

$$\therefore \quad \sum_{w'' \in W(\alpha_3,\ldots,\alpha_l)} f(w_{\alpha_1} * w_{\alpha_2} w'') = \sum_{w'' \in W(\alpha_3,\ldots,\alpha_l)} f(*w_{\alpha_1} w_{\alpha_2} w'')$$

$$= \sum_{w'' \in W(\alpha_3,\ldots,\alpha_l)} f(w_{\alpha_1} w_{\alpha_2} w'' *)$$

$$= \sum_{w'' \in W(\alpha_3,\ldots,\alpha_l)} f(w_{\alpha_1} w_{\alpha_2} w''),$$

which ensures

$$S_{l-1}(t) = \sum_{w' \in W(\alpha_2,\ldots,\alpha_l)} f(w_{\alpha_1} w') = \sum_{w'' \in W(\alpha_3,\ldots,\alpha_l)} f(w_{\alpha_1} w'')$$

$$+ \sum_{w'' \in W(\alpha_3,\ldots,\alpha_l)} f(w_{\alpha_1} w_{\tau'} w'')$$

$$+ d_{l-2} \sum_{w'' \in W(\alpha_3,\ldots,\alpha_l)} f(w_{\alpha_1} w_{\alpha_2} w''). \tag{8.41}$$

Choosing α_1, $\tau' = 2\alpha_2 + \cdots + 2\alpha_{l-1} + \alpha_l$, $\alpha_3, \ldots, \alpha_l$ to be a basis in V, we easily find*

$$w_{\tau'} = \begin{pmatrix} 1 & 1 & \\ 0 & -1 & \\ & & I \end{pmatrix}, \quad w_{\alpha_1} w_{\tau'} = \begin{pmatrix} -1 & -1 & \\ 2 & 1 & \\ & & I \end{pmatrix}, \quad w'' = \begin{pmatrix} 1 & 0 & \\ 0 & 1 & * \cdots * \\ & & * \end{pmatrix},$$

---

* It seems there are some errors in the expressions of $w_{\alpha_1} w_{\tau'}$, $w_{\alpha_2} w''$, and $w_{\alpha_1} w_{\alpha_2} w''$, but the main result is still true. (Translator)

provided $w'' \in W(\alpha_3, \ldots, \alpha_l)$. Hence, this yields

$$\mathrm{tr}_V \, w_{\alpha_1} w_{\tau'} w'' = \mathrm{tr}_{V(\alpha_3, \ldots, \alpha_l)} w'', \qquad \mathrm{tr}_V \, w_{\alpha_1} w'' = \mathrm{tr}_{V(\alpha_3, \ldots, \alpha_l)} w'';$$

that is,

$$\sum_{w'' \in W(\alpha_3, \ldots, \alpha_l)} f(w_{\alpha_1} w'') = F_{l-2}(t), \qquad \sum_{w'' \in W(\alpha_3, \ldots, \alpha_l)} f(w_{\alpha_1} w_{\tau'} w'') = F_{l-2}(t).$$

If we choose $\alpha_1, \ldots, \alpha_l$ as a basis in V, then w_{α_1}, w_{α_2}, $w'' \in W(\alpha_3, \ldots, \alpha_l)$ amount to

$$\begin{pmatrix} -1 & 0 & \\ 0 & 1 & \\ & & I \end{pmatrix}, \begin{pmatrix} 1 & 1 & 0 \\ 0 & -1 & 0 \\ 0 & 1 & 1 \\ & & & I \end{pmatrix}, \begin{pmatrix} 1 & 0 & 0 & \cdots & 0 \\ 0 & 1 & * & \cdots & * \\ 0 & 0 & x_3 & & * \\ \vdots & \vdots & & \ddots & \\ 0 & 0 & * & & x_1 \end{pmatrix},$$

$$w_{\alpha_2} w'' = \begin{pmatrix} 1 & 1 & & & \\ 0 & -1 & & * & \\ 0 & 1 & x_3 & & \\ \vdots & & 0 & & \ddots \\ \vdots & & & & \\ 0 & 0 & * & & x_l \end{pmatrix}, \qquad w_{\alpha_1} w_{\alpha_2} w'' = \begin{pmatrix} -1 & -1 & & & \\ 1 & 0 & & * & \\ 0 & 1 & x_3 & & * \\ \vdots & \vdots & & \ddots & \\ 0 & 0 & * & & x_l \end{pmatrix},$$

$$\therefore \quad \mathrm{tr}_V \, w_{\alpha_1} w_{\alpha_2} w'' = \mathrm{tr}_{V(\alpha_2, \ldots, \alpha_l)} W_{\alpha_1} W''.$$

Hence, it follows that

$$\sum_{w'' \in W(\alpha_3, \ldots, \alpha_l)} f(w_{\alpha_1} w_{\alpha_2} w'') = \sum_{w'' \in W(\alpha_3, \ldots, \alpha_l)} f(w_{\alpha_2} w'') = S_{l-2}(t).$$

Equation (8.41) becomes

$$S_{l-1}(t) = 2F_{l-2}(t) + d_{l-2} S_{l-2}(t),$$

and

$$F_l(t) = (t + t^{-1}) F_{l-1}(t) + d_{l-1} S_{l-1}(t)$$

$$= (t + t^{-1}) F_{l-1}(t) + 2d_{l-1} F_{l-2}(t) + d_{l-1} d_{l-2} S_{l-2}(t).$$

The same argument gives

$$F_{l-1}(t) = (t + t^{-1})F_{l-2}(t) + d_{l-2}S_{l-2}(t),$$

the substitution of which into the preceding equality gives

$$F_l(t) = (t + t^{-1} + d_{l-1})F_{l-1}(t) + d_{l-1}(2 - t - t^{-1})F_{l-2}(t).$$

Since $d_l = d_{l-1} + 2$, the preceding equality can be written as

$$F_l - d_l F_{l-1} = t^{-1}(1 - t)^2(F_{l-1} - d_{l-1}F_{l-2}).$$

Set $\zeta = t^{-1}(1 - t)^2$. The recursion allows

$$F_l - d_l F_{l-1} = \zeta^{l-1}(F_1 - d_1 F_0),$$

where $F_0 = 1$, and $F_1 = t + t^{-1}$ and $F_1 - d_1 F_0 = t^{-1} + t - 2 = \zeta$ hold.

$$\therefore \quad F_l = d_l F_{l-1} + \zeta^l = \zeta^l + d_l \zeta^{l-1} + d_l d_{l-1} F_{l-2}$$

$$= \cdots = \zeta^l + d_l \zeta^{l-1} + d_l d_{l-1} \zeta^{l-2} + \cdots + d_l d_{l-1} \cdots d_1.$$

Equation (8.35) is proved.

Case 3. $W(D_l)$. As mentioned before, if we realize $W(C_l)$ to be a signed permutation group, then $W(D_l)$ can be regarded as the subgroup $W^+(C_l)$ of $W(C_l)$ consisting of all positive permutations. In $W(C_l)$, w_τ corresponds to the permutation $(\lambda_1, \ldots, \lambda_l) \to (-\lambda_1, \ldots, \lambda_l)$. Thus

$$\sum_{w \in W(C_l)} f(w) = \sum_{w \in W^+(C_l)} f(w) + \sum_{w \in W^+(C_l)} f(w_\tau w).$$

Now we investigate the relation between tr $w_\tau w$ and tr w when $w \in W^+(C_l)$:

$$w = \begin{pmatrix} 1 & 0 & \cdots & 0 \\ 0 & & & \\ \vdots & & w' & \\ 0 & & & \end{pmatrix}, \qquad w_\tau w = \begin{pmatrix} -1 & 0 & \cdots & 0 \\ 0 & & & \\ \vdots & & w' & \\ 0 & & & \end{pmatrix},$$

$$\text{tr } w_\tau w = -1 + \text{tr } w', \qquad w' \in W^+(C_{l-1}),$$

$$W = \begin{pmatrix} -1 & 0 & \cdots & 0 \\ 0 & & & \\ \vdots & & w' & \\ 0 & & & \end{pmatrix}, \qquad w_\tau w = \begin{pmatrix} 1 & 0 & \cdots & 0 \\ 0 & & & \\ \vdots & & w' & \\ 0 & & & \end{pmatrix},$$

$$\text{tr } w_\tau w = 1 + \text{tr } w', \qquad w' \in W^-(C_{l-1}).$$

$$w = \begin{pmatrix} 0 & * & \cdots & * \\ * & & & \\ \vdots & & * & \\ * & & & \end{pmatrix};$$

that is, $w \in W^+(C_l)$ but not in

$$\begin{pmatrix} 1 & \\ & W^+(C_{l-1}) \end{pmatrix} \cup \begin{pmatrix} -1 & \\ & W^-(C_{l-1}) \end{pmatrix}.$$

Then $\text{tr } w_\tau w = \text{tr } w$. Therefore, we have

$$\sum_{w \in W^+(C_l)} f(w_\tau w) = t^{-1} \sum_{w' \in W^+(C_{l-1})} f(w') + t \sum_{w' \in W^-(C_{l-1})} f(w')$$

$$+ \sum_{w' \in W(C_l)} f(w') - t \sum_{w' \in W^+(C_{l-1})} f(w')$$

$$- t^{-1} \sum_{w' \in W^-(C_{l-1})} f(w'), \qquad (8.42)$$

because $W(D_l) = W^+(C_l)$. Write $F_l(t) = \sum_{w \in W(D_l)} f(w)$, $C_l(t) = \sum_{w \in W(C_l)} f(w)$. Then

$$\sum_{w \in W^-(C_l)} f(w) = C_l(t) - F_l(t),$$

Equation (8.42) turns out to be

$$C_l(t) = F_l(t) + t^{-1}F_{l-1}(t) + t[C_{l-1}(t) - F_{l-1}(t)]$$

$$+ F_l(t) - tF_{l-1}(t) - t^{-1}[C_{l-1}(t) - F_{l-1}(t)]$$

$$= 2F_l(t) + 2(t^{-1} - t)F_{l-1}(t) + (t - t^{-1})C_{l-1}(t).$$

$$\therefore \quad C_l(t) - 2F_l(t) = (t - t^{-1})[C_{l-1}(t) - 2F_{l-1}(t)]$$

$$= \cdots = (t - t^{-1})^{l-1}[C_1(t) - 2F_1(t)]$$

$$= -(t - t^{-1})^l,$$

because it is easy to check $C_l(t) = t + t^{-1}$, $F_1(t) = t$. Thus we confirm that

$$F_l(t) = \tfrac{1}{2}[C_l(t) + (t - t^{-1})^l]$$

$$= \tfrac{1}{2}[(t - t^{-1})^l + \xi^l + 2l\xi^{l-1} + 2ld_{l-1}\xi^{l-2} + \cdots + 2ld_{l-1} \cdots d_1]$$

$$= \tfrac{1}{2}[(t - t^{-1})^l + \xi^l] + l\xi^{l-1} + ld_{l-1}\xi^{l-2} + \cdots + ld_{l-1} \cdots d_1$$

$$= \tfrac{1}{2}(t - t^{-1})^l + \xi^l + d_l\xi^{l-1} + d_ld_{l-1}\xi^{l-2} + \cdots + d_ld_{l-1} \cdots d_1,$$

where the expression (8.35) of $C_l(t)$ is adapted with $\xi = t^{-1}(1 - t)^2$. Thus (8.36) is proved, and the proof of the theorem is complete.   □

References

1. A. Borel *et al.*, *Seminar on Algebraic Groups and Related Finite Groups*, Lecture Notes in Math., vol. 131, Springer-Verlag, New York (1970).
2. R. Bott, An application of the Morse theory to the topology of Lie groups, *Bull. Soc. Math. France* **84**, 251–282 (1956).
3. R. W. Carter, *Simple Groups of Lie Type*, Wiley, New York (1972).
4. R. W. Carter, Conjugacy classes in the Weyl group, *Compositio Math.* **25**, 1–20 (1972).
5. L. Solomon, The orders of finite Chevalley groups, *J. Algebra* **3**, 376–393 (1966).
6. R. Steinberg, *Lectures on Chevalley groups*, Yale University (1967).
7. Wan Ze-xian, *Lie Algebras*, Science Press, Beijing (1964) (in Chinese).

9

The Trace Formula of the Weyl Group Representations of the Symmetric Group

Let g be a complex simple Lie algebra, h its Cartan subalgebra, and Σ and $\Pi = \{\beta_1, \ldots, \beta_l\}$ the root system of g and a basis of the root system, respectively. It is known that the Weyl group $W(g)$ is a finite group generated by reflections [3]

$$w_{\beta_i}: \xi \in h, \qquad \xi \to \xi - \frac{2(\xi, \beta_i)}{(\beta_i, \beta_i)} \beta_i.$$

The polynomial ring on h is denoted by S. For $w \in W$, $f \in S$, define the operator

$$wf(x) = f(w^{-1}x).$$

Then S becomes a W-module. $f \in S$ is said to be W-invariant if for any $w \in W$, $wf = f$. The subring of S consisting of all invariant polynomials is denoted by I. A basic fact on I is the famous Chevalley theorem: $\exists I_i \in I$, $i = 1, \ldots, l$, such that $I = R[I_1, \ldots, I_l]$; i.e., I is a real polynomial ring with basis $I_1, \ldots, I_l$ (for example, see Ref. 4). If we denote the subspace of homogeneous polynomials with degree d in S by S_d, $I_d = S_d \cap I$, then I_d is a W-invariant submodule. The dimension $\dim I_d$ of I_d; i.e., the number of maximal linear independent, W-invariant homogeneous polynomials with degree d is given by Molien's formula [3]:

$$\frac{1}{|W|} \sum_{w \in W} \frac{1}{\det(1 - tw)} = \frac{1}{\prod_{i=1}^{l}(1 - t^{d_i})} = \sum_{d=0}^{\infty} (\dim I_d) t^d, \qquad (9.1)$$

where d_i is the degree of I_i; i.e., $d_i = \deg I_i$, $i = 1, \ldots, l$. It is a set of basic invariants of g. It is also called the Poincaré indices of g. This is because the Poincaré polynomial of g is $p_g(t) = (1 + t^{2d_i-1}) \cdots (1 + t^{2d_i-1})$.

The purpose of this chapter is to generalize (9.1), regarded as a formula on the representation of the finite group W. To do this, we give (9.1) an interpretation of representation theory: $f \to wf$ gives a representation of W on S_d, with base space S_d. In general, this representation is reducible. Hence it decomposes into the sum of irreducible representations. If we denote the identity representation of W by π_0, the multiplicity of π_0 in the preceding representation by N_0^d, then (9.1) can be rewritten as

$$\sum_{d=0}^{\infty} N_0^d t^d = \frac{1}{|W|} \sum_{w \in W} \frac{1}{\det(1 - tw)} = \prod_{i=1}^{l} \frac{1}{(1 - t^{d_i})}. \tag{9.2}$$

It is natural to ask if there is any formula similar to (9.2) for the other irreducible representation π ($\neq \pi_0$) of W. In Section 9.1 we give a formula similar to (9.2) for any irreducible representation π—called the trace formula of Weyl groups.

Until now, we have never seen any discussion on the representation of the Weyl group of simple algebras from a general view point, so we cannot give a function similar to the right-hand side of (9.2) for a general W and its irreducible representations. We only consider this problem for the most important Weyl group $W(A_l)$ (it is isomorphic to the symmetric group S_{l+1} of degree $l + 1$) and give some results, including

$$\frac{1}{|W|} \sum_{w \in W(A_l)} \frac{\mathrm{tr}(w)}{\det(1 - tw)} = \frac{t}{(1 - t)(1 - t^2) \cdots (1 - t^{l-1})(1 - t^{l+1})},$$

etc. These identities solve the problem of the decomposition of S as a representation space of the symmetric group. It is certainly significant for the representation theory of the symmetric groups. This is the content of Section 9.2.

The author would like to express his hearty thanks to Wan Ze-xian for his reading this manuscript and for his help.

9.1. The Trace Formula of Weyl Groups

The meanings of g, h, W, S, and S_d are as before. As a finite group, it is known that W has only finite number of nonequivalent irreducible representations, denoted by $\pi_0, \pi_1, \ldots, \pi_q$. To fix the notation, we denote by π_0 the identity representation $w \to 1$ and by π_1 the alternating representation $w \to \det w$.

As the representation space of W, S_d, on which w acts by $w: f \to wf = f(w^{-1}x)$, is reducible in general. It can be decomposed into the direct sum of irreducible representations. We denote by N_j^d the multiplicity of π_j ($j = 0, 1, \ldots, q$) in this representation, by x_j the character of π_j, and by n_j the dimension of π_j.

Consider the operator T_φ on S_d:

$$T_\varphi = \frac{1}{|W|} \sum_{w \in W} \varphi(w) w, \tag{9.3}$$

where $|W|$ is the order of W and $\varphi(w)$ is any function on W. In the following, we investigate the trace of T_φ on S_d.

Choose any $w \in W$. It is known that the elementary divisors of w are all simple, and its eigenvalues are all the roots of unity. Denote the eigenvalues of w^{-1} by $\lambda_1, \ldots, \lambda_l$, and the corresponding eigenvectors by $e_1, \ldots, e_l$; i.e.,

$$w^{-1} e_i = \lambda_i e_i.$$

Choose $e_1, \ldots, e_l$ to be the basis vectors of the space h. The coordinate corresponding to e_i is denoted by y_i. Then $\{y_1^{i_1} \cdots y_l^{i_l} | \sum i_k = d\}$ is a basis of S_d. It is clear that

$$w: y_1^{i_1} \cdots y_l^{i_l} \to \lambda_1^{i_1} \cdots \lambda_l^{i_l} y_1^{i_1} \cdots y_l^{i_l}.$$

Thus

$$\operatorname{tr} T_\varphi = \frac{1}{|W|} \sum_{w \in W} \sum_{\sum i_k = d} \varphi(w) \lambda_1^{i_1} \cdots \lambda_l^{i_l}$$

$$= \frac{1}{|W|} \sum_{w \in W} \varphi(w) \sum_{\sum i_k = d} \lambda_1^{i_1} \cdots \lambda_l^{i_l}.$$

But $\sum_{\sum i_k = d} \lambda_1^{i_1} \cdots \lambda_l^{i_l}$ is exactly the coefficient of t^d in the expansion of

$$\frac{1}{(1 - t\lambda_1) \cdots (1 - t\lambda_l)} = \frac{1}{\det(1 - tw)}.$$

In the following, by $\{g(t)\}_d$ we denote the coefficient of t^d in the expansion of the power series $g(t)$. Thus

$$\operatorname{tr} T_\varphi = \frac{1}{|W|} \sum_{w \in W} \varphi(w) \left\{ \frac{1}{\det(1 - tw)} \right\}_d$$

$$= \frac{1}{|W|} \left\{ \sum_{w \in W} \frac{\varphi(w)}{\det(1 - tw)} \right\}_d. \tag{9.4}$$

On the other hand, the reduced decomposition of S_d is

$$S_d = N_0^d H_0 \oplus N_1^d H_1 \oplus \cdots \oplus N_q^d H_q,$$

where by H_i we denote the reduced submodule in S_d. The restriction of $w: f \rightarrow wf$ on H_i is equivalent to π_i, tr $\pi_i = x_i$. Then the trace of T_φ on S_d is

$$\frac{1}{|W|} \sum_{w \in W} \sum_{i=0}^{q} \varphi(w) N_i^d x_i(w). \tag{9.5}$$

For the irreducible representation π_j, by constructing the power series $F_j(t) = \sum_{d=0}^{\infty} N_j^d t^d$ we have the following theorem.

THEOREM 9.1 (*The trace formula of Weyl groups*). *For any function* $\varphi(w)$ *on* W, *we have*

$$\frac{1}{|W|} \sum_{w \in W} \frac{\varphi(w)}{\det(1 - tw)} = \sum_{i=0}^{q} F_j(t) \frac{1}{|W|} \sum_{w \in W} \varphi(w) x_j(w), \tag{9.6}$$

where $x_j(w)$ *is the character of the irreducible representation* π_j ($j = 0, \ldots, q$), *which are the nonequivalent irreducible representations of* W.

PROOF. By (9.4) and (9.5), for any nonnegative integer d we have

$$\frac{1}{|W|} \left\{ \sum_{w \in W} \frac{\varphi(w)}{\det(1 - tw)} \right\}_d = \frac{1}{|W|} \sum_{w \in W} \sum_{j=0}^{q} N_j^d \varphi(w) x_j(w).$$

Multiplying both sides by t^d and taking the summation over all d, we have (9.6). The theorem is proved. $\square$

COROLLARY 9.2. *In* (9.6) *we take* $\varphi(w) = 1$. *By the orthogonality of the characters* (*note that for the Weyl groups,* $x_j(w)$ *are real*), *we have, when* $j \neq 0$, $(1/|W|) \sum_{w \in W} x_j(w) = 0$ *and* $(1/|W|) \sum_{w \in W} 1 = 1$. *Thus* (9.6) *becomes*

$$\frac{1}{|W|} \sum_{w \in W} \frac{1}{\det(1 - tw)} = F_0(t). \tag{9.7}$$

By definition, $F_0(t) = \sum_{d=0}^{\infty} N_0^d t^d$. By Chevalley's theorem, $I = R[I_1, \ldots, I_l]$. The subspace I_d of invariant d-degree homogeneous polynomials has a basis $\{I_1^{i_1} \cdots I_l^{i_l} | \sum_{k=1}^{l} i_k d_k = d\}$. Its dimension, $N_0^d = \dim I_d$, is equal to the number of nonnegative integer solutions $(i_1, \ldots, i_l)$ of the equation $\sum_{k=1}^{l} i_k d_k = d$, i.e., the coefficient $\{1/(1 - t^{d_1}) \cdots (1 - t^{d_l})\}_d$ of t^d in the expansion of $1/(1 - t^{d_1}) \cdots (1 - t^{d_l})$; hence

$$F_0(t) = \frac{1}{(1 - t^{d_1}) \cdots (1 - t^{d_l})}. \tag{9.8}$$

Combining (9.7) and (9.8), we obtain Molien's formula.

COROLLARY 9.3. *Choose* $\varphi(w) = x_j(w)$ *in* (9.6). *By the orthogonality of the characters,*

$$\frac{1}{|W|} \sum_{w \in W} x_i(w) x_j(w) = \delta_{ij},$$

(9.6) *becomes*

$$F_j(t) = \sum_{d=0}^{\infty} N_j^d t^d = \frac{1}{|W|} \sum_{w \in W} \frac{x_j(w)}{\det(1 - tw)}. \tag{9.9}$$

Equation (9.9) can be regarded as a generalization of Molien's formula to the irreducible representation π_j. For general π_j's, how to determine $F_j(t)$ will be an interesting problem.

It is known that the number of conjugacy classes in a finite group is the same as the number of nonequivalent irreducible representations. By W_i $(i = 0, 1, \ldots, q)$ we denote the different conjugacy classes in W. At all elements of a conjugacy class, $\det(1 - tw)$ takes the same value. We set $f_i(t) = \det(1 - tw)$, $w \in W_i$.

THEOREM 9.4. *The following identity holds:*

$$\frac{1}{f_i(t)} = \sum_{j=0}^{q} x_j(W_i) F_j(t), \qquad i = 0, 1, \ldots, q. \tag{9.10}$$

In particular, we have

$$\frac{1}{(1 - t)^l} = \sum_{j=0}^{q} n_j F_j(t), \qquad n_j = \dim \pi_j. \tag{9.11}$$

PROOF. In (9.6), put

$$\varphi(w) = \begin{cases} 1, & w \in W_i, \\ 0, & \text{otherwise.} \end{cases}$$

Then the left-hand side of (9.6) is $(|W_i|/|W|)(1/f_i(t))$, and the right-hand side is

$$\sum_{j=0}^{q} F_j(t) \frac{1}{|W|} \sum_{w \in W_i} x_j(w) = \sum_{j=0}^{q} F_j(t) \frac{|W_i|}{|W|} x_j(w_i).$$

Thus (9.10) is proved. In (9.10), put $i = 0$ and $W_0 = (e)$, where e is the identity of W. Since $x_j(e) = \dim \pi_j = n_j$ and $f_0(t) = (1-t)^l$, we get (9.11).
$\square$

DEFINITION 9.5. A polynomial $f \in S$ is called an alternating polynomial if $wf = (\det w)f$.

By Σ^+ we denote the set of all positive roots of g. For $x \in h$ and $r \in \Sigma^+$, (r, x) is a 1-form of x, where $(\ ,\ )$ is the Killing form of g.

LEMMA 9.6 [5]. f is an alternating polynomial if and only if it is a product of a W-invariant polynomial and $\prod_{r \in \Sigma^+} (x, r)$.

PROOF. Step 1. First we prove that $\prod_{r \in \Sigma^+} (x, r)$ is alternating. Choose any reflection w_{α_i}, $\alpha_i \in \Pi$,

$$w_{\alpha_i} \prod_{r \in \Sigma^+} (x, r) = \prod_{r \in \Sigma^+} (w_{\alpha_i} x, r) = \prod_{r \in \Sigma^+} (x, w_{\alpha_i} r).$$

It is well known that w_{α_i} changes α_i to $-\alpha_i$ and permutes the other positive roots. Thus

$$\prod_{r \in \Sigma^+} (w_{\alpha_i} r, x) = - \prod_{r \in \Sigma^+} (r, x).$$

But any element in W can be generated by w_{α_i} $(i = 1, \ldots, l)$; hence

$$w \prod_{r \in \Sigma^+} (x, r) = (\det w) \prod_{r \in \Sigma^+} (x, r);$$

i.e. $\prod_{r \in \Sigma^+} (x, r)$ is alternating.

Step 2. If $f(x)$ is an alternating polynomial (i.e., $f(wx) = (\det w)f(x)$), then for each $r \in \Sigma^+$ we have $f(w_r x) = -f(x)$. Then if x belongs to the hyperplane $(x, r) = 0$, we have $w_r x = x$. Thus $f(x) = -f(x)$ implies $f(x) = 0$ (i.e., $(x, r)|f(x)$), for all $r \in \Sigma^+$. Since (x, r) are all irreducible for $r \in \Sigma^+$, we have,

$$\prod_{r \in \Sigma^+} (x, r)|f(x);$$

i.e.,

$$f(x) = g(x) \prod_{r \in \Sigma^+} (x, r), \qquad g \in S.$$

Since f and $\prod (x, r)$ are both alternating, g is W-invariant. The lemma is proved.   □

THEOREM 9.7. *If we denote the alternating representation of W by π_i: $w \to \det w$, we have*

$$F_1(t) = \frac{1}{|W|} \sum_{w \in W} \frac{\det w}{\det (1 - tw)} = \frac{t^N}{(1 - t^{d_1}) \cdots (1 - t^{d_l})}, \qquad (9.12)$$

where N is the number of all positive roots in Σ.

PROOF. By Lemma 9.6, an alternating homogeneous polynomial differs from a W-invariant polynomial only by an N-degree homogeneous factor $\prod_{r \in \Sigma^+} (x, r)$. Thus

$$N_1^{d+N} = N_0^d \qquad \text{and} \qquad \sum N_0^d t^d = \frac{1}{\prod (1 - t^{d_i})};$$

so

$$F_1(t) = \sum_{d=0}^{\infty} N_1^{d+N} t^{d+B} = t^N \sum_{d=0}^{\infty} N_0^d t^d = \frac{t^N}{\prod (1 - t^{d_i})}. \qquad □$$

9.2. The Symmetric Group

To determine the expressions $F_j(t)$ introduced in the preceding section for the Weyl group of any simple Lie algebra and its irreducible representation π_j seems to be a difficult problem. The reason is that, though the structure and representations of Weyl groups were investigated individually, yet they were rarely treated by a general method. (In Ref. 3 the equivalence classes were discussed by a general method similar to the Dynkin diagram.) In this section, we concentrate our discussion on $F_j(t)$ of the Weyl group of the simple algebra A_l. The reason is that $W(A_l)$ is isomorphic to the symmetric group S_{l+1} of $l+1$ letters and the symmetric groups are known to be the most important both in theory and applications.

The classical facts about the representations of symmetric groups mentioned in the following can be found in Ref. 1. By S_{l+1} we denote the permutation group of the letters $1, 2, \ldots, l+1$. Its conjugacy classes are denoted by $J(\alpha_1, \ldots, \alpha_{l+1})$, where $\alpha_i \geq 0$, $1\alpha_1 + 2\alpha_2 + \cdots + (l+1)\alpha_{l+1} = l+1$. There are

$$n(\alpha_1, \ldots, \alpha_{l+1}) = \frac{(l+1)!}{\alpha_1! \cdots \alpha_{l+1}! \, 1^{\alpha_1} 2^{\alpha_2} \cdots (l+1)^{\alpha_{l+1}}}.$$

elements in $J(\alpha_1, \ldots, \alpha_{l+1})$.

Note that $W(A_l)$ is isomorphic to S_{l+1} [6]. Let $\Pi = \{\beta_1, \ldots, \beta_l\}$ be a basis of the root system of A_l. The reflection w_{β_i} determined by β_i is simply denoted by w_i. Then the isomorphic mapping between $W(A_l)$ and S_{l+1} is $w_i \leftrightarrow (i, i+1)$, where $(i, i+1)$ is the permutation of interchanging i and $i+1$ and keeping other letters fixed. For a cycle of length i, $(1, 2, \ldots, i) \in S_{l+1}$, the corresponding element in $W(A_l)$ is $(1, 2, \ldots, i) = (1, 2)(2, 3) \cdots (i-1, i) \leftrightarrow w_1 w_2 \cdots w_{i-1}$. Thus we obtain

$$(1, 2, \ldots, i_1)(i_1 + 1, \ldots, i_1 + i_2)$$

$$\cdots (i_1 + i_2 + \cdots + i_{k-1} + 1, \ldots, i_1 + i_2 + \cdots + i_k)$$

$$\leftrightarrow w_1 \cdots w_{i_1-1} w_{i_1+1} \cdots w_{i_1+i_2-1} \cdots w_{i_1+\cdots+i_{k-1}+1} \cdots w_l,$$

$$\sum_{s=1}^{k} i_s = l+1. \quad (9.13)$$

LEMMA 9.8

$$\det (1 - t w_1 \cdots w_k) = (1 - t)^{l-k}(1 + t + \cdots + t^k). \quad (9.14)$$

PROOF. $w_1 \cdots w_k$ is a Coxeter element in the linear subspace $V(\beta_1, \ldots, \beta_k)$ spanned by $\beta_1, \ldots, \beta_k$. Its characteristic polynomial on $V(\beta_1, \ldots, \beta_k)$ is known to be $1 + t + \cdots + t^k$, and its characteristic polynomial on $V(\beta_{k+1}, \ldots, \beta_l)$ is clearly $(1 - t)^{l-k}$. But we have $V = V(\beta_1, \ldots, \beta_k) \oplus (\beta_{k+1}, \ldots, \beta_l)$. Thus (9.14) is proved.   □

If we denote the isomorphic image of the conjugacy class $J(\alpha_1, \ldots, \alpha_{l+1}) \subset S_{l+1}$ in $W(A_l)$ again by $J(\alpha_1, \ldots, \alpha_{l+1})$, then we have

LEMMA 9.8. *If* $w \in J(\alpha_1, \ldots, \alpha_{l+1}) \subset W(A_l)$, *then*

$$\det(1 - tw) = (1 - t)^{\alpha_1 + \cdots + \alpha_{l+1} - 1}(1 + t)^{\alpha_2}(1 + t + t^2)^{\alpha_3}$$

$$\cdots (1 + t + \cdots + t^l)^{\alpha_{l+1}}. \tag{9.15}$$

PROOF. Any $w \in J(\alpha_1, \ldots, \alpha_{l+1})$ corresponds to an element in S_{l+1} which is the product of α_1's 1-cycles, α_2's 2-cycles, $\ldots$, α_{l+1}'s $(l + 1)$-cycles. Here a k-cycle means that the length of the cycle is k. By (9.13), each k-cycle corresponds to a Coxeter element with $k - 1$ factors, $w_{j+1} \cdots w_{j+k-1}$. This term contributes a term $(1 + t + \cdots + t^{k-1})$ to $\det(1 - tw)$, by Lemma 9.8. This proves (9.15). □

COROLLARY 9.9. *If* $w \in J(\alpha_1, \ldots, \alpha_{l+1})$, *then* $\operatorname{tr} w = \alpha_1 - 1$. (9.16)

PROOF. Take the coefficient of t in (9.15). □

Before proving the following theorem, we recall that the Poincaré indices of A_l are $d_1 = 2, d_2 = 3, \ldots, d_l = l + 1$.

THEOREM 9.10. *The following identity is valid:*

$$\frac{1}{|W(A_l)|} \sum_{w \in W(A_l)} \frac{\operatorname{tr} w}{\det(1 - tw)}$$

$$= \frac{t}{(1 - t)(1 - t^2) \cdots (1 - t^l)^{\wedge}(1 - t^{l+1})}, \tag{9.17}$$

where $^{\wedge}$ *means this term does not occur in the expression.*

PROOF. It is known that $|W(A_l)| = (l + 1)!$. By Lemma 9.8 and Corollary 9.9,

$$\frac{1}{|W(A_l)|} \sum_{w \in W(A_l)} \frac{\operatorname{tr} w}{\det(1 - tw)}$$

$$= \frac{1}{(l + 1)!} \sum_{\sum_{i=1}^{l+1}; \alpha_i \geq 0 \ i\alpha_i = l+1} n(\alpha_1, \ldots, \alpha_{l+1})$$

$$\times \frac{\alpha_1 - 1}{(1 - t)^{\alpha_1 + \cdots + \alpha_{l+1} - 1}(1 + t)^{\alpha_2} \cdots (1 + t + \cdots + t^l)^{\alpha_{l+1}}}$$

$$= \sum_{\sum \alpha_i \geq 0, i\alpha_i = l+1} \frac{1}{\alpha_1! \cdots \alpha_{l+1}! \, 1^{\alpha_1} \cdots (l+1)^{\alpha_{l+1}}}$$

$$\times \frac{\alpha_1 - 1}{(1-t)^{\alpha_1 + \cdots + \alpha_{l+1} - 1}(1+t)^{\alpha_2} \cdots (1+t+\cdots+t^l)^{\alpha_{l+1}}}$$

$$= \sum_{\sum \alpha_i \geq 0, i\alpha_i = l+1} \frac{1}{\alpha_1! \cdots \alpha_{l+1}! \, 1^{\alpha} \cdots (l+1)^{\alpha_{l+1}}}$$

$$\times \frac{\alpha_1}{(1-t)^{\alpha_1 + \cdots + \alpha_{l+1} - 1}(1+t)^{\alpha_2} \cdots (1+t+\cdots+t^l)^{\alpha_{l+1}}}$$

$$- \frac{1}{|W(A_l)|} \sum_{w \in W(A_l)} \frac{1}{\det(1-tw)}$$

$$= \sum_{\substack{\alpha_1 > 0, \, \alpha_{l+1} = 0 \\ \sum i\alpha_i = l+1}} \frac{1}{(\alpha_1 - 1)! \, \alpha_2! \cdots \alpha_l! \, 1^{\alpha_1} \cdots l^{\alpha_l}}$$

$$\times \frac{1}{(1-t)(1-t)^{(\alpha_1 - 1) + \alpha_2 + \cdots + \alpha_l - 1}(1+t)^{\alpha_2} \cdots (1+t+\cdots+t^{l-1})^{\alpha_l}}$$

$$- \frac{1}{\prod_{k=2}^{l+1}(1-t^k)}$$

$$= \frac{1}{l!} \sum_{\substack{(\alpha_1 - 1) + 2\alpha_2 + \cdots + l\alpha_l = l \\ \alpha_1 \geq 1, \, \alpha_i \geq 0}} \frac{n(\alpha_1 - 1, \alpha_2, \ldots, \alpha_l)}{1-t}$$

$$\times \frac{1}{(1-t)^{(\alpha_1 - 1) + \alpha_2 + \cdots + \alpha_l - 1}(1+t) \cdots (1+t+\cdots+t^{l-1})^{\alpha_l}}$$

$$- \frac{1}{\prod_{k=2}^{l+1}(1-t^k)}$$

$$= \frac{1}{1-t} \frac{1}{|W(A_{l-1})|} \sum_{w \in W(A_{l-1})} \frac{1}{\det(1-tw)} - \frac{1}{\prod_{k=2}^{l+1}(1-t^k)}$$

$$= \frac{1}{(1-t)(1-t^2)\cdots(1-t^l)} - \frac{1}{(1-t^2)\cdots(1-t^{l+1})}$$

$$= \frac{t(1 - t^l)}{(1 - t)(1 - t^2) \cdots (1 - t^l)(1 - t^{l+1})}$$

$$= \frac{t}{(1 - t)(1 - t^2) \cdots (1 - t^l)^{\wedge}(1 - t^{l+1})}. \qquad \square$$

In the following we discuss general irreducible representations of S_{l+1}. It is well known that each irreducible representation of S_{l+1} is given by a Young tableau $m_1 \geq m_2 \geq \cdots \geq m_\nu > 0$, where $\nu \leq l + 1$, $\sum_{i=1}^{\nu} m_i = l + 1$. The irreducible representation determined by the tableau $m_1 \geq m_2 \geq \cdots \geq m_\nu > 0$ is denoted by $\pi(m_1, \ldots, m_\nu)$ and its character by $\chi^{(m_1, \ldots, m_\nu)}$, whose value on the class $J(\alpha_1, \ldots, \alpha_{l+1})$ is expressed by $\chi^{(m_1, \ldots, m_\nu)}_{(\alpha_1, \ldots, \alpha_{l+1})}$. Concerning $\chi^{(m_1, \ldots, m_\nu)}_{(\alpha_1, \ldots, \alpha_{l+1})}$ we have the following Gamba formula [1], pp. 197–201:

$$\chi^{(m_1, \ldots, m_\nu)}_{(\alpha_1, \ldots, \alpha_{l+1})} = \sum_{\substack{k_1 + 2k_2 + \cdots pk_p \leq p \\ k_i \geq 0, \, p = m_2 + \cdots + m_\nu}} [m_2, \ldots, m_\nu]_{1^{k_1} \cdots p^{k_p}} (0)$$

$$\times \binom{\alpha_1}{k_1} \cdots \binom{\alpha_p}{k_p}, \qquad (9.18)$$

where $\binom{\alpha_i}{k_i}$ are binomial coefficients, $[m_2, \ldots, m_\nu]_{1^{k_1} \cdots p^{k_p}}(0)$ are Gamba's coefficients. The details of computing these coefficients can be found in Ref. 1 and are omitted here. For example, it is not difficult to compute

$$\chi^{(l, 1)}_{(\alpha_1, \ldots, \alpha_{l+1})} = \alpha_1 - 1, \qquad (9.19)$$

$$\chi^{(l-1, 1, 1)}_{(\alpha_1, \ldots, \alpha_{l+1})} = \tfrac{1}{2}\alpha_1(\alpha_1 - 1) - \alpha_1 - \alpha_2 + 1, \qquad (9.20)$$

etc. By Lemma 9.8 and its corollaries, it is not difficult to check that $\chi^{(l, 1)}_{(\alpha_1, \ldots, \alpha_{l+1})}(w) = \mathrm{tr}\, w$, $\chi^{(l-1, 1, 1)}_{(\alpha_1, \ldots, \alpha_{l+1})}(w) = \sigma_2(w)$. Here $\sigma_2(w)$ is the coefficient of t^2 in the expansion of $\det(1 - tw)$; i.e., $\sigma_2(w) = \sum_{i<j} \lambda_i \lambda_j$, λ_i ($i = 1, \ldots, l$) are the eigenvalues of w. In other words, the irreducible representation of S_{l+1} corresponding to the Young tableau $(l, 1, 0, \ldots, 0)$ is exactly the irreducible representation $w \to W$ of $W(A_l)$. $\pi(l - 1, 1, 1)$ corresponds to the irreducible representation $w \to w \otimes_s w$ of $W(A_l)$, where by $\otimes_s$ we denote the symmetric direct product of matrices.

In the following we simply write $\alpha_{(k_1,\ldots,k_p)}^{(m_2,\ldots,m_\nu)}$ instead of Gamba's coefficient $[m_2,\ldots,m_\nu]_1{}^{k_i}\ldots_p{}^{k_p}(0)$. Note that formally it does not depend on m_1. To find the polynomial $F_{(m_1,\ldots,m_\nu)}(t)$ corresponding to the irreducible representation $\pi(m_1,\ldots,m_\nu)$, using (9.9) and putting $p = m_2 + \cdots + m_\nu$, we have

$$F_{(m_1,\ldots,m_\nu)}(t)$$

$$= \frac{1}{(l+1)!} \sum_{w \in W(A_l)} \frac{\chi^{(m_1,\ldots,m_\nu)}(w)}{\det(1-tw)}$$

$$= \frac{1}{(l+1)!} \sum_{\substack{\alpha_1+2\alpha_2+\cdots+(l+1)\alpha_{l+1}=l+1 \\ \alpha_i \geq 0}} n(\alpha_1,\ldots,\alpha_{l+1})$$

$$\times \frac{\chi_{(\alpha_1,\ldots,\alpha_{l+1})}^{(m_1,\ldots,m_\nu)}}{(1-t)^{\alpha_1+\cdots+\alpha_{l+1}-1}(1+t)^{\alpha_2}\cdots(1+t+\cdots+t^l)^{\alpha_{l+1}}}$$

$$= \frac{1}{(l+1)!} \sum_{(\alpha)} n(\alpha_1,\ldots,\alpha_{l+1})$$

$$\times \sum_{\substack{k_1+2k_2+\cdots+pk_p \leq p \\ k_i \geq 0}} \frac{\alpha_{(k_1,\ldots,k_p)}^{(m_2,\ldots,m_\nu)}\binom{\alpha_1}{k_1}\cdots\binom{\alpha_p}{k_p}}{(1-t)^{\alpha_1-1}(1-t^2)^{\alpha_2}\cdots(1-t^{l+1})^{\alpha_{l+1}}}$$

$$= \sum_{(\alpha)}\sum_{(k)} \frac{\left[\alpha_{(k_1,\ldots,k_p)}^{(m_2,\ldots,m_\nu)}\alpha_1(\alpha_1-1)\cdots(\alpha_1-k_1+1)\cdots\alpha_p(\alpha_p-1)\cdots(\alpha_p-k_p+1)\right]}{\alpha_1!\cdots\alpha_{l+1}!\,1^{\alpha_1}\cdots(l+1)^{\alpha_{l+1}}k_1!\cdots k_p!\,(1-t)^{\alpha_1-1}(1-t^2)^{\alpha_2}\cdots(1-t^{l+1})^{\alpha_{l+1}}}$$

$$= \sum_{(k)}\sum_{(\alpha)} \frac{\left[\alpha_{(k_1,\ldots,k_p)}^{(m_2,\ldots,m_\nu)}\alpha_1(\alpha_1-1)\cdots(\alpha_1-k_1+1)\cdots\alpha_p(\alpha_p-1)\cdots(\alpha_p-k_p+1)\right]}{\alpha_1!\cdots\alpha_{l+1}!\,k_1!\cdots k_p!\,1^{\alpha_1}\cdots(l+1)^{\alpha_{l+1}}(1-t)^{\alpha_1-1}(1-t^2)^{\alpha_2}\cdots(1-t^{l+1})^{\alpha_{l+1}}}.$$

In the summation $\sum_{(\alpha)}$, when $\alpha_i = 0, 1, \ldots, k_1-1, \ldots, \alpha_p = 0, 1, \ldots, k_p-1$ the result is zero. Thus the preceding expression is equal to

$$\sum_{(k)} \frac{\alpha_{(k_1,\ldots,k_p)}^{(m_2,\ldots,m_\nu)}}{k_1!\cdots k_p!} \sum_{\substack{\alpha_i \geq k_i\,(i=1,\ldots,p) \\ \sum_{i=1}^{l+1} i\alpha_i = l+1}} [(\alpha_1-k_1)!\cdots(\alpha_p-k_p)!\,\alpha_{p+1}!\cdots\alpha_{l+1}!$$

$$\times (1-t)^{\alpha_1-1}(1-t^2)^{\alpha_2}\cdots(1-t^{l+1})^{\alpha_{l+1}}]^{-1}$$

$$= \sum_{(k)} \frac{\alpha^{(m_2, \ldots, m_\nu)}_{(k_1, \ldots, k_p)}}{k_1! \cdots k_p! \, 1^{k_1} \cdots p^{k_p} \cdot (1-t)^{k_1} \cdots (1-t^p)^{k_p}}$$

$$\times \sum_{\substack{\alpha_i - k_i \geq 0 \, (1 \leq i \leq p) \\ \sum_{i=1}^{p} i(\alpha_i - k_i) + (p+1)\alpha_{p+1} \\ + \cdots + (l+1)\alpha_{l+1} = l+1 - \sum_{i=1}^{p} ik_i}} \frac{n(\alpha_1 - k_1, \ldots, \alpha_p - k_p, \alpha_{p+1}, \ldots, \alpha_l)}{[(l+1 - \sum_{i=1}^{p} ik_i)! \, (1-t)^{\alpha_1 - k_1 - 1}(1-t^2)^{\alpha_2 - k_2}}$$
$$\cdots (1-t^p)^{\alpha_p - k_p}(1-t^{p+1})^{\alpha_{p+1}} \cdots (1-t^{l+1})^{\alpha_{l+1}}]$$

$$(9.21)$$

By Molien's formula, the inner summation is equal to

$$\frac{1}{|W(A_{l-\sum_{i=1}^{p} ik_i})|} \sum_{w \in W(A_{l-\sum_{i=1}^{p} ik_i})} \frac{1}{\det(1-tw)}$$

$$= \frac{1}{(1-t^2)(1-t^3) \cdots (1-t^{l+1-\sum_{i=1}^{p} ik_i})}$$

Suppose that $j \leq p = m_2 + \cdots + m_\nu$. Put

$$B_j^{(m_2, \ldots, m_\nu)}(t) = \sum_{\substack{ik_1 + \cdots + pk_p = j \\ k_i \geq 0}} \frac{\alpha^{(m_2, \ldots, m_\nu)}_{(k_1, \ldots, k_p)}}{k_1! \cdots k_p! \, 1^{k_1} \cdots p^{k_p}}$$

$$\times \frac{1}{(1-t)^{k_1} \cdots (1-t^p)^{k_p}}. \qquad (9.22)$$

Then (9.21) can be written as

$$F_{(m_1, \ldots, m_\nu)}(t) = \sum_{j=0}^{p} \frac{B_j^{(m_2, \ldots, m_\nu)}(t)}{(1-t^2)(1-t^3) \cdots (1-t^{l+1-j})}. \qquad (9.23)$$

In summary, we get the following theorem.

THEOREM 9.11. *Let* $\pi(m_1, \ldots, m_\nu)$ *be the irreducible representation of* $W(A_l)$ *corresponding to the Young tableau* $m_1 \geq m_2 \geq \cdots \geq m_\nu > 0$, $N^d_{(m_1, \ldots, m_\nu)}$ *the multiplicity of this representation occurring in the irreducible decomposition of the* $W(A_l)$*-submodule* S_d, *and* $p = m_2 + \cdots + m_\nu$. *Then*

$$F_{(m_1, \ldots, m_\nu)}(t) = \sum_{d=0}^{\infty} N^d_{(m_1, \ldots, m_\nu)} t^d = \frac{1}{|W(A_l)|} \sum_{w \in W(A_l)} \frac{\chi^{(m_1, \ldots, m_\nu)}(w)}{\det(1-tw)}$$

$$= \sum_{j=0}^{p} \frac{B_j^{(m_2, \ldots, m_\nu)}(t)}{(1-t^2) \cdots (1-t^{l+1-j})}, \qquad (9.24)$$

where $B_j^{(m_2, \ldots, m_\nu)}(t)$ *is given by* (9.22).

EXAMPLE 9.12. We have pointed out that $\chi^{(l, 1)} = \operatorname{tr} w = \alpha_1 - 1$, so that

$$F_{(l, 1)}(t) = \frac{t}{(1 - t)(1 - t^2) \cdots (1 - t^l)^\wedge (1 - t^{l+1})}$$

is given by (9.17).

For $\chi^{(l-1, 1, 1)}$, by (9.20),

$$\chi^{(l-1, 1, 1)}_{(\alpha_1, \ldots, \alpha_{l+1})} = \tfrac{1}{2}\alpha_1(\alpha_1 - 1) - \alpha_1 - \alpha_2 + 1 = \sigma_2(w),$$

where $p = 1 + 1 = 2$, we can compute

$$B_0^{(l-1, 1, 1)}(t) = 1$$

$$B_1^{(l-1, 1, 1)}(t) = -\frac{1}{1 - t},$$

$$B_2^{(l-1, 1, 1)}(t) = \frac{1}{2}\frac{1}{(1 - t)^2} - \frac{1}{1 - t^2}.$$

By (9.24) we can compute

$$F_{(l-1, 1, 1)}(t) = \frac{1}{(1 - t^2) \cdots (1 - t^{l+1})} - \frac{1}{(1 - t)(1 - t^2) \cdots (1 - t^l)}$$

$$+ \frac{1}{2}\left[\frac{1}{(1 - t)^2} - \frac{1}{(1 - t^2)}\right]\frac{1}{(1 - t^2) \cdots (1 - t^{l-1})}$$

$$= \frac{t^3}{(1 - t^2)(1 - t)(1 - t^2) \cdots (1 - t^{l-1})^\wedge (1 - t^l)^\wedge (1 - t^{l+1})}.$$

This can also be written as follows.

COROLLARY 9.13. *We have the identity*

$$\frac{1}{|W(A_l)|}\sum_{w \in W(A_l)} \frac{\sigma_2(w)}{\det(1 - tw)}$$

$$= \frac{t^3}{(1 - t^2)(1 - t)(1 - t^2) \cdots (1 - t^{l-1})^\wedge (1 - t^l)^\wedge (1 - t^{l+1})}. \tag{9.25}$$

Theorem 9.11 gives the expression of $F_{(m_1,\ldots,m_\nu)}(t)$, but the computation of Gamba's coefficient is rather complicated. A simpler method is still needed. In the following we will give a recursion formula for computing $F_{(m_1,\ldots,m_\nu)}(t)$.

By Π we denote the basis of the root system of $W(A_l)$. For β_i ($i = 1,\ldots,j$) $\in \Pi$, we denote the subgroup generated by w_{β_i} ($i = 1,\ldots,j$) by $W(\beta_1,\ldots,\beta_j)$. Let $f(w)$ be a class function on W; i.e., it satisfies $f(ww'w^{-1}) = f(w')$, $w, w' \in W$. Then by Ref. 7, we have

$$\sum_{w \in W(A_l)} f(w) = \sum_{w' \in W(\beta_2,\ldots,\beta_l)} f(w') + l \times \sum_{w' \in W(\beta_3,\ldots,\beta_l)} f(w_{\beta_1}w')$$

$$+ l(l-1) \sum_{w' \in W(\beta_4,\ldots,\beta_l)} f(w_{\beta_1}w_{\beta_2}w') + l(l-1)\cdots(l-k+1)$$

$$\times \sum_{\substack{w' \in W(\beta_{k+2},\ldots,\beta_l) \\ +l!f(w_{\beta_1}\cdots w_{\beta_l})}} f(w_{\beta_1}\cdots w_{\beta_k}w') + \cdots$$

$$\cdots + l(l-1)\cdots 2f(w_{\beta_1}\cdots w_{\beta_{l-1}}) \tag{9.26}$$

Obviously, $W(\beta_{k+2},\ldots,\beta_l)$ is isomorphic to the symmetric group. $S(k+2,\ldots,l+1)$ of the letters $k+2,\ldots,l+1$, which is a subgroup of $S_{l+1} = S(1, 2,\ldots,l+1)$. We explained before that $w_{\beta_1}\cdots w_{\beta_i}$ corresponds to the cycle $(1, 2,\ldots,i+1)$ in S_{l+1}. If $w = w'w''$, where $w' = w_{\beta_1}\cdots w_{\beta_{h-1}}$ corresponds to the cycle $(1, 2,\ldots,h)$ and $w'' \in W(\beta_{h+1},\ldots,\beta_l)$, then for the character $\chi^{(m_1,\ldots,m_\nu)}$ we have [1, p. 193]

$$\chi^{(m_1,\ldots,m_\nu)}(w) = \chi^{(m_1,\ldots,m_\nu)}(w'w'') = \chi^{(m_1-h, m_2,\ldots,m_\nu)}(w'')$$

$$+ \chi^{(m_1, m_2-h,\ldots,m_\nu)}(w'') + \cdots + \chi^{(m_1,\ldots,m_\nu-h)}(w''), \tag{9.27}$$

where $\chi^{(m_1,\ldots,m_i-h,\ldots,m_\nu)}$ is the character of an irreducible representation of the symmetric group of degree $l+1-h$. But note that $\chi^{(m_1,\ldots,m_i-h,\ldots,\mu_\nu)}$ may not satisfy the decreasing condition $m_1 \geq m_2 \geq \cdots \geq m_i - h \geq m_{i+1} \geq \cdots \geq m_\nu$. In the following, a character not satisfying the decreasing condition is also called a character. If $m_i < m_{i+1}$, then interchange m_i and m_{i+1} in the following manner:

$$\chi^{(m_1,\ldots,m_i, m_{i+1},\ldots, m_\nu)} = -\chi^{(m_1,\ldots, m_{i+1}-1, m_i+1,\ldots,m_\nu)}. \tag{9.28}$$

For example, $\chi^{(-1, 2)} = -\chi^{(1, 0)}$, $\chi^{(0, 1)} = -\chi^{(0, 1)}$ (hence $\chi^{(0, 1)} = 0$), etc. If the superscript interchanged still spoils the decreasing condition, then we repeat this procedure until the condition is satisfied. If the last number is negative, then this character is considered to be zero [1, p. 194]. Clearly, if $\sum_{i=1}^{\nu} m_i < 0$, then $\chi^{(m_1, \ldots, m_\nu)} = 0$. Here we emphasize that the character mentioned below is always understood in this way.

Since

$$F_{(m_1, \ldots, m_\nu)}(t) = \frac{1}{|W|} \sum_{w \in W(A_l)} \frac{\chi^{(m_1, \ldots, m_\nu)}(w)}{\det (1 - tw)},$$

the subscript of $F_{(m_1, \ldots, m_\nu)}(t)$ is also understood in the same way. In the following we simply write $\langle m_1, \ldots, m_\nu \rangle$ instead of $F_{(m_1, \ldots, m_\nu)}(t)$. For example,

$$\langle l + 1 \rangle = \frac{1}{(1 - t^2) \cdots (1 - t^{l+1})},$$

etc. If $m_i < m_{i+1}$, then by (9.28),

$$\langle m_1, \ldots, m_i, m_{i+1}, \ldots, m_\nu \rangle = -\langle m_1, \ldots, m_{i+1} - 1, m_i + 1, \ldots, m_\nu \rangle.$$

For example, $\langle -1, 2 \rangle = -\langle 1, 0 \rangle = -\langle 1 \rangle$, $\langle 0, 1 \rangle = 0$, etc.

In (9.26), put

$$f(w) = \frac{1}{|W|} \frac{\chi^{(m_1, \ldots, m_\nu)}(w)}{\det (1 - tw)}.$$

Then for $w = w_{\beta_1} \cdots w_{\beta_k} w'$ and $w' \in W(\beta_{k+2}, \ldots, \beta_l)$,

$$\det (1 - tw) = (1 - t)(1 + t + \cdots + t^k) \det (1 - tw'),$$

where $w' \in W(A_{l-k-1})$. By (9.27),

$$f(w_{\beta_1} \cdots w_{\beta_k} w') = \frac{1}{|W|} \sum_{i=1}^{\nu} \frac{\chi^{(m_1, \ldots, m_i - k - 1, \ldots, m_\nu)}(w')}{(1 - t^{k+1}) \det (1 - tw')}. \tag{9.29}$$

Thus (9.26) becomes

$$\frac{1}{(l + 1)!} \sum_{w \in W(A_l)} \frac{\chi^{(m_1, \ldots, m_\nu)}(w)}{\det (1 - tw)}$$

$$
= \frac{1}{(l+1)!} \sum_{i=1}^{\nu} \sum_{w \in W(A_{l-1})} \frac{\chi^{(m_1, \ldots, m_i-1, \ldots, m_\nu)}(w)}{(1-t) \det(1-tw)}
$$

$$
+ \frac{l}{(l+1)!} \sum_{i=1}^{\nu} \sum_{w \in W(A_{l-2})} \frac{\chi^{(m_1, \ldots, m_i-2, \ldots, m_\nu)}(w)}{(1-t^2) \det(1-tw)} + \cdots
$$

$$
+ \frac{l(l-1) \cdots (l-k+1)}{(l+1)!} \sum_{i=1}^{\nu} \sum_{w \in W(A_{l-k-1})} \frac{\chi^{(m_1, \ldots, m_i-k-1, \ldots, m_\nu)}(w)}{(1-t^{k+1}) \det(1-tw)}
$$

$$
+ \cdots + \frac{l(l-2) \cdots 2}{(l+1)!} \sum_{i=1}^{\nu} \frac{\chi^{(m_1, \ldots, m_i-l, \ldots, m_\nu)}(w)}{1-t^l}
$$

$$
+ \frac{l!}{(l+1)!} \frac{1-t}{1-t^{l+1}} \sum_{i=1}^{\nu} \chi^{(m_1, \ldots, m_i-l-1, \ldots, m_\nu)}(w). \tag{9.30}
$$

Since $\sum_{i=1}^{\nu} m_i = l+1$, the last two terms are zero except for $\nu = 1$ (or $m_1 = l+1$) and $\chi^{(1,0,\ldots,0)} = 1$, $\chi^{(0,0,\ldots,0)} = 1$. If we define $\langle 1 \rangle = 1$, $\langle 0 \rangle = 1-t$, then (9.30) can be written as

$$
(l+1)\langle m_1, \ldots, m_\nu \rangle = \frac{1}{1-t} \sum_{i=1}^{\nu} \langle m_1, \ldots, m_i - 1, \ldots, m_\nu \rangle
$$

$$
+ \frac{1}{1-t^2} \sum_{i=1}^{\nu} \langle m_1, \ldots, m_i - 2, \ldots, m_\nu \rangle
$$

$$
+ \cdots + \frac{1}{1-t^l} \sum_{i=1}^{\nu} \langle m_1, \ldots, m_i - l, \ldots, m_\nu \rangle
$$

$$
+ \frac{1}{1-t^{l+1}} \sum_{i=1}^{\nu} \langle m_1, \ldots, m_i - l - 1, \ldots, m_\nu \rangle. \tag{9.31}
$$

Especially, when $\nu = 1$ or $m_1 = l+1$, (9.31) becomes

$$
(l+1)\langle l+1 \rangle = \frac{1}{1-t} \langle l \rangle + \frac{1}{1-t^2} \langle l-1 \rangle
$$

$$
+ \cdots + \frac{1}{1-t^l} \langle 1 \rangle + \frac{1}{1-t^{l+1}} \langle 0 \rangle, \tag{9.32}
$$

where

$$\langle k \rangle = F_{(k,\,0,\ldots,0)} = \frac{1}{(1-t^2)\cdots(1-t^k)} \qquad (k \geq 2), \langle 1 \rangle = 1, \langle 0 \rangle = 1 - t.$$

In the set $\{\langle m_1, \ldots, m_{l+1}\rangle\}$ we introduce an order $<$:

DEFINITION 9.14. $\langle m_1, \ldots, m_{l+1}\rangle < \langle n_1, \ldots, n_{l+1}\rangle$ if $\sum m_i < \sum n_i$ or $\sum m_i = \sum n_i$, and the first nonzero component in $\langle m_1 - n_1, m_2 - n_2, \ldots, m_{l+1} - n_{l+1}\rangle$ is positive.

The main result in this section is the following.

THEOREM 9.15. *For* $\langle m_1, \ldots, m_\nu \rangle$, *we have the following expression*:

$$\langle m_1, \ldots, m_\nu \rangle = (\langle m_1 \rangle, \langle m_1 + 1 \rangle, \ldots, \langle m_1 + \nu - 1 \rangle)$$

$$\times \begin{pmatrix} \langle 0 \rangle & \langle 1 \rangle & \cdots & \langle \nu - 1 \rangle \\ & & & \vdots \\ & & & \langle 1 \rangle \\ & & & \langle 0 \rangle \end{pmatrix}^{-1} \begin{pmatrix} \langle 0, & m_2, \ldots, m_\nu \rangle \\ \langle -1, & m_2, \ldots, m_\nu \rangle \\ \vdots \\ \langle -\nu + 1, & m_2, \ldots, m_\nu \rangle \end{pmatrix}. \quad (9.33)$$

Since $\langle 0, m_2, \ldots, m_\nu \rangle < \langle m_1, \ldots, m_\nu \rangle$, *etc.*, (9.33) *is a recursion formula*.

PROOF. *Step 1.* It is easy to check that when $m_1 \leq -\nu$, $\langle m_1, m_2, \ldots, m_\nu \rangle = 0$. Now choose $\alpha_1(t), \alpha_2(t), \ldots, \alpha_\nu(t)$ satisfying

$$\begin{aligned} \langle 0, & m_2, \ldots, m_\nu \rangle = \alpha_1(t)\langle 0 \rangle + \alpha_2(t)\langle 1 \rangle + \cdots + \alpha_\nu(t)\langle \nu - 1 \rangle, \\ \langle -1, & m_2, \ldots, m_\nu \rangle = \alpha_2(t)\langle 0 \rangle + \cdots + \alpha_\nu(t)\langle \nu - 2 \rangle, \\ \vdots \quad & \qquad\qquad \vdots \\ \langle -\nu + 1, & m_2, \ldots, m_\nu \rangle = \alpha_\nu(t)\langle 0 \rangle. \end{aligned} \quad (9.34)$$

Since $\langle 0 \rangle = 1 - t \neq 0$, (9.34) has a unique solution

$$\begin{pmatrix} \alpha_1(t) \\ \vdots \\ \alpha_\nu(t) \end{pmatrix} = \begin{pmatrix} \langle 0 \rangle & \langle 1 \rangle & \cdots & \langle \nu - 1 \rangle \\ & & & \vdots \\ & & & \langle 1 \rangle \\ & & & \langle 0 \rangle \end{pmatrix}^{-1} \begin{pmatrix} \langle 0, & m_2, \ldots, m_\nu \rangle \\ \langle -1, & m_2, \ldots, m_\nu \rangle \\ \vdots \\ \langle -\nu + 1, & m_2, \ldots, m_\nu \rangle \end{pmatrix}.$$

$$(9.35)$$

In other words, we have shown that when $m_1 \leq 0$, there exist $\alpha_1(t), \ldots, \alpha_\nu(t)$ such that

$$\langle m_1, m_2, \ldots, m_\nu \rangle = \alpha_1(t)\langle m_1 \rangle + \alpha_2(t)\langle m_1 + 1 \rangle + \cdots + \alpha_\nu(t)\langle m_1 + \nu - 1 \rangle,$$

$$(9.36)$$

where $\alpha_1(t), \ldots, \alpha_\nu(t)$ depend only on $m_2, \ldots, m_\nu$ but not m_1.

Step 2. Then we prove that (9.36) is also true for any m_1 by induction. Assume that (9.36) is established for all labels $< \langle m_1 + 1, \ldots, m_\nu \rangle$, then we need only show that, similarly, it is also true for $\langle m_1 + 1, m_2, \ldots, m_\nu \rangle$, where $\alpha_1(t), \ldots, \alpha_\nu(t)$ is also given by (9.35).

By (9.31) (put $N = \sum_{i=1}^\nu m_i + 1$),

$$N\langle m_1 + 1, m_2, \ldots, m_\nu \rangle = \sum_{k=1}^{l+1} \frac{\langle m_1 + 1 - k, m_2, \ldots, m_\nu \rangle}{1 - t^k}$$

$$+ \sum_{k=1}^{l+1} \frac{\langle m_1 + 1, m_2 - k, \ldots, m_\nu \rangle}{1 - t^k} + \cdots$$

$$+ \cdots + \sum_{k=1}^{l+1} \frac{\langle m_1 + 1, m_2, \ldots, m_\nu - k \rangle}{1 - t^k}. \qquad (9.37)$$

By the inductive hypothesis, when $k \geq 1$,

$$\langle m_1 + 1 - k, m_2, \ldots, m_\nu \rangle = \alpha_1(t)\langle m_1 + 1 - k \rangle + \cdots + \alpha_\nu(t)\langle m_1 + k \rangle,$$

so

$$\sum_{k=1}^{l+1} \frac{\langle m_1 + 1 - k, m_2, \ldots, m_\nu \rangle}{1 - t^k}$$

$$= \alpha_1(t) \sum_{j=0}^{l} \frac{\langle m_1 - j \rangle}{1 - t^{j+1}} + \cdots + \alpha_\nu(t) \sum_{j=0}^{l} \frac{\langle m_1 + \nu - 1 - j \rangle}{1 - t^{j+1}}$$

$$= (m_1 + 1)\alpha_1(t)\langle m_1 + 1 \rangle + \cdots + (m_1 + \nu)\alpha_\nu(t)\langle m_1 + \nu \rangle. \qquad (9.38)$$

The last step comes from (9.32).

Concerning other terms in (9.37), for instance, $\langle m_1 + 1, m_2 - k, \ldots, m_\nu \rangle$ $(k \geq 1)$. By definition of the order of the label set, they are all $< \langle m_1, \ldots, m_\nu \rangle$. Thus by the inductive hypothesis, we have

$$\sum_{k=1}^{l+1} \frac{\langle m_1 + 1, m_2 - k, \ldots, m_\nu \rangle}{1 - t^k} = b_1(t)\langle m_1 + 1 \rangle + \cdots + b_\nu(t)\langle m_1 + \nu \rangle,$$

$$\sum_{k=1}^{l+1} \frac{\langle m_1 + 1, m_2, \ldots, m_\nu - k \rangle}{1 - t^k} = c_1(t)\langle m_1 + 1 \rangle + \cdots + c_\nu(t)\langle m_1 + \nu \rangle,$$

where $b_1(t), \ldots, b_\nu(t); \cdots; c_1(t), \ldots, c_\nu(t)$ are all independent of m_1. Thus (9.37) becomes

$$N\langle m_1 + 1, m_2, \ldots, m_\nu \rangle = ((m_1 + 1)\alpha_1(t) + b_1(t) + \cdots + c_1(t))\langle m_1 + 1 \rangle$$

$$+ \cdots + ((m_1 + \nu)\alpha_\nu(t) + b_\nu(t)$$

$$+ \cdots + c_\nu(t))\langle m_1 + \nu \rangle.$$

Thus, we need only show that

$$\frac{1}{N}((m_1 + 1)\alpha_1(t) + b_1(t) + \cdots + c_1(t)) = \alpha_1(t), \ldots,$$

$$\frac{1}{N}((m_1 + \nu)\alpha_\nu(t) + b_\nu(t) + \cdots + c_\nu(t)) = \alpha_\nu(t).$$

Note that $N = m_1 + 1 + m_2 + \cdots + m_\nu$. It remains to show that

$$b_1(t) + \cdots + c_1(t) = (m_2 + \cdots + m_\nu)\alpha_1(t)$$

$$b_\nu(t) + \cdots + c_\nu(t) = (m_2 + \cdots + m_\nu + 1 - \nu)\alpha_\nu(t). \tag{9.39}$$

Both sides of (9.39) are independent of m_1. By the inductive hypothesis, (9.36) is true for $m_1 - 1$ and m_1. Using a similar procedure passing from $m_1 - 1$ to m_1, we get (9.40). Thus (9.40) is established. In other words, (9.36) is proved, and $\alpha_1(t), \ldots, \alpha_\nu(t)$ in (9.36) are given by (9.35). Then Theorem 9.15 is proved. $\quad\square$

Theorem 9.15 gives a simpler method for computing $\langle m_1, \ldots, m_\nu \rangle$. Here are some examples.

EXAMPLE 9.16. Find $\langle s, 1 \rangle$. By (9.33),

$$\langle s, 1 \rangle = (\langle s \rangle, \langle s + 1 \rangle) \begin{pmatrix} \langle 0 \rangle & \langle 1 \rangle \\ & \langle 0 \rangle \end{pmatrix}^{-1} \begin{pmatrix} \langle 0, 1 \rangle \\ \langle -1, 1 \rangle \end{pmatrix},$$

where $\langle 0, 1 \rangle = 0$, $\langle -1, 1 \rangle = -\langle 0 \rangle = -(1 - t)$,

$$\begin{pmatrix} \langle 0 \rangle & \langle 1 \rangle \\ & \langle 0 \rangle \end{pmatrix}^{-1} = \frac{1}{1 - t} \begin{pmatrix} 1 & -1/(1 - t) \\ 0 & 1 \end{pmatrix}.$$

Hence

$$
\langle s, 1 \rangle = \frac{1}{1-t} (\langle s \rangle, \langle s+1 \rangle) \begin{pmatrix} 1 & -1/(1-t) \\ 0 & 1 \end{pmatrix} \begin{pmatrix} 0 \\ -1+t \end{pmatrix}
$$

$$
= \frac{1}{1-t} \langle s \rangle - \langle s+1 \rangle = \frac{t}{(1-t)(1-t^2) \cdots (1-t^s)^\wedge (1-t^{s+1})}.
$$

This is exactly (9.17).

EXAMPLE 9.17. Find $\langle s, k \rangle$.

$$
\langle s, k \rangle = (\langle s \rangle, \langle s+1 \rangle) \begin{pmatrix} \langle 0 \rangle & \langle 1 \rangle \\ & \langle 0 \rangle \end{pmatrix}^{-1} \begin{pmatrix} \langle 0, k \rangle \\ \langle -1, k \rangle \end{pmatrix}
$$

$$
= \frac{1}{1-t} (\langle s \rangle, \langle s+1 \rangle) \begin{pmatrix} 1 & -1/(1-t) \\ 0 & 1 \end{pmatrix} \begin{pmatrix} \langle 0, k \rangle \\ \langle -1, k \rangle \end{pmatrix}.
$$

But

$$
\langle 0, k \rangle = -\langle k-1, 1 \rangle = \langle k \rangle - \frac{1}{1-t} \langle k-1 \rangle; \qquad \langle -1, k \rangle = -\langle k-1 \rangle.
$$

Thus

$$
\langle s, k \rangle = \frac{1}{1-t} (\langle s \rangle, \langle s+1 \rangle) \begin{pmatrix} 1 & -1/(1-t) \\ 0 & 1 \end{pmatrix} \begin{pmatrix} \langle k \rangle - [1/(1-t)]\langle k-1 \rangle \\ -\langle k-1 \rangle \end{pmatrix}
$$

$$
= \frac{1}{1-t} (\langle s \rangle\langle k \rangle - \langle s+1 \rangle\langle k-1 \rangle)
$$

$$
= \frac{t^k}{(1-t)(1-t^2) \cdots (1-t^{s+1-k})^\wedge \cdots (1-t^{s+1})(1-t^2) \cdots (1-t^k)}.
$$

EXAMPLE 9.18. Find $\langle s, 1, 1 \rangle$.

$$
\langle s, 1, 1 \rangle = (\langle s \rangle, \langle s+1 \rangle, \langle s+2 \rangle) \begin{pmatrix} \langle 0 \rangle & \langle 1 \rangle & \langle 2 \rangle \\ & \langle 0 \rangle & \langle 1 \rangle \\ & & \langle 0 \rangle \end{pmatrix}^{-1} \begin{pmatrix} \langle 0, 1, 1 \rangle \\ \langle -1, 1, 1 \rangle \\ \langle -2, 1, 1 \rangle \end{pmatrix},
$$

$$
\begin{pmatrix} \langle 0 \rangle & \langle 1 \rangle & \langle 2 \rangle \\ & \langle 0 \rangle & \langle 1 \rangle \\ & & \langle 0 \rangle \end{pmatrix}^{-1} = \frac{1}{1-t} \begin{pmatrix} 1 & -1/(1-t) & t/[(1-t)(1-t^2)] \\ & 1 & -1/(1-t) \\ & & 1 \end{pmatrix},
$$

$$
\langle 0, 1, 1 \rangle = 0, \quad \langle -1, 1, 1 \rangle = 0, \quad \langle -2, 1, 1 \rangle = -\langle 0, 1, 1 \rangle = \langle 0 \rangle = 1 - t.
$$

By substituting, we obtain

$$\langle s, 1, 1 \rangle = \frac{t}{(1-t)(1-t^2)} \langle s \rangle - \frac{1}{1-t} \langle s+1 \rangle + \langle s+2 \rangle$$

$$= \frac{t^3}{(1-t^2)(1-t)(1-t^2) \cdots (1-t^s)^\wedge (1-t^{s+1})^\wedge (1-t^{s+2})}.$$

EXAMPLE 9.19. For symmetric groups of lower degree (e.g., $W(A_2) \simeq S_3$ and $W(A_3) \simeq S_4$), all their irreducible representations and corresponding $\langle m_1, \ldots, m_\nu \rangle$ are listed as follows. For $W(A_2) \simeq S_3$:

Irreducible representations	$\langle m_1, \ldots, m_\nu \rangle$
$(3, 0, 0)$	$\langle 3 \rangle = \dfrac{1}{(1-t^2)(1-t^3)}$
$(2, 1, 0)$	$\langle 2, 1 \rangle = \dfrac{t}{(1-t)(1-t^3)}$
$(1, 1, 1)$	$\langle 1, 1, 1 \rangle = \dfrac{t^3}{(1-t^2)(1-t^3)}$

For $W(A_3) \simeq S_4$:

Irreducible representations	$\langle m_1, \ldots, m_\nu \rangle$
$(4, 0, 0, 0)$	$\langle 4 \rangle = \dfrac{1}{(1-t^2)(1-t^3)(1-t^4)}$
$(3, 1, 0, 0)$	$\langle 3, 1 \rangle = \dfrac{t}{(1-t)(1-t^2)(1-t^4)}$
$(2, 2, 0, 0)$	$\langle 2, 2 \rangle = \dfrac{t^2}{(1-t^2)^2(1-t^3)}$
$(2, 1, 1, 0)$	$\langle 2, 1, 1 \rangle = \dfrac{t^3}{(1-t)(1-t^2)(1-t^4)}$
$(1, 1, 1, 1)$	$\langle 1, 1, 1, 1 \rangle = \dfrac{t^6}{(1-t^2)(1-t^3)(1-t^4)}.$

References

1. H. Boerner, *Representations of Groups*, North-Holland, Amsterdam (1963).
2. R. W. Carter, Simple groups of Lie type, Wiley-International, London and New York (1972).
3. R. W. Carter, Conjugacy classes in the Weyl groups, *Compositio Math.*, vol. 25, 1972.
4. C. Chevalley, Invariants of finite groups generated by reflections, *Am. J. Math.* 77 (1955).
5. L. Solomon, The orders of finite Chevalley groups, *J. Algebra* 3 (1966).
6. Ze-xian Wan, *Lie Algebras*, Science Press, Beijing (1964) (in Chinese).
7. Zhong Jia-Qing, On the sum of class functions of Weyl groups, *Acta Math. Sinica* 23 (1980) (in Chinese) (Chapter 8 of this volume).

10

Some Types of Nonsymmetric Homogeneous Domains

Current research in the classification theory of bounded homogeneous domains is mainly concerned with Siegel domains. Pyateckii-Shapiro gave the first example of nonsymmetric bounded homogeneous domains in 1959. Based on that, he introduced the concept of Siegel domains, and proved the following fundamental result [3]: any bounded homogeneous domain is analytically equivalent to an affine homogeneous Siegel domain of the first or second kind. Up to now, the research on Siegel domains has yielded many results. But these works are mainly on the algebraic constructions of automorphism groups of the Siegel domains; the results on properties of geometry and function theory are not numerous.

The authors believe that the construction of some explicit examples of nonsymmetric Siegel domains (especially if these examples are very close to the symmetric domains) and the comparison of these with symmetric domains would be good for the research of general Siegel domains. From this point of view, the four explicit examples of affine homogeneous Siegel domains of the second type introduced by Pyateckii-Shapiro [4] are important. They are equal to the four symmetric classical domains plus a "tail," and their bases are self-dual cones.

Starting with some types of non-self-dual cones, we construct some new types of nonsymmetric Siegel domains. They are the Siegel domains of the first kind and are subspaces of symmetric classical domains. They are quite close to the symmetric classical domains. Besides these, this chapter

---

Joint work with Yin Wei-Ping.

also gives the proofs of non-self-dual cones and nonsymmetric domains. Other results about these domains will be published elsewhere.

The authors thank Professor Lu Qi-Keng for his support and help. His paper [5] gave us much inspiration.

10.1. Non-Self-Dual Cones

An open set $V \in \mathbb{R}^n$ is said to be a convex cone if $x \in V$, $\lambda > 0$, implies $\lambda x \in V$ and if $x, y \in V$, implies $\lambda x + \mu y \in V$, where $\lambda \geq 0$, $\mu \geq 0$, and $\lambda + \mu = 1$. We always suppose that V does not contain any straight line.

For a convex cone V and a Euclidean metric (determined by the inner product (x, y)), the set

$$V^* = \{a | (a, x) > 0, \forall x \in \bar{V} - \{0\}\}$$

is also a convex cone and is called the dual cone of V. If for certain Euclidean metrics we have $V = V^*$, then V is called a self-dual cone.

A linear transformation group which preserves V is denoted by $G(V)$. If $G(V)$ operates transitively on V, then V is called affine homogeneous and $G(V)$ is called the motion group of V.

It is proved in Ref. 7 that there are five different classes of affine homogeneous self-dual cones under the affine transformations. For the non-self-dual cones, it gives an example of dimensional 5. In this section, we will introduce some new types of affine homogeneous cones, give the motion groups of them, and prove that they are non-self-dual and simply transitive, i.e., a subgroup of motions acts simply transitively on it.

10.1.1. Definition of V_{II}

Consider $m \times m$ real symmetric square matrices Y with the following block form:

$$Y = Y', \qquad m_1 + \cdots + m_s = m,$$

$$Y = \begin{pmatrix} Y_{11} & Y_{12} & \cdots & Y_{1s} \\ Y_{21} & Y_{22} & \cdots & Y_{2s} \\ \cdots & \cdots & \cdots & \cdots \\ Y_{s1} & Y_{s2} & \cdots & Y_{ss} \end{pmatrix} \begin{matrix} m_1 \\ m_2 \\ \vdots \\ m_s \end{matrix} \qquad (10.1)$$
$$\quad\; m_1 \quad\; m_2 \quad \cdots \quad m_s$$

Take $1 = k_1 < k_2 < \cdots < k_l = s$. Let

$$Y_{ij} = 0 \qquad i \neq k_1, \ldots, k_l, j > i, \text{ or } j \neq k_1, \ldots, k_l, i > j. \qquad (10.2)$$

Then such a collection is denoted by

$$S\begin{pmatrix} m_1, \ldots, m_s \\ k_1, \ldots, k_l \end{pmatrix}.$$

For example, $(k_1, k_2) = (1, s)$. Then

$$Y = \begin{pmatrix} Y_{11} & Y_{12} & \cdots & Y_{1s} \\ Y_{21} & Y_{22} & & 0 \\ \vdots & & \ddots & \\ Y_{sl} & 0 & & Y_{ss} \end{pmatrix}, \qquad Y = Y', \qquad (10.3)$$

$s = 5$, $(k_1, k_2, k_3) = (1, 3, 5)$. Then

$$Y = \begin{pmatrix} Y_{11} & Y_{12} & Y_{13} & Y_{14} & Y_{15} \\ Y_{21} & Y_{22} & 0 & 0 & 0 \\ Y_{31} & 0 & Y_{33} & Y_{34} & Y_{35} \\ Y_{41} & 0 & Y_{43} & Y_{44} & 0 \\ Y_{51} & 0 & Y_{53} & 0 & Y_{55} \end{pmatrix}, \qquad Y = Y'. \qquad (10.4)$$

DEFINITION 10.1.

$$V_{II}\begin{pmatrix} m_1, \ldots, m_s \\ k_1, \ldots, k_l \end{pmatrix} = \left\{ Y > 0 \,\middle|\, Y \in S\begin{pmatrix} m_1, \ldots, m_s \\ k_1, \ldots, k_l \end{pmatrix} \right\}. \qquad (10.5)$$

Obviously, the dimension of V_{II} is

$$\tfrac{1}{2} \sum_{i=1}^{s} m_i(m_i + 1) + m_{k_1}(m_{k_1+1} + \cdots + m_s) + \cdots + m_{k_{l-1}}(m_{k_{l-1}+1} + \cdots + m_s).$$

$$(10.6)$$

THEOREM 10.2. *The cone*

$$V_{II}\begin{pmatrix} m_1, \ldots, m_s \\ k_1, \ldots, k_l \end{pmatrix}$$

is affine homogeneous and simply transitive.

PROOF. *Step 1.* For

$$V_{II}\begin{pmatrix} m_1, \ldots, m_s \\ k_1, \ldots, k_l \end{pmatrix},$$

we consider the transformation

$$Y \to AYA', \tag{10.7}$$

where

$$A = \begin{pmatrix} A_{11} & A_{12} & \cdots & A_{1s} \\ 0 & A_{22} & & 0 \\ \vdots & & 0 & \\ 0 & & & A_{ss} \end{pmatrix} \begin{matrix} m_1 \\ m_2 \\ \vdots \\ m_s \end{matrix}, \qquad A_{ii} = \begin{pmatrix} a_{11}^{(i)} & a_{12}^{(i)} & \cdots & a_{1m_i}^{(i)} \\ 0 & a_{22}^{(i)} & & * \\ \vdots & & 0 & \\ 0 & & & a_{m_i m_i}^{(i)} \end{pmatrix}.$$

$$\begin{matrix} m_1 & m_2 & \cdots & m_s \end{matrix}$$

$$\tag{10.8}$$

Obviously, transformation (10.7) maps

$$S\binom{m_1,\ldots,m_s}{1,s}$$

onto itself. In order to prove that the

$$S\binom{m_1,\ldots,m_s}{k_1,\ldots,k_l}$$

is transitive, it suffices to prove that for any

$$Y_0 \in V_{II}\binom{m_1,\ldots,m_s}{1,s},$$

there exists A with the form of (10.8), such that $AA' = Y_0$; that is,

$$\begin{pmatrix} A_{11} & \cdots & A_{1s} \\ & \ddots & 0 \\ 0 & & A_{ss} \end{pmatrix} \begin{pmatrix} A'_{11} & & 0 \\ \vdots & \ddots & \\ A'_{1s} & 0 & A'_{ss} \end{pmatrix} = \begin{pmatrix} \sum_{i=1}^{s} A_{1i}A'_{1i} & A_{12}A'_{22} & \cdots & A_{1s}A'_{ss} \\ A_{22}A'_{12} & A_{22}A'_{22} & & 0 \\ \vdots & & \ddots & \\ A_{ss}A'_{1s} & 0 & & A_{ss}A'_{ss} \end{pmatrix}$$

$$= \begin{pmatrix} Y_{11}^0 & Y_{12}^0 & \cdots & Y_{1s}^0 \\ Y_{21}^0 & Y_{22}^0 & & 0 \\ \vdots & & \ddots & \\ Y_{s1}^0 & 0 & & Y_{ss}^0 \end{pmatrix}.$$

It is well known that we can take A_{kk} to be of the form (10.8), such that $A_{kk}A'_{kk} = Y^0_{kk}$ $(1 < k \leq s)$. Let $A_{1k} = Y^0_{1k}A'^{-1}_{kk}$; i.e., $A_{1k}A'_{kk} = Y^0_{1k}$. So, we need only

$$Y^0_{11} - A_{12}A'_{12} - \cdots - A_{1s}A'_{1s} > 0;$$

i.e.,

$$Y^0_{11} - Y^0_{12}(Y^0_{22})^{-1}(Y^0_{12})' - \cdots - Y^0_{1s}(Y^0_{ss})^{-1}(Y^0_{1s})' > 0. \qquad (10.9)$$

Using the well-known identity $(\det S_{22} \neq 0)$

$$\begin{pmatrix} I & -S_{12}S^{-1}_{22} \\ 0 & I \end{pmatrix}\begin{pmatrix} S_{11} & S_{12} \\ S'_{12} & S_{22} \end{pmatrix}\begin{pmatrix} I & 0 \\ -S^{-1}_{22}S'_{12} & I \end{pmatrix} = \begin{pmatrix} S_{11} - S_{12}S^{-1}_{22}S'_{12} & 0 \\ 0 & S_{22} \end{pmatrix}$$

$$(10.10)$$

and letting $S_{11} = Y^0_{11}$, $S_{12} = (Y^0_{12}, \ldots, Y^0_{1s})$, $S_{22} = [Y^0_{22}, \ldots, Y^0_{ss}]$, we have (10.9). So, we take A_{11} of the form (1.7), such that

$$A_{11}A'_{11} = Y^0_{11} - Y^0_{12}(Y^0_{22})^{-1}(Y^0_{12})' - \cdots - Y^0_{1s}(Y^0_{ss})^{-1}(Y^0_{1s})'.$$

Hence, we prove that

$$V_{II}\begin{pmatrix} m_1, \ldots, m_s \\ 1, s \end{pmatrix}$$

is homogeneous. It is well known that if a positive definite symmetric matrix $Y = AA'$, where A is an upper-triangular matrix and the diagonal elements of A are positive, then A is unique. So if we take positive values for the diagonal elements of A_{ii}, then we get the simple transitive group of

$$V_{II}\begin{pmatrix} m_1, \ldots, m_s \\ 1, s \end{pmatrix}.$$

Step 2. For

$$V_{II}\begin{pmatrix} m_1, \ldots, m_s \\ 1, k_2, \ldots, k_{l-1}, s \end{pmatrix}$$

we have

$$Y = Y', \quad Y = \begin{pmatrix} Y_{11} & \cdots & Y_{1s} \\ Y_{21} & \cdots & Y_{2s} \\ \cdots & \cdots & \cdots \\ Y_{s1} & \cdots & Y_{ss} \end{pmatrix}, \quad Y_{ij} = 0 \ (i \neq 1, k_2, \ldots, k_{l-1}, s; i \neq j).$$

Consider $Y \to AYA'$, where

$$A = \begin{pmatrix} A_{11} & \cdots & A_{1s} \\ & \ddots & * \\ 0 & & A_{ss} \end{pmatrix}, \quad A_{ij} = 0 \quad (i \neq 1, k_2, \ldots, k_{l-1}, s; i < j),$$

$$A_{ii} = \begin{pmatrix} a_{11}^{(i)} & \cdots & a_{1m_i}^{(i)} \\ & \ddots & * \\ 0 & & a_{m_i m_i}^{(i)} \end{pmatrix}. \tag{10.11}$$

Such matrices form a group, which is denoted by

$$G_{II} \begin{pmatrix} m_1, \ldots, m_s \\ k_1, \ldots, k_l \end{pmatrix}.$$

Now, we prove that G_{II} acts transitively on V_{II}. In order to do this, it suffices to prove that for any $Y_0 \in V_{II}$, there exists $A \in G_{II}$ such that $AA' = Y_0$. Let

$$Y_0 = \begin{pmatrix} \tilde{Y}_{11} & \tilde{Y}_{12} \\ \tilde{Y}'_{12} & \tilde{Y}_{22} \end{pmatrix},$$

where

$$\tilde{Y}_{II} \in V_{11} \begin{pmatrix} m_1, \ldots, m_{k_{l-1}} - 1 \\ k_1, \ldots, k_{l-2}, k_{l-1} - 1 \end{pmatrix},$$

$$\tilde{Y}_{22} = \begin{pmatrix} Y_{k_{l-1}k_{l-1}} & \cdots & Y_{k_{l-1}s} \\ \vdots & 0 & \\ & & 0 \\ Y'_{k_{l-1}s} & & Y_{ss} \end{pmatrix}, \qquad \tilde{Y}_{12} = \begin{pmatrix} Y_{1k_{l-1}} & \cdots & Y_{1s} \\ 0 & \cdots & 0 \\ Y_{k_2k_{l-1}} & \cdots & Y_{k_2s} \\ 0 & \cdots & 0 \\ * & \cdots & * \\ \cdots & \cdots & \cdots \\ Y_{k_{l-2}k_{l-1}} & \cdots & Y_{k_{l-2}s} \\ 0 & \cdots & 0 \end{pmatrix}. \qquad (10.12)$$

Let

$$A = \begin{pmatrix} \tilde{A}_{11} & \tilde{A}_{12} \\ 0 & \tilde{A}_{22} \end{pmatrix},$$

where

$$\tilde{A}_{11} \in G_{II}\begin{pmatrix} m_1, \ldots, m_{k_{l-1}} - 1 \\ k_1, \ldots, k_{l-2}, k_{l-1} - 1 \end{pmatrix}, \qquad \tilde{A}_{11} = \begin{pmatrix} A_{1k_{l-1}} & \cdots & A_{1s} \\ \cdots & \cdots & \cdots \\ A_{k_{l-1}-1k_{l-1}} & \cdots & A_{k_{l-1}-1s} \end{pmatrix},$$

$$A_{ij} = 0 \quad (i \neq 1, k_2, \ldots, k_{l-2})$$

$$\tilde{A}_{22} = \begin{pmatrix} A_{k_{l-1}k_{l-1}} & \cdots & A_{k_{l-1}s} \\ & \ddots & 0 \\ 0 & & A_{ss} \end{pmatrix}.$$

If $AA' = Y_0$, then, $\tilde{A}_{22}\tilde{A}'_{22} = \tilde{Y}_{22}$, $\tilde{A}_{12}\tilde{A}'_{22} = \tilde{Y}_{12}$, $\tilde{A}_{11}\tilde{A}'_{11} + \tilde{A}_{12}\tilde{A}'_{12} = \tilde{Y}_{11}$. But from Step 1, there exists $\tilde{A}_{22}$ of the above form, such that $\tilde{A}_{22}\tilde{A}'_{22} = \tilde{Y}_{22}$. Let $\tilde{A}_{12} = \tilde{Y}_{12}\tilde{A}'^{-1}_{22}$. (It is easy to see that $\tilde{A}^{-1}_{22}$, $\tilde{Y}_{22}\tilde{A}^{-1}_{22}$ and $\tilde{Y}_{12}$ have the same form as $\tilde{A}_{22}$.)

From the identity (10.10), we know that

$$\tilde{Y}_{11} - \tilde{A}_{12}\tilde{A}'_{12} = \tilde{Y}_{11} - \tilde{Y}_{12}\tilde{Y}^{-1}_{22}\tilde{Y}'_{12} > 0.$$

But

$$\tilde{A}_{12}\tilde{A}'_{12} \in S\begin{pmatrix} m_1, \ldots, m_{k_{l-1}} - 1 \\ k_1, \ldots, k_{l-1}, k_{l-1} - 1 \end{pmatrix},$$

and $\tilde{Y}_{11} - \tilde{Y}_{12}\tilde{Y}_{22}'^{-1}\tilde{Y}_{12}'$ also has the same form but the degree is lower. So, by induction, we can obtain

$$\tilde{A}_{11} \in G_{II}\begin{pmatrix} m_1, \ldots, m_{k_{l-1}} - 1 \\ k_1, \ldots, k_{l-2}, k_{l-1} - 1 \end{pmatrix},$$

such that $\tilde{A}_{11}\tilde{A}_{11}' = \tilde{Y}_{11} - \tilde{Y}_{12} - \tilde{Y}_{22}^{-1}\tilde{Y}_{12}'$. Hence, we have

$$A = \begin{pmatrix} \tilde{A}_{11} & \tilde{A}_{12} \\ 0 & \tilde{A}_{22} \end{pmatrix} \in G_{II}\begin{pmatrix} m_1, \ldots, m_s \\ k_1, \ldots, k_l \end{pmatrix} \quad \text{and} \quad AA' = Y_0.$$

If we take positive values for the diagonal elements of A_{kk} in (10.11), then we get the simply transitive motion group.   □

REMARK 10.3.   If $m_1 = m_2 = m_3 = 1$, then

$$V_{II}\begin{pmatrix} 1 & 1 & 1 \\ & 1 & 3 \end{pmatrix} = \left\{ Y = \begin{pmatrix} y_{11} & y_{12} & y_{13} \\ y_{12} & y_{22} & 0 \\ y_{13} & 0 & y_{33} \end{pmatrix} \middle| Y > 0 \right\},$$

which is the non-self-dual cone introduced by Vinberg [7].

REMARK 10.4.   If $m_1 = m_2 = \cdots = m_3 = 1$, then

$$V_{II}\begin{pmatrix} 1, \ldots, 1 \\ 1, n \end{pmatrix} = \left\{ Y = \begin{pmatrix} y_1 & y_2 & \cdots & y_n \\ y_2 & t_2 & & \mathbf{0} \\ \vdots & & \ddots & \\ y_n & \mathbf{0} & & t_n \end{pmatrix} \middle| Y > 0 \right\},$$

that is,

$$t_2 \cdots t_n\left(y_1 - \frac{y_2^2}{t_2} - \cdots - \frac{y_n^2}{t_n}\right) > 0,$$

$$t_2 > 0,$$

$$\vdots$$

$$t_n > 0.$$

Let $t_2 = t_3 = \cdots = t_n = t$. Then we have

$$y_1 t - y_2^2 - \cdots - y_n^2 > 0, \qquad t > 0.$$

This is the cone V_4 introduced by Pyateckii-Shapiro [4].

10.1.2. Definition of V_I

Consider $m \times m$ Hermitian matrices H possessing the following block form:

$$H = \bar{H}', \qquad H = \begin{pmatrix} H_{11} & H_{12} & \cdots & H_{1s} \\ H_{21} & H_{22} & \cdots & H_{2s} \\ \cdots & \cdots & \cdots & \cdots \\ H_{s1} & H_{s2} & \cdots & H_{ss} \end{pmatrix} \begin{matrix} m_1 \\ m_2 \\ \vdots \\ m_s \end{matrix}, \quad m = m_1 + m_2 + \cdots + m_s.$$
$$\begin{matrix} m_1 & m_2 & \cdots & m_s \end{matrix} \qquad\qquad (10.13)$$

Take $1 = k_1 < k_2 < \cdots < k_l = s$. Let

$$H_{ij} = 0 \qquad (i \neq k_1, \ldots, k_l; \ i \neq j).$$

The set of such matrices is denoted by

$$H\begin{pmatrix} m_1, \ldots, m_s \\ k_1, \ldots, k_l \end{pmatrix}.$$

DEFINITION 10.5.

$$V_I\begin{pmatrix} m_1, \ldots, m_s \\ k_1, \ldots, k_l \end{pmatrix} = \left\{ H > 0 \,\Big|\, H \in H\begin{pmatrix} m_1, \ldots, m_s \\ k_1, \ldots, k_l \end{pmatrix} \right\}. \qquad (10.14)$$

Obviously,

$$\dim V_I = \sum_{i=1}^{s} m_i^2 + 2m_{k_1}(m_{k_1+1} + m_{k_1+2} + \cdots + m_s)$$

$$+ \cdots + 2m_{k_{l-1}}(m_{k_{l-1}+1} + \cdots + m_s).$$

For example,

$$H \in V_I\begin{pmatrix} m_1, \ldots, m_s \\ 1, s \end{pmatrix}.$$

Then

$$H = \bar{H}', \qquad H = \begin{pmatrix} H_{11} & H_{12} & \cdots & H_{1s} \\ H_{21} & H_{22} & & 0 \\ \vdots & & \ddots & \\ H_{s1} & 0 & & H_{ss} \end{pmatrix} > 0.$$

THEOREM 10.6. *The cone*

$$V_I \begin{pmatrix} m_1, \ldots, m_s \\ k_1, \ldots, k_l \end{pmatrix}$$

is affine homogeneous and simply transitive.

PROOF. Similar to Theorem 10.2. But here the motion group is of the form $H \to AH\bar{A}'$,

$$A = \begin{pmatrix} A_{11} & A_{12} & \cdots & A_{1s} \\ & 0 & & * \\ & & & A_{ss} \end{pmatrix}, \qquad A_{ii} = \begin{pmatrix} a_{11}^{(i)} & \cdots & a_{1m_i}^{(i)} \\ 0 & & * \\ & & a_{m_im_i}^{(i)} \end{pmatrix},$$

$$A_{ij} = 0 \qquad (i \neq k_1, \ldots, k_l; j > i). \tag{10.15}$$

This group is denoted by

$$G_I \begin{pmatrix} m_1, \ldots, m_s \\ k_1, \ldots, k_l \end{pmatrix}.$$

In order to prove that G_I acts transitively on V_I, except for changing identity (10.10) to the identity

$$\begin{pmatrix} I & -H_{12}H_{22}^{-1} \\ 0 & I \end{pmatrix} \begin{pmatrix} H_{11} & H_{12} \\ \bar{H}'_{12} & H_{22} \end{pmatrix} \begin{pmatrix} I & 0 \\ -(H_{12}H_{22}^{-1})' & I \end{pmatrix}$$

$$= \begin{pmatrix} H_{11} - H_{12}H_{22}^{-1}\bar{H}'_{12} & 0 \\ 0 & H_{22} \end{pmatrix}, \tag{10.16}$$

the rest of the proof is same as for Theorem 10.2. And if we take positive values for the diagonal elements of A_{ii}, then we get a simply transitive group with the form of (10.15). $\qquad\square$

10.1.3. Definition of V_{III}

Consider the $2m \times 2m$ Hermitian matrix and skew-symmetric matrix J:

$$H = \bar{H}', \qquad H = \begin{pmatrix} H_{11} & \cdots & H_{1s} \\ H_{21} & \cdots & H_{2s} \\ \cdots & \cdots & \cdots \\ H_{s1} & \cdots & H_{ss} \end{pmatrix} \begin{matrix} 2m_1 \\ 2m_2 \\ \vdots \\ 2m_s \end{matrix}, \qquad m = 2m_1 + \cdots + 2m_s,$$
$$\phantom{H = \bar{H}', \qquad H =} \begin{matrix} 2m_1 & \cdots & 2m_s \end{matrix}$$

$$J = \begin{pmatrix} J_{m_1} & 0 \\ 0 & J_{m_s} \end{pmatrix}, \qquad J_{m_i} = \begin{pmatrix} j & 0 \\ 0 & j \end{pmatrix}_{2m_i \times 2m_i}, \qquad j = \begin{pmatrix} 0 & 1 \\ -1 & 0 \end{pmatrix}.$$

Take $1 = k_1 < k_2 < \cdots < k_l = s$, such that

$$H_{ij} = 0 \quad (i \neq k_1, \ldots, k_l; i \neq j) \qquad \text{and} \qquad JH = \bar{H}J. \qquad (10.17)$$

The set of such matrices H is denoted by

$$J\begin{pmatrix} m_1, \ldots, m_s \\ k_1, \ldots, k_l \end{pmatrix}.$$

DEFINITION 10.7.

$$V_{III}\begin{pmatrix} m_1, \ldots, m_s \\ k_1, \ldots, k_l \end{pmatrix} = \left\{ H > 0 \,\middle|\, H \in J\begin{pmatrix} m_1, \ldots, m_s \\ k_1, \ldots, k_l \end{pmatrix} \right\}.$$

Obviously,

$$\dim V_{III} = \sum_{i=1}^{s} 2m_i^2 + 4m_1(m_2 + \cdots + m_s) + 4m_{k_2}(m_{k_2+1} + \cdots + m_s)$$

$$+ \cdots + 4m_{k_{l-1}}(m_{k_{l-1}+1} + \cdots + m_s).$$

In order to prove that V_{III} is affine transitive, we first prove the following lemma.

LEMMA 10.8. *The group* $\{H \to AH\bar{A}', JA = \bar{A}J\}$ *acts transitively on the cone* $\{H = \bar{H}' > 0, JH = \bar{H}J\}$, *where*

$$J = \begin{pmatrix} j & 0 \\ 0 & j \end{pmatrix}, \qquad A = \begin{pmatrix} A_{11} & \cdots & A_{1n} \\ 0 & & * \\ & & A_{nn} \end{pmatrix},$$

A_{ii} *are* 2×2 *upper-triangular matrices and* $j = \begin{pmatrix} 0 & 1 \\ -1 & 0 \end{pmatrix}$.

PROOF. The proof of this lemma can be found in Ref. 6. But we give a simple proof here, and from the proof we can see how to construct the simply transitive subgroup.

We need only prove that for any $H > 0$, $JH = \bar{H}J$, there exists A with the above conditions such that $A\bar{A}' = H$. First we prove the case 2×2. For this case

$$H = \begin{pmatrix} h_{11} & h_{12} \\ \bar{h}_{12} & h_{22} \end{pmatrix}, \quad j = \begin{pmatrix} 0 & 1 \\ -1 & 0 \end{pmatrix};$$

$$jH = \bar{H}j \quad \text{i.e.,} \quad H = \begin{pmatrix} h_{11} & 0 \\ 0 & h_{11} \end{pmatrix},$$

so we can take

$$A = \begin{pmatrix} \sqrt{h_{11}} & 0 \\ 0 & \sqrt{h_{11}} \end{pmatrix}.$$

In the case $2n \times 2n$, $H = (H_{ij})$, H_{ij} is a 2×2 matrix, and from $JH = \bar{H}J$ we have $jH_{ij} = \bar{H}_{ij}j$. So, from the 2×2 case, there exists A_{nn} such that $jA_{nn} = \bar{A}_{nn}j$ and $A_{nn}\bar{A}'_{nn} = H_{nn}$.

Let $A_{1n} = H_{1n}\bar{A}'^{-1}_{nn}$. Then we have

$$jA_{1n} = jH_{1n}\bar{A}'^{-1}_{nn} = \bar{H}_{1n}j\bar{A}'^{-1}_{nn} = \bar{H}_{1n}A'^{-1}_{nn}j = \bar{A}_{1n}j.$$

From the form (10.16), we have $H_{n-1\,n-1} - A_{1n}\bar{A}'_{1n} > 0$ and

$$j(H_{n-1\,n-1} - A_{1n}\bar{A}'_{1n}) = \overline{(H_{n-1\,n-1} - A_{1n}\bar{A}'_{1n})}j.$$

From the 2×2 case again, there exists an upper-triangular matrix (in fact it is a diagonal matrix) such that

$$jA_{n-1\,n-1} = \bar{A}_{n-1\,n-1}j, \quad A_{n-1}\bar{A}'_{n-1} = H_{n-1\,n-1}.$$

Proceeding in this way, we can complete the proof of lemma. From this we can see that if we take positive values for the diagonal elements of A, we can get a simply transitive group.   □

THEOREM 10.9. *The group*

$$G_{III}\begin{pmatrix} m_1, \ldots, m_s \\ k_1, \ldots, k_l \end{pmatrix}$$

acts transitively on V_{III}. G_{III} has a simply transitive subgroup and has the following form:

$$H \in V_{III}, \qquad H \to A H \bar{A}',$$

$$A = \begin{pmatrix} A_{11} & A_{12} & \cdots & A_{1s} \\ 0 & A_{22} & \cdots & A_{2s} \\ \vdots & & \ddots & * \\ 0 & 0 & & A_{ss} \end{pmatrix}, \qquad A_{ij} = 0 \quad (i \ne k_1, \ldots, k_l; \, i < j).$$

$$(10.18)$$

$JA = \bar{A}J$, J has the form (10.17), and A_{ii} is an upper-triangular matrix.

PROOF. For any $F \in G_{III}$, we have $F(V_{III}) = V_{III}$, and

$$J(AH\bar{A}') = \bar{A}JH\bar{A}' = \bar{A}\bar{H}J\bar{A}' = -\bar{A}\bar{H}(\bar{A}J)' = -\bar{A}\bar{H}(JA)' = \overline{AH\bar{A}'J}.$$

Step 1. First, we prove that

$$G_{III}\begin{pmatrix} m_1, \ldots, m_s \\ 1, s \end{pmatrix}$$

acts transitively on

$$V_{III}\begin{pmatrix} m_1, \ldots, m_s \\ 1, s \end{pmatrix}.$$

It suffices to prove that for any

$$H_0 \in V_{III}\begin{pmatrix} m_1, \ldots, m_s \\ 1, s \end{pmatrix},$$

there exists

$$A \in G_{III}\begin{pmatrix} m_1, \ldots, m_s \\ 1, s \end{pmatrix},$$

such that

$$A\bar{A}' = H_0. \tag{10.19}$$

$$A = \begin{pmatrix} A_{11} & A_{12} & \cdots & A_{1s} \\ 0 & A_{22} & & 0 \\ \vdots & & \ddots & \\ 0 & 0 & & A_{ss} \end{pmatrix}, \quad H_0 = \begin{pmatrix} H_{11} & H_{12} & \cdots & H_{1s} \\ \bar{H}'_{12} & H_{22} & & 0 \\ \vdots & & \ddots & \\ \bar{H}'_{1s} & & & H_{ss} \end{pmatrix}.$$

But (10.19) can be written

$$A_{11}\bar{A}'_{11} + \cdots + A_{1s}\bar{A}'_{1s} = H_{11}, \qquad A_{11}\bar{A}'_{1i} = H_{1i}, \qquad H_{ii} = A_{ii}\bar{A}'_{ii} \quad (i > 1).$$

From the lemma, there exists A_{ii} such that $J_{m_i}A_{ii} = \bar{A}_{ii}J_{m_i}$ and

$$A_{ii}\bar{A}'_{ii} = H_{ii}. \tag{10.20}$$

Let $A_{1i} = H_{1i}\bar{A}'^{-1}_{ii}$. From $J_{m_1}H_{1i} = \bar{H}_{1i}J_{m_i}$ we know that

$$J_{m_1}A_{1i} = \bar{A}_{1i}J_{m_i}. \tag{10.21}$$

From (10.16) we have

$$H_{11} - A_{12}\bar{A}'_{12} - \cdots - A_{1s}\bar{A}'_{1s} = H_{11} - H_{12}H^{-1}_{22}\bar{H}'_{12} - \cdots - H_{1s}H^{-1}_{ss}\bar{H}'_{1s} > 0.$$

And from (10.20) and (10.21), we have

$$J_{m_1}(H_{11} - A_{12}\bar{A}'_{12} - \cdots - A_{1s}\bar{A}'_{1s}) = \overline{(H_{11} - A_{12}\bar{A}'_{12} - \cdots - A_{1s}\bar{A}'_{1s})}J_{m_1}.$$

Then from the lemma, we get A_{11} such that

$$J_{m_1}A_{11} = \bar{A}_{11}J_{m_1}, \qquad A_{11}\bar{A}'_{11} = H_{11} - A_{12}\bar{A}'_{12} - \cdots - A_{1s}\bar{A}'_{1s}.$$

Hence, the matrix

$$A = \begin{pmatrix} A_{11} & A_{12} & \cdots & A_{1s} \\ 0 & A_{22} & & 0 \\ \vdots & & \ddots & \\ 0 & 0 & & A_{ss} \end{pmatrix}$$

satisfies the conditions of the theorem.

Step 2. In the general case

$$V_{III}\begin{pmatrix} m_1, \ldots, m_s \\ k_1, \ldots, k_l \end{pmatrix},$$

it still suffices to prove that for any $H \in V_{III}$, there exists $A \in G_{III}$, such that $A\bar{A}' = H$.

$$H = \begin{pmatrix} \tilde{H}_{11} & \tilde{H}_{12} \\ \tilde{H}_{21} & \tilde{H}_{22} \end{pmatrix}, \qquad \tilde{H}_{11} \in G_{III}\begin{pmatrix} m_1, \ldots, m_{k_{l-1}-1} \\ k_1, \ldots, k_{l-2}, k_{l-1} - 1 \end{pmatrix},$$

$$\tilde{H}_{22} = \begin{pmatrix} H_{k_{l-1}k_{l-1}} & \cdots & & H_{k_{l-1}s} \\ & H_{k_{l-1}+1\,k_{l-1}+1} & & 0 \\ \bar{H}'_{k_{l-1}s} & & 0 & H_{ss} \end{pmatrix},$$

$$\tilde{H}_{12} = \begin{pmatrix} H_{1k_{l-1}} & \cdots & H_{1s} \\ \cdots & \cdots & \cdots \\ H_{k_{l-1}-1\,k_{l-1}} & \cdots & H_{k_{l-1}-1\,s} \end{pmatrix},$$

$$H_{ij} = 0 \qquad (i \neq k_1, \ldots, k_{l-2})$$

$$A = \begin{pmatrix} \tilde{A}_{11} & \tilde{A}_{12} \\ 0 & \tilde{A}_{22} \end{pmatrix}, \qquad \tilde{A}_{22} = \begin{pmatrix} A_{k_{l-1}k_{l-1}} & \cdots & A_{k_{l-1}s} \\ 0 & & 0 \\ \vdots & 0 & \\ 0 & & A_{ss} \end{pmatrix},$$

$$\tilde{A}_{11} \in G_{III}\begin{pmatrix} m_1, \ldots, m_{k_{l-1}-1} \\ k_1, \ldots, k_{l-2}, k_{l-1} - 1 \end{pmatrix},$$

$$\tilde{A}_{12} = \begin{pmatrix} A_{1k_{l-1}} & \cdots & A_{1s} \\ \cdots & \cdots & \cdots \\ A_{k_{l-1}-1\,k_{l-1}} & \cdots & A_{k_{l-1}-1\,s} \end{pmatrix}, \qquad A_{ij} = 0 \quad (i \neq k_1, \ldots, k_{l-2}). \quad (10.22)$$

From Step 1, there exists $\tilde{A}_{22}$ of the form (10.22), such that $\tilde{A}_{22}\bar{A}'_{22} = \tilde{H}_{22}$ and

$$J\tilde{A}_{22} = \bar{\tilde{A}}_{22}J. \tag{10.23}$$

Let $\tilde{A}_{12} = \tilde{H}_{12}\bar{\tilde{A}}'^{-1}_{22}$. Then we have $J\tilde{A}_{12} = \bar{\tilde{A}}_{12}J$. Because

$$\tilde{H}_{11} - \tilde{H}_{12}\tilde{H}^{-1}_{22}\bar{\tilde{H}}'_{12} > 0.$$

By induction, we can get

$$\tilde{A}_{11} \in G_{III}\begin{pmatrix} m_1, \ldots, m_{k_{l-1}-1} \\ k_1, \ldots, k_{l-2}, k_{l-1} - 1 \end{pmatrix},$$

such that $\tilde{A}_{11}\bar{A}'_{11} = \tilde{H}_{11} - \tilde{H}_{12}\tilde{H}^{-1}_{22}\bar{\tilde{H}}'_{12}.$

Hence,

$$A = \begin{pmatrix} \tilde{A}_{11} & \tilde{A}_{12} \\ 0 & \tilde{A}_{22} \end{pmatrix} \in G_{III}\begin{pmatrix} m_1, \ldots, m_s \\ k_1, \ldots, k_l \end{pmatrix}$$

satisfies the conditions of the theorem. And if we take positive values for the diagonal elements of A_{ii}, we get the simply transitive subgroup.   ☐

10.1.4. Non-self-duality

In this section we prove that V_I, V_{II}, V_{III} are non-self-dual cones. Here, we will use the results of Ref. 1. There it is proved that any affine homogeneous convex cone can be obtained by finitely many steps of S-operations and C-operations [1].

DEFINITION 10.10. From the affine transitive convex cone $V \in \mathbb{R}$, we can get a real Siegel domain by the following method: take a V-homogeneous bilinear form F (for the definition see Ref. 4), which is defined on $\mathbb{R}^m$. Then

$$S(V) = \{(y, t) \in \mathbb{R}^{n+m} \mid y - F(t, t) \in V\}$$

is called a real Siegel domain associated with V and F. And the operation from $V \to S(V)$ is called an S-operation. Because F is V-homogeneous, $S(V)$ is affine homogeneous.

DEFINITION 10.11. For any affine transitive real Siegel domain $S \in \mathbb{R}^{n+m}$, there exists a convex cone $C(S) \in \mathbb{R}^{n+m+1}$:

$$C(S) = \{(y, t, \gamma) \in \mathbb{R}^n \times \mathbb{R}^m \times \mathbb{R}^1 \mid y\gamma - F(t, t) \in V, \gamma > 0\}. \tag{10.24}$$

The operation from $S \to C(S)$ is called a C-operation.

$C(S)$ is an affine transitive convex cone. That fact was proved in Ref. 1, and the following important result was also proved in Ref. 1.

THEOREM 10.12. *Any affine homogeneous convex cone V can be obtained from the cone $V^{(1)} = \{y \mid y > 0\}$ by a finite number of S-operations and C-operations. And each S-operation (or C-operation) is unique (up to an affine mapping).*

$$V^{(1)} \underset{S}{\to} S^{(2)} = S(V^{(1)}) \underset{C}{\to} V^{(2)} = C(S^{(2)}) \underset{S}{\to} S^{(3)}$$

$$= S(V^{(2)}) \to \cdots \to S^{(l)} \underset{C}{\to} V^{(l)} = C(S^{(l)}) = V. \tag{10.25}$$

The sequence (10.25) is called a generating sequence of V. It is easily seen that by the S-operation $V^{(i)} \to S^{(i+1)}$, the dimension increases to $n_i = \dim \mathbb{R}_i^{n_i}$, and the $V^{(i)}$-bilinear form $F^{(i)}$ is defined on $\mathbb{R}_i^{n_i}$; but by the C-operation $S^{(i)} \to V^{(i)}$, the dimension increases only by one. Let $V^{(i)} \subset \mathbb{R}_{ii}^1$. Then (10.25) can be written in terms of the underlying spaces:

$$\mathbb{R}_{11}^1 \xrightarrow{S} \mathbb{R}_{11}^1 \times \mathbb{R}_2^{n_2} \xrightarrow{C} \mathbb{R}_{11}^1 \times \mathbb{R}_2^{n_2} \times \mathbb{R}_{22}^1 \to \cdots \xrightarrow{C} \mathbb{R}_{11}^1 \times \mathbb{R}_2^{n_2} \times \cdots \times \mathbb{R}_l^{n_l} \times \mathbb{R}_{ll}^1.$$

$$(10.26)$$

Because $F^{(j)}$ has meaning on $\mathbb{R}_{11}^1 \times \mathbb{R}_2^{n_2} \times \mathbb{R}_{22}^1 \times \cdots \times \mathbb{R}_{j-1\,j-1}'$, then $F^{(j)}|_{\mathbb{R}_{1i}^1}$ induces a bilinear form $F_i^{(j)}$ on $\mathbb{R}_{ii}^1$ ($i \leq j-1$). Here $\mathbb{R}_{ii}^1 = \mathbb{R}^+$. Let $F^{(j)}$ be defined on $\mathbb{R}_j^{n_j}$ and $\mathbb{R}_{ij}$ be the positive definite subspace of $F_i^{(j)}$ on $\mathbb{R}_i^{n_j}$. Then we have (see Ref. 1)

$$\mathbb{R}_j^{n_j} = \mathbb{R}_{1j} \times \mathbb{R}_{2j} \times \cdots \times \mathbb{R}_{j-1\,j}.$$

$$(10.27)$$

Let $n_{ij} = \dim \mathbb{R}_{ij}$ ($i < j$). Then n_{ij} is an affine invariant of V. And iff every n_{ij} ($i < j$) is equal to each other, then V is a self-dual cone. So, we will give the generating sequences of V_I, V_{II}, and V_{III}. Then we can see the non-self-dual property.

For simplicity, we only discuss the case for $s = 3$, $k_1 = 1$, $k_2 = 3$. The other cases are the same.

Case 1. The generating sequence of

$$V_{II}\begin{pmatrix} m_1, m_2, m_3 \\ 1, 3 \end{pmatrix}$$

is

$$Y = \begin{pmatrix} Y_{11} & Y_{12} & Y_{13} \\ Y_{12}' & Y_{22} & 0 \\ Y_{13}' & 0 & Y_{33} \end{pmatrix} > 0, \qquad Y_{ij} = (y_{kl}^{(i,j)}).$$

First, the generating sequence of the cone $\{Y_{11} > 0\}$ is of the form

$$V^{(1)} = \{y_{11}^{(11)} > 0\} \xrightarrow{S} S^{(2)} = \{y_{11}^{(11)} - (y_{12}^{(11)})^2 > 0\} \xrightarrow{C} V^{(2)}$$

$$= \left\{ \begin{matrix} y_{11}^{(11)} y_{22}^{(11)} - (y_{12}^{(11)})^2 \\ y_{22}^{(11)} > 0 \end{matrix} \right., \text{ i.e., } \begin{pmatrix} y_{11}^{(11)} & y_{12}^{(11)} \\ y_{12}^{(11)} & y_{22}^{(11)} \end{pmatrix} > 0 \right\}$$

$$\xrightarrow{s} S^{(3)} = \left\{ \begin{pmatrix} y_{11}^{(11)} & y_{12}^{(11)} \\ y_{12}^{(11)} & y_{22}^{(11)} \end{pmatrix} - \begin{pmatrix} y_{13}^{(11)} \\ y_{23}^{(11)} \end{pmatrix} (y_{13}^{(11)} y_{23}^{(11)}) > 0 \right\} \xrightarrow{c} V^{(3)}$$

$$= \left\{ \begin{aligned} & \begin{pmatrix} y_{11}^{(11)} & y_{12}^{(11)} \\ y_{12}^{(11)} & y_{22}^{(11)} \end{pmatrix} y_{33}^{(11)} - \begin{pmatrix} y_{13}^{(11)} \\ y_{23}^{(11)} \end{pmatrix} (y_{13}^{(11)} y_{23}^{(11)}) > 0; \\ & y_{33}^{(11)} > 0 \end{aligned} \right.$$

i.e.,

$$\begin{pmatrix} y_{11}^{(11)} & y_{12}^{(11)} & y_{13}^{(11)} \\ y_{12}^{(11)} & y_{22}^{(11)} & y_{23}^{(11)} \\ y_{13}^{(11)} & y_{23}^{(11)} & y_{33}^{(11)} \end{pmatrix} > 0 \right\} \xrightarrow{s} * \to * \cdots * \to V^{(m_1)} = \{Y_{11} > 0\}.$$

Here, we used the identity

$$\begin{pmatrix} I & -ya' \\ 0 & 1 \end{pmatrix} \begin{pmatrix} Y & a' \\ a & y \end{pmatrix} \begin{pmatrix} I & 0 \\ -y'a & 1 \end{pmatrix} = \begin{pmatrix} Y - (1/y)a'a & 0 \\ 0 & y \end{pmatrix}. \quad (10.28)$$

Continuing, we have

$$V^{(m_1)} \xrightarrow{s} S^{(m_1+1)} \xrightarrow{c} \cdots \xrightarrow{c} V^{(m_1+m_2)} = \begin{pmatrix} Y_{11} & Y_{12} \\ Y'_{12} & Y_{22} \end{pmatrix} > 0.$$

Further, we have

$$V^{(m_1+m_2)} = \begin{pmatrix} Y_{11} & Y_{12} \\ Y'_{12} & Y_{22} \end{pmatrix} > 0$$

$$\xrightarrow{s} \begin{pmatrix} Y_{11} & Y_{12} \\ Y'_{12} & Y_{22} \end{pmatrix} - \begin{pmatrix} y_{11}^{(13)} \\ \vdots \\ y_{1m_1}^{(13)} \\ 0 \\ \vdots \\ 0 \end{pmatrix} (y_{11}^{(13)} \cdots y_{1m_1}^{(13)} 0 \cdots 0) > 0$$

$$\rightarrow \begin{cases} \begin{pmatrix} Y_{11} & Y_{12} \\ Y'_{12} & Y_{22} \end{pmatrix} y_{11}^{(13)} - \begin{pmatrix} y_{11}^{(13)} \\ \vdots \\ y_{1m_1}^{(13)} \\ 0 \\ \vdots \\ 0 \end{pmatrix} (y_{11}^{(13)} \cdots y_{1m_1}^{(13)} 0 \cdots 0) > 0 \\[20pt] y_{11}^{(33)} > 0; \end{cases}$$

that is,

$$\left(\begin{matrix} Y_{11} & Y_{12} & \begin{pmatrix} y_{11}^{(13)} \\ \vdots \\ y_{1m_1}^{(13)} \end{pmatrix} \\ Y'_{12} & Y_{22} & 0 \\ (y_{11}^{(13)} \cdots y_{1m_1}^{(13)}) & 0 & y_{11}^{(33)} \end{matrix} \right) > 0 \Bigg\}$$

$$\rightarrow \cdots \xrightarrow{C} V^{(m_1+m_2+m_3)} = \begin{pmatrix} Y_{11} & Y_{12} & Y_{13} \\ Y'_{12} & Y_{22} & 0 \\ Y'_{13} & 0 & Y_{33} \end{pmatrix} > 0.$$

And the $V^{(j)}$-bilinear forms are

$$F^{(2)} = (y_{12}^{(11)})^2, \qquad F^{(3)} = \begin{pmatrix} y_{13}^{(11)} \\ y_{23}^{(11)} \end{pmatrix} (y_{13}^{(11)} y_{23}^{(11)}), \ldots,$$

$$F^{(m_1+m_2)} = \begin{pmatrix} y_{1m_2}^{(12)} \\ \vdots \\ y_{m_1 m_2}^{(12)} \\ \vdots \\ y_{m_2-1 m_2}^{(22)} \end{pmatrix} (y_{1m_2}^{(12)} \cdots y_{m_1 m_2}^{(12)} \cdots y_{m_2-1 m_2}^{(22)}),$$

$$F^{(m_1+m_2+1)} = \begin{pmatrix} y_{11}^{(13)} \\ \vdots \\ y_{m_1 1}^{(13)} \\ 0 \\ \vdots \\ 0 \end{pmatrix} (y_{11}^{(13)} \cdots y_{m_1 1}^{(13)} 0 \cdots 0), \ldots,$$

$$F^{(m_1+m_2+m_3)} = \begin{pmatrix} y_{1m_3}^{(13)} \\ \vdots \\ y_{m_1m_3}^{(13)} \\ 0 \\ \vdots \\ 0 \end{pmatrix} (y_{1m_3}^{(13)} \cdots y_{m_1m_3}^{(13)} 0 \cdots 0).$$

For example,

$$F^{(3)} = \begin{pmatrix} y_{13}^{(11)} y_{13}^{(11)} & y_{13}^{(11)} y_{23}^{(11)} \\ y_{13}^{(11)} y_{23}^{(11)} & y_{23}^{(11)} y_{23}^{(11)} \end{pmatrix},$$

so $n_{23} = n_{13} = 1$.

For the same reason, if $j \le m_1 + m_2$ we have $n_{ij} = 1$ $(i < j)$. If $j = m_1 + m_2 + 1$, from the expression of $F^{(m_1+m_2+1)}$, we have

$$n_{i, m_1+m_2+1} = 1 \quad (i \le m_1), \qquad n_{i, m_1+m_2+1} = 0 \quad (m_1 < i < m_1 + m_2 + 1).$$

Also for the same reason, we have

$$n_{ij} = 1 \quad (i \le m_1, j > m_1 + m_2),$$
$$n_{ij} = 0 \quad (m_1 < i < m_1 + m_2 + 1, j > m_1 + m_2).$$

So n_{ij} are not equal to each other, and hence the

$$V_{II}\begin{pmatrix} m_1, m_2, m_3 \\ 1, 3 \end{pmatrix}$$

is a non-self-dual cone. Similarly,

$$V_{II}\begin{pmatrix} m_1, \ldots, m_s \\ k_1, \ldots, k_l \end{pmatrix}$$

is also a non-self-dual cone (if $l < s$).

Case 2. The generating sequence of

$$V_I\begin{pmatrix} m_1, m_2, m_3 \\ 1, 3 \end{pmatrix}$$

is

$$V^{(1)} =. \{h_{11} > 0\} \xrightarrow{S} \{h_{11} - h_{12}\bar{h}_{12} > 0\} \xrightarrow{C} V^{(2)}$$

$$= \left\{ \begin{pmatrix} h_{11} & h_{12} \\ \bar{h}_{12} & h_{22} \end{pmatrix} > 0 \right\} \xrightarrow{S} * \to \cdots \to V^{(m_1 + m_2)}$$

$$= \begin{pmatrix} H_{11} & H_{12} \\ \bar{H}'_{12} & H_{22} \end{pmatrix} > 0 \xrightarrow{S} \begin{pmatrix} H_{11} & H_{12} \\ \bar{H}'_{12} & H_{22} \end{pmatrix} - \begin{pmatrix} * \\ 0 \end{pmatrix} (\overline{*}' \quad 0) > 0$$

$$\xrightarrow{C} \begin{pmatrix} H_{11} & H_{12} & * \\ \bar{H}'_{12} & H_{22} & 0 \\ \overline{*}' & 0 & h_{11}^{(33)} \end{pmatrix} > 0 \xrightarrow{S} * \to \cdots \xrightarrow{C} V^{(m_1 + m_2 + m_3)}$$

$$= \begin{pmatrix} H_{11} & H_{12} & H_{13} \\ \bar{H}'_{12} & H_{22} & 0 \\ \bar{H}'_{13} & 0 & H_{33} \end{pmatrix} > 0.$$

So, we have

$$n_{ij} = 2 \quad (j \le m_1 + m_2, j > 1); \qquad n_{ij} = 0 \quad (i > m_1, j > m_1 + m_2).$$

Hence

$$V_I \begin{pmatrix} m_1, m_2, m_3 \\ 1, 3 \end{pmatrix}$$

is a non-self-dual cone. If $l < s$, for the same reason,

$$V_I \begin{pmatrix} m_1, \ldots, m_s \\ k_1, \ldots, k_l \end{pmatrix}$$

is a non-self-dual cone.

   Case 3. The generating sequence of

$$V_{III} \begin{pmatrix} m_1, m_2, m_3 \\ 1, 3 \end{pmatrix}.$$

is

$$H = \begin{pmatrix} \tilde{H}_{11} & \tilde{H}_{12} & \tilde{H}_{13} \\ \tilde{H}'_{12} & H_{22} & 0 \\ \tilde{H}'_{13} & 0 & H_{33} \end{pmatrix}, \quad H = \bar{H}', \quad JH = \bar{H}'J, \quad J = \begin{pmatrix} j & \\ & \ddots & \\ & & j \end{pmatrix},$$

$$j = \begin{pmatrix} 0 & 1 \\ -1 & 0 \end{pmatrix}.$$

Let

$$\tilde{H}_{11} = \begin{pmatrix} H_{11}^{(11)} & \cdots & H_{1m_1}^{(11)} \\ \cdots & \cdots & \cdots \\ \overline{H_{1m_1}^{(11)}} & \cdots & H_{m_1 m_1}^{(11)} \end{pmatrix} \begin{matrix} 2 \\ \vdots \\ 2 \end{matrix}$$

$$2 \quad \cdots \quad 2$$

Then $J\tilde{H}_{11} = \overset{\approx}{\tilde{H}}_{11}J$; i.e., $JH_{ij}^{(11)} = \bar{H}_{ij}^{(11)}j$. So the generating sequence of

$$H_{ii}^{(11)} = \begin{pmatrix} h_{ii}^{(11)} & 0 \\ 0 & h_{ii}^{(11)} \end{pmatrix}, \quad H_{ij}^{(11)} = \begin{pmatrix} a & b \\ -\bar{b} & a \end{pmatrix}, \quad i \neq j,$$

is

$$V^{(1)} = \{ h_{11}^{(11)} > 0; \text{ i.e., } H_{11}^{(11)} > 0 \} \overset{s}{\to} H_{11}^{(11)} - H_{12}^{(11)} \bar{H}_{12}'^{(11)} > 0 \overset{c}{\to} V^{(2)}$$

$$= \begin{cases} h_{12}^{(11)} H_{11}^{(11)} - H_{12}^{(11)} \bar{H}_{12}'^{(11)} > 0 \\ h_{22}^{(11)} > 0; \end{cases}$$

i.e.,

$$\begin{pmatrix} H_{11}^{(11)} & H_{12}^{(11)} \\ \bar{H}_{12}'^{(11)} & H_{22}^{(11)} \end{pmatrix} > 0 \} \overset{s}{\to} \cdots \overset{c}{\to} V^{(m_1 + m_2)}$$

$$= \left\{ \begin{pmatrix} \tilde{H}_{11} & \tilde{H}_{12} \\ \tilde{H}'_{12} & \tilde{H}_{22} \end{pmatrix} > 0 \right\} \overset{s}{\to} \begin{pmatrix} \tilde{H}_{11} & \tilde{H}_{12} \\ \tilde{H}'_{12} & \tilde{H}_{22} \end{pmatrix}$$

$$- \begin{pmatrix} H_{11}^{(13)} \\ \vdots \\ H_{m_1}^{(13)} \\ 0 \\ \vdots \\ 0 \end{pmatrix} (\bar{H}_{11}'^{(13)} \cdots \bar{H}_{m_1 1}'^{(13)} 0 \cdots 0) > 0 \overset{c}{\to} \begin{pmatrix} \tilde{H}_{11} & \tilde{H}_{12} & * \\ \tilde{H}'_{12} & \tilde{H}_{22} & 0 \\ * & 0 & H_{11}^{(33)} \end{pmatrix} > 0$$

$$\rightarrow \cdots \rightarrow V^{(m_1+m_2+m_3)} = \left\{ \begin{pmatrix} \tilde{H}_{11} & \tilde{H}_{12} & \tilde{H}_{13} \\ \tilde{H}'_{12} & \tilde{H}_{22} & 0 \\ \tilde{H}'_{13} & 0 & \tilde{H}_{33} \end{pmatrix} > 0 \right\}.$$

Here, the $V^{(i)}$-bilinear form is

$$F^{(2)} = H_{12}^{(11)} \bar{H}'^{(11)}_{12} = \begin{pmatrix} a\bar{a} + b\bar{b} & * \\ * & a\bar{a} + b\bar{b} \end{pmatrix},$$

so $n_{12} = 4$, and the others are similar to this. Finally we have

$$n_{ij} = 4 \quad (i < j, j \le m_1 + m_2); \qquad n_{ij} = 0 \quad (i < j, i > m_1, j > m_1 + m_2).$$

So, if $l < s$, then

$$V_{III}\begin{pmatrix} m_1, \ldots, m_s \\ k_1, \ldots, k_l \end{pmatrix}$$

is a non-self-dual cone. $\qquad\qquad\Box$

From Cases 1 to 3, we have the following theorem.

THEOREM 10.13. *If $l < s$, then*

$$V_I\begin{pmatrix} m_1, \ldots, m_s \\ k_1, \ldots, k_l \end{pmatrix}, \quad V_{II}\begin{pmatrix} m_1, \ldots, m_s \\ k_1, \ldots, k_l \end{pmatrix}, \quad and \quad V_{III}\begin{pmatrix} m_1, \ldots, m_s \\ k_1, \ldots, k_l \end{pmatrix}$$

are non-self-dual cones.

At the end of this section, we would like to add some words about V_{III}. If we use the language of quaternions, then V_{III} has the same form as V_I and V_{II}. Because, for any quaternion number $\sigma = a + ib + jc + kd = (a + ib) + (c + id)j = z_1 + z_2 j$, where z_1, z_2 are complex numbers. Then

$$\sigma \leftrightarrow \begin{pmatrix} z_1 & z_2 \\ -\bar{z}_2 & \bar{z}_1 \end{pmatrix} \qquad (10.29)$$

is an isomorphism. If H denotes a quaternion square matrix $H = H_1 + H_2 j$, where H_1, H_2 are complex square matrices, then

$$H \leftrightarrow \begin{pmatrix} H_1 & H_2 \\ -\bar{H}_2 & \bar{H}_1 \end{pmatrix} \qquad (10.30)$$

is an isomorphism between the quaternion matrix ring and the $2n \times 2n$ special matrix ring. It is well known that $\bar{\sigma} = a - ib - jc - kd$ is the conjugate of σ. Then the conjugate matrix of H is

$$\bar{H} = \begin{pmatrix} \bar{H}_1 & -H_2 \\ \bar{H}_2 & H_1 \end{pmatrix}.$$

We can define the adjoint matrix and adjoint operator $'$ as usual. Then

$$H^* = \bar{H}' = \overline{(H')} \leftrightarrow \begin{pmatrix} \bar{H}_1' & -H_2' \\ \bar{H}_2' & H_1' \end{pmatrix}.$$

Obviously, we have

$$(AB)^* = B^* A^*. \tag{10.31}$$

So, for any $1 \times n$ quaternion matrix z and quaternionic Hermitian square matrix H ($H^* = H$), we have

$$zHz^* = (zHz^*)^* \in \mathbb{R}. \tag{10.32}$$

Hence, we have the following definition: A quaternion square matrix H is called positive definite, if for any quaternary $1 \times n$ matrix $z \neq 0$, we always have

$$zHz^* > 0. \tag{10.33}$$

If

$$H = \begin{pmatrix} \sigma_{11} & \sigma_{12} & \cdots & \sigma_{1n} \\ \bar{\sigma}_{12} & \sigma_{22} & \cdots & \sigma_{2n} \\ \cdots & \cdots & \cdots & \cdots \\ \bar{\sigma}_{1n} & \bar{\sigma}_{2n} & \cdots & \sigma_{nn} \end{pmatrix} = H^*,$$

where σ_{ij} are quaternionic numbers, from (10.29) we have

$$\sigma_{ij} \leftrightarrow \begin{pmatrix} a_{ij} & b_{ij} \\ -b_{ij} & \bar{a}_{ij} \end{pmatrix},$$

then

$$
H \leftrightarrow \begin{pmatrix} \begin{pmatrix} a_{11} & b_{11} \\ -\bar{b}_{11} & \bar{a}_{11} \end{pmatrix} \cdots \begin{pmatrix} a_{1n} & b_{1n} \\ -\bar{b}_{1n} & \bar{a}_{1n} \end{pmatrix} \\ \cdots \quad \cdots \quad \cdots \quad \cdots \quad \cdots \\ \begin{pmatrix} \bar{a}_{1n} & -b_{1n} \\ \bar{b}_{1n} & \bar{a}_{1n} \end{pmatrix} \cdots \begin{pmatrix} a_{nn} & b_{nn} \\ -\bar{b}_{nn} & \bar{a}_{nn} \end{pmatrix} \end{pmatrix} = \begin{pmatrix} H_{11} & H_{12} & \cdots & H_{1n} \\ \bar{H}'_{12} & H_{22} & \cdots & H_{2n} \\ \cdots & \cdots & \cdots & \cdots \\ \bar{H}'_{1n} & \bar{H}'_{2n} & \cdots & H_{nn} \end{pmatrix} = \tilde{H}.
$$

Because (10.29) and (10.30) are holomorphisms then following the definition of (10.33), we know that the quaternionic Hermitian square matrix $H > 0$ iff $\tilde{H} > 0$ (in $\mathbb{C}$). And by the special form of $\tilde{H}$, we easily have

$$
J\tilde{H} = \tilde{H}J, \qquad J = \begin{pmatrix} j & \\ & \ddots \\ & & j \end{pmatrix}_{2n \times 2n}, \qquad j = \begin{pmatrix} 0 & 1 \\ -1 & 0 \end{pmatrix}. \tag{10.34}
$$

This condition can also determine the form of $\tilde{H}$ completely. And (10.34) is just the condition used in the definition of the cone V_{III}. □

10.2. New Types of Nonsymmetric Homogeneous Domains

10.2.1. The Definitions

Based on some types of non-self-dual cones introduced in Section 10.1 and according to the construction of Siegel domains, we introduce some new types of nonsymmetric homogeneous domains. And the special cases are just the same types of nonsymmetric domains introduced in Ref. 4.

Let

$$
C\begin{pmatrix} m_1, \ldots, m_s \\ k_1, \ldots, k_l \end{pmatrix}
$$

denote matrices of the following form:

$$
Z = \begin{pmatrix} Z_{11} & Z_{12} & \cdots & Z_{1s} \\ Z_{21} & Z_{22} & \cdots & Z_{2s} \\ \cdots & \cdots & \cdots & \cdots \\ Z_{s1} & Z_{s2} & \cdots & Z_{ss} \end{pmatrix} \begin{matrix} m_1 \\ m_2 \\ \vdots \\ m_s \end{matrix} \,,
$$

$$
Z_{ij} = 0, \quad i \neq k_1, \ldots, k_l, \quad j > i, \quad \text{or} \quad j \neq k_1, \ldots, k_l, \quad i > j. \tag{10.35}
$$

where $1 = k_1 < k_2 < \cdots < k_l = s$. If $1 = s$, then the matrix has the usual meaning. If $l = 2$, then a matrix

$$Z \in C\binom{m_1, \ldots, m_s}{1, s}$$

has the form

$$Z = \begin{pmatrix} Z_{11} & Z_{12} & \cdots & Z_{1s} \\ Z_{21} & Z_{22} & & \\ \vdots & & 0 & \\ Z_{s1} & 0 & & Z_{ss} \end{pmatrix}.$$

DEFINITION 10.14.

$$S_I = \left\{ Z \in C\binom{m_1, \ldots, m_s}{k_1, \ldots, k_l} \,\middle|\, I_m Z \in V_I\binom{m_1, \ldots, m_s}{k_1, \ldots, k_l} \right\},$$

$$S_{II} = \left\{ Z \in C\binom{m_1, \ldots, m_s}{k_1, \ldots, k_l} \,\middle|\, I_m Z \in V_{II}\binom{m_1, \ldots, m_s}{k_1, \ldots, k_l} \right\},$$

$$S_{III} = \left\{ Z \in C\binom{2m_1, \ldots, 2m_s}{k_1, \ldots, k_l} \,\middle|\, I_m Z \in V_{III}\binom{m_1, \ldots, m_s}{k_1, \ldots, k_l}, JZ = Z'J \right\}.$$

Here, if $l = s$, then S_I, S_{II}, S_{III} are symmetric classical domains as usual.

THEOREM 10.15. *Domains* S_I, S_{II}, S_{III} *are homogeneous and analytic equivalent to bounded domains.*

PROOF. In fact, any $Z \in S_I$ (or S_{II}, S_{III}) can be written as $Z = X + iY$, and $Y \in V_I$ (or V_{II}, V_{III}). But V_I (or V_{II}, V_{III}) is affine transitive (see Theorems 10.2, 10.6, and 10.9), from the fundamental fact of the theory of Siegel domains [4]. We know that S_I (or S_{II}, S_{III}) is homogeneous.

Because S_I is a submanifold of the classical domain $R_I = \{Z \,|\, (Z - \bar{Z}')/2i > 0\}$, and R_I is analytic equivalent to a bounded domain. So S_I is analytic equivalent to a bounded domain. For the same reason, S_{II} and S_{III} are also analytic equivalent to bounded domains. Now, we use cones V_I, V_{II}, and V_{III} to construct Siegel domains of type II. There are many methods to do this, but we only take the classical one:

Let

$$U_1 = \begin{pmatrix} U_{11} & U_{12} & \cdots & U_{1s} \\ 0 & U_{22} & \cdots & U_{2s} \\ \vdots & & \ddots & \vdots \\ 0 & 0 & & U_{ss} \end{pmatrix} \begin{matrix} m_1 \\ m_2 \\ \vdots \\ m_s \end{matrix}, \qquad U_2 = \begin{pmatrix} V_{11} & V_{12} & \cdots & V_{1s} \\ 0 & V_{22} & \cdots & V_{2s} \\ \vdots & & \ddots & \vdots \\ 0 & 0 & & V_{ss} \end{pmatrix} \begin{matrix} m_1 \\ m_2 \\ \vdots \\ m_s \end{matrix},$$

$$\qquad n_1 \quad n_2 \quad \cdots \quad n_s \qquad\qquad\qquad p_1 \quad p_2 \quad \cdots \quad p_s$$

$$\tag{10.36}$$

$$U_{ij} = 0, \quad V_{ij} = 0 \qquad (i \neq k_1, \ldots, k_l; j > i).$$

And let $U = (U_1, U_2)$, then we have the V_I-Hermitian form (U, V):

$$F(U, V) = U_1 \bar{V}'_1 + \bar{V}_2 U'_2. \tag{10.37}$$

DEFINITION 10.16. The domain

$$W_I = \left\{ (Z, U) \,\middle|\, \frac{1}{i}(Z - \bar{Z}') - U_1 \bar{U}'_1 - \bar{U}'_2 U'_2 > 0, \right.$$

$$\left. Z \in C\begin{pmatrix} m_1, \ldots, m_s \\ k_1, \ldots, k_l \end{pmatrix}, U = (U_1, U_2) \text{ has the form of (10.36)} \right\}$$

$$\tag{10.38}$$

Let

$$U = \begin{pmatrix} U_{11} & U_{12} & \cdots & U_{1s} \\ 0 & U_{22} & \cdots & U_{2s} \\ \vdots & & \ddots & \vdots \\ 0 & 0 & & U_{ss} \end{pmatrix} \begin{matrix} m_1 \\ m_2 \\ \vdots \\ m_s \end{matrix}, \qquad U_{ij} = 0 \quad (i \neq k_1, \ldots, k_l; j > i),$$

$$\qquad n_1 \quad n_2 \quad \cdots \quad n_s$$

$$\tag{10.39}$$

and V_{II}-Hermitian form $F(U, V) = U\bar{V}' + \bar{V}U'$. Then

domain $W_{II} = \left\{ (Z, U) \,\middle|\, \frac{1}{i}(Z - \bar{Z}') - U\bar{U}' - \bar{U}U' > 0, Z = Z', \right.$

$$\left. Z \in C\begin{pmatrix} m_1, \ldots, m_s \\ k_1, \ldots, k_l \end{pmatrix}, U \text{ has the form of (10.39)} \right\}.$$

$$\tag{10.40}$$

Let U have the form of (10.39), and V_{III}-Hermitian form $F(U, V) = U\bar{V}' + J\bar{V}U'J$, then

$$\text{domain } W_{III} = \left\{ (Z, U) \left| \frac{1}{i}(Z - \bar{Z}') - U\bar{U}' - J\bar{U}U'J > 0, \quad JZ = Z'J, \right. \right.$$

$$\left. Z \in C\left(\begin{matrix} 2m_1, \ldots, 2m_s \\ k_1, \ldots, k_l \end{matrix} \right), \; U \text{ has the form of (10.39)} \right\}.$$

$$(10.41)$$

Here, if $l = s$, then they are the three types of nonsymmetric domains introduced in Ref. 4.

THEOREM 10.17. W_I, W_{II}, and W_{III} are homogeneous domains, and analytic equivalent to bounded domains.

PROOF. According to the basic fact of Siegel domains, it suffices to prove that the V-Hermitian forms of W_I, W_{II}, and W_{III} are V-transitive (see Ref. 4). For V_I, V_{II}, V_{III}, their motion groups are, respectively, as follows:

$$H \to AH\bar{A}', \qquad Y \to AYA', \qquad H \to AH\bar{A}'.$$

In W_I, let $U_1 \to AU_1$, $U_2 \to \bar{A}U_2$, $A \in G_I$. Then

$$U_1\bar{V}_1' + \bar{V}_2U_2' \to A(U_1\bar{V}_1' + \bar{V}_2U_2')\bar{A}'.$$

In W_{II}, let $U \to AU$, $A \in G_{II}$. Then

$$U\bar{V}' + \bar{V}U' \to AU\bar{V}'A' + A\bar{V}U'A' = A(U\bar{V}' + \bar{V}U')A'.$$

In W_{III}, let $U \to AU$, $A \in G_{III}$. Then $JA = \bar{A}J$, and

$$U\bar{V}' + J\bar{V}U'J \to AU\bar{V}'\bar{A}' + J\bar{A}\bar{V}U'\bar{A}'J = A[U\bar{V}' + J\bar{V}U'J']\bar{A}'.$$

So, the V-Hermite forms corresponding to the W_I, W_{II}, and W_{III}, are V-transitive. Hence W_I, W_{II}, W_{III} are homogeneous domains. Because they are submanifolds of the three types of nonsymmetric domains introduced in Ref. 4 which are analytically equivalent to bounded domains [4], so W_I, W_{II}, W_{III} are analytically equivalent to bounded domains.

10.2.2. Proof of Nonsymmetry

In this section, we prove that if $l < s$, then S_I, S_{II}, S_{III}, W_I, W_{II}, W_{III} are nonsymmetric. If $l = s$, it is well known that S_I, S_{II}, S_{III} are symmetric, and W_I, W_{II}, W_{III} are the same as the three types of nonsymmetric domains introduced in Ref. 4. So, the famous nonsymmetric domains constructed by Pyateckii-Shapiro are only special cases here.

We start with the following lemma.

LEMMA 10.18. *Suppose $\mathscr{D}$ is a bounded domain, $0 \in \mathscr{D}$. If φ_0 is the involution of $\mathscr{D}$ at point 0, and G_0 denotes the stability subgroup at point 0 for the holomorphic automorphism of $\mathscr{D}$, then for any $\varphi \in G_0$, we have*

$$\varphi\varphi_0 = \varphi_0\varphi, \tag{10.42}$$

and if $\mathscr{D}_\varphi$, the set of fixed points of φ, is a domain, then φ_0 is also an involution of $\mathscr{D}_\varphi$ at point 0 [4].

The involution φ_0 has an expression at point 0, as follows:

$$\varphi_0: \begin{cases} w_j = -z_j + \text{terms of high degree}, \\ i = 1, 2, \ldots, n. \end{cases} \tag{10.43}$$

We use this lemma to prove that

THEOREM 10.19. *The domain*

$$S_{II}\begin{pmatrix} 1, 1, 1 \\ 1, 3 \end{pmatrix} = \left\{ Z \, \middle| \, \frac{1}{2i}(Z - \bar{Z}') > 0, \quad Z = \begin{pmatrix} z_1 & z_2 & z_3 \\ z_2 & t_2 & 0 \\ z_3 & 0 & t_3 \end{pmatrix} \right\} \tag{10.44}$$

is a nonsymmetric domain.

First, we prove the lemma.

LEMMA 10.20. *The stability subgroup of $S_{II}(\begin{smallmatrix} 1; 1, 1 \\ 1, 3 \end{smallmatrix})$ at point iI contains the following mapping:*

$$Z \to W = \begin{pmatrix} w_1 & w_2 & w_3 \\ w_2 & \xi_2 & 0 \\ w_3 & 0 & \xi_3 \end{pmatrix},$$

$$w_1 = z_1 - c_2(a_2 + c_2 t_2)^{-1} z_2^2 - c_3(a_3 + c_3 t_3)^{-1} z_3^2,$$

$$w_j = z_j(a_j + c_j t_j)^{-1} \qquad (j = 2, 3), \tag{10.45}$$

$$\xi_j = (a_j + c_j t_j)^{-1}(-c_j + a_j t_j),$$

where a_j, c_j are real and satisfy

$$\begin{pmatrix} a_j & -c_j \\ c_j & a_j \end{pmatrix}\begin{pmatrix} 0 & 1 \\ -1 & 0 \end{pmatrix}\begin{pmatrix} a_j & -c_j \\ c_j & a_j \end{pmatrix}' = \begin{pmatrix} 0 & 1 \\ -1 & 0 \end{pmatrix}. \tag{10.46}$$

PROOF. In fact, the mapping (10.45) can be written as

$$W = (A + ZC)^{-1}(-C + ZA),$$

where

$$A = \begin{pmatrix} 1 & & \\ & a_2 & \\ & & a_3 \end{pmatrix}, \qquad C = \begin{pmatrix} 0 & 0 & 0 \\ 0 & c_2 & 0 \\ 0 & 0 & c_3 \end{pmatrix}$$

and condition (10.46) is

$$\begin{pmatrix} A & -C \\ C & A \end{pmatrix}\begin{pmatrix} 0 & I \\ -I & 0 \end{pmatrix}\begin{pmatrix} A & -C \\ C & A \end{pmatrix}' = \begin{pmatrix} 0 & I \\ -I & 0 \end{pmatrix}. \tag{10.47}$$

But it is easy to see that $(Z - \bar{Z})/2i > 0$ becomes $(W - \bar{W}')/2i > 0$ under condition (10.47).

For the simplification, $S_{II}(^{1\,1\,1}_{1\,3})$ becomes S'_{II} under a parallel motion:

$$S'_{II} = \left\{ Z \,|\, I + \frac{1}{2i}(Z - \bar{Z}') > 0, \quad Z = \begin{pmatrix} z_1 & z_2 & z_3 \\ z_2 & t_2 & 0 \\ z_3 & 0 & t_3 \end{pmatrix} \right\}.$$

The mapping (10.45) has the following expression at point $0 \in S'_{II}$:

$$w_1 = z_1 - \frac{c_2 z_2^2}{a_2 + ic_2} - \frac{c_2 z_3^2}{a_3 + ic_3} + \frac{c_2^2 z_2^2 t_2}{(a_2 + ic_2)^2} + \frac{c_3^2 z_3^2 t_3}{(a_3 + ic_3)^2} + \cdots,$$

$$w_j = \frac{z_j}{a_j + ic_j} - \frac{c_j}{(a_j + ic_j)^2} z_j t_j + \frac{c_j^2}{(a_j + ic_j)^3} z_j t_j^2 + \cdots, \qquad j = 2, 3, \tag{10.48}$$

$$\xi_j = \frac{a_j - ic_j}{a_j + ic_j} t_j - \frac{c_j(a_j - ic_j)}{(a_j + ic_j)^2} t_j^2 + \frac{c_j^2(a_j - ic_j)}{(a_j + ic_j)^3} t_j^3 + \cdots, \qquad j = 2, 3.$$

We also use the following fact: the involution of the classical domain $(Z - \bar{Z})/2i > 0$, $Z = Z'$ at point iI is $Z \to -Z^{-1}$. And the involution of the domain $\{I + (Z - \bar{Z}')/2i > 0, Z = Z'\}$ at point 0 is

$$W = -Z - iZ^2 + Z^3 + \cdots. \qquad \square \quad (10.49)$$

Now, we prove Theorem 10.19. If the domain is symmetric, and its involution at point 0 is φ_0, from Lemma 10.18, we have $\varphi_0: W = -Z + \cdots$. Consider the following element φ_1, which belongs to the stability subgroup at 0:

$$\varphi_1: Z \to \begin{pmatrix} 1 & & \\ & 1 & \\ & & -1 \end{pmatrix}, \quad Z = \begin{pmatrix} 1 & & \\ & 1 & \\ & & -1 \end{pmatrix} = \begin{pmatrix} z_1 & z_2 & -z_3 \\ z_2 & t_2 & 0 \\ -z_3 & 0 & t_3 \end{pmatrix}.$$

The fixed set of φ_1 is denoted by $\mathscr{D}_{\varphi_1}$. Obviously,

$$\mathscr{D}_{\varphi_1} = \{z_3 = 0\} \times \left\{ 1 + \frac{1}{2i}(t_3 - \bar{t}_3) > 0 \right\}$$

$$\times \left\{ I + \frac{1}{2i} \left[\begin{pmatrix} z_1 & z_2 \\ z_2 & t_2 \end{pmatrix} - \overline{\begin{pmatrix} z_1 & z_2 \\ z_2 & t_2 \end{pmatrix}}' \right] > 0 \right\}.$$

From Lemma 10.18 and (10.49), we have

$$w_3|_{z_3=0} = 0,$$

$$\xi_3|_{z_3=0} = -t_3 - it_3^2 + t_3^3 + \cdots,$$

$$\begin{pmatrix} w_1 & w_2 \\ w_2 & \xi_2 \end{pmatrix}_{z_3=0} \qquad\qquad (10.50)$$

$$= -\begin{pmatrix} z_1 & z_2 \\ z_2 & t_2 \end{pmatrix} - i\begin{pmatrix} z_1^2 + z_2^2 & z_2(z_1 + t_2) \\ z_2(z_1 + t_2) & z_2^2 + t_2^2 \end{pmatrix},$$

$$+ \begin{pmatrix} (z_1^2 + z_2^2)z_1 + z_2^2(z_1 + t_2) & * \\ (z_1^2 + z_2^2)z_2 + z_2 t_2(z_1 + t_2), & z_2^2(z_1 + t_2) + t_2(z_2^2 + t_2^2) \end{pmatrix}$$

$$+ \text{high-degree terms}.$$

For the same reason, we have

$$w_2|_{z_2=0} = 0,$$

$$\xi_2|_{z_2=0} = -t_2 - it_2^2 + t_2^3 + \cdots,$$

$$\begin{pmatrix} w_1 & w_3 \\ w_3 & \xi_3 \end{pmatrix}_{z_2=0}$$

$$= -\begin{pmatrix} z_1 & z_3 \\ z_3 & t_3 \end{pmatrix} - i\begin{pmatrix} z_1^2 + z_3^2 & z_3(z_1 + t_3) \\ z_3(z_1 + t_3) & z_3^2 + t_3^2 \end{pmatrix} \tag{10.50}$$

$$+ \begin{pmatrix} (z_1^2 + z_3^2)z_1 + (z_1 + t_3)z_3^2 & * \\ (z_1^2 + z_3^2)z_3 + (z_1 + t_3)z_3t_3, & z_3^2(z_1 + t_3) + t_3(z_3^2 + t_3^2) \end{pmatrix}$$

+ high-degree terms.

From the above, every term in the expression of φ_0 except the terms having the factor z_2z_3 can be determined completely, because any terms which do not contain z_2z_3 can appear in the form (10.50) or (10.51). So, we have

$$\varphi_0: \begin{cases} w_1 = -z_1 - i(z_1^2 + z_3^2 + z_3^2) + az_2z_3 + (z_1^3 + 2z_1z_2^2 + 2z_1z_3^2 + z_2^2t_2 + z_3^2t_3) \\ \qquad + z_2z_3\Sigma_1z + \text{terms of degree more than 4,} \\[6pt] w_2 = -z_2 - iz_2(z_1 + t_2) + bz_2z_3 + (z_1^2z_2 + z_2^3 + z_2t_2^2 + z_1z_2t_2) \\ \qquad + z_2z_3\Sigma_2z + \text{terms of degree more than 4,} \\[6pt] w_3 \text{ similar to } w_2, \\[6pt] \xi_2 = -t_2 - i(t_2^2 + z_2^2) + cz_2z_3 + (z_1z_2^2 + 2z_2^2t_2 + z_2^3) + z_2z_3\Sigma_3z \\ \qquad + \text{terms of degree more than four,} \\[6pt] \xi_3 \text{ similar to } \xi_2, \end{cases} \tag{10.52}$$

where $\Sigma_1z, \Sigma_2z, \Sigma_3z$ are linear terms:

$$\Sigma_iz = a_iz_1 + b_iz_2 + c_iz_3 + d_it_1 + e_it_2.$$

Equation (10.48) is denoted by φ. Then from $\varphi\varphi_0 = \varphi_0\varphi$, we can prove that

$$a = b = c = 0. \tag{10.53}$$

For example, in order to get $b = 0$, we have

$$\varphi\varphi_0 \colon w_2 = \frac{1}{a_2 + ic_2}[-z_2 - i(z_1 + t_2)z_2 + bz_2z_3 + \cdots]$$

$$-\frac{c_2z_2t_2}{(a_2 + ic_2)^2} + \text{high-degree terms,}$$

$$\varphi_0\varphi \colon w_2 = -\left[\frac{z_2}{a_2 + ic_2} - \frac{c_2z_2t_2}{(a_2 + ic_2)^2} + \cdots\right] - i\frac{z_2}{a_2 + ic_2}$$

$$\times \left(z_1 + \frac{a_2 - ic_2}{a_2 + ic_2}t_2\right) + \frac{bz_2z_3}{(a_2 + ic_2)(a_3 + ic_3)}$$

$$+ \text{high-degree terms.}$$

From (10.42), compare the coefficients of z_3z_2, and observe that a_3, c_3 can be varied arbitrarily. Then we can get $b = 0$. Based on (10.53), we calculate $\Sigma_2 z$.

$$\varphi\varphi_0 \colon w_2 = \frac{1}{a_2 + ic_2}[-z_2 - iz_2(z_1 + t_2) + (z_1^2z_2 + z_2^3 + z_2t_2^2 + z_1z_2t_2) + z_2z_3\Sigma_2 z]$$

$$+ (-1)\frac{c_2}{(a_2 + ic_2)^2}[(-z_2 - i(z_1 + t_2)z_2)(-t_2 - i(t_2^2 + z_2^2))]$$

$$-\frac{c_2^2}{(a_2 + ic_2)^3}z_2t_2^2 + \text{high-degree terms.}$$

$$\varphi_0\varphi \colon w_2 = -\left[\frac{z_2}{a_2 + ic_2} - \frac{c_2z_2t_2}{(a_2 + ic_2)^2} + \frac{c_2^3}{(a_2 + ic_2)^3}z_2t_2^2 + \cdots\right]$$

$$+ \frac{ic_3z_2z_3^2}{(a_2 + ic_2)(a_3 + ic_3)} + \frac{z_2z_3}{(a_2 + ic_2)(a_3 + ic_3)}\Sigma_2\varphi(z)$$

$$+ \text{the terms with degree three but not containing } z_2z_3$$

$$+ \text{high-degree terms,}$$

where $\Sigma_2\varphi(z)$ denotes the linear terms of $z \overset{\varphi}{\to} \varphi(z)$. Comparing the coefficients of $z_2z_3\Sigma_2 z$, we see that we have $\Sigma_2 z = \lambda z_3$ and

$$\lambda = \frac{ic_3}{a_3 + ic_3} + \frac{\lambda}{(a_3 + ic_3)^2}, \quad \text{i.e.} \quad \lambda = \frac{ic_3(a_3 + ic_3)}{(a_3 + ic_3)^2 - 1}.$$

This fact contradicts that λ is a constant. So φ_0 is not an involution, and we complete the proof.   □

THEOREM 10.21. $S_{III}\binom{2;2;2}{1,3}$ is a nonsymmetric domain, where

$$S_{III}\binom{2,2,2}{1,3} = \left\{ Z \left| \frac{1}{2i}(Z - \bar{Z}') > 0, JZ = Z'J, J = \begin{pmatrix} j & & \\ & j & \\ & & j \end{pmatrix}, \right. \right.$$

$$\left. j = \begin{pmatrix} 0 & 1 \\ -1 & 0 \end{pmatrix}, \quad Z = \begin{pmatrix} Z_{11} & Z_{12} & Z_{13} \\ Z_{21} & Z_{22} & 0 \\ Z_{31} & 0 & Z_{33} \end{pmatrix} \right\}$$

and Z_{ij} are 2×2 square matrices.

PROOF. Because $JZ = Z'J$, Z has the following form:

$$Z = \begin{pmatrix} z_1 & 0 & z_4 & z_5 & z_8 & z_9 \\ 0 & z_1 & z_6 & z_7 & z_{10} & z_{11} \\ z_7 & -z_5 & z_2 & 0 & 0 & 0 \\ -z_6 & z_4 & 0 & z_2 & 0 & 0 \\ z_{11} & -z_9 & 0 & 0 & z_3 & 0 \\ -z_{10} & z_8 & 0 & 0 & 0 & z_3 \end{pmatrix}$$

Let

$$Z_1 = \begin{pmatrix} z_1 & 0 & z_4 & z_5 \\ 0 & z_1 & z_6 & z_7 \\ z_7 & -z_5 & z_2 & 0 \\ -z_6 & z_4 & 0 & z_2 \end{pmatrix}.$$

Then $S_{III}\binom{2;2;2}{1,3}$ can be written as

$$\frac{1}{2i}(Z_1 - \bar{Z}_1')y_3 + \frac{1}{4}\begin{pmatrix} z_8 - \bar{z}_{11} & z_9 + \bar{z}_{10} \\ z_{10} + \bar{z}_9 & z_{11} - \bar{z}_8 \end{pmatrix}\begin{pmatrix} z_{11} - \bar{z}_8 & -(z_9 + \bar{z}_{10}) \\ -(z_{10} + \bar{z}_9) & (z_8 - \bar{z}_{11}) \end{pmatrix} > 0,$$

$$y_3 = \frac{1}{2i}(z_3 - \bar{z}_3).$$

From this, we see that this domain has an analytic automorphism of the following form:

$$z_8 \to z_{11}, \quad z_{11} \to z_8, \quad z_9 \to z_{10}, \quad z_{10} \to z_9, \quad z_j = z_j \quad (j = 1, \ldots, 7).$$

By the same reasoning it also has the following analytic automorphism:

$$z_1 \to z_1, \quad z_2 \to z_2, \quad z_3 \to z_3, \quad z_4 \to z_6, \quad z_6 \to z_4, \quad z_5 \to z_7,$$

$$z_7 \to z_5, \quad z_8 \to z_{11}, \quad z_{11} \to z_8, \quad z_9 \to z_{10}, \quad z_{10} \to z_9.$$

The set of fixed points of this automorphism is

$$\left\{ Z = \begin{vmatrix} z_1 & 0 & z_4 & z_5 & z_8 & z_9 \\ 0 & z_1 & z_5 & z_4 & z_9 & z_8 \\ z_4 & -z_5 & z_2 & 0 & 0 & 0 \\ -z_5 & z_4 & 0 & z_2 & 0 & 0 \\ z_8 & -z_9 & 0 & 0 & z_3 & 0 \\ -z_9 & z_8 & 0 & 0 & 0 & z_3 \end{vmatrix} \Big| \frac{1}{2i}(Z - \bar{Z}') > 0 \right\},$$

which admits the analytic automorphism: $Z \to Z'$. Because $JZ' = -JZ'JJ = -JJZJ = ZJ$, and the set of fixed points of mapping $Z \to Z'$ is

$$\left\{ Z | Z = \begin{vmatrix} z_1 & 0 & z_4 & 0 & z_8 & 0 \\ 0 & z_1 & 0 & z_4 & 0 & z_8 \\ z_4 & 0 & z_2 & 0 & 0 & 0 \\ 0 & z_4 & 0 & z_2 & 0 & 0 \\ z_8 & 0 & 0 & 0 & z_3 & 0 \\ 0 & z_8 & 0 & 0 & 0 & z_3 \end{vmatrix} = Z' \right\};$$

i.e., the domain

$$\frac{1}{2i}\left[\begin{pmatrix} z_1 & z_4 & z_8 \\ z_4 & z_2 & 0 \\ z_8 & 0 & z_3 \end{pmatrix} - \begin{pmatrix} \bar{z}_1 & \bar{z}_4 & \bar{z}_8 \\ \bar{z}_4 & \bar{z}_2 & 0 \\ \bar{z}_8 & 0 & \bar{z}_3 \end{pmatrix} \right] > 0.$$

From Theorem 10.19 we know that the above domain is nonsymmetric. So the proof is completed. □

Now, we prove the two main results in this section.

THEOREM 10.22. *Suppose $l < s$. Then S_I, S_{II}, S_{III} are nonsymmetric.*

PROOF. First, we consider

$$S_{II} = \left\{ Z \left| \frac{1}{2i}(Z - \bar{Z}) > 0, \quad Z = Z', \quad Z \in C\begin{pmatrix} m_1, \ldots, m_s \\ k_1, \ldots, k_l \end{pmatrix} \right\}.$$

Because, $l < s$, we have a number j satisfying the following condition:

$$j \notin \{k_1, k_2, \ldots, k_l\}, \quad \text{but} \quad 1, 2, \ldots, j-1 \in \{k_1, k_2, \ldots, k_l\}.$$

Obviously, $1 < j < s$, and

$$Z \doteq \begin{pmatrix} Z_{11} \cdots\cdots\cdots\cdots\cdots Z_{1s} \\ \quad * \cdots\cdots\cdots * \\ \quad\quad\quad Z_{jj}\cdots 0\cdots 0 \\ \quad\quad\quad\quad\quad 0 \quad 0 \\ Z'_{1s} \quad * \quad 0 \quad\quad Z_{ss} \end{pmatrix},$$

so we consider the analytic automorphism

$$Z \to I_0 Z I'_0, \qquad I_0 = \begin{pmatrix} \overset{\leftarrow\; j\; \longrightarrow}{} \\ I \\ \quad -I \\ \quad\quad \ddots \\ \quad\quad\quad -I \\ \quad\quad\quad\quad I \\ \quad\quad\quad\quad\quad -I \\ \quad\quad\quad\quad\quad\quad \ddots \\ \quad\quad\quad\quad\quad\quad\quad I \end{pmatrix}.$$

The set of fixed points of this automorphism is

$$\left\{ \frac{1}{2i}\left[\begin{pmatrix} Z_{11} & Z_{1j} & Z_{1s} \\ Z'_{1j} & Z_{jj} & 0 \\ Z'_{1s} & 0 & Z_{ss} \end{pmatrix} - \overline{\begin{pmatrix} Z_{11} & Z_{1j} & Z_{1s} \\ Z'_{1j} & Z_{jj} & 0 \\ Z'_{1s} & 0 & Z_{ss} \end{pmatrix}} \right] > 0 \right\}$$

$\times$ some symmetric domains.

The first domain in the above space has the following analytic automorphism:

$$\begin{pmatrix} Z_{11} & Z_{1j} & Z_{1s} \\ Z'_{1j} & Z_{jj} & 0 \\ Z'_{1s} & 0 & Z_{ss} \end{pmatrix} \to I_* \begin{pmatrix} Z_{11} & Z_{1j} & Z_{1s} \\ Z'_{1j} & Z_{jj} & 0 \\ Z'_{1s} & 0 & Z_{ss} \end{pmatrix} I'_*,$$

$$I_* = \left(\begin{array}{ccc} \left(\begin{smallmatrix} 1 & & & \\ & -1 & & \\ & & \ddots & \\ & & & -1 \end{smallmatrix}\right) & & \\ & \left(\begin{smallmatrix} 1 & & & \\ & -1 & & \\ & & \ddots & \\ & & & -1 \end{smallmatrix}\right) & \\ & & \left(\begin{smallmatrix} 1 & & & \\ & -1 & & \\ & & \ddots & \\ & & & -1 \end{smallmatrix}\right) \end{array} \right) \begin{array}{l} m_1 \\[2em] m_j. \\[2em] m_s \end{array}$$

The set of fixed points of this automorphism is

$$\left\{ \begin{pmatrix} z_{11}^{(11)} & & 0 & z_{11}^{(1j)} & & 0 & z_{11}^{(1s)} & 0 \\ & \ddots & & & & & & \\ 0 & & z_{m_1 m_1}^{(11)} & & z_{11}^{(jj)} & & 0 & \\ z_{11}^{(1j)} & & & \ddots & & & & \\ 0 & & & & z_{m_j m_j}^{(jj)} & & z_{11}^{(ss)} & \\ z_{11}^{(1s)} & & 0 & & & z_{11}^{(ss)} & \ddots & \\ 0 & & & & & & & z_{m_s m_s}^{(ss)} \end{pmatrix} \right\};$$

$$\left\{ \frac{1}{2i}\left[\begin{pmatrix} z_{11}^{(11)} & z_{11}^{(1j)} & z_{11}^{(1s)} \\ z_{11}^{(1j)} & z_{11}^{(jj)} & 0 \\ z_{11}^{(1s)} & 0 & z_{11}^{(ss)} \end{pmatrix} - \overline{\begin{pmatrix} z_{11}^{(11)} & z_{11}^{(1j)} & z_{11}^{(1s)} \\ z_{11}^{(1j)} & z_{11}^{(jj)} & 0 \\ z_{11}^{(1s)} & 0 & z_{11}^{(ss)} \end{pmatrix}} \right] > 0 \right\}$$

$\times$ some upper half-planes.

From Theorem 10.19 the above domain is nonsymmetric, so S_{ll} is a nonsymmetric domain.

Second, we consider

$$S_I = \left\{ Z \left| \frac{1}{2i}(Z - \bar{Z}') > 0, Z \in C\begin{pmatrix} m_1, \ldots, m_s \\ k_1, \ldots, k_l \end{pmatrix} \right. \right\}.$$

Then the set of fixed points of the automorphism $Z \to Z'$ is S_{II}. But S_{II} is a nonsymmetric domain. From Lemma 10.18 we know that S_I is non-symmetric.

Finally, we consider

$$S_{III} = \left\{ Z \left| \frac{1}{2i}(Z - \bar{Z}') > 0, JZ = Z'J, l < s, Z \in C\begin{pmatrix} 2m_1, \ldots, 2m_s \\ k_1, \ldots, k_l \end{pmatrix} \right. \right\}.$$

Suppose that $\gamma \in (1, 2, \ldots, s)$ and $\gamma \notin (k_1, \ldots, k_l)$, consider the automorphism $Z \to I_0 Z I_0'$, where

$$I_0 = \begin{pmatrix} I & & & & & & & \\ & -I & & & & & & \\ & & \ddots & & & & & \\ & & & -I & & & & \\ & & & & I \cdots\cdots\cdots\cdots\cdots\cdots \\ & & & & & -I & & \\ & & & & & & \ddots & \\ & & & & & & & -I \\ & & & & & & & & I \end{pmatrix} \Big\downarrow r.$$

Obviously, $I_0 J = J I_0$. But the set of fixed points of this automorphism is the product of some symmetric domains and the following domain:

$$\left\{ Z \left| \frac{1}{2i} \left[\begin{pmatrix} Z_{11} & Z_{1r} & Z_{1s} \\ Z_{r1} & Z_{rr} & 0 \\ Z_{s2} & 0 & Z_{ss} \end{pmatrix} - \overline{\begin{pmatrix} Z_{11} & Z_{1r} & Z_{1s} \\ Z_{r1} & Z_{rr} & 0 \\ Z_{s1} & 0 & Z_{ss} \end{pmatrix}}' \right] > 0, JZ = Z'J \right. \right\}.$$

Consider the analytic automorphism of this domain:

$$\begin{pmatrix} Z_{11} & Z_{1r} & Z_{1s} \\ Z_{r1} & Z_{rr} & 0 \\ Z_{s1} & 0 & Z_{ss} \end{pmatrix} \to I_1 \begin{pmatrix} Z_{11} & Z_{1r} & Z_{1s} \\ Z_{r1} & Z_{rr} & 0 \\ Z_{s1} & 0 & Z_{ss} \end{pmatrix} I_s'$$

$$I_1 = \begin{pmatrix} \begin{pmatrix} \begin{pmatrix}1&0\\0&1\end{pmatrix} & & & \\ & -\begin{pmatrix}1&0\\0&1\end{pmatrix} & & \\ & & \ddots & \\ & & & -\begin{pmatrix}1&0\\0&1\end{pmatrix} \end{pmatrix} & & & \\ & \begin{pmatrix} \begin{pmatrix}1&0\\0&1\end{pmatrix} & & & \\ & -\begin{pmatrix}1&0\\0&1\end{pmatrix} & & \\ & & \ddots & \\ & & & -\begin{pmatrix}1&0\\0&1\end{pmatrix} \end{pmatrix} & & \\ & & \begin{pmatrix} \begin{pmatrix}1&0\\0&1\end{pmatrix} & & & \\ & -\begin{pmatrix}1&0\\0&1\end{pmatrix} & & \\ & & \ddots & \\ & & & -\begin{pmatrix}1&0\\0&1\end{pmatrix} \end{pmatrix} \end{pmatrix} \begin{matrix} 2m_1 \\ \\ 2m_r \\ \\ 2m_s \end{matrix}$$

The set of fixed points of this automorphism is the product of some symmetric domains and following domain:

$$\left\{ Z \,\middle|\, \frac{1}{2i}[Z - \bar{Z}'] > 0,\ JZ = Z'J,\ Z = \begin{pmatrix} z_{11}^{(11)} & z_{12}^{(11)} & z_{11}^{(1r)} & z_{12}^{(1r)} & z_{11}^{(1s)} & z_{12}^{(1s)} \\ z_{21}^{(11)} & z_{22}^{(11)} & z_{21}^{(1r)} & z_{21}^{(1r)} & z_{21}^{(1s)} & z_{22}^{(1s)} \\ z_{11}^{(r1)} & z_{12}^{(r1)} & z_{11}^{(rr)} & z_{12}^{(rr)} & 0 & 0 \\ z_{21}^{(r1)} & z_{22}^{(r1)} & z_{21}^{(rr)} & z_{22}^{(rr)} & 0 & 0 \\ z_{11}^{(s1)} & z_{12}^{(s1)} & 0 & 0 & z_{11}^{(ss)} & z_{12}^{(ss)} \\ z_{21}^{(s1)} & z_{22}^{(s1)} & 0 & 0 & z_{21}^{(ss)} & z_{22}^{(ss)} \end{pmatrix} \right\}.$$

From Theorem 10.21, the above domain is a nonsymmetric domain. So, S_{III} is also a nonsymmetric domain, and Theorem 10.22 follows.

THEOREM 10.23. *Suppose $l < s$. Then W_I, W_{II}, W_{III} are nonsymmetric domains.*

PROOF. W_I, W_{II}, W_{III} have the following form:

$$\{Z \,|\, (1/2i)(Z - \bar{Z}') - F(U, U) > 0\},$$

where Z, F satisfy some other conditions. The W's admit an analytic automorphism of the form $Z \to Z$, $U \to -U$. But the set of fixed points of this automorphism is S_I, S_{II}, S_{III}. If $l < s$, by Theorem 10.22 S_I, S_{II}, S_{III} are nonsymmetric domains. So, Theorem 10.23 follows. □

10.2.3. Generalizations

Based on V_I, V_{II}, V_{III}, we can consider other cones and domains. Since the latter lose the classical meaning, we will only discuss V_{II}.

For

$$Y \in V_{II} \begin{pmatrix} m_1, \ldots, m_s \\ k_1, \ldots, k_l \end{pmatrix},$$

Y has the form

$$Y = \begin{pmatrix} Y_{11} & Y_{12} & \cdots & Y_{1s} \\ Y'_{12} & Y_{22} & \cdots & Y_{2s} \\ \vdots & \vdots & \ddots & \\ Y'_{1s} & Y'_{2s} & & Y_{ss} \end{pmatrix}, \qquad Y_{ij} = 0 \quad (i \neq k_1, \ldots, k_l, j > i),$$

where $Y_{11}, Y_{22}, \ldots, Y_{ss}$ are symmetric matrices. If we divide them into some groups $Y_{i_1 i_1}, \ldots, Y_{i_g i_g}; Y_{j_1 j_1}, \ldots, Y_{j_p j_p}; \cdots$, the elements of the same group are equal to each other: $Y_{i_1 i_1} = Y_{i_2 i_2} = \cdots = Y_{i_g i_g}; Y_{j_1 j_1} = Y_{j_2 j_2} = \cdots = Y_{j_p j_p}; \cdots$. Then they are also homogeneous cones, and their affine motion group is

$$Y \to AYA', \quad A = \begin{pmatrix} A_{11} & A_{12} & \cdots & A_{1s} \\ 0 & A_{22} & \cdots & A_{2s} \\ \vdots & & \ddots & \\ 0 & & & A_{ss} \end{pmatrix}, \quad A_{ij} = 0, \quad i \neq k_1, \ldots, k_l; j > i,$$

and

$$A_{i_1 i_1} = A_{i_2 i_2} = \cdots = A_{i_q i_q}; \qquad A_{j_1 j_1} = A_{j_2 j_2} = \cdots = A_{j_p j_p}; \cdots.$$

We can prove that these cones are non-self-dual in most cases, but they also contain some self-dual cones.

From such cones, we can get Siegel domains of first and second kinds. In each explicit situation, using the results of Ref. 4 and the last section, it is not difficult to determine if they are symmetric.

Among these, we consider

$$V_{IV}: \begin{pmatrix} Y_{11} & Y_{12} & \cdots & Y_{1s} \\ Y'_{12} & Y_{22} & & \\ \vdots & & \ddots & 0 \\ Y'_{1s} & 0 & & Y_{ss} \end{pmatrix} > 0.$$

The simplest one is $V_4: y_1 y_2 - y_3^2 - \cdots - y_n^2 > 0$, $y_2 > 0$.

From V_{IV}, we can get Siegel domains, the simplest one is the symmetric classical domains of the fourth type. If the degree of Y_{22} is not less than 2, then the corresponding Siegel domains are nonsymmetric.

Take V_4 for the base, and let $F(U, U) = U\bar{U}' + \bar{U}U'$,

$$U = \begin{pmatrix} v & u \\ 0 & u_1 T_1 \\ \vdots & \vdots \\ 0 & u_1 T_n \end{pmatrix}.$$

Then $F(U, U) \in V_4 \Rightarrow T_i \bar{T}'_j + T_i \bar{T}'_i = 2\delta_{ij}I$.

Let

$$Z = \begin{pmatrix} z_1 & z_3 & \cdots & & z_{n+2} \\ z_3 & z_2 & & 0 & \\ \vdots & & 0 & \ddots & \\ z_{n+2} & & & & z_2 \end{pmatrix}.$$

Then

$$\frac{1}{2i}(Z - \bar{Z}) - \begin{pmatrix} v & u \\ 0 & u_1 T_1 \\ \vdots & \vdots \\ 0 & u_1 T_n \end{pmatrix} \overline{\begin{pmatrix} v & u \\ 0 & u_1 T_1 \\ \vdots & \vdots \\ 0 & u_1 T_n \end{pmatrix}}' - \overline{\begin{pmatrix} v & u \\ 0 & u_1 T_1 \\ \vdots & \vdots \\ 0 & u_1 T_n \end{pmatrix}} \begin{pmatrix} v & u \\ 0 & u_1 T_1 \\ \vdots & \vdots \\ 0 & u_1 T_n \end{pmatrix}' > 0$$

is the nonsymmetric domain of the fourth type introduced in Ref. 4. □

References

1. S. G. Gindikin, Analysis on the homogeneous domains, *UMN* **19**, 3-92 (1964).
2. I. I. Pyateckii-Shapiro, On a problem of E. Cartan, *DAN* **124** (2) 272-273 (1959).
3. I. I. Pyateckii-Shapiro, Generalized upper halfplanes in the theory of many complex variables, Transactions of the Stockholm Mathematical Congress (1963).
4. I. I. Pyateckii-Shapiro, *Automorphic Functions and the Geometry of Classical Domains*, Gordon and Breach, New York (1969).
5. Lu Qi-Keng, A class of the homogeneous complex manifolds, *Acta Math. Sinica* **12**, 229-249 (1962).
6. Lu Qi-Keng, *Classical Domains and Classical Manifolds*, Shanghai Science and Technology Publishers (1963).
7. E. B. Vinberg, Homogeneous cones, *DAN* **133** (1) (1960).

11

The Extension Spaces of Nonsymmetric Classical Domains

We have introduced some new types of nonsymmetric classical domains in Ref. 4, whose special cases are the examples of Ref. 3. In this chapter we discuss "extension spaces" for these domains. To define "extension space," we introduce infinite distance points. In one complex variable, the Gauss plane can be compactified by introducing the unique infinite distance point, which is very convenient in many problems. In several complex variables, extension spaces for the four types of classical symmetric domains are well known. But for nonsymmetric domains, there is only Ref. 2.

The extension spaces of bounded homogeneous domains are the complex manifolds, which are homogeneous manifolds in general, such that the bounded homogeneous domains can be realized as the "super disk" in it. Super disks come from the work in matrix geometry by Hua Loo-Keng. For example, we can introduce homogeneous coordinates in the Gauss plane such that this plane is extended to a complex projective space of dimension 1: $\mathfrak{Z} = (z_1, z_2), (z_1, z_2) \neq (0, 0)$, $\mathfrak{Z}_1$ equivalent to $\mathfrak{Z}_2$ iff $\mathfrak{Z}_1 = \lambda \mathfrak{Z}_2$, $\lambda \neq 0$. In the coordinate neighborhood $z_1 \neq 0$, the local coordinates can be defined by $z_1^{-1} z_2 = z$, and the point of $z_1 = 0$ is called ∞. Obviously, the "∞" consists of only one point. Now, let $H = [1, -1]$. Then the set

$$\{\mathfrak{Z} \mid \mathfrak{Z} H \bar{\mathfrak{Z}}' > 0\} \tag{11.1}$$

is called the *super disk H*. The set (11.1) is contained in the coordinate neighborhood $z_1 \neq 0$. Using local coordinates, the set (11.1) is

$$\{z \mid 1 - |z|^2 > 0\}. \tag{11.2}$$

Joint work with Yin Wei-Ping.

It is well known that an appropriate extension space is very useful for research in the homogeneous domain itself. For example, if the extension space is chosen appropriately (for a domain, its extension space is not unique), then the identity component of the analytic automorphism group of this transitive domain can be obtained by the motion group of its extension space restricted to the super disk. That fact has been proved for symmetric classical domains. It is also true for some nonsymmetric domains.

We thank Professor Hua Loo Keng for his interest and encouragement. We also thank Professor Lu Qi-Keng for his support and advice. His paper [2] is the model of this chapter.

11.1. The Method of Lu

We realize the extension spaces of the nonsymmetric domains in Ref. 4 by using the homogeneous complex manifolds $\mathcal{B}(\gamma_1, \ldots, \gamma_p; s_1, \ldots, s_p)$ introduced in Ref. 2. The manifold $\mathcal{B}(\gamma_1, \ldots, \gamma_p; s_1, \ldots, s_p)$ is constructed as follows. In the set $E(\gamma_1, \ldots, \gamma_p; s_1, \ldots, s_p) =$

$$
E = \left\{ \mathscr{L} \mid \mathscr{L} = \begin{pmatrix} \mathscr{L}_{11} & \mathscr{L}_{12} & \cdots & \mathscr{L}_{1p} \\ 0 & \mathscr{L}_{22} & \cdots & \mathscr{L}_{2p} \\ \vdots & & \ddots & \vdots \\ 0 & & \mathbf{0} & \mathscr{L}_{pp} \end{pmatrix} \begin{matrix} r_1 \\ r_2, \\ \\ r_p \end{matrix} \quad \text{rank } \mathscr{L}_{ii} = \gamma_i \right\},
$$
$$
\qquad\qquad r_1 + s_1 \quad r_2 + s_2 \quad \cdots \quad r_p + s_p \qquad\qquad (11.3)
$$

we introduce the equivalence relation

$$
\mathscr{L}_1 \sim \mathscr{L}_2 \quad \text{iff} \quad \mathscr{L}_2 = P\mathscr{L}_1,
$$

where

$$
P = \begin{pmatrix} P_{11} & P_{12} & \cdots & P_{1p} \\ & P_{22} & \cdots & P_{2p} \\ & & \ddots & \\ & \mathbf{0} & & P_{pp} \end{pmatrix} \begin{matrix} r_1 \\ r_2, \\ \\ r_p \end{matrix} \qquad \det P \neq 0. \qquad (11.4)
$$
$$
\qquad\qquad r_1 \quad r_2 \quad \cdots \quad r_p
$$

The equivalence class, which contains $\mathscr{L}$, is denoted by $\pi(\mathscr{L})$. The set of all $\pi(\mathscr{L})$ is denoted by $\mathcal{B}(\gamma_1, \ldots, \gamma_p; s_1, \ldots, s_p)$. It is proved in Ref. 2 that $\mathcal{B}(\gamma_1, \ldots, \gamma_p; s_1, \ldots, s_p)$ is a homogeneous complex manifold; in general, it is noncompact. Its local coordinates can be given as follows: the coordinate neighborhood $M(\alpha)$, where $\alpha = (\alpha_{11}, \ldots, \alpha_{1\gamma_1}, \ldots, \alpha_{p1}, \ldots, \alpha_{p\gamma_p})$, consists of such $\mathscr{L}$, of which the subdeterminant $\mathscr{L}_{ii}^1$ is nonsingular and $\mathscr{L}_{ii}^1$ consists of the α_{i1}th, $\ldots$, $\alpha_{i\gamma_i}$th columns of $\mathscr{L}_{ii}$. Then we have

$$
\mathscr{L} =
\begin{pmatrix}
(\mathscr{L}_{11}^1\mathscr{L}_{11}^2)P(\alpha_1) & (\mathscr{L}_{12}^1\mathscr{L}_{12}^2)P(\alpha_2) & \cdots & (\mathscr{L}_{1p}^1\mathscr{L}_{1p}^2)P(\alpha_p) \\
 & (\mathscr{L}_{22}^1\mathscr{L}_{22}^2)P(\alpha_2) & \cdots & (\mathscr{L}_{2p}^1\mathscr{L}_{2p}^2)P(\alpha_p) \\
 & & \ddots & \vdots \\
 & & & (\mathscr{L}_{pp}^1\mathscr{L}_{pp}^2)P(\alpha_p)
\end{pmatrix}
$$

$$
=
\begin{pmatrix}
\mathscr{L}_{11}^1 & \cdots & \mathscr{L}_{1p}^1 & \mathscr{L}_{11}^2 & \cdots & \mathscr{L}_{1p}^2 \\
 & \ddots & \vdots & & \ddots & \vdots \\
 & & \mathscr{L}_{pp}^1 & & & \mathscr{L}_{pp}^2
\end{pmatrix}
NP(\alpha) = (Z_1, Z_2)NP(\alpha),
$$

$$(11.5)$$

where N is a fixed permutation matrix, and $P(\alpha) = [P(\alpha_1), \ldots, P(\alpha_p)]$ is also a permutation matrix, but $P(\alpha_i)$ is dependent on $\alpha_i = (\alpha_{i1}, \ldots, \alpha_{i\gamma_i})$. So the local coordinates in $M(\alpha)$ can be defined by

$$
\pi(\mathscr{L}) \xrightarrow{\theta(\alpha_1, \ldots, \alpha_p)} Z_1^{-1}Z_2 = Z,
$$

$$
Z =
\begin{pmatrix}
Z_{11} & Z_{12} & \cdots & Z_{1p} \\
0 & Z_{22} & \cdots & Z_{2p} \\
\vdots & & \ddots & \vdots \\
0 & & 0 & Z_{pp}
\end{pmatrix}
\begin{matrix} r_1 \\ r_2 \\ \vdots \\ r_p \end{matrix}
$$

$$
 \begin{matrix} s_1 & s_2 & \cdots & s_p \end{matrix}
$$

$$(11.6)$$

Hence, (11.5) can be written as $\mathscr{L} = P(I, Z)NP(\alpha)$, where P is a nonsingular square matrix.

Obviously, the complex manifold $\mathscr{B}(\gamma_1, \ldots, \gamma_p; s_1, \ldots, s_p)$ admits the motion group $G(\gamma_1, \ldots, \gamma_p; s_1, \ldots, s_p)$:

$$
\mathscr{L} = \mathscr{L}Q, \qquad Q =
\begin{pmatrix}
Q_{11} & Q_{12} & \cdots & Q_{1p} \\
 & Q_{22} & \cdots & Q_{2p} \\
 & & & \vdots \\
 & & 0 & Q_{pp}
\end{pmatrix}
\begin{matrix} r_1 + s_1 \\ r_2 + s_2, \\ \vdots \\ r_p + s_p \end{matrix}
\qquad \det Q \neq 0,
$$

$$
\phantom{\mathscr{L} = \mathscr{L}Q, \qquad Q =} \begin{matrix} r_1 + s_1 & r_2 + s_2 & \cdots & r_p + s_p \end{matrix}
$$

$$(11.7)$$

and $G(\gamma_1, \ldots, \gamma_p; s_1, \ldots, s_p)$ acts transitively on $\mathscr{B}(\gamma_1, \ldots, \gamma_p; s_1, \ldots, s_p)$ (see Ref. 2). In the same manner as in Ref. 2 we introduce complex submanifolds $\mathscr{B}_J(\gamma_1, \ldots, \gamma_p; s_1, \ldots, s_p)$. Suppose J is an $(m+n) \times (m+n)$ nonsingular symmetric or skew-symmetric matrix, where $m = \gamma_1 + \cdots + \gamma_p$, $n = s_1 + \cdots + s_p$. Then the set of all $\pi(\mathscr{Z})$, which satisfies the condition $\mathscr{Z}J\bar{\mathscr{Z}}' = 0$, is a submanifold and will be denoted by $\mathscr{B}_J(\gamma_1, \ldots, \gamma_p; s_1, \ldots, s_p)$.

Then we introduce the super disk $\mathscr{B}_H$. Here H is a nonsingular Hermite matrix with degree $m + n$. The set of all $\pi(\mathscr{Z})$ with the property

$$\mathscr{Z}H\bar{\mathscr{Z}}' > 0 \tag{11.8}$$

is called the super disk defined by H. It may be empty. If it is not empty, then $\mathscr{B}_H$ is an open set in $\mathscr{B}(\gamma_1, \ldots, \gamma_p; s_1, \ldots, s_p)$. And $\mathscr{B}_H \cap \mathscr{B}_J = \mathscr{B}_{HJ}$ is called the super disk in $\mathscr{B}_J$, or $\mathscr{B}_J$ is the extension space of $\mathscr{B}_{JH}$.

Now, we take $m = n$, $r = r_1 + \cdots + r_q$, $s = s_1 + \cdots + s_q$, $r_{q+1} + \cdots + r_{q+l} = n - r - s$, $s_{q+1} + \cdots + s_{q+l} = n - r - s$, $r_{q+l+1} + \cdots + r_p = s$, $s_{q+l+1} + \cdots + s_p = r$. And

$$H_0 = \begin{pmatrix} 0 & 0 & 0 & 0 & 0 & iI^{(r)} \\ 0 & 0 & 0 & 0 & iI^{(n-r-s)} & 0 \\ 0 & 0 & I^{(s)} & 0 & 0 & 0 \\ 0 & 0 & 0 & -I^{(s)} & 0 & 0 \\ 0 & -iI^{(n-r-s)} & 0 & 0 & 0 & 0 \\ -iI^{(r)} & 0 & 0 & 0 & 0 & 0 \end{pmatrix}.$$

Let $H = N'H_0N$. Then N is defined by (11.5). As in Ref. 2 all of the points of $\mathscr{B}_H$ are contained in the neighborhood $M(\alpha)$, where $\alpha = (1, 2, \ldots, \gamma_1; 1, 2, \ldots, \gamma_2, \ldots, 1, 2, \ldots, \gamma_p)$. In the local coordinates of $M(\alpha)$, the points of $\mathscr{B}_H$ can be written as

$$\begin{pmatrix} I & 0 & 0 & R_{11} & R_{12} & R_{13} \\ 0 & I & 0 & 0 & R_{22} & R_{23} \\ 0 & 0 & I & 0 & 0 & R_{33} \end{pmatrix} H_0 \begin{pmatrix} I & & & R_{11} & R_{12} & R_{13} \\ & I & & 0 & R_{22} & R_{23} \\ & & I & 0 & 0 & R_{33} \end{pmatrix}' > 0,$$

$$\tag{11.9}$$

where

$$R_{11} = \begin{pmatrix} Z_{11} & Z_{12} & \cdots & Z_{1q} \\ 0 & Z_{22} & \cdots & Z_{2q} \\ \vdots & & \ddots & \vdots \\ 0 & & & Z_{qq} \end{pmatrix} \begin{matrix} r_1 \\ r_2, \\ \vdots \\ r_q \end{matrix}$$

$$\qquad\quad s_1 \quad\ s_2 \quad \cdots \quad s_q$$

$$R_{12} = \begin{pmatrix} Z_{1\,q+1} & \cdots & Z_{1\,q+l} \\ \vdots & \ddots & \vdots \\ Z_{q\,q+1} & & Z_{q\,q+l} \end{pmatrix} \begin{matrix} r_1 \\ \vdots \\ r_p \end{matrix} \quad ,$$
$$\begin{matrix} s_{q+1} & \cdots & s_{q+l} \end{matrix}$$

$$R_{13} = \begin{pmatrix} Z_{1\,q+l+1} & \cdots & & Z_{1p} \\ \vdots & \ddots & & \vdots \\ Z_{q\,q+l+1} & \cdots & & Z_{qp} \end{pmatrix} \begin{matrix} r_1 \\ \vdots \\ r_p \end{matrix} \quad ,$$
$$\begin{matrix} s_{q+l+1} & \cdots & s_p \end{matrix}$$

$$R_{22} = \begin{pmatrix} Z_{q+1\,q+1} & \cdots & Z_{q+1\,q+l} \\ & \ddots & \vdots \\ & 0 & Z_{q+l\,q+l} \end{pmatrix} \begin{matrix} r_{q+1} \\ \vdots \\ r_{q+l} \end{matrix} \quad ,$$
$$\begin{matrix} s_{q+1} & \cdots & s_{q+l} \end{matrix}$$

$$R_{23} = \begin{pmatrix} Z_{q+1\,q+l+1} & \cdots & & Z_{q+1\,p} \\ \vdots & \ddots & & \vdots \\ Z_{q+1\,q+l+1} & \cdots & & Z_{q+1\,p} \end{pmatrix} \begin{matrix} r_{q+1} \\ \vdots \\ r_{q+l} \end{matrix} \quad ,$$
$$\begin{matrix} s_{q+l+1} & \cdots & s_p \end{matrix}$$

$$R_{33} = \begin{pmatrix} Z_{q+l+1\,q+l+1} & \cdots & & Z_{q+l+1\,p} \\ & \ddots & & \vdots \\ & 0 & & Z_{pp} \end{pmatrix} \begin{matrix} r_{q+l+1} \\ \vdots \\ r_p \end{matrix} \quad .$$
$$\begin{matrix} s_{q+l+1} & \cdots & s_p \end{matrix}$$

Here if we take $r_1 = s_{q+l+1}, r_2 = s_{q+l+2}, \ldots, r_q = s_p; s_1 = r_{q+l+1}, \ldots, s_q = r_p;$ $r_{q+1} = s_{q+1}, \ldots, r_{q+l} = s_{q+l}$, and $q = p - q - l$. Let

$$J = N'J_0N,$$

$$J_0 = \begin{pmatrix} 0 & 0 & 0 & 0 & 0 & I^{(r)} \\ 0 & 0 & 0 & 0 & I^{(n-r-s)} & 0 \\ 0 & 0 & 0 & I^{(s)} & 0 & 0 \\ 0 & 0 & -I^{(s)} & 0 & 0 & 0 \\ 0 & -I^{(n-r-s)} & 0 & 0 & 0 & 0 \\ -I^{(r)} & 0 & 0 & 0 & 0 & 0 \end{pmatrix} .$$

Then the points of $\mathcal{B}_H \cap \mathcal{B}_J$ can be written as

$$
\begin{pmatrix} I & & & R_{11} & R_{12} & R_{13} \\ & I & & 0 & R_{22} & R_{23} \\ & & I & 0 & 0 & R_{33} \end{pmatrix} J_0 \begin{pmatrix} I & & & R_{11} & R_{12} & R_{13} \\ & I & & 0 & R_{22} & R_{23} \\ & & I & 0 & 0 & R_{33} \end{pmatrix}' = 0,
$$

$$
\begin{pmatrix} I & & & R_{11} & R_{12} & R_{13} \\ & I & & 0 & R_{22} & R_{23} \\ & & I & 0 & 0 & R_{33} \end{pmatrix} H_0 \overline{\begin{pmatrix} I & & & R_{11} & R_{12} & R_{13} \\ & I & & 0 & R_{22} & R_{23} \\ & & I & 0 & 0 & R_{33} \end{pmatrix}}' > 0.
$$

$$(11.10)$$

From the equality, we have $R_{13} = R_{13}'$, $R_{12} = R_{23}'$, $R_{22} = R_{22}'$, $R_{11} = R_{33}'$, and from the block form of R_{ij}, we must have

$$
R_{22} = \begin{pmatrix} Z_{q+1\,q+1} & & \\ & \ddots & 0 \\ 0 & & Z_{q+l\,q+l} \end{pmatrix}, \qquad R_{11} = R_{33}' = \begin{pmatrix} Z_{11} & & \\ & \ddots & 0 \\ 0 & & Z_{qq} \end{pmatrix}.
$$

By a simple calculation, (11.10) has the form

$$
\frac{1}{i}\left[\begin{pmatrix} R_{13} & R_{12} \\ R_{12}' & R_{22} \end{pmatrix} - \begin{pmatrix} \overline{R_{13}} & \overline{R_{12}} \\ R_{12}' & R_{22} \end{pmatrix} \right] - \begin{pmatrix} R_{11} \\ 0 \end{pmatrix} (\bar{R}_{11}' \quad 0) - \begin{pmatrix} \bar{R}_{11} \\ 0 \end{pmatrix} (R_{11}' \quad 0) > 0.
$$

$$(11.11)$$

In view of the form of R_{22}, we know that the above domain is one of the nonsymmetric domains W_{II} constructed in Ref. 4. We can see that it may be very difficult to realize all of the nonsymmetric domains introduced in Ref. 4 by using only $\mathcal{B}(\gamma_1, \ldots, \gamma_p; s_1, \ldots, s_p)$.

11.2. New Methods

11.2.1. Some Homogeneous Manifolds

In this section, new types of homogeneous complex manifolds will be introduced.

Consider the set of matrices with the following properties:

$$
\mathcal{Z} = \begin{pmatrix} \mathcal{Z}_{11} & \mathcal{Z}_{12} & \cdots & \mathcal{Z}_{1p} \\ & \mathcal{Z}_{22} & \cdots & \mathcal{Z}_{2p} \\ & & \ddots & \vdots \\ 0 & & & \mathcal{Z}_{pp} \end{pmatrix} \begin{matrix} r_1 \\ r_2, \\ \vdots \\ r_p \end{matrix} \qquad \text{rank } \mathcal{Z}_{ii} = r_i, \quad (11.12)
$$

$$
\begin{matrix} r_1 + s_1 & r_2 + s_2 & \cdots & r_p + s_p \end{matrix}
$$

Take $1 < k_1 < k_2 < \cdots < k_l < p$, $\rho_1, \rho_2, \ldots, \rho_l$ such that $k_i < \rho_i \leq p$. If the matrix in (11.12) has the form

$$\mathcal{L}_{k_1, k_1+1} = \mathcal{L}_{k_1, k_1+2} = \cdots = \mathcal{L}_{k_1, \rho_1} = 0,$$

$$\mathcal{L}_{k_2, k_2+1} = \mathcal{L}_{k_2, k_2+2} = \cdots = \mathcal{L}_{k_2, \rho_2} = 0,$$

$$\cdots \tag{11.13}$$

$$\mathcal{L}_{k_l, k_l+1} = \mathcal{L}_{k_l, k_l+2} = \cdots = \mathcal{L}_{k_l, \rho_l} = 0,$$

then it is called a (k, ρ)-matrix. For example, if $p = 5$, $k_1 = 2$, $k_2 = 3$, $\rho_1 = 3$, and $\rho_2 = 4$, then

$$\mathcal{L} = \begin{pmatrix} \mathcal{L}_{11} & \mathcal{L}_{12} & \mathcal{L}_{13} & \mathcal{L}_{14} & \mathcal{L}_{15} \\ 0 & \mathcal{L}_{22} & 0 & \mathcal{L}_{24} & \mathcal{L}_{25} \\ 0 & 0 & \mathcal{L}_{33} & 0 & \mathcal{L}_{35} \\ 0 & 0 & 0 & \mathcal{L}_{44} & \mathcal{L}_{45} \\ 0 & 0 & 0 & 0 & \mathcal{L}_{55} \end{pmatrix}. \tag{11.14}$$

The set of such matrices with the above properties is denoted by

$$E\begin{pmatrix} \gamma_1, \ldots, \gamma_p & k_1, \ldots, k_l \\ s_1, \ldots, s_p & \rho_1, \ldots, \rho_l \end{pmatrix},$$

which for simplicity will be denoted by

$$E\begin{pmatrix} \gamma & k \\ s & \rho \end{pmatrix}.$$

Now, we introduce an equivalence relation in it.

$$\mathcal{L}_1 \sim \mathcal{L}_2 \quad \text{iff} \quad \mathcal{L}_2 = P\mathcal{L}_1, \qquad \det P \neq 0,$$

$$P = \begin{pmatrix} P_{11} & P_{12} & \cdots & P_{1p} \\ 0 & P_{22} & \cdots & P_{2p} \\ \vdots & & \ddots & \vdots \\ 0 & & & P_{pp} \end{pmatrix}, \tag{11.15}$$

$$P_{k_1 k_1+1} = P_{k_1 k_1+2} = \cdots = P_{k_1 \rho_1} = 0,$$

$$\cdots,$$

$$P_{k_l k_l+1} = P_{k_l k_l+2} = \cdots = P_{k_l \rho_l} = 0.$$

This is an equivalence relation. The important point is that the P's with the form of (11.15) form a matrix group. It is not difficult to prove this, but from the following example we can understand the general case. For example, when $k_1 = 2$, we have

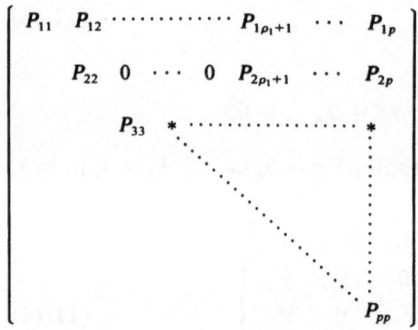

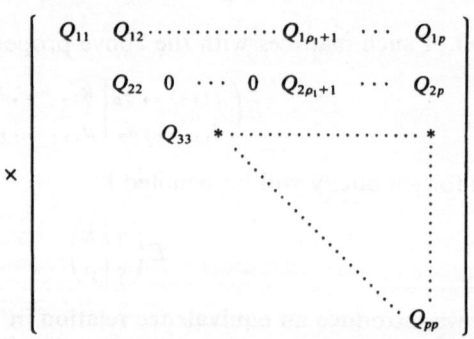

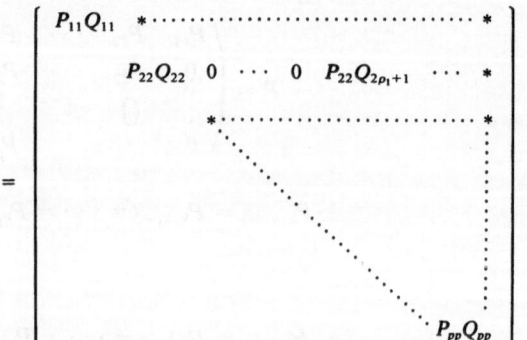

,

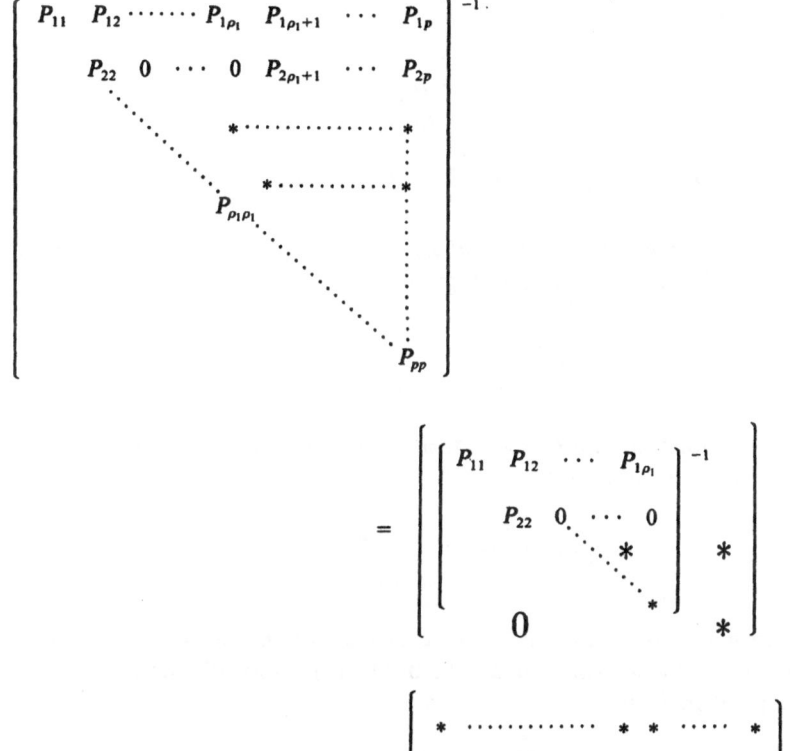

The equivalence class, which contains $\mathscr{L}$, is denoted by $\pi(\mathscr{L})$. The set of all $\pi(\mathscr{L})$ is denoted by

$$\mathscr{B}\begin{pmatrix} \gamma_1, \ldots, \gamma_p & k_1, \ldots, k_l \\ s_1, \ldots, s_p & \rho_1, \ldots, \rho_l \end{pmatrix}.$$

We thus have the next theorem.

THEOREM 11.1.

$$\mathscr{B}\left(\begin{matrix} \gamma_1, \ldots, \gamma_p \\ s_1, \ldots, s_p \end{matrix}\middle|\begin{matrix} k_1, \ldots, k_l \\ \rho_1, \ldots, \rho_l \end{matrix}\right)$$

is a homogeneous complex manifold and is noncompact if $p > 1$.

PROOF. *Step 1.* Let $\pi : \mathscr{X} \to \pi(\mathscr{X})$ be the projection where the topology of $\pi(\mathscr{X})$ is induced by the topology of

$$E\left(\begin{matrix} \gamma_1, \ldots, \gamma_p \\ s_1, \ldots, s_p \end{matrix}\middle|\begin{matrix} k_1, \ldots, k_l \\ \rho_1, \ldots, \rho_l \end{matrix}\right).$$

So π is an open mapping. Second, π is a continuous mapping. In fact, we take an arbitrary open set

$$O \in \mathscr{B}\left(\begin{matrix} \gamma_1, \ldots, \gamma_p \\ s_1, \ldots, s_p \end{matrix}\middle|\begin{matrix} k_1, \ldots, k_l \\ \rho_1, \ldots, \rho_l \end{matrix}\right).$$

According to the definition, there exists an open set B in E such that $\pi(B) = O$. We also take P of the form (11.15) arbitrarily. Then PB is also an open set in E, so

$$\pi^{-1}(0) = \sum_P PB = \text{open set},$$

where P ranges over all of the P with the form of (11.15). So π is a continuous open mapping.

Step 2. In

$$\mathscr{B}\left(\begin{matrix} \gamma_1, \ldots, \gamma_p \\ s_1, \ldots, s_p \end{matrix}\middle|\begin{matrix} k_1, \ldots, k_l \\ \rho_1, \ldots, \rho_l \end{matrix}\right)$$

we define the coordinate neighborhood system $\{M(\alpha)|\alpha = (\alpha_1, \ldots, \alpha_p),$ $\alpha_1 = \alpha_{11}, \ldots, \alpha_{1\gamma_1}; \cdots; \alpha_p = \alpha_{p1}, \ldots, \alpha_{p\gamma_p}\}$ as follows:

$$\mathscr{X} \in E\left(\begin{matrix} \gamma_1, \ldots, \gamma_p \\ s_1, \ldots, s_p \end{matrix}\middle|\begin{matrix} k_1, \ldots, k_l \\ \rho_1, \ldots, \rho_l \end{matrix}\right),$$

of which the subdeterminants consisting of the α_{i1}th, $\ldots$, $\alpha_{i\gamma_i}$th columns of

$\mathscr{L}_{ii}$ are nonsingular. If $\pi(\mathscr{L}) \in M(\alpha)$, then

$$\mathscr{L} = \begin{pmatrix} (\mathscr{L}_{11}^1\mathscr{L}_{11}^2)P(\alpha_1) & \cdots & \cdots & (\mathscr{L}_{1p}^1\mathscr{L}_{1p}^2)P(\alpha_p) \\ & (\mathscr{L}_{22}^1\mathscr{L}_{22}^2)P(\alpha_2) & \cdots & (\mathscr{L}_{2p}^1\mathscr{L}_{2p}^2)P(\alpha_p) \\ & & \ddots & \\ & & & (\mathscr{L}_{pp}^1\mathscr{L}_{pp}^2)P(\alpha_p) \end{pmatrix}$$

$$= \begin{pmatrix} \mathscr{L}_{11}^1 & \mathscr{L}_{12}^1 & \cdots & \mathscr{L}_{1p}^1 & \mathscr{L}_{11}^2 & \cdots & Z_{1p}^2 \\ & \mathscr{L}_{22}^1 & \cdots & \mathscr{L}_{2p}^1 & & \ddots & \vdots \\ & & \ddots & \vdots & & & \\ & & & \mathscr{L}_{pp}^1 & & & \mathscr{L}_{pp}^2 \end{pmatrix} NP(\alpha) = (Z_1, Z_2)NP(\alpha),$$

$$(11.16)$$

where Z_1, Z_2 are (k, ρ)-matrices, N is a fixed permutation matrix, $P(\alpha) = [P(\alpha_1), \ldots, P(\alpha_p)]$, and $P(\alpha_i)$ is a permutation matrix that depends on $(\alpha_{i1}, \ldots, \alpha_{i\gamma_i})$.

The local coordinates in $M(\alpha)$ are defined by

$$\pi(\mathscr{L}) \xrightarrow{\theta(\alpha)} Z_1^{-1}Z_2 = Z, \qquad (11.17)$$

where Z is still a (k, ρ)-matrix; that is

$$Z = \begin{pmatrix} Z_{11} & Z_{12} & \cdots & Z_{1p} \\ 0 & Z_{22} & \cdots & Z_{2p} \\ \vdots & & 0 & \ddots & \vdots \\ 0 & & & Z_{pp} \end{pmatrix} \begin{matrix} r_1 \\ r_2 \\ \vdots \\ r_p \end{matrix} \qquad \begin{matrix} Z_{k_1k_1+1} = \cdots = Z_{k_1\rho_1} = 0, \\ \cdots, \\ Z_{k_lk_l+1} = \cdots = Z_{k_l\rho_l} = 0. \end{matrix}$$

$$\begin{matrix} s_1 & s_2 & \cdots & s_p \end{matrix} \qquad (11.18)$$

Step 3. Hausdorff property. Suppose $\pi(\mathscr{L}_1) \neq \pi(\mathscr{L}_2)$. If $\mathscr{L}_1$, $\mathscr{L}_2$ are in the same coordinate neighborhood, and Z^1, Z^2 are their local coordinates, respectively, then $Z^1 \neq Z^2$, and we can get two neighborhoods which do not intersect each other but they contain Z^1 and Z^2, respectively. If $\mathscr{L}_1$, $\mathscr{L}_2$ are not in the same coordinate neighborhood, suppose $\mathscr{L}_1 \in M(\alpha_0)$,

$$\alpha_0 = \begin{pmatrix} 1, 2, \ldots, r_1 \\ 1, 2, \ldots, r_2 \\ \cdots, \\ 1, 2, \ldots, r_p \end{pmatrix}, \qquad \mathscr{L}_1 \sim (I, Z^1)N.$$

Because $\mathcal{L} \to \pi(\mathcal{L})$ is an open continuous mapping, if K is a closed set in $\mathcal{B}$ then $\pi^{-1}(K)$ is also a closed set. Near Z^1, we choose two open sets U_1, U_2 containing the point (I, Z^1), such that $\pi(\mathcal{L}_1) \in U_1 \subset \bar{U}_1 \subset U_2 \subset \bar{U}_2 \subset M(\alpha_0)$. Then $\pi^{-1}(U_1)$ is an open set in $E(^{\gamma}_s|^{k}_{\rho})$ and $\pi^{-1}(U_1)$ contains $\mathcal{L}_1$. Consider $E - \pi^{-1}(\bar{U}_2)$. From the above we know that $\pi^{-1}(\bar{U}_2)$ is a closed set, so $E - \pi^{-1}(\bar{U}_2)$ is an open set, and obviously $\mathcal{L}_2$ is in $E - \pi^{-1}(\bar{U}_2)$. So U_1 and $\pi(E - \pi^{-1}(\bar{U}_2))$ are two open sets and contain $\pi(\mathcal{L}_1)$, $\pi(\mathcal{L}_2)$, respectively.

Step 4. Holomorphism. Suppose $\pi(\mathcal{L}) \in M(\alpha) \cap M(\beta)$, $\alpha \neq \beta$. Then from (11.16) we have

$$\mathcal{L} \sim (I, Z)NP(\alpha) = R(I, W)NP(\beta),$$

$$R(I, W) = (I, Z)NP(\alpha)P^{-1}(\beta)N^{-1} = (I, Z)\begin{pmatrix} A & B \\ C & D \end{pmatrix}.$$

Because $\det R \neq 0$, $\det (A + ZC) \neq 0$. Hence

$$W = (A + ZC)^{-1}(B + ZB) \tag{11.19}$$

is holomorphic in Z.

Step 5. Homogeneity. The motion group is

$$\mathcal{L} \to \mathcal{L}Q,$$

$$Q = \begin{pmatrix} Q_{11} & Q_{12} & \cdots & Q_{1p} \\ & Q_{22} & \cdots & Q_{2p} \\ & & \ddots & \vdots \\ 0 & & & Q_{pp} \end{pmatrix} \begin{matrix} r_1 + s_1 \\ r_2 + s_2 \\ \vdots \\ r_p + s_p \end{matrix} \quad \begin{matrix} Q_{k_1 k_1+1} = \cdots = Q_{k_1 \rho_1} = 0, \\ Q_{k_2 k_2+1} = \cdots = Q_{k_2 \rho_2} = 0, \\ \vdots \\ Q_{k_l k_l+1} = \cdots = Q_{k_l \rho_l} = 0. \end{matrix}$$

$$r_1 + s_1 \quad r_2 + s_2 \quad \cdots \quad r_p + s_p$$

$$\tag{11.20}$$

For any $\pi(\mathcal{L}_1) \neq \pi(\mathcal{L}_2)$, we would like to find Q with the form of (11.20) such that $\mathcal{L}_1 Q = \mathcal{L}_2$. We use the inductive method for p. If $p = 1$, we get the ordinary Grassmann manifold and, of course, the assertion is true. Now suppose it is true for $p - 1$. Let

$$\mathcal{L}_1 = \begin{pmatrix} \tilde{\mathcal{L}}_1 & \tilde{\mathcal{L}}_{12} \\ 0 & \mathcal{L}_{pp} \end{pmatrix}, \qquad \mathcal{L}_2 = \begin{pmatrix} \tilde{W}_1 & \tilde{W}_{12} \\ 0 & W_{pp} \end{pmatrix}.$$

Then $\tilde{\mathscr{X}}_1$, $\tilde{\mathscr{W}}_1$ belong to the case of $p-1$. From the induction, there exist

$$\tilde{Q}_1 = \begin{pmatrix} Q_{11} & \cdots & Q_{1\,p-1} \\ & \ddots & \\ & & Q_{p-1\,p-1} \end{pmatrix}, \qquad \begin{matrix} Q_{k_1 k_1+1} = \cdots = Q_{k_1 \rho_1} = 0, \\ \cdots, \\ Q_{k_l k_l+1} = \cdots = Q_{k_l \rho_l} = 0, \end{matrix}$$

and Q_{pp} such that

$$\tilde{\mathscr{X}}_1 \tilde{Q}_1 = \tilde{\mathscr{W}}_1, \qquad \mathscr{X}_{pp} Q_{pp} = \mathscr{W}_{pp}.$$

Thus,

$$Q = \begin{pmatrix} \tilde{Q}_1 & \tilde{Q}_{12} \\ 0 & Q_{pp} \end{pmatrix}$$

satisfies $\mathscr{X}_1 Q = \mathscr{X}_2$ iff

$$\tilde{\mathscr{X}}_1 \tilde{Q}_{12} = \tilde{\mathscr{W}}_{12} - \tilde{\mathscr{X}}_{12} Q_{pp}, \tag{11.21}$$

where $\tilde{\mathscr{X}}_{12}$, $\tilde{\mathscr{W}}_{12}$ have the same form:

$$\tilde{\mathscr{X}}_{12} = \left.\begin{pmatrix} \mathscr{X}_{1p} \\ \mathscr{X}_{2p} \\ \vdots \\ \mathscr{X}_{p-1\,p} \end{pmatrix}\right\vert \begin{matrix} r_1 \\ r_2 \\ \vdots \\ r_{p-1} \end{matrix}, \qquad \tilde{\mathscr{W}}_{12} = \left.\begin{pmatrix} \mathscr{W}_{1p} \\ \vdots \\ \mathscr{W}_{p-1\,p} \end{pmatrix}\right\vert \begin{matrix} r_1 \\ \vdots \\ r_{p-1} \end{matrix},$$

$$r_p + s_p \qquad\qquad\qquad r_p + s_p$$

among $\{\mathscr{X}_{ip}\}$ and $\{\mathscr{W}_{ip}\}$ there may be some entries equal to zero. The $\tilde{Q}_{12}$ satisfying (11.21) also has the above form. If $\mathscr{X}_{ip} = 0$ (then $\mathscr{W}_{ip} = 0$), and the elements of the ith row of $\tilde{\mathscr{X}}_1$ are equal to zero except $\mathscr{X}_{ii}$. We keep those rows of $\tilde{\mathscr{X}}_1$, among whose elements only $\mathscr{X}_{ii} \neq 0$, and also keep the corresponding columns (except the $i = p - 1$), then we obtain the matrix

$$\tilde{\mathscr{X}}_1 \to \begin{pmatrix} \mathscr{X}_{11} & \mathscr{X}_{1*} & \cdots & \mathscr{X}_{1\,p-1} \\ & \mathscr{X}_{**} & \cdots & * \\ & & \ddots & \vdots \\ & & & \mathscr{X}_{p-1\,p-1} \end{pmatrix}.$$

Deleting the zero elements from the columns at the right-hand side of

(11.21), then (11.21) takes the following form:

$$
\begin{pmatrix}
\mathcal{Z}_{11} & \mathcal{Z}_{1^*} & \cdots & \mathcal{Z}_{1\,p-1} \\
 & \mathcal{Z}_{**} & \cdots & * \\
 & & \ddots & \vdots \\
 & & & \mathcal{Z}_{p-1\,p-1}
\end{pmatrix}
\begin{pmatrix}
Q_{1p} \\
* \\
\vdots \\
Q_{p-1\,p}
\end{pmatrix}
=
\begin{pmatrix}
* \\
\vdots \\
*
\end{pmatrix}.
\tag{11.22}
$$

Based on the following lemma and usual methods of linear algebra, we can easily solve (11.22). Finally, we add some zeros to the solution $(Q'_{1p}, *, \ldots, *, Q_{p-1\,p})$ of (11.22) to get $\tilde{Q}_{12}$, and thus complete our proof.   □

LEMMA 11.2. *For any* $m \times n$ $(n \geq m)$ *matrix* $\mathcal{Z}$ *with rank* m *and any* $m \times p$ *matrix* $\mathcal{W}$, *there exists an* $n \times p$ *matrix* A *such that*

$$
\mathcal{Z}A = \mathcal{W}.
\tag{11.23}
$$

PROOF. In fact, because the Grassmann manifold is homogeneous, there exist two nonsingular matrices R and B of degree m and n, respectively, such that

$$
\mathcal{Z} = R(I, 0)B = (R, 0)B = (I, 0)\begin{pmatrix} R & 0 \\ 0 & I \end{pmatrix}B
$$

$$
= (I, 0)B_1,
$$

where B_1 is a nonsingular matrix of degree n. So (11.23) becomes $(I, 0)B_1 A = \mathcal{W}$. We take an $n \times p$ matrix A_1 such that the first m rows are $\mathcal{W}$. Then $A = B_1^{-1}A_1$ satisfies the condition (11.23). The lemma is proved.

Now, we return to the group (11.20). The action of (11.20) is holomorphic. In fact, for $\mathcal{Z} \to \mathcal{Z}Q$, if $\pi(\mathcal{Z}) \in M(\alpha)$, $\pi(\mathcal{Z}Q) \in M(\beta)$, then $\mathcal{Z}Q \sim (I, \mathcal{W})NP(\beta)$ and simultaneously $\mathcal{Z}Q \sim (I, Z)NP(\alpha)Q$. So

$$
R(I, \mathcal{W})NP(\beta) = (I, Z)NP(\alpha)Q,
$$

$$
R(I, \mathcal{W}) = (I, Z)NP(\alpha)QP^{-1}(\beta)N^{-1} = (I, Z)\begin{pmatrix} A & B \\ C & D \end{pmatrix}.
$$

But $\det R \neq 0$, so $\det(A + ZC) \neq 0$, and hence

$$
\mathcal{W} = (A + ZC)^{-1}(B + ZD)
\tag{11.24}
$$

is holomorphic in Z.

Step 6. If $p > 1$, then $\mathcal{B}$ is noncompact. The proof is the same as Ref. 2. Theorem 11.1 is proved.   □

Using

$$\mathscr{B}\left(\begin{array}{c}\gamma_1,\ldots,\gamma_p \\ s_1,\ldots,s_p\end{array}\middle|\begin{array}{c}k_1,\ldots,k_l \\ \rho_1,\ldots,\rho_l\end{array}\right)$$

and appropriate submanifolds, we can get the extension spaces of many nonsymmetric domains introduced in Ref. 4. First, we give an example. From this example, we can understand the general constructions.

EXAMPLE 11.3. Determine the extension space of the domain

$$\left\{(Z-U)\,\middle|\,\frac{1}{i}(Z-\bar{Z}')-U\bar{U}'-\bar{U}U'>0\right\},$$

where

$$Z=Z'=\begin{pmatrix} Z_{11} & Z_{12} & Z_{13} & Z_{14} & Z_{15} \\ Z'_{12} & Z_{22} & 0 & 0 & 0 \\ Z'_{13} & 0 & Z_{33} & Z_{34} & Z_{35} \\ Z'_{14} & 0 & Z'_{34} & Z_{44} & 0 \\ Z'_{15} & 0 & Z'_{35} & 0 & Z_{55} \end{pmatrix}, \qquad U=\begin{pmatrix} U_{11} & U_{12} \\ 0 & 0 \\ U_{31} & 0 \\ 0 & 0 \\ 0 & 0 \end{pmatrix}.$$

We take

$$\mathscr{B}\left(\begin{array}{c}\gamma_1,\ldots,\gamma_7 \\ s_1,\ldots,s_7\end{array}\middle|\begin{array}{c}2,4 \\ 6,5\end{array}\right).$$

Let $\pi(\mathscr{L})\in M(\alpha_0)$, where

$$\alpha_0=\begin{pmatrix} 1,2,\ldots,r_1 \\ 1,2,\ldots,r_2 \\ \cdots, \\ 1,2,\ldots,r_p \end{pmatrix},$$

$$\mathscr{L}=\begin{pmatrix} \mathscr{L}_{11} & \mathscr{L}_{12} & \mathscr{L}_{13} & \mathscr{L}_{14} & \mathscr{L}_{15} & \mathscr{L}_{16} & \mathscr{L}_{17} \\ & \mathscr{L}_{22} & 0 & 0 & 0 & 0 & \mathscr{L}_{27} \\ & & \mathscr{L}_{33} & \mathscr{L}_{34} & \mathscr{L}_{35} & \mathscr{L}_{36} & \mathscr{L}_{37} \\ & & & \mathscr{L}_{44} & 0 & \mathscr{L}_{46} & \mathscr{L}_{47} \\ & & & & \mathscr{L}_{55} & \mathscr{L}_{56} & \mathscr{L}_{57} \\ & & & & & \mathscr{L}_{66} & \mathscr{L}_{67} \\ & & & & & & \mathscr{L}_{77} \end{pmatrix}$$

$$\sim \begin{pmatrix} (I, Z_{11}) & (0, Z_{12}) & (0, Z_{13}) & (0, Z_{14}) & & \cdots & (0, Z_{17}) \\ & (I, Z_{22}) & 0 & \cdots & & 0 & (0, Z_{27}) \\ & & (I, Z_{33}) & & & & (0, Z_{37}) \\ & & & (I, Z_{44}), & 0, & * & * \\ & & & & & \cdots & \\ & & & & & & (I, Z_{77}) \end{pmatrix}$$

$$= \begin{pmatrix} (I, Z_{11}) & (I, Z_{13}) & (0, Z_{14}) & (0, Z_{15}) & (0, Z_{17}) & (0, Z_{12}) & (0, Z_{16}) \\ (0,0) & (0,0) & (0,0) & (0, Z_{27}) & (I, Z_{22}) & (0,0) \\ (I, Z_{33}) & (0, Z_{34}) & (0, Z_{35}) & (0, Z_{37}) & (0,0) & (0, Z_{36}) \\ & (I, Z_{44}) & (0,0) & (0, Z_{47}) & (0,0) & (0, Z_{46}) \\ & & (I, Z_{55}) & (0, Z_{57}) & (0,0) & (0, Z_{56}) \\ & & & (0, Z_{67}) & (0,0) & (I, Z_{66}) \\ & & & (I, Z_{77}) & (0,0) & (0,0) \end{pmatrix} N_1$$

$$= \begin{pmatrix} I & & & & & Z_{13} & Z_{11} & Z_{14} & Z_{15} & Z_{17} & Z_{12} & Z_{16} \\ & I & & & & 0 & 0 & 0 & 0 & Z_{27} & Z_{22} & 0 \\ & & I & & & Z_{33} & 0 & Z_{34} & Z_{35} & Z_{37} & 0 & Z_{36} \\ & & & I & & 0 & 0 & Z_{44} & 0 & Z_{47} & 0 & Z_{46} \\ & & & & I & 0 & 0 & 0 & Z_{55} & Z_{57} & 0 & Z_{56} \\ & & & & & I & 0 & 0 & 0 & 0 & Z_{67} & 0 & Z_{66} \\ & & & & & & I & 0 & 0 & 0 & 0 & Z_{77} & 0 & 0 \end{pmatrix} N_2 N_1$$

$$= \begin{pmatrix} I & & & R_{11} & R_{12} & R_{13} \\ & I & & 0 & R_{22} & R_{23} \\ & & I & 0 & 0 & R_{33} \end{pmatrix} N_2 N_1, \tag{11.25}$$

where N_1, N_2 are two fixed permutation matrices. Let $N_3 = N_2 N_1$.

$$H_0 = \begin{pmatrix} 0 & 0 & 0 & 0 & 0 & iI \\ 0 & 0 & 0 & 0 & iI & 0 \\ 0 & 0 & I & 0 & 0 & 0 \\ 0 & 0 & 0 & -I & 0 & 0 \\ 0 & -iI & 0 & 0 & 0 & 0 \\ -iI & 0 & 0 & 0 & 0 & 0 \end{pmatrix}$$

$$J_0 = \begin{pmatrix} 0 & 0 & 0 & 0 & 0 & I \\ 0 & 0 & 0 & 0 & I & 0 \\ 0 & 0 & 0 & I & 0 & 0 \\ 0 & 0 & -I & 0 & 0 & 0 \\ 0 & -I & 0 & 0 & 0 & 0 \\ -I & 0 & 0 & 0 & 0 & 0 \end{pmatrix}.$$

Take $H = N_3' H_0 N_3$, $J = N_3' J_0 N_3$. Then

$$\mathscr{B}_H \begin{pmatrix} \gamma_1, \ldots, \gamma_7 \mid 2, 4 \\ s_1, \ldots, s_7 \mid 6, 5 \end{pmatrix} = \{\pi(\mathscr{L}) \mid \mathscr{L} H \bar{\mathscr{L}}' > 0\}$$

has $\pi(\mathscr{L}) \in M(\alpha_0)$, α_0 as above. So the points of $\mathscr{B}_H$ have the local coordinates (11.25). Now, consider $\mathscr{B}_H \cap \mathscr{B}_J$:

$$\mathscr{L} H \bar{\mathscr{L}}' > 0 \qquad \text{and} \qquad \mathscr{L} J \bar{Z}' = 0.$$

The equality says:

$$\begin{pmatrix} I & & & R_{11} & R_{12} & R_{13} \\ & I & & 0 & R_{22} & R_{23} \\ & & I & 0 & 0 & R_{33} \end{pmatrix} J_0 \begin{pmatrix} I & & & R_{11} & R_{12} & R_{13} \\ & I & & 0 & R_{22} & R_{23} \\ & & I & 0 & 0 & R_{33} \end{pmatrix}' = 0,$$

i.e., $R_{13} = R_{13}'$, $R_{12} = R_{23}'$, $R_{33} = R_{11}'$, $R_{22} = R_{22}'$. And the inequality says:

$$\begin{pmatrix} I & & & R_{11} & R_{12} & R_{13} \\ & I & & 0 & R_{22} & R_{23} \\ & & I & 0 & 0 & R_{33} \end{pmatrix} H_0 \begin{pmatrix} I & & & R_{11} & R_{12} & R_{13} \\ & I & & '0 & R_{22} & R_{23} \\ & & I & 0 & 0 & R_{33} \end{pmatrix}' > 0,$$

that is,

$$\frac{1}{i}\left[\begin{pmatrix} R_{13} & R_{12} \\ R_{23} & R_{22} \end{pmatrix} - \overline{\begin{pmatrix} R_{13} & R_{12} \\ R_{23} & R_{22} \end{pmatrix}}'\right] - \begin{pmatrix} R_{11} \\ 0 \end{pmatrix}(\bar{R}_{11}' \quad 0) - \begin{pmatrix} \bar{R}_{33}' \\ 0 \end{pmatrix}(R_{33} \quad 0) > 0, \tag{11.26}$$

where

$$R_{13} = R_{13}' = \begin{pmatrix} Z_{17} & Z_{12} & Z_{16} \\ Z_{27} & Z_{22} & 0 \\ Z_{37} & 0 & Z_{36} \end{pmatrix},$$

$$R_{12} = R'_{23} = \begin{pmatrix} Z_{14} & Z_{15} \\ 0 & 0 \\ Z_{34} & Z_{35} \end{pmatrix} = \begin{pmatrix} Z_{47} & 0 & Z_{46} \\ Z_{57} & 0 & Z_{56} \end{pmatrix}',$$

$$R_{22} = R'_{22} = \begin{pmatrix} Z_{44} & 0 \\ 0 & Z_{55} \end{pmatrix},$$

$$R_{11} = R'_{33} = \begin{pmatrix} Z_{13} & Z_{11} \\ 0 & 0 \\ Z_{33} & 0 \end{pmatrix} = \begin{pmatrix} Z_{67} & 0 & Z_{66} \\ Z_{77} & 0 & 0 \end{pmatrix}'.$$

Substituting these into (11.26) and changing variables, we have

$$\frac{1}{i} \left[\begin{pmatrix} W_{11} & W_{12} & W_{13} & W_{14} & W_{15} \\ W'_{12} & W_{22} & 0 & 0 & 0 \\ W'_{13} & 0 & W_{33} & W_{34} & W_{35} \\ W'_{14} & 0 & W'_{34} & W_{44} & 0 \\ W'_{15} & 0 & W'_{35} & 0 & W_{55} \end{pmatrix} - \begin{pmatrix} \overline{W_{11}} & \overline{W_{12}} & \overline{W_{13}} & \overline{W_{14}} & \overline{W_{15}} \\ \overline{W'_{12}} & \overline{W_{22}} & 0 & 0 & 0 \\ \overline{W'_{13}} & 0 & \overline{W_{33}} & \overline{W_{34}} & \overline{W_{35}} \\ \overline{W'_{14}} & 0 & \overline{W'_{34}} & \overline{W_{44}} & 0 \\ \overline{W'_{5}} & 0 & \overline{W'_{35}} & 0 & \overline{W_{55}} \end{pmatrix}' \right]$$

$$- \begin{pmatrix} U_{11} & U_{12} \\ 0 & 0 \\ U_{31} & 0 \\ 0 & 0 \\ 0 & 0 \end{pmatrix} \begin{pmatrix} \overline{U_{11}} & \overline{U_{12}} \\ 0 & 0 \\ \overline{U_{31}} & 0 \\ 0 & 0 \\ 0 & 0 \end{pmatrix}' - \begin{pmatrix} \overline{U_{11}} & \overline{U_{12}} \\ 0 & 0 \\ \overline{U_{31}} & 0 \\ 0 & 0 \\ 0 & 0 \end{pmatrix} \begin{pmatrix} U_{11} & U_{12} \\ 0 & 0 \\ U_{31} & 0 \\ 0 & 0 \\ 0 & 0 \end{pmatrix}' > 0.$$

This is the domain given by the present example. It can be realized as a super disk $\mathscr{B}_{J \cap H}$ in $\mathscr{B}_J$, where J and H are the same as before. This domain is of the type W_{II} introduced in Ref. 4.

REMARK. If we only take $\mathscr{B}_H$, then we have

$$\frac{1}{i} \left[\begin{pmatrix} R_{13} & R_{12} \\ R_{23} & R_{22} \end{pmatrix} - \begin{pmatrix} \overline{R_{13}} & \overline{R_{12}} \\ \overline{R_{23}} & \overline{R_{22}} \end{pmatrix}' \right] - \begin{pmatrix} R_{11} \\ 0 \end{pmatrix} (\bar{R}'_{11} \quad 0) - \begin{pmatrix} \bar{R}'_{33} \\ 0 \end{pmatrix} (R_{33} \quad 0) > 0,$$

(11.27)

where R_{ij} are defined by (11.25), and (11.27) is of the type W_I in Ref. 4.

11.2.2. The New Construction

Now, we give a general method to generalize the above example. We still consider W_{II}:

$$\frac{1}{i}[Z - \bar{Z}] - U\bar{U}' - \bar{U}U' > 0, \qquad Z = Z', \tag{11.28}$$

where

$$Z = \begin{pmatrix} Z_{11} & Z_{12} & \cdots & Z_{1s} \\ Z_{21} & Z_{22} & \cdots & Z_{2s} \\ & & \vdots & \\ Z_{s1} & Z_{s2} & \cdots & Z_{ss} \end{pmatrix}, \qquad Z_{ij} = 0, \quad i < j, i \neq k_1, \ldots, k_l,$$

$$1 = k_1 < k_2 < \cdots < k_l < s.$$

For convenience, we take $(k_1, \ldots, k_l) = (1, 3, \ldots, 2l - 1)$. The rest of the cases are like this one. So U of (11.28) is of the form

$$U = \begin{pmatrix} U_{11} & U_{12} \cdots \cdots \cdots \cdots \cdots U_{1\,l-1} & U_{1l} \\ 0 & 0 \cdots \cdots \cdots \cdots \cdots 0 & 0 \\ U_{31} & U_{32} \cdots \cdots \cdots \cdots U_{3\,l-1} & 0 \\ 0 & 0 \cdots \cdots \cdots \cdots \cdots 0 & 0 \\ U_{51} & U_{52} \quad \cdots \quad U_{5\,l-2} \quad 0 & 0 \\ \cdots \cdots \cdots \cdots \cdots \cdots \cdots \cdots \cdots \cdots \\ U_{2\,l-11} & 0 \cdots \cdots \cdots \cdots \cdots 0 \\ 0 \cdots \cdots \cdots \cdots \cdots \cdots 0 \end{pmatrix}.$$

Consider the matrix

$$\mathscr{A}(Z, U, V) =$$

$$(11.29)$$

We change the columns of $\mathscr{A}(Z, U, V)$ as follows: the columns, which contain

$$U_{1l}, Z_{22}, U_{3\,l-1}, Z_{44}, U_{5\,l-2}, Z_{66}, \ldots, U_{2\,l-1,1}, Z_{2l\,2l}, Z_{2l+12l+1}, \ldots,$$

$$Z_{ss}, Z_{11}, Z_{33}, \ldots, Z_{2l-12l-1}, \tag{11.30}$$

respectively, are arranged according to the order of (11.30), and the elements of the diagonal $U_{1l}, Z_{22}, \ldots, Z_{ss}, V_{1l}, V_{3\,l-1}, \ldots, V_{2\,l-11}$ are changed by $(I, U_{1l}), (I, Z_{22}), \ldots, (I, V_{2\,l+1})$, respectively, and the rest of the elements of the diagonal $U_{12}, Z_{12}, \ldots$ are changed by $(0, U_{12}), (0, Z_{12}), \ldots$, respectively. Then

$$
\mathcal{A}(Z, U, V) =
\begin{bmatrix}
(I, U_{11}) & (0, Z_{12}) & (0, U_{1l-1}) & (0, Z_{14}) & \cdots & (0, U_{11}) & (0, Z_{12l}) & \cdots & (0, Z_{1s}) & (0, Z_{12l-1}) & \cdots & (0, Z_{11}) \\
 & (I, Z_{22}) & 0 & \cdots\cdots\cdots\cdots\cdots\cdots\cdots\cdots\cdots\cdots\cdots\cdots\cdots\cdots\cdots & 0 & (0, Z_{21}) \\
 & & (I, U_{3l-1}) & (0, Z_{34}) & \cdots & (0, U_{31}) & (0, Z_{32l}) & \cdots & (0, Z_{3s}) & (0, Z_{32l-1}) & \cdots & (0, Z_{31}) \\
 & & & (I, Z_{44}) & 0 \cdots\cdots\cdots\cdots\cdots\cdots\cdots\cdots\cdots\cdots\cdots & 0 & (0, Z_{41}) \\
 & & & & (0, U_{2l-11}) & (0, Z_{2l-12l}) \cdots\cdots\cdots\cdots\cdots\cdots\cdots\cdots & (0, Z_{2l-11}) \\
 & & & & & (I, Z_{2l2l}) & 0 & \cdots & 0 & (0, Z_{2l2l-1}) & \cdots & (0, Z_{2l1}) \\
 & & & & & & 0 & \cdots & 0 & (0, Z_{2l+12l-1}) & & (0, Z_{2l+11}) \\
 & & & & & & & & 0 & \vdots & \\
 & & & & & & & & (1, Z_{ss}) & (0, Z_{ss2l-1}) & \cdots & (0, Z_{s1}) \\
 & & & & & & & & & (I, V_{1l}) & \cdots & (0, V_{11}) \\
 & & & & & & & & & & & (I, V_{2l-11})
\end{bmatrix}
N
$$

(11.31)

In terms of homogeneous coordinates, (11.31) can be written as

$$
\begin{bmatrix}
\mathcal{Z}_{11} & \mathcal{Z}_{12} & \mathcal{Z}_{13} & \cdots & \mathcal{Z}_{12l} & \cdots & \mathcal{Z}_{1s} & \cdots & \cdots & \mathcal{Z}_{1\,s+l} \\
 & \mathcal{Z}_{22} & 0 & \cdots & 0 & \cdots & 0 & \cdots & 0 & \mathcal{Z}_{2\,s+l} \\
 & & \mathcal{Z}_{33} & \cdots & \mathcal{Z}_{32l} & \cdots & \mathcal{Z}_{3s} & \cdots & \cdots & \mathcal{Z}_{3\,s+l} \\
 & & & \mathcal{Z}_{44} & 0 \cdots 0 & \cdots & 0 & \cdots & 0 & Z_{4\,s+l} \\
 & & & & \cdots\cdots & \cdots\cdots\cdots\cdots & \\
 & & & & \mathcal{Z}_{2l2l} & 0 \cdots 0 & \mathcal{Z}_{2l\,s+1} & \cdots & \mathcal{Z}_{2l\,s+l} \\
 & & & & & 0 \cdots 0 & \mathcal{Z}_{2l+1\,s+1} & & \mathcal{Z}_{2l+1\,s+l} \\
 & & & & & 0 & & & \\
 & & & & & & \mathcal{Z}_{ss} & \mathcal{Z}_{ss+1} & \mathcal{Z}_{ss+l} \\
 & & & & & & & \mathcal{Z}_{s+1s+1} & \mathcal{Z}_{s+1s+l} \\
 & & & & & & & & \mathcal{Z}_{s+ls+l}
\end{bmatrix}
$$

So, we consider the manifold

$$\mathscr{B}\begin{pmatrix} r_1,\ldots,r_{s+l} & 2, & 4, & 2l, & 2l+1,\ldots,s \\ s_1,\ldots,s_{s+l} & s+l-1, & s+l-1,\ldots,s, & s,\ldots,s \end{pmatrix}.$$

Let $\mathscr{Z}$ have the form of the above matrix, then we have

$$\mathscr{Z} = R\begin{pmatrix} (I, U_{11}) & (0, Z_{12}) & \cdots & & (0, Z_{11}) \\ & (I, Z_{22}) & 0 & \cdots & 0 & (0, Z_{21}) \\ & & 0 & & \ddots & \vdots \\ & & & & & (I, V_{2\,l-11}) \end{pmatrix} = R\mathscr{A}(Z, U, V)N.$$

(11.32)

Let $H = N'H_0N$, $J = N'J_0N$, where N is defined by (11.31), (11.32),

$$H_0 = \begin{pmatrix} & & & & & & & iI \\ & & & & & & iI & \\ & & & & & I & & \\ & & & & -I & & & \\ & & & -iI & & & & \\ & -iI & & & & & & \\ -iI & & & & & & & \end{pmatrix},$$

$$J_0 = \begin{pmatrix} & & & & & & & I \\ & & & & & & I & \\ & & & & & I & & \\ & & & & -I & & & \\ & & & -I & & & & \\ & & -I & & & & & \\ & -I & & & & & & \end{pmatrix}.$$

Then for $\mathscr{Z} \in \mathscr{B}_J \cap \mathscr{B}_H$, we have

$$\mathscr{A}(Z, U, V)NHN'\overline{\mathscr{A}(Z, U, V)}' > 0, \tag{11.33}$$

$$\mathscr{A}(Z, U, V)NHN'\mathscr{A}'(Z, U, V) = 0. \tag{11.34}$$

Then (11.29) can be written as

$$\mathscr{A}(Z, U, V) = \begin{pmatrix} I & & & U & Z_{12} & Z_{11} \\ & I & & 0 & Z_{22} & Z_{23} \\ & & I & 0 & 0 & V \end{pmatrix}.$$

Then from (11.34), we have

$$Z_{11} = Z'_{11}, \quad Z_{22} = Z'_{22}, \quad Z_{12} = Z'_{23}, \quad U = V'.$$

But (11.33) is

$$\frac{1}{i}(Z - \bar{Z}') - U\bar{U}' - \bar{V}'V > 0,$$

where

$$Z = \begin{pmatrix} Z_{11} & Z_{12} \\ Z_{23} & Z_{22} \end{pmatrix}.$$

Using the symmetric condition as above, we have

$$\frac{1}{i}(Z - \bar{Z}) - U\bar{U}' - \bar{U}U' > 0, \tag{11.35}$$

where

$$Z = \begin{bmatrix} Z_{11} & Z_{12} & Z_{13} & Z_{14} & \cdots\cdots & Z_{12l-1} & \cdots\cdots & Z_{1s} \\ Z'_{12} & Z_{22} & 0 & \cdots\cdots\cdots\cdots & 0 & \cdots\cdots\cdots & 0 \\ \vdots & 0 & Z_{33} & \cdots\cdots\cdots & Z_{32l-1} & \cdots\cdots & Z_{3s} \\ \vdots & \vdots & & Z_{44} & 0\cdots 0 & \cdots\cdots\cdots & 0 \\ \vdots & \vdots & \vdots & 0 & \cdots\cdots\cdots\cdots\cdots & \\ \vdots & \vdots & \vdots & \vdots & & Z_{2l-12l-1} & \cdots\cdots & Z_{2l-1s} \\ \vdots & \vdots & \vdots & \vdots & \vdots & Z_{2l2l} & 0\cdots\cdots 0 \\ \vdots & \vdots & \vdots & \vdots & \vdots & 0 & \\ Z'_{1s} & 0 & Z'_{3s} & 0 & \vdots & Z'_{2l-1s} & 0 & Z_{ss} \end{bmatrix},$$

$$U = \begin{bmatrix} U_{11} & U_{12} & \cdots & U_{1l-2} & U_{1l-1} & U_{1l} \\ 0 & \cdots\cdots\cdots\cdots\cdots\cdots\cdots\cdots\cdots & 0 \\ U_{31} & U_{32} \cdots\cdots\cdots\cdots\cdots U_{3l-1} & 0 \\ 0 & \cdots\cdots\cdots\cdots\cdots\cdots\cdots\cdots & 0 \\ U_{51} & U_{52} & \cdots & U_{5l-2} & 0 & 0 \\ 0 & \cdots\cdots\cdots\cdots\cdots\cdots\cdots & 0 \\ & \cdots\cdots\cdots\cdots\cdots\cdots\cdots\cdots\cdots \\ U_{2l-11} & 0 \cdots\cdots\cdots\cdots\cdots\cdots & 0 \\ 0 & 0\cdots\cdots\cdots\cdots\cdots\cdots & 0 \end{bmatrix}.$$

11.2.3. Generalizations

In this section we consider the construction of more general extension spaces. Consider the k sets of square matrices $L_1, L_2, \ldots, L_k$. Each L_i is a subgroup of the linear group. That is, if $A, B \in L_i$, then $AB \in L_i$, $A^{-1} \in L_i$. For $A_i \in L_i$, take A_{ij} such that

$$A = \begin{pmatrix} A_1 & A_{12} & \cdots & A_{1k} \\ & A_2 & \cdots & A_{2k} \\ & & \ddots & \\ \mathbf{0} & & & A_k \end{pmatrix}$$

also is a group. Such an $A_{ij}(i < j)$ is said to be conjugate to $L_1, \ldots, L_k$. For example, if $L_1, \ldots, L_k$ are upper triangular matrices, then any A_{ij} is conjugate to $L_1, \ldots, L_k$. For example, again if

$$L_1 = \left\{ \begin{pmatrix} * & * & * \\ & * & 0 \\ & & * \end{pmatrix} \right\}, \quad L_2 = \left\{ \begin{pmatrix} * & * & * \\ & * & 0 \\ & & * \end{pmatrix} \right\},$$

and A_{12} has the form

$$
\begin{pmatrix} * & * & * \\ 0 & 0 & 0 \\ * & * & * \end{pmatrix} \quad \text{or} \quad \begin{pmatrix} * & * & * \\ 0 & 0 & 0 \\ 0 & 0 & 0 \end{pmatrix}
$$

then A_{12} is conjugate to L_1, L_2. In fact, matrices with the form

$$
A = \begin{pmatrix} * & * & * & * & * & * \\ & * & 0 & 0 & 0 & 0 \\ & & * & * & * & * \\ & & & * & * & * \\ & & & & * & 0 \\ & & & & & * \end{pmatrix} \quad \text{or} \quad \begin{pmatrix} * & * & * & * & * & * \\ & * & 0 & 0 & 0 & 0 \\ & & * & 0 & 0 & 0 \\ & & & * & * & * \\ & & & & * & 0 \\ & & & & & * \end{pmatrix}
$$

form a subgroup of the linear group. For example, for

$$
L_1 = \left\{ \begin{pmatrix} * & * & * \\ & * & 0 \\ & & * \end{pmatrix} \right\}, \quad L_2 = \left\{ \begin{pmatrix} * & 0 & 0 \\ * & * & 0 \\ * & 0 & * \end{pmatrix} \right\},
$$

then

$$
A_{12} = \begin{pmatrix} * & * & * \\ * & * & 0 \\ * & 0 & * \end{pmatrix}
$$

is conjugate to L_1, L_2. Because, if

$$
\begin{pmatrix} A_1 & A_{12} \\ 0 & A_2 \end{pmatrix} \begin{pmatrix} B_1 & B_{12} \\ 0 & B_2 \end{pmatrix} = \begin{pmatrix} C_1 & C_{12} \\ 0 & C_2 \end{pmatrix},
$$

then

$$
C_{12} = \begin{pmatrix} * & * & * \\ * & * & 0 \\ * & 0 & * \end{pmatrix}.
$$

Now, for the subgroup L of the matrix group with degree m and the subgroup $P(L)$ of the group consisting of all permutation matrices, we associate a matrix set $\tilde{L}$ with the following property:

> For $\mathcal{Z} \in \tilde{L}$, there exists $N \in P(L)$, such that
>
> $\mathcal{Z} = (\mathcal{Z}_1, \mathcal{Z}_2) N$, $\mathcal{Z}_1 \in L$, and for any $R \in L$,   (11.36)
>
> we always have $R\mathcal{Z} \in \tilde{L}$.

For example

$$L = \left\{ \begin{pmatrix} Z_{11} & * & * \\ & Z_{22} & 0 \\ & & Z_{33} \end{pmatrix} \right\}.$$

Then

$$\tilde{L} = \begin{pmatrix} \mathcal{Z}_{11} & \mathcal{Z}_{12} & \mathcal{Z}_{13} \\ 0 & \mathcal{Z}_{22} & 0 \\ 0 & 0 & \mathcal{Z}_{33} \end{pmatrix},$$

where rank $\mathcal{Z}_{li}$ = the number of rows of $\mathcal{Z}_{li}$. For the same reason,

$$L = \left\{ \begin{pmatrix} Z_{11} & 0 & 0 \\ * & Z_{22} & 0 \\ * & 0 & Z_{33} \end{pmatrix} \right\}.$$

Then

$$\tilde{L} = \left\{ \begin{pmatrix} \mathcal{Z}_{11} & 0 & 0 \\ \mathcal{Z}_{21} & \mathcal{Z}_{22} & 0 \\ \mathcal{Z}_{31} & 0 & \mathcal{Z}_{33} \end{pmatrix} \right\}.$$

Now, for the system $\tilde{L}_1, \ldots, \tilde{L}_k$ explained as above, we construct the following matrix set $E(\tilde{L}_1, \ldots, \tilde{L}_k)$:

$$\mathcal{Z} = \begin{pmatrix} \tilde{\mathcal{Z}}_{11} & \tilde{\mathcal{Z}}_{12} & \cdots & \tilde{\mathcal{Z}}_{1k} \\ & \tilde{\mathcal{Z}}_{22} & \cdots & \tilde{\mathcal{Z}}_{2k} \\ & & \ddots & \vdots \\ & \mathbf{0} & & \tilde{\mathcal{Z}}_{kk} \end{pmatrix},$$ (11.37)

where $\tilde{\mathfrak{X}}_{ii} \in \tilde{L}_i$ ($i = 1, \ldots, k$), and $\tilde{\mathfrak{X}}_{12}, \ldots, \tilde{\mathfrak{X}}_{1k}, \ldots, \tilde{\mathfrak{X}}_{k-1\,k}$ is conjugate to $\tilde{L}_1, \ldots, \tilde{L}_k$. From the definition of $\tilde{L}_i$ (see (11.36)), we have

$$\mathfrak{X} = \begin{pmatrix} (\tilde{\mathfrak{X}}_{11}^1, \tilde{\mathfrak{X}}_{11}^2) N_1 & (\tilde{\mathfrak{X}}_{12}^1, \tilde{\mathfrak{X}}_{12}^2) N_2 & \cdots & (\tilde{\mathfrak{X}}_{1k}^1, \tilde{\mathfrak{X}}_{1k}^2) N_k \\ & (\tilde{\mathfrak{X}}_{22}^1, \tilde{\mathfrak{X}}_{22}^2) N_2 & \cdots & (\tilde{\mathfrak{X}}_{2k}^1, \tilde{\mathfrak{X}}_{2k}^2) N_k \\ & & \ddots & \vdots \\ & & & (\tilde{\mathfrak{X}}_{kk}^1, \tilde{\mathfrak{X}}_{kk}^2) N_k \end{pmatrix}, \quad N_i \in P(L_i).$$

We introduce the equivalence relation in $E(\tilde{L}_1, \ldots, \tilde{L}_k)$:

$$\mathfrak{X}_1 \sim \mathfrak{X}_2 \Leftrightarrow \mathfrak{X}_2 = \tilde{P}\mathfrak{X}_1, \qquad \tilde{P} = \begin{pmatrix} \tilde{P}_{11} & \tilde{P}_{12} & \cdots & \tilde{P}_{1k} \\ & \tilde{P}_{22} & \cdots & \tilde{P}_{2k} \\ & & \ddots & \vdots \\ & & & \tilde{P}_{kk} \end{pmatrix}, \quad (11.38)$$

where $\tilde{P}_{ii} \in L_i$, and $\tilde{P}_{12}, \ldots, \tilde{P}_{1k}, \ldots, \tilde{P}_{k-1\,k}$ have the same forms as the $\tilde{Z}_{12}^1, \ldots, \tilde{Z}_{k-1\,k}^1$. That is, the $\tilde{P}_{12}, \ldots, \tilde{P}_{k-1\,k}$ are conjugate to $L_1, \ldots, L_k$.

The equivalence class of $\mathfrak{X}$ is denoted by $\pi(\mathfrak{X})$. The set of all $\pi(\mathfrak{X})$ is denoted by $\mathcal{B}(\tilde{L}_1, \ldots, \tilde{L}_k)$. Then we have the next result.

THEOREM 11.4. $\mathcal{B}(\tilde{L}_1, \ldots, \tilde{L}_k)$ *is a complex manifold.*

PROOF. *Step 1.* Because the rank of $\mathfrak{X}_{ii}$ is maximum, $\mathfrak{X}$, being a set of matrices in complex Euclidean space, is an open set. The projection $\pi : \mathfrak{X} \to \pi(\mathfrak{X})$ is an open mapping. If O is an open set in $\mathcal{B}(\tilde{L}_1, \ldots, \tilde{L}_k)$, then $\pi^{-1}(O) = \sum_{\tilde{P}} \tilde{P}O$ is an open set, where $\tilde{P}$ includes all of the matrices defined by (11.38), so π is continuous.

Step 2. The local coordinate system is symbolized by $N_1, \ldots, N_k$, where $N_i \in P(L_i)$. $\mathfrak{X}$ belongs to the coordinate neighborhood $M(N_1, \ldots, N_k)$, if $\tilde{\mathfrak{X}}_{ii} = (\tilde{\mathfrak{X}}_{ii}^1, \tilde{\mathfrak{X}}_{ii}^2) N_i$ and $\tilde{\mathfrak{X}}_{ii}^1 \in L_i$. The local coordinates in $M(N_1, \ldots, N_k)$ are defined by

$$\begin{pmatrix} \tilde{\mathfrak{X}}_{11}^1 & \cdots & \tilde{\mathfrak{X}}_{1k}^1 \\ & \ddots & \vdots \\ 0 & & \tilde{\mathfrak{X}}_{kk}^1 \end{pmatrix}^{-1} \begin{pmatrix} \tilde{\mathfrak{X}}_{11}^2 & \cdots & \tilde{\mathfrak{X}}_{1k}^2 \\ & \ddots & \vdots \\ 0 & & \tilde{\mathfrak{X}}_{kk}^2 \end{pmatrix} = \tilde{\mathfrak{X}}. \quad (11.39)$$

So

$$\mathfrak{X} = \begin{pmatrix} \tilde{\mathfrak{X}}_{11}^1 & \cdots & \tilde{\mathfrak{X}}_{1k}^1 \\ & \ddots & \vdots \\ 0 & & \tilde{\mathfrak{X}}_{kk}^1 \end{pmatrix} (I, \tilde{Z}) N_0 N, \quad (11.40)$$

where N_0 is a fixed permutation matrix and $N = [N_1, \ldots, N_k]$. Because $\tilde{\mathfrak{X}}_{ii} \in \tilde{L}_i$ (see definition (11.36)), we have

$$(I, \tilde{Z})N_0 N = \begin{pmatrix} (I, \tilde{Z}_{11}) & & \cdots & & (0, \tilde{Z}_{1k}) \\ & (I, \tilde{Z}_{22}) & \cdots & & (0, \tilde{Z}_{2k}) \\ & & & \ddots & & \vdots \\ & \mathbf{0} & & & (I, \tilde{Z}_{kk}) \end{pmatrix} \in E(\tilde{L}_1, \ldots, \tilde{L}_k).$$

The holomorphicity of the transform of local coordinates can be proved as in Theorem 11.1.

Step 3. The Hausdorff property can be proved as in Theorem 11.1. So Theorem 11.4 is proved. $\qquad\square$

If we do not add some conditions to $L_1, \ldots, L_k$ and $\tilde{L}_1, \ldots, \tilde{L}_k$, then it is difficult to see that $\mathcal{B}(\tilde{L}_1, \ldots, \tilde{L}_k)$ is homogeneous. But as we will discuss in the following, usually $\mathcal{B}(\tilde{L}_1, \ldots, \tilde{L}_k)$ is homogeneous. We can get the extension spaces for almost all nonsymmetric domains introduced in Ref. 4 from $\mathcal{B}(\tilde{L}_1, \ldots, \tilde{L}_k)$. Only for S_I, S_{II}, S_{III} will we have some changes based on $\mathcal{B}(\tilde{L}_1, \ldots, \tilde{L}_k)$.

Consider $E_0(\tilde{L}_1, \ldots, \tilde{L}_k)$, which is similar to $E(\tilde{L}_1, \ldots, \tilde{L}_k)$:

$$\mathfrak{X} = \begin{pmatrix} \tilde{\mathfrak{X}}_{11} & \tilde{\mathfrak{X}}_{12} & \cdots & \tilde{\mathfrak{X}}_{1k} \\ & \tilde{\mathfrak{X}}_{22} & \cdots & \tilde{\mathfrak{X}}_{2k} \\ & \mathbf{0} & \ddots & \vdots \\ & & & \tilde{\mathfrak{X}}_{kk} \end{pmatrix},$$

where $\tilde{\mathfrak{X}}_{ii} \in \tilde{L}_i$ $(i \neq 1, k)$, and $\tilde{\mathfrak{X}}_{11}, \tilde{\mathfrak{X}}_{kk}$ are as follows:

$$\tilde{\mathfrak{X}}_{11} = (\tilde{\mathfrak{X}}_{11}^1, 0)N_1, \quad N_1 \in P(L_1), \quad \tilde{\mathfrak{X}}_{11}^1 \in L_1,$$
$$\tilde{\mathfrak{X}}_{kk} = (\tilde{\mathfrak{X}}_{kk}^1, 0)N_k, \quad N_k \in P(L_k), \quad \tilde{\mathfrak{X}}_{kk}^1 \subset L_k. \tag{11.41}$$

$E_0(\tilde{L}_1, \ldots, \tilde{L}_k)$ with the equivalence relation

$$\mathfrak{X}_1 \sim \mathfrak{X}_2 \Leftrightarrow \mathfrak{X}_2 = \tilde{P}\mathfrak{X}_1, \quad \tilde{P} = \begin{pmatrix} \tilde{P}_{11} & \tilde{P}_{12} & \cdots & \tilde{P}_{1k} \\ & \tilde{P}_{22} & \cdots & \tilde{P}_{2k} \\ & \mathbf{0} & \ddots & \vdots \\ & & & \tilde{P}_{kk} \end{pmatrix}, \quad P_{ii} \in L_i, \tag{11.42}$$

becomes a complex manifold. The proof is the same as in Theorem 11.4, and we denote it by $\mathcal{B}_0(\tilde{L}_1, \ldots, \tilde{L}_k)$. In the following cases, $\mathcal{B}_0$ is homogeneous.

Now, we use $\mathscr{B}(\tilde{L}_1, \ldots, \tilde{L}_k)$ and $\mathscr{B}_0(\tilde{L}_1, \ldots, \tilde{L}_k)$ to construct the extension spaces for most of the nonsymmetric domains introduced in Ref. 4. We will use some typical examples. From the examples, we can see all the important points for general cases.

EXAMPLE 11.5. Find the extension space for

$$W_I = \left\{ (Z, U, V) \,\Big|\, \frac{1}{2i}(Z - \bar{Z}') - U\bar{U}' - \bar{V}V' > 0 \right\},$$

where

$$Z = \begin{pmatrix} Z_{11} & Z_{12} & \cdots & Z_{1s} \\ Z_{21} & Z_{22} & \cdots & Z_{2s} \\ & & \vdots & \\ Z_{s1} & Z_{s2} & \cdots & Z_{ss} \end{pmatrix}, \quad \begin{matrix} Z_{k_i j} = 0 \\ Z_{j k_i} = 0 \end{matrix}, \quad j > k_i, 1 < k_1 < k_2 < \cdots < k_l < s,$$

$$U = \begin{pmatrix} U_{11} & U_{12} & \cdots & U_{1s} \\ & U_{22} & \cdots & U_{2s} \\ & 0 & \ddots & \vdots \\ & & & U_{ss} \end{pmatrix}, \quad V = \begin{pmatrix} V_{11} & V_{12} & \cdots & V_{1s} \\ & V_{22} & \cdots & V_{2s} \\ & 0 & \ddots & \vdots \\ & & & V_{ss} \end{pmatrix}, \quad (11.43)$$

$$U_{k_i j} = 0, \; V_{k_i j} = 0, \quad j > k_i, i = 1, 2, \ldots, l.$$

Here we omit the degrees of block matrices in (11.43), because that is inessential in our discussion.

Take the manifold $\mathscr{B}(\tilde{L}_1, \tilde{L}_2)$, where

$$L_1 = \begin{pmatrix} *_{11} & *_{12} & \cdots & *_{1s} \\ & *_{22} & \cdots & *_{2s} \\ & 0 & \ddots & \vdots \\ & & & *_{ss} \end{pmatrix}, \quad *_{k_i j} = 0, \quad j > k_i,$$

$$i = 1, 2, \ldots, l, \det(*_{ii}) \neq 0,$$

$$L_2 = \begin{pmatrix} *_{11} & & & \\ \vdots & *_{22} & & 0 \\ & & \ddots & \\ *_{s1} & *_{s2} & & *_{ss} \end{pmatrix}, \quad *_{j k_i} = 0, \quad j > k_i,$$

$$i = 1, 2, \ldots, l, \det(*_{ii}) \neq 0.$$

Obviously, L_1, L_2 are groups, and the following matrix is conjugate to L_1, L_2:

$$\begin{pmatrix} *_{11} & \cdots & *_{1s} \\ \vdots & \ddots & \vdots \\ *_{s1} & \cdots & *_{ss} \end{pmatrix}, \qquad *_{k_ij} = 0, \ *_{jk_i} = 0, \quad j > k_i, \ i = 1, 2, \ldots, l.$$

The complex manifold $\mathscr{B}(\tilde{L}_1, \tilde{L}_2)$ can be written as follows:

$$\mathscr{X} = \begin{pmatrix} \tilde{\mathscr{X}}_{11} & \tilde{\mathscr{X}}_{12} \\ 0 & \tilde{\mathscr{X}}_{22} \end{pmatrix},$$

$$\tilde{\mathscr{X}}_{11} = \begin{pmatrix} \mathscr{X}_{11} & \mathscr{X}_{12} & \cdots & \mathscr{X}_{1s} \\ & \mathscr{X}_{22} & \cdots & \mathscr{X}_{2s} \\ & 0 & \ddots & \vdots \\ & & & \mathscr{X}_{ss} \end{pmatrix} \in \tilde{L}_1, \quad \tilde{\mathscr{X}}_{22} = \begin{pmatrix} w_{11} & & & 0 \\ & w_{22} & & \\ \vdots & & \ddots & \\ w_{s1} & w_{s2} & \cdots & w_{ss} \end{pmatrix} \in \tilde{L}_2,$$

$$\mathscr{X}_{k_ij} = 0, j > k_i, \qquad\qquad w_{jk_i} = 0, j > k_i$$

$$\tilde{\mathscr{X}}_{12} = \begin{pmatrix} x_{11} & \cdots & x_{1s} \\ \vdots & \ddots & \vdots \\ x_{s1} & \cdots & x_{ss} \end{pmatrix}, \qquad x_{k_ij} = 0, \quad x_{jk_i} = 0, j > k_i.$$

In the first coordinate neighborhood we have

$$\mathscr{X} \sim \left(\left(\begin{pmatrix} I, & \begin{matrix} U_{11} & \cdots & U_{1s} \\ & \ddots & \vdots \\ 0 & & U_{ss} \end{matrix} \end{pmatrix} \begin{pmatrix} 0, & \begin{matrix} Z_{11} & \cdots & Z_{1s} \\ \vdots & \ddots & \vdots \\ Z_{s1} & \cdots & Z_{ss} \end{matrix} \end{pmatrix} \\ (0, 0) \qquad\qquad \begin{pmatrix} I, & \begin{matrix} V_{11} \\ \vdots & \ddots & 0 \\ V_{s1} & \cdots & V_{ss} \end{matrix} \end{pmatrix} \right) N, \right. \tag{11.44}$$

where $U_{k_ij} = 0$, $V_{jk_i} = 0$, $Z_{k_ij} = 0$, $Z_{jk_i} = 0$, $j > k_i$, $i = 1, 2, \ldots, l$. Equation (11.44) also can be written as

$$\mathscr{X} = R\begin{pmatrix} I & 0 & U & Z \\ 0 & I & 0 & V' \end{pmatrix} N_0,$$

where N_0 is a fixed permutation matrix. Let $h = N_0' H_0 N_0$,

$$H_0 = \begin{pmatrix} 0 & 0 & 0 & iI \\ 0 & I & 0 & 0 \\ 0 & 0 & -I & 0 \\ -iI & 0 & 0 & 0 \end{pmatrix}.$$

Then

$$\mathcal{B}_H = \left\{ (Z, U, V) \,\Big|\, \frac{1}{i}(Z - \bar{Z}') - U\bar{U}' - \bar{V}V' > 0 \right\}.$$

So the complex manifold $\mathcal{B}(\tilde{L}_1, \tilde{L}_2)$ is the extension space of the domain (11.43).

Now we prove that the manifold $\mathcal{B}(\tilde{L}_1, \tilde{L}_2)$ is homogeneous. Its motion group

$$\mathcal{L} = \begin{pmatrix} \tilde{\mathcal{L}}_{11} & \tilde{\mathcal{L}}_{12} \\ 0 & \tilde{\mathcal{L}}_{22} \end{pmatrix} \to \mathcal{L}Q, \qquad Q = \begin{pmatrix} \tilde{Q}_{11} & \tilde{Q}_{12} \\ 0 & \tilde{Q}_{22} \end{pmatrix}, \tag{11.45}$$

$$\tilde{Q}_{11} = \begin{pmatrix} Q_{11} & \cdots & Q_{1s} \\ & \ddots & \vdots \\ 0 & & Q_{ss} \end{pmatrix}, \quad \tilde{Q}_{22} = \begin{pmatrix} P_{11} & & 0 \\ \vdots & \ddots & \vdots \\ P_{s1} & \cdots & P_{ss} \end{pmatrix},$$

$$\tilde{Q}_{12} = \begin{pmatrix} R_{11} & \cdots & R_{1s} \\ \vdots & & \vdots \\ R_{s1} & \cdots & R_{ss} \end{pmatrix}$$

satisfies $Q_{k_i j} = 0, P_{jk_i} = 0, R_{k_i j} = 0, R_{jk_i} = 0, j > k_i \, (i = 1, 2, \ldots, l), \det Q_{ii} \neq 0,$ $\det P_{ii} \neq 0 \, (i = 1, 2, \ldots, s).$

In order to prove that the group (11.45) acts transitively on $\mathcal{B}(\tilde{L}_1, \tilde{L}_2)$, it suffices to prove that for any two points $\mathcal{L}$, $w \in E(\tilde{L}_1, \tilde{L}_2)$, there exists Q with the form (11.45) such that

$$\mathcal{L}Q = w. \tag{11.46}$$

In block form, (11.46) can be written as

$$\tilde{\mathcal{L}}_{11}\tilde{Q}_{11} = \tilde{w}_{11}, \qquad \tilde{\mathcal{L}}_{22}\tilde{Q}_{22} = \tilde{w}_{22}, \qquad \tilde{\mathcal{L}}_{11}\tilde{Q}_{12} + \tilde{\mathcal{L}}_{12}\tilde{Q}_{22} - \tilde{w}_{12}.$$

We can solve the first equation from Theorem 11.1. By the same reasoning we can also solve the second equation. So the problem is to find $\tilde{Q}_{12}$ such that

$$\tilde{\mathscr{X}}_{11}\tilde{Q}_{12} = \tilde{w}_{12} - \tilde{\mathscr{X}}_{12}\tilde{Q}_{22}, \tag{11.47}$$

where $\tilde{\mathscr{X}}_{12}, \tilde{\mathscr{X}}_{11}, \tilde{w}_{12}, \tilde{Q}_{12}, \tilde{Q}_{22}$ are as in (11.45). In general, using the inductive method we can solve (11.47), but it is too complicated. We use the following simple example to explain the essential points of the proof.

For example, if

$$\tilde{\mathscr{X}}_{11} = \begin{pmatrix} \mathscr{X}_{11} & \mathscr{X}_{12} & \mathscr{X}_{13} \\ 0 & \mathscr{X}_{22} & 0 \\ 0 & 0 & \mathscr{X}_{33} \end{pmatrix}, \qquad \tilde{Q}_{12} = \begin{pmatrix} Q_{11} & Q_{12} & Q_{13} \\ Q_{21} & Q_{22} & 0 \\ Q_{31} & 0 & Q_{33} \end{pmatrix},$$

$$\tilde{w}_{12} - \tilde{\mathscr{X}}_{12}\tilde{Q}_{22} = \begin{pmatrix} *_{11} & *_{12} & *_{13} \\ *_{21} & *_{22} & 0 \\ *_{31} & 0 & *_{33} \end{pmatrix},$$

then (11.47) becomes

$$(\mathscr{X}_{33}Q_{31}, 0, \mathscr{X}_{33}Q_{33}) = (*_{31}, 0, *_{33}),$$

$$(\mathscr{X}_{22}Q_{21}, \mathscr{X}_{22}Q_{22}, 0) = (*_{21}, *_{22}, 0),$$

$$(\mathscr{X}_{11}\mathscr{X}_{12}\mathscr{X}_{13})\begin{pmatrix} Q_{11} \\ Q_{21} \\ Q_{31} \end{pmatrix} = *_{11}, \qquad (\mathscr{X}_{11}, \mathscr{X}_{12}, \mathscr{X}_{13})\begin{pmatrix} Q_{12} \\ Q_{22} \\ 0 \end{pmatrix} = *_{12},$$

$$(\mathscr{X}_{11}, \mathscr{X}_{12}, \mathscr{X}_{13})\begin{pmatrix} Q_{13} \\ 0 \\ Q_{33} \end{pmatrix} = *_{13},$$

because the rank of $\mathscr{X}_{11}$, $\mathscr{X}_{22}$, and $\mathscr{X}_{33}$ is equal to the number of rows in each. Using Lemma 11.2 we can get the solution $\tilde{Q}_{12}$ immediately. So $\mathscr{B}(\tilde{L}_1, \tilde{L}_2)$ is a homogeneous complex manifold.

REMARK 11.6. Here the example is for W_I. If we want to find the extension spaces of W_{II}, W_{III}, then we need to also discuss $\mathscr{B}_J(\tilde{L}_1, \tilde{L}_2)$, where J is an appropriate skew-symmetric matrix.

EXAMPLE 11.7. Find the extension space of the domain

$$\frac{1}{2i}\left[\begin{pmatrix} Z_{11} & Z_{12} & \cdots & Z_{1n} \\ Z'_{12} & Z_{22} & & 0 \\ \vdots & & \ddots & \\ Z'_{1n} & 0 & & Z_{nn} \end{pmatrix} - \overline{\begin{pmatrix} Z_{11} & Z_{12} & \cdots & Z_{1n} \\ Z'_{12} & Z_{22} & & 0 \\ \vdots & & \ddots & \\ Z'_{1n} & 0 & & Z_{nn} \end{pmatrix}}\right] > 0.$$

Construct the manifold

$$\mathscr{L} = \begin{bmatrix} \mathscr{L}_{11} & \mathscr{L}_{12} \cdots\cdots\cdots\cdots \mathscr{L}_{1\,n+1} \\ & \mathscr{L}_{22}|0 \cdots\cdots 0 \; Z_{2\,n+1} \\ & \quad \mathscr{L}_{33} \; 0 \cdots 0 \; \mathscr{L}_{3\,n+1} \\ & \qquad \ddots \qquad \vdots \\ \mathbf{0} & \qquad\qquad \mathscr{L}_{n+1\,n+1} \end{bmatrix} \begin{matrix} 1 \\ 1 \\ 1 \\ \vdots \\ 1 \end{matrix}$$
$$\qquad\qquad\quad 2 \quad\; 2 \cdots\cdots\cdots 2$$

where $\mathscr{L}_{11} = (\mathscr{L}_{11}^1, 0)$, $\mathscr{L}_{n+1\,n+1} = (\mathscr{L}_{n+1\,n+1}^1, 0)$, $\mathscr{L}_{11} \neq 0$, $\mathscr{L}_{n+1\,n+1}^1 \neq 0$, and $\mathscr{L}_{ii}$ are the 1×2 matrices, and their rank is equal to 1. The equivalence relation is

$$Z_1 \sim Z_2 \quad \text{if} \quad Z_2 = PZ_1, \qquad P = \begin{bmatrix} p_{11} & p_{12} \cdots\cdots\cdots\cdots\cdots p_{1\,n+1} \\ & p_{22} \; 0 \; \cdots \; 0 \; p_{2\,n+1} \\ & \quad \ddots \qquad\qquad \vdots \\ & \qquad\qquad\qquad\quad p_{n+1\,n+1} \end{bmatrix}.$$

This is the complex manifold $\mathscr{B}_0(\tilde{L}_1, \tilde{L}_2, \tilde{L}_3)$ expounded above, where L_1, L_3 are nonzero complex numbers, and L_2 is the subgroup consisting of the matrices

$$\begin{bmatrix} * & 0 & \cdots & 0 \\ & * & \cdots & 0 & 0 \\ & \mathbf{0} & \ddots & \vdots & \vdots \\ & & & * & 0 \\ & & & & * \end{bmatrix}.$$

Now, we come back to (11.48),

$$
\mathscr{L} =
\begin{bmatrix}
\mathscr{L}^1_{11} & \mathscr{L}^1_{12} \cdots\cdots\cdots \mathscr{L}^1_{1\,n+1} & 0 & \mathscr{L}^2_{12} \cdots\cdots\cdots\cdots \mathscr{L}^2_{1\,n+1} \\[6pt]
& \mathscr{L}^1_{22}\ 0 \cdots 0\ \mathscr{L}^1_{2\,n+1} & & \mathscr{L}^2_{22}\ 0 \cdots 0\ \mathscr{L}^2_{2\,n+1} \\[6pt]
& \ddots \qquad\vdots & & \ddots \qquad\vdots \\[6pt]
& \mathscr{L}^1_{n+1\,n+1} & & 0
\end{bmatrix}
N,
$$

and take $H = N' H_0 N$,

$$
H_0 =
\begin{pmatrix}
 & & & & i \\
 & & & iI & \\
 & & 1 & & \\
 & & & -1 & \\
 & -iI & & & \\
 -i & & & &
\end{pmatrix}
\begin{matrix} 1 \\ n \\ 1 \\ 1 \\ n \\ 1 \end{matrix}.
$$

Then it is easy to prove that the subdeterminant, which is formed by the first $n+1$ columns of $\mathscr{L}N'$, is not equal to zero, and the super disk $\mathscr{B}_H = \{\pi(\mathscr{L}) \,|\, \mathscr{L}H\bar{\mathscr{L}}' > 0\}$ is contained in the first coordinate neighborhood. In local coordinates, this is

$$
\begin{bmatrix}
1 & & & & 0 & Z_{12} & \cdots\cdots & Z_{1\,n+1} \\
 & \ddots & & & & Z_{22}\ 0 \cdots 0\ Z_{2\,n+1} \\
 & & \ddots & & & & \ddots \\
 & & & & & & \ Z_{nn}\ Z_{n\,n+1} \\
 & & & 1 & & & & 0
\end{bmatrix}
H_0
$$

$$
\begin{bmatrix}
1 & & & & 0 & Z_{12} & \cdots\cdots & Z_{1\,n+1} \\
 & \ddots & & & & Z_{22}\ 0 \cdots 0\ Z_{2\,n+1} \\
 & & \ddots & & & & \ddots \\
 & & & & & & \ Z_{nn}\ Z_{n\,n+1} \\
 & & & 1 & & & & 0
\end{bmatrix}'
> 0;
$$

i.e.,

$$\frac{1}{2i}\left[\begin{pmatrix} Z_{1\,n+1} & Z_{12} & \cdots & Z_{1n} \\ Z_{2\,n+1} & Z_{22} & & \\ \vdots & & 0 & \\ Z_{n\,n+1} & 0 & & Z_{nn} \end{pmatrix} - \overline{\begin{pmatrix} Z_{1\,n+1} & Z_{12} & \cdots & Z_{1n} \\ Z_{2\,n+1} & Z_{22} & & \\ \vdots & & 0 & \\ Z_{n\,n+1} & 0 & & Z_{nn} \end{pmatrix}}' \right] > 0.$$

If consider $\mathcal{B}^0_{HJ}$, H is defined by (11.49) and J is defined by

$$J = N'J_0N, \qquad J_0 = \begin{pmatrix} & & & & & 1 & 1 \\ & & & & I & & n \\ & & & 1 & & & 1 \\ & & -1 & & & & 1 \\ & -I & & & & & n \\ -1 & & & & & & 1 \end{pmatrix}.$$

Then from $\mathscr{Z}J\mathscr{Z}' = 0$, we have

$$(Z_{12}, \ldots, Z_{1n}) = (Z_{2n+1}, \ldots, Z_{n\,n+1}),$$

so $\mathcal{B}^0_J$ is the extension space.

EXAMPLE 11.8. Find the extension space of $S_I(S_{II}, S_{III})$.

$$\frac{1}{2i}(Z - \bar{Z}') > 0, \qquad Z = \begin{pmatrix} Z_{11} & \cdots & Z_{1s} \\ \vdots & \ddots & \vdots \\ Z_{s1} & & Z_{ss} \end{pmatrix}, \tag{11.49}$$

$$Z_{k_ij} = 0, \quad Z_{jk_i} = 0, j > k_i, 1 < k_1 < k_2 < \cdots < k_l < s.$$

We can suppose that $k_l = s - 1$; otherwise, $Z_{s-1\,s-1}, Z_{s-1\,s}, Z_{s\,s-1}, Z_{ss}$ can be merged into one block. Let $j_0 \in \{1, 2, \ldots, s - 1\}$, $j_0 \notin \{k_1, \ldots, k_l\}$, but $j_0 + i \in \{k_1, \ldots, k_l\}$ $(i = 1, 2, \ldots, s - j_0 - 1)$; that is,

$$Z = \begin{bmatrix} Z_{11} & \cdots\cdots\cdots\cdots\cdots\cdots\cdots\cdots\cdots\cdots & Z_{1s} \\ & Z_{22} & \cdots\cdots\cdots\cdots\cdots\cdots & * \\ & & \ddots & \cdots\cdots\cdots\cdots & \\ & & & Z_{j_0 j_0} & \cdots\cdots\cdots & Z_{j_0 s} \\ & & & & Z_{j_0+1\,j_0+1} & 0 & \cdots & 0 \\ & & & & & \ddots & & \vdots \\ & & & & & & \ddots & 0 \\ Z_{s1} & * & & Z_{sj_0} & 0 & \cdots\cdots & Z_{ss} \end{bmatrix}$$

Now, we take the complex manifold $\mathcal{B}_0(\tilde{L}_1, \tilde{L}_2, \tilde{L}_3)$, where

$$
L_1 = \begin{pmatrix} *_{11} & & & *_{1j_0} \\ & *_{22} & \cdots & *_{2j_0} \\ & & \ddots & \\ \mathbf{0} & & & *_{j_0 j_0} \end{pmatrix}, \quad *_{kj} = 0, \quad j > k_i, \quad k_i < j_0,
$$

$$
L_2 = \begin{bmatrix} *_{i_0+1\, j_0+1} & \cdots\cdots\cdots & 0 \\ \vdots & \ddots & \vdots \\ \vdots & & \ddots & 0 \\ 0 & \cdots\cdots\cdots\cdots & *_{ss} \end{bmatrix},
$$

$$
L_3 = \begin{bmatrix} \tilde{*}_{11} & 0 & \cdots\cdots & 0 \\ \vdots & \tilde{*}_{22} & & \vdots \\ \vdots & & \ddots & \vdots \\ \tilde{*}_{j_0 1} & \tilde{*}_{j_0 2} & \cdots\cdots & \tilde{*}_{j_0 j_0} \end{bmatrix}, \quad \tilde{*}_{jk_i} = 0, \quad j > k_i, \quad k_i < j_0.
$$

If the square matrix, which is the left part of $\mathscr{L}N'$, is not singular, then

$$
\mathscr{L} \sim \begin{pmatrix} I & & & 0 & \tilde{Z}_{12} & \tilde{Z}_{11} \\ & I & & 0 & \tilde{Z}_{22} & \tilde{Z}_{21} \\ & & I & 0 & 0 & 0 \end{pmatrix} N, \tag{11.50}
$$

where

$$
\begin{pmatrix} \tilde{Z}_{11} & \tilde{Z}_{12} \\ \tilde{Z}_{21} & \tilde{Z}_{22} \end{pmatrix} = (11.49), \qquad \tilde{Z}_{22} = \begin{bmatrix} Z_{j_0+1\, j_0+1} & \cdots\cdots\cdots & 0 \\ \vdots & \ddots & \vdots \\ \vdots & & \ddots & \vdots \\ 0 & \cdots\cdots\cdots\cdots & Z_{ss} \end{bmatrix}.
$$

Using H, which we have used many times, we obtain

$$
\frac{1}{2i}(Z - \bar{Z}') > 0,
$$

where Z is the same as (11.49). So, $\mathcal{B}_0$ is the extension space of (11.49). If we consider $\mathcal{B}_J^0 \cap \mathcal{B}_H^0$, and

$$J = N'J_0N, \qquad J_0 = \begin{pmatrix} & & & & & & I \\ & & & & & I & \\ & & & & I & & \\ & & & -I & & & \\ & & -I & & & & \\ & -I & & & & & \end{pmatrix},$$

then we obtain

$$S_{II} : \frac{1}{2i}(Z - \bar{Z}) > 0, \qquad Z = Z',$$

where Z is the same as (11.49).

If we take

$$J_0 = \begin{pmatrix} & & & & & K \\ & & & & K & \\ & & & I & & \\ & & -I & & & \\ & -K & & & & \\ -K & & & & & \end{pmatrix},$$

$$K = \begin{bmatrix} \begin{pmatrix} 0 & 1 \\ -1 & 0 \end{pmatrix} & & \\ & \ddots & \\ & & \begin{pmatrix} 0 & 1 \\ -1 & 0 \end{pmatrix} \end{bmatrix},$$

then we obtain

$$S_{III} : \frac{1}{2i}(Z - \bar{Z}') > 0, \qquad JZ = Z'J, \quad J = \left[\begin{pmatrix} 0 & 1 \\ -1 & 0 \end{pmatrix}, \dots, \begin{pmatrix} 0 & 1 \\ -1 & 0 \end{pmatrix} \right].$$

EXAMPLE 11.9. Find the extension space of the following domain W_I (W_{II}, W_{III}):

$$\frac{1}{i}(Z - \bar{Z}') - U\bar{U}' - \bar{V}V' > 0, \qquad (11.51)$$

where

$$Z = \begin{pmatrix} Z_{11} & \cdots & Z_{1s} \\ \vdots & \ddots & \vdots \\ Z_{s1} & & Z_{ss} \end{pmatrix},$$

$Z_{k_i j} = 0, j > k_i, Z_{j k_i} = 0, j > k_i, 1 < k_1 < \cdots < k_l < s,$

$$U = \begin{pmatrix} U_{11} & U_{12} & \cdots & U_{1q} \\ 0 & U_{22} & \cdots & U_{2q} \\ \vdots & & \ddots & \vdots \\ 0 & & & U_{qq} \\ 0 & & & 0 \end{pmatrix}, \qquad V = \begin{pmatrix} V_{11} & V_{12} & \cdots & V_{1q} \\ 0 & V_{22} & \cdots & V_{2q} \\ \vdots & & \ddots & \vdots \\ 0 & & & V_{qq} \\ 0 & & & 0 \end{pmatrix},$$

$q < s$, $U_{k_i j} = 0$, $V_{k_i j} = 0$, $j > k_i$, $k_i \leq q$. We can suppose $k_l = s - 1$, and let p satisfy the conditions

$$p \in \{1, 2, \ldots, s - 1\}, \quad p \notin \{k_1, \ldots, k_l\}, \quad p + i \in \{k_1, \ldots, k_l\},$$

$$i = 1, 2, \ldots, s - 1 - p.$$

Then we can construct the extension spaces in two cases: $q > p$ or $q \leq p$. For these two cases, we use two simple examples to explain the construction of extension spaces. But the method can be used easily for general cases.

Case 1. Suppose

$$Z = \begin{pmatrix} Z_{11} & Z_{12} & Z_{13} \\ Z_{21} & Z_{22} & 0 \\ Z_{31} & 0 & Z_{33} \end{pmatrix}, \quad U = \begin{pmatrix} u_{11} & u_{12} \\ 0 & u_{22} \\ 0 & 0 \end{pmatrix}, \quad V = \begin{pmatrix} v_{11} & v_{12} \\ 0 & v_{22} \\ 0 & 0 \end{pmatrix}. \quad (11.52)$$

Then the extension space is

$$\mathscr{Z} = \begin{pmatrix} \mathscr{Z}_{11} & \mathscr{Z}_{12} & \mathscr{Z}_{13} & \mathscr{Z}_{14} & \mathscr{Z}_{15} \\ 0 & \mathscr{Z}_{22} & 0 & \mathscr{Z}_{24} & \mathscr{Z}_{25} \\ 0 & 0 & \mathscr{Z}_{33} & \mathscr{Z}_{34} & 0 \\ 0 & 0 & 0 & \mathscr{Z}_{44} & 0 \\ 0 & 0 & 0 & 0 & \mathscr{Z}_{55} \end{pmatrix},$$

where $\mathscr{X}_{ij}$ are 1×2 matrices and the rank of $\mathscr{X}_{ii}$ is 1. The equivalence relation is if $\mathscr{X}_2 = P\mathscr{X}_1$, then $\mathscr{X}_1 \sim \mathscr{X}_2$, where

$$P = \begin{pmatrix} p_{11} & p_{12} & p_{13} & p_{14} & p_{15} \\ 0 & p_{22} & 0 & p_{24} & p_{25} \\ 0 & 0 & p_{33} & p_{34} & 0 \\ 0 & 0 & 0 & p_{44} & 0 \\ 0 & 0 & 0 & 0 & p_{55} \end{pmatrix}, \qquad \det P \neq 0.$$

If we use the language of $\mathscr{B}(\tilde{L}_1, \tilde{L}_2, \tilde{L}_3)$, then

$$L_1 = \left\{ \begin{pmatrix} * & * \\ 0 & * \end{pmatrix} \right\}, \quad L_2 = \{(*)\}, \quad L_3 = \left\{ \begin{pmatrix} * & 0 \\ * & * \end{pmatrix} \right\}.$$

If we use H (or J, K) again, then we obtain the extension space.

Case 2. Suppose

$$Z = \begin{pmatrix} Z_{11} & Z_{12} & Z_{13} & Z_{14} & Z_{15} \\ Z_{21} & Z_{22} & 0 & 0 & 0 \\ Z_{31} & 0 & Z_{33} & Z_{34} & Z_{35} \\ Z_{41} & 0 & Z_{43} & Z_{44} & 0 \\ Z_{51} & 0 & Z_{53} & 0 & Z_{55} \end{pmatrix},$$

$$U = \begin{pmatrix} u_{11} & u_{12} \\ 0 & u_{22} \\ 0 & 0 \\ 0 & 0 \\ 0 & 0 \end{pmatrix}, \quad V = \begin{pmatrix} v_{11} & v_{12} \\ 0 & v_{22} \\ 0 & 0 \\ 0 & 0 \\ 0 & 0 \end{pmatrix}.$$

(11.53)

Its extension space is

$$\mathscr{X} = \begin{pmatrix} \mathscr{X}_{11} & \mathscr{X}_{12} & 0 & \mathscr{X}_{14} & \mathscr{X}_{15} & \mathscr{X}_{16} & \mathscr{X}_{17} & \mathscr{X}_{18} \\ & \mathscr{X}_{22} & 0 & 0 & 0 & \mathscr{X}_{26} & \mathscr{X}_{27} & 0 \\ & & \mathscr{X}_{33} & \mathscr{X}_{34} & \mathscr{X}_{35} & \mathscr{X}_{36} & 0 & \mathscr{X}_{38} \\ & & & \mathscr{X}_{44} & 0 & \mathscr{X}_{46} & 0 & \mathscr{X}_{48} \\ & & & & \mathscr{X}_{55} & \mathscr{X}_{56} & 0 & \mathscr{X}_{58} \\ & & & & & \mathscr{X}_{66} & 0 & 0 \\ & & & & & & \mathscr{X}_{76} & \mathscr{X}_{77} & 0 \\ & & & & & & 0 & 0 & \mathscr{X}_{88} \end{pmatrix},$$

where the Z_{ij} are 1×2 matrices, rank $\mathscr{X}_{ii} = 1$, $\mathscr{X}_{33} = (\mathscr{X}_{33}^1, 0)$, $\mathscr{X}_{88} = (\mathscr{X}_{88}^1, 0)$.

The equivalence relation is as follows: if $\mathcal{Z}_2 = P\mathcal{Z}_1$, then $\mathcal{Z}_1 \sim \mathcal{Z}_2$, where

$$
P = \begin{pmatrix}
p_{11} & p_{12} & 0 & p_{14} & p_{15} & p_{16} & p_{17} & p_{18} \\
 & p_{22} & 0 & 0 & 0 & p_{26} & p_{27} & 0 \\
 & & p_{33} & p_{34} & p_{35} & p_{36} & 0 & p_{38} \\
 & & & p_{44} & 0 & p_{46} & 0 & p_{48} \\
 & & & & p_{55} & p_{56} & 0 & p_{58} \\
 & & & & & p_{66} & 0 & 0 \\
 & & & & & p_{76} & p_{77} & 0 \\
 & & & & & 0 & 0 & p_{88}
\end{pmatrix}, \quad \det P \neq 0.
$$

Using $\mathcal{B}_0(\tilde{L}_1, \tilde{L}_2, \tilde{L}_3)$, we get

$$
L_1 = \left\{ \begin{pmatrix} * & * \\ 0 & * \end{pmatrix} \right\}, \quad L_2 = \{(*)\}, \quad L_3 = \left\{ \begin{pmatrix} * & 0 \\ 0 & * \end{pmatrix} \right\},
$$

$$
L_4 = \left\{ \begin{pmatrix} * & 0 \\ * & * \end{pmatrix} \right\}, \quad L_5 = \{(*)\}.
$$

But $\tilde{L}_2$ is of the form $(\mathcal{Z}^1_{33}, 0)$ and $\tilde{L}_5$ is of the form $(\mathcal{Z}^1_{88}, 0)$. The local coordinates of the above points (in the first coordinate neighborhood) are

$$
\begin{pmatrix}
u_{11} & u_{12} & 0 & Z_{14} & Z_{15} & Z_{11} & Z_{12} & Z_{13} \\
 & u_{22} & 0 & 0 & 0 & Z_{21} & Z_{22} & 0 \\
 & & 0 & Z_{34} & Z_{35} & Z_{31} & 0 & Z_{33} \\
 & & & Z_{44} & 0 & Z_{41} & 0 & Z_{43} \\
 & & & & Z_{55} & Z_{51} & 0 & Z_{53} \\
 & & & & & v_{11} & 0 & 0 \\
 & & & & & v_{12} & v_{22} & 0 \\
 & & & & & 0 & 0 & 0
\end{pmatrix}.
$$

Then using H again, (11.53) can be realized as a super disk of $\mathcal{B}_0(\tilde{L}_1, \ldots, \tilde{L}_5)$.

EXAMPLE 11.10. Now, we can easily obtain the extension space of the following domain:

$$\frac{1}{i}(Z - \bar{Z}) - U\bar{U}' - \bar{U}U' > 0,$$

$$Z = Z',$$

$$Z = \begin{pmatrix} Z_{11} & \cdots & Z_{1q} & \cdots & Z_{1s} \\ \vdots & & \vdots & & \vdots \\ Z_{q1} & \cdots & Z_{qq} & \cdots & Z_{qs} \\ \vdots & & \vdots & & \vdots \\ Z_{s1} & \cdots & Z_{sq} & \cdots & Z_{ss} \end{pmatrix} \qquad U = \begin{pmatrix} U_{11} & \cdots & U_{1q} \\ \vdots & & \vdots \\ 0 & \cdots & U_{qq} \\ 0 & \cdots & 0 \\ \vdots & & \vdots \\ 0 & \cdots & 0 \end{pmatrix}. \qquad (11.54)$$

We take complex manifold $\mathcal{B}(\tilde{L}_1, \tilde{L}_2, \tilde{L}_3)$, where

$$L_1 = \left\{ \begin{pmatrix} *_{11} & \cdots & *_{1q} \\ & \ddots & \vdots \\ & & *_{qq} \end{pmatrix} \right\}, \quad L_2 = \left\{ \begin{pmatrix} *_{q+1\,q+1} & \cdots & *_{q+1\,s} \\ \vdots & \ddots & \vdots \\ *_{s\,q+1} & \cdots & *_{ss} \end{pmatrix} \right\},$$

$$L_3 = \left\{ \begin{pmatrix} \tilde{*}_{11} & & 0 \\ \vdots & \ddots & \\ \tilde{*}_{q1} & & \tilde{*}_{qq} \end{pmatrix} \right\}.$$

That is,

$$\mathcal{Z} = \begin{bmatrix} \mathcal{Z}_{11} & \cdots & \mathcal{Z}_{1q} & \mathcal{Z}_{1\,q+1} & \cdots\cdots & \mathcal{Z}_{1s} & \cdots\cdots\cdots & \mathcal{Z}_{1\,s+q} \\ & \ddots & \vdots & \vdots & & \vdots & & \vdots \\ & & \mathcal{Z}_{qq} & \vdots & & \vdots & & \vdots \\ & & & \mathcal{Z}_{q+1\,q+1} & \cdots & \mathcal{Z}_{q+1\,s} & \cdots\cdots & \mathcal{Z}_{q+1\,s+q} \\ & & & \vdots & & \vdots & & \vdots \\ & & & \mathcal{Z}_{s\,q+1} & \cdots\cdots & \mathcal{Z}_{ss} & \cdots\cdots & \mathcal{Z}_{ss+q} \\ & & & & & \mathcal{Z}_{s+1\,s+1} & \cdots & 0 \\ & & & & & \vdots & \ddots & \vdots \\ & & & & & \mathcal{Z}_{s+q\,s+1} & \cdots & \mathcal{Z}_{s+q\,s+q} \end{bmatrix}.$$

The equivalence relation is as follows: if $\mathscr{L}_2 = P\mathscr{L}_1$, then $\mathscr{L}_1 \sim \mathscr{L}_2$, where

$$P = \begin{pmatrix} P_1 & * & * \\ & P_2 & * \\ & & P_3 \end{pmatrix}, \qquad P_i \in L_i.$$

Let

$$\mathscr{L} = R\begin{pmatrix} I & & U & \tilde{Z}_{12} & \tilde{Z}_{11} \\ & I & 0 & \tilde{Z}_{22} & \tilde{Z}_{21} \\ & I & 0 & 0 & V \end{pmatrix} N,$$

where N is a permutation matrix. Suppose

$$H = N'H_0N, \qquad J = N'J_0N,$$

$$H_0 = \begin{pmatrix} & & & & & iI \\ & & & & iI & \\ & & I & & & \\ & & & -I & & \\ & -iI & & & & \\ -iI & & & & & \end{pmatrix},$$

$$J_0 = \begin{pmatrix} & & & & & I \\ & & & & I & \\ & & I & & & \\ & & & -I & & \\ & -I & & & & \\ -I & & & & & \end{pmatrix}.$$

Then (11.54) can be realized as $\mathscr{B}_H(\tilde{L}_1, \tilde{L}_2, \tilde{L}_3) \cap \mathscr{B}_J(\tilde{L}_1, \tilde{L}_2, \tilde{L}_3)$. So $\mathscr{B}_J(\tilde{L}_1, \tilde{L}_2, \tilde{L}_3)$ is the extension space of (11.54).

References

1. Lu Qi-Keng. *Classical Manifolds and Classical Domains*, Shanghai Science and Technology, Shanghai (1963).
2. Lu Qi-Keng. A class of homogeneous complex manifolds, *Acta Math., Sinica* **12**, 229–249 (1962).
3. I. I. Pyateckii-Shapiro, *Automorphic Functions and the Geometry of Classical Domains*, Gordon and Breach, New York (1969).
4. Zhong Jia-Qing and Yin Wei-Ping. Some types of nonsymmetric homogeneous domains, *Acta Math. Sinica* **24**, 587–613 (1981) (Chapter 10 of this volume).

12

Cohomology of Extension Spaces for Classical Domains

12.1. Introduction

The Grassmann manifolds are considered to be the most important and canonical examples in compact Kähler manifolds. The Grassmann manifold $G(m + n, n)$ consists of the set of all n-dimensional linear subspaces in $\mathbb{C}^{m+n}$, and can be realized as

$$G(m + n, n) = \{\mathfrak{Z} \,|\, \mathfrak{Z} \text{ an } m \times (m + n) \text{ matrix, rank } \mathfrak{Z} = m,$$

$$\mathfrak{Z}_1 \sim \mathfrak{Z}_2 \Leftrightarrow \mathfrak{Z}_1 = Q\mathfrak{Z}_2, \det Q \neq 0\}. \tag{12.1}$$

Let

$$H = \begin{pmatrix} I^{(m)} & 0 \\ 0 & -I^{(n)} \end{pmatrix}.$$

By using local coordinates, the set

$$\{\mathfrak{Z} \in G(m + n, n) \,|\, \mathfrak{Z}H\bar{\mathfrak{Z}}' > 0\} \tag{12.2}$$

given by (see Ref. 7, pp. 53-54)

$$R_{\mathrm{I}} : I - Z\bar{Z}' > 0, \quad Z \text{ symmetric, being an } m \times n \text{ complex matrix,}$$

is just the first classical domain in several complex variables [6]. According to Professor L. K. Hua, the Grassmann manifold $G(m + n, n)$ is called the extension space of R_1.

In $G(2n, n)$, we define

$$G_{II}(2n, n) = \left\{ \mathfrak{Z} \mid \mathfrak{Z} \in G(2n, n), \mathfrak{Z} \begin{pmatrix} 0 & I^{(n)} \\ -I^{(n)} & 0 \end{pmatrix} \mathfrak{Z}' = 0 \right\}, \quad (12.3)$$

$$G_{III}(2n, n) = \left\{ \mathfrak{Z} \mid \mathfrak{Z} \in G(2n, n), \mathfrak{Z} \begin{pmatrix} 0 & I^{(n)} \\ I^{(n)} & 0 \end{pmatrix} \mathfrak{Z}' = 0 \right\}, \quad (12.4)$$

which are, respectively, compact Kähler manifolds of dimension $\frac{1}{2}n(n + 1)$ and $\frac{1}{2}n(n - 1)$, and are homogeneous [7, pp. 14–24].

We define the "hyperball" of $G_{II}(2n, n)$ and $G_{III}(2n, n)$ to be

$$\left\{ \mathfrak{Z} \mid \mathfrak{Z} \in G_{II}(2n, n), \mathfrak{Z} \begin{pmatrix} I^{(n)} & 0 \\ 0 & -I^{(n)} \end{pmatrix} \bar{\mathfrak{Z}}' > 0 \right\},$$

$$\left\{ \mathfrak{Z} \mid \mathfrak{Z} \in G_{III}(2n, n), \mathfrak{Z} \begin{pmatrix} I^{(n)} & 0 \\ 0 & -I^{(n)} \end{pmatrix} \bar{\mathfrak{Z}}' > 0 \right\}.$$

By using local coordinates, they can be represented as

$$R_{II}: I - Z\bar{Z} > 0, \qquad Z = Z', \quad (12.5)$$

$$R_{III}: I - Z\bar{Z} > 0, \qquad Z = -Z', \quad (12.6)$$

Z denoting $n \times n$ complex matrices and Z' the transpose of Z. As is well known, (12.5) and (12.6) are the second and the third symmetric classical domains. Consequently, as closed submanifolds $G_{II}(2n, n)$ and $G_{III}(2n, n)$ of the Grassmannian, they are also called the extension spaces of R_{II} and R_{III}, respectively.

A lot of work has been done on the properties of geometry and function theory for extension spaces, but little on their topological structure. In this respect, the cohomology ring of the extension space $G(m + n, n)$ of R_1 is well known, and can be given by the Schubert calculus [5]. Considering that $G(m + n, n)$ is homogeneous, its cohomology can also be studied by the method of Lie algebras. For general compact Kähler manifolds, the newly developed methods of studying their cohomology are based on Bott's residual formula [3, 5]. In 1967, by using zero points of a holomorphic vector field of a manifold, Bott first constructed a formula for the Chern numbers of the manifold. On the basis of this, Carrell and Lieberman proved the following important result in Refs. 3 and 4.

THEOREM 12.1 (*Carrell–Lieberman*). *Let M be an n-dimensional compact Kähler manifold, V a holomorphic vector field on M, and Z the set of zero points of V. Then*

1. *If $Z \neq \emptyset$, then $H^{p,q}(M) = 0$, for $p \neq q$.*
2. *If Z is finite and nondegenerate, let $H^0(M, \theta_Z)$ be the ring of complex functions on Z, and then $H^0(M, \theta_Z)$ has a filtration:*

$$H^0(M, \theta_Z) = F_n \supset F_{n-1} \supset \cdots \supset F_1 \supset F_0 = \mathbb{C},$$

$$F_i \cdot F_j \subseteq F_{i+j}, \tag{12.7}$$

such that

$$F_p / F_{p-1} \cong H^{2p}(M, \mathbb{C}). \tag{12.8}$$

Using the above theorem, Carrell [2] rewrote the classical result of the Schubert calculus concerning the cohomology of $G(m + n, n)$.

This chapter also aims at calculating the cohomology of $G_{II}(2n, n)$ and $G_{III}(2n, n)$ by means of the Carrell-Lieberman theorem. In comparison with the situation in Grassmannians, no classical result of the Schubert calculus is needed as background. And a set of bases of the cohomology group for $G_{II}(2n, n)$ and $G_{III}(2n, n)$ is given. As a corollary, if

$$P(t) = \sum b_p t^p \qquad (b_p = \dim H^p(M, \mathbb{C}), \text{ the } p\text{th Betti number})$$

denotes the Poincaré polynomial, then we get

$$P_{G_{II}(2n,n)}(t) = (1 + t^2)(1 + t^4) \cdots (1 + t^{2n}), \tag{12.9}$$

$$P_{G_{III}(2n,n)}(t) = 2(1 + t^2)(1 + t^4) \cdots (1 + t^{2n-2}). \tag{12.10}$$

12.2. The Case of $G_{II}(2n, n)$

To work out the cohomology of $G_{II}(2n, n)$, we first give a sketch of the main results of Ref. 4 here. Suppose M signifies a compact Kähler manifold of dimension m, and v is a holomorphic vector field on M. Let Ω^p denote the sheaf of germs of holomorphic p-forms on M with finite and nondegenerate zero set. $\Omega^0 = \theta$ stands for the sheaf of germs of holomorphic functions on M.

Suppose $E \to M$ is a θ-module vector bundle on M of rank k. E is said to be v-equivariant if v can be lifted to a map $\tilde{v}: E \to E$ such that

$$\tilde{v}(f \cdot s) = v(f) \cdot s + f \cdot \tilde{v}(s).$$

Here f and s are sections of θ and E, respectively.

Let $u = \{U_\alpha\}$ be a Leray cover of M, and D an analytic connection of E (see Ref. 1).

$$D = \{U_\alpha, D_\alpha\}, \qquad D_\alpha: E|_{U_\alpha} \to E \times \Omega^1|_{U_\alpha},$$

Thus, $D_\alpha \in \Gamma(U_\alpha, \text{Hom}(E, E) \otimes \Omega^1)$. Evidently, $\{K_{\alpha\beta} = D_\alpha - D_\beta, U_\alpha \cap U_\beta\}$ is a cocycle in the Čech cohomology $H^1(u, \text{Hom}(E, E) \otimes \Omega^1)$. Therefore, $\{K_{\alpha\beta}\} \in H^1(U, \text{Hom}(E, E) \otimes \Omega^1)$ is called the Atiyah–Chern class of E, denoted $C(E)$.

As a holomorphic vector field on M, v induces naturally a contraction map $i(v): \Omega^p \to \Omega^{p-1}$. Since $C(E) \in H^1(u, \text{Hom}(E, E) \otimes \Omega^1)$, we have

$$i(v)C(E) \in C^1(u, \text{Hom}(E, E)).$$

According to Ref. 4, from the v-equivariance of the vector bundle E, if $L_\alpha = \tilde{v} - i(v)D_\alpha$, then $L_\alpha \in \Gamma(U_\alpha, \text{Hom}(E, E))$ and $L = \{U_\alpha, L_\alpha\} \in C^0(u, \text{Hom}(E, E))$, satisfying

$$i(v)C(E) = i(v)k_{\alpha\beta} = L_\beta - L_\alpha$$

on $U_\alpha \cap U_\beta$. Hence $i(v)C(E) \in H^1(M, \text{Hom}(E, E))$.

Now from the double complex (δ being the coboundary operator in Čech cohomology)

$$C^p(u, \text{Hom}(E, E) \otimes \Omega^q) \xrightarrow{\delta} C^{p+1}(u, \text{Hom}(E, E) \otimes \Omega^q),$$

$$C^p(u, \text{Hom}(E, E) \otimes \Omega^q) \xrightarrow{i(v)} C^p(u, \text{Hom}(E, E) \otimes \Omega^{q-1}),$$

we can form a total complex

$$K^\cdot = \{K^0 \to K^1 \to K^2 \to \cdots\},$$

where $K^j = \bigoplus_{p-q=j} C^p(u, \text{Hom}(E, E) \otimes \Omega^q)$ and the coboundary operator $K^j \to K^{j+1}$ is $\delta + (-1)^p i(v)$. Let $H^0(K)$ be the cohomology group of K^0. It was proved in Ref. 4 that $H^0(K)$ is isomorphic to $H^0(M, \theta_\mathscr{Z})$,

$$\varphi: H^0(K) \cong H^0(M, \theta_\mathscr{Z}), \tag{12.11}$$

where Z means the set of zero points of v, Z being assumed to be finite and nondegenerate and $\theta_Z = \theta/i(v)\Omega^1$. It is easy to see that $H^0(M, \theta_Z)$ is the ring of all complex functions on Z. Strictly speaking, the isomorphism φ in (12.11) is induced by the composition of projections φ_1 and φ_2, where

$$\varphi_1 \colon K^j = C^j(u, \theta) \oplus C^{j+1}(u, \Omega^1) \oplus \cdots \oplus C^{j+m}(u, \Omega^m) \to C^j(u, \theta),$$

$$\varphi_2 \colon \theta \to \theta/i(v)\Omega^1 = \theta_Z.$$

Since $\{K_{\alpha\beta}\} \in C^1(u, \mathrm{Hom}\,(E, E) \otimes \Omega^1)$, and $L \in C^0(u, \mathrm{Hom}\,(E, E))$, it is easy to see that $\{K_{\alpha\beta} \oplus L_\alpha\}$ is a cocycle in $C^0(K)$; i.e.,

$$\{C(E) \oplus L\} \in H^0(K) \cong H^0(M, \theta_Z).$$

Denote the cohomology class of $\{C(E) \oplus L\}$ by $\tilde{C}(E)$.

Under the map φ, $\tilde{C}(E)$ is determined by its values $\tilde{v}_z$ on Z (i.e., $\tilde{v}_z \in H^0(Z, \mathrm{Hom}\,(E, E))$.

Now suppose σ_d is the d-degree elementary symmetric homogeneous polynomial of $k \times k$ matrices (k is the rank of E). Then Ref. 4 proves that under the isomorphism φ, we shall have

$$\varphi(\sigma_d(\tilde{v}_Z)) = \sigma_d(C(E)). \tag{12.12}$$

Furthermore, if $H^0(M, \theta_Z) = F_n \supset F_{n-1} \supset \cdots \supset F_0 = \mathbb{C}$ is the filtration shown by (12.7), we have isomorphisms

$$\varphi \colon F_d/F_{d-1} \to H^d(M, \Omega^d),$$

$$\sigma_d(\tilde{v}_z) \in F_d/F_{d-1}, \qquad \sigma_d(C(E)) \in H^d(M, \Omega^d). \tag{12.13}$$

By the Dolbeaut Theorem, $\sigma_d(C(E))$ represents the dth Chern class $C_d(E)$ of E.

In short, to give a description of $H^{d,d}(M, C)$, we may start by constructing a v-equivariant vector bundle E, find $\tilde{C}(E)$, and restrict it to Z (the set of zero points of v) to get $\sigma_d(\tilde{v}_z)$. We then have an element in F_d/F_{d-1}. Similarly, we can find a base of F_d/F_{d-1}. By the isomorphism $H^d(M, \Omega^d) \cong H^{d,d}(M, C)$, we get a basis of $H^{d,d}(M, C)$.

Now choose $M = G_{\mathrm{II}}(2n, n)$, and let $I = \{1 \le i_1 < i_2 < \cdots < i_n \le 2n\}$. For $\mathfrak{Z} \in G_{\mathrm{II}}(2n, n)$, we use Z_I to denote the submatrix of $\mathfrak{Z}$ consisting of the $(i_1, \ldots, i_n)$th columns of $\mathfrak{Z}$. If $U_I = \{\mathfrak{Z} \in G_{\mathrm{II}}(2n, n) | \det Z_I \neq 0\}$, then $\{U_I\}$ is an open cover of $G_{\mathrm{II}}(2n, n)$ for all I. Let E be the restriction to $G_{\mathrm{II}}(2n, n)$ of the universal bundle of $G(2n, n)$. E refers to a holomorphic vector bundle of rank n, and on $U_I \cap U_J$, the transition function is written as $\{f_{IJ} = Z_I \cdot Z_J^{-1}, \ U_I \cap U_J\}$. Just as the Grassmannian, E is v-equivariant for any holomorphic vector field v on $G_{II}(2n, n)$ [4, pp. 269–270]. Analogous to Ref. 2, the Atiyah–Chern class of E gives

$$C(E) = \{df_{IJ} \cdot f_{IJ}^{-1}\}, \tag{12.14}$$

and

$$df_{IJ} \cdot f_{IJ}^{-1} = d(Z_I \cdot Z_J^{-1}) \cdot Z_J Z_I^{-1} = dZ_I \cdot Z_I^{-1} - Z_I Z_J^{-1} \, dZ_J \cdot Z_I^{-1}$$

$$= dZ_I \cdot Z_I^{-1} - f_{IJ}(dZ_J \cdot Z_J^{-1})f_{IJ}^{-1},$$

$$i(v)C(E) = i(v)(dZ_I \cdot Z_I^{-1}) - f_{IJ}i(v)(dZ_J \cdot Z_J^{-1})f_{IJ}^{-1}, \tag{12.15}$$

$$\tilde{C}(E) = df_{IJ} \cdot f_{IJ}^{-1} - i(v)(dZ_I \cdot Z_I^{-1}) \xrightarrow{\varphi} (-1)i(v)(dZ_I \cdot Z_I^{-1})\big|_{Z \cap U_I}$$

(see Ref. 4, pp. 269–270).

Now let us fix a holomorphic vector field of $G_{II}(2n, n)$ and determine its zero points. On $G_{II}(2n, n)$, a holomorphic vector field can be generated by the one-parameter group

$$\mathfrak{Z} \to \mathfrak{Z} \, e^{tN}, \qquad NJ + JN' = 0, \qquad J = \begin{pmatrix} 0 & I \\ -I & 0 \end{pmatrix}$$

where N is a $2n \times 2n$ complex matrix. Suppose $N = (\eta_{ij})_{1 \le i,j \le 2n}$ and then

$$\mathfrak{Z} \, e^{tN} = (w_{ij}) = (Z_{ij}) \, e^{tN} = (Z_{ij})(I + tN + o(t^2))$$

$$= \mathfrak{Z} + t\left(\sum_{k=1}^{2n} Z_{ik}\eta_{kj} \right) + o(t^2).$$

As a consequence, the holomorphic vector field generated by N leads to

$$v = \sum_{\substack{1 \le i \le n \\ 1 \le j \le 2n}} \frac{\partial w_{ij}}{\partial t}\bigg|_{t=0} \cdot \frac{\partial}{\partial Z_{ij}} = \sum_{\substack{1 \le i \le n \\ 1 \le j \le 2n}} \left(\sum_{k=1}^{2n} Z_{ik}\eta_{kj} \right) \frac{\partial}{\partial Z_{ij}}. \tag{12.16}$$

LEMMA 12.2. *For a matrix N, if $NJ + JN' = 0$ and λ is an eigenvalue of N, then $-\lambda$ is an eigenvalue of N, too.*

PROOF. Suppose x is an eigenvector of N corresponding to λ; i.e., $xN = \lambda x$. Since

$$N = -JN'J' = -JN'J^{-1},$$

we have $\lambda x = xN = -xJN'J^{-1}$, $-\lambda(xJ) = (xJ)N'$, and $N(xJ)' = -\lambda(xJ)'$. Therefore $-\lambda$ is also an eigenvalue. $\qquad\square$

From this we may conclude that the eigenvalues of N appear in pairs with opposite signs. We may choose N to be

$$N = \begin{pmatrix} \lambda_1 & & & & & \\ & \ddots & & & & \\ & & \lambda_n & & & \\ & & & -\lambda_1 & & \\ & & & & \ddots & \\ & & & & & -\lambda_n \end{pmatrix}, \qquad \lambda_i \neq 0, \quad i = 1,\ldots,n, \qquad (12.17)$$

and then $e_i = (0,\ldots,1,\ldots,0)$ stands for an eigenvector corresponding to λ_i and

$$e_{n+i} = (0,\ldots,\overset{n+1}{1},\ldots,0)$$

stands for an eigenvector corresponding to $-\lambda_i$. Choose arbitrarily n vectors among $I = \{i_1,\ldots,i_n\}$ $(1 \leq i_1 < \cdots < i_n \leq 2n)$ to form

$$\mathcal{Z}_I = \begin{pmatrix} e_{i_1} \\ \vdots \\ e_{i_n} \end{pmatrix}.$$

We claim $\mathcal{Z}_I \in G_{\mathrm{II}}(2n, n)$ if and only if for each i, i and $n+i$ are not simultaneously contained in I, which can be proved by

$$\mathcal{Z}_I \in G_{\mathrm{II}}(2n, n) \Leftrightarrow e_{i_k}\begin{pmatrix} 0 & I \\ -I & 0 \end{pmatrix} e'_{i_j} = 0.$$

Since $\mathcal{Z}$ is the homogeneous coordinate of $G_{\mathrm{II}}(2n, n)$, $\mathcal{Z} \sim Q\mathcal{Z}$, all permutation of the rows in $\mathcal{Z}_I$ will give the same point in $G_{\mathrm{II}}(2n, n)$. Therefore, $\{\mathcal{Z}_I\}$ determines $1 + \binom{n}{1} + \cdots + \binom{n}{n} = 2^n$ points in $G_{\mathrm{II}}(2n, n)$.

Now we choose any one of these points, such as $\mathcal{Z}_I$:

$$\mathcal{Z}_I e^{tN} = \begin{pmatrix} e_{i_1} \\ \vdots \\ e_{i_n} \end{pmatrix} e^{tN} = \begin{pmatrix} e^{\pm t\lambda_{i_1}} & & \\ & \ddots & \\ & & e^{\pm t\lambda_{i_n}} \end{pmatrix}\begin{pmatrix} e_{i_1} \\ \vdots \\ e_{i_n} \end{pmatrix} \sim \begin{pmatrix} e_{i_1} \\ \vdots \\ e_{i_n} \end{pmatrix},$$

$\mathcal{Z}_I$ is a fixed point of $\mathcal{Z} \to \mathcal{Z} \, e^{tN}$ and, thus, a zero point of the vector field v. In other words, the degree of the zero set Z in the holomorphic vector field v determined by (12.17) amounts to $\deg(Z) = 2^n$.

Let $I = (i_1, \ldots, i_k) \subset (1, \ldots, n)$ $(0 \le k \le n)$ and $I' = (j_1, \ldots, j_{n-k})$ be the complement of I. Then points of Z can be written as

$$Z = \{e_{I,I'}\}, \qquad e_{I,I'} = \begin{pmatrix} e_{i_1} \\ \vdots \\ e_{i_k} \\ e_{n+j_1} \\ \vdots \\ e_{n+j_{n-k}} \end{pmatrix}.$$

We claim v is nondegenerate at every point of Z. Take $I = (1, \ldots, n)$ as an example,

$$e_{I,I'} = \begin{pmatrix} \overset{\ulcorner n \urcorner}{1} & & & 0 & \cdots & 0 \\ & \ddots & & \vdots & & \vdots \\ & & 1 & 0 & \cdots & 0 \end{pmatrix},$$

and take the local coordinates at this point to be $w = Z_1^{-1} Z_2$ with $\mathcal{Z} = (Z_1, Z_2)$. In these coordinates, $e_{I,I'}$ corresponds to the point $w = 0$. With respect to this coordinate, the one-parameter subgroup can be written as

$$w \to (A + wC)^{-1}(B + wD) = \begin{pmatrix} e^{-t\lambda_1} & & \\ & \ddots & \\ & & e^{-t\lambda_n} \end{pmatrix} w \begin{pmatrix} e^{-t\lambda_1} & & \\ & \ddots & \\ & & e^{-t\lambda_n} \end{pmatrix},$$

$$\begin{pmatrix} A & B \\ C & D \end{pmatrix} = \begin{pmatrix} e^{t\lambda_1} & & & & & \\ & \ddots & & & & \\ & & e^{t\lambda_n} & & & \\ & & & e^{-t\lambda_1} & & \\ & & & & \ddots & \\ & & & & & e^{-t\lambda_n} \end{pmatrix}.$$

From this the nondegeneracy is easy to derive. To sum up, we have

$$\dim H^0(G_{11}(2n, n), \theta_Z) = 2^n.$$

The notations $I = (i_1, \ldots, i_k)$ and $J = (n + j_1, \ldots, n + j_{n-k})$ are adopted as above. Let

$$U_{II'} = \{\mathfrak{Z} \in G_{II}(2n, n) \mid \mathfrak{Z} = (\mathfrak{Z}_1, \ldots, \mathfrak{Z}_{2n}),$$

$$\det(\mathfrak{Z}_{i_1}, \ldots, \mathfrak{Z}_{i_k}, \mathfrak{Z}_{n+j_1}, \ldots, \mathfrak{Z}_{n+j_{n-k}}) \neq 0\},$$

where $\mathfrak{Z}_i$ is the ith column of $\mathfrak{Z}$, and $U_{II'}$ is a neighborhood of $e_{I,I'}$. According to (12.15), we have to find $i(v)(dZ_{I,I'} \cdot Z_{I,I'}^{-1})|_{e_{I,I'}}$. By direct computation, we get

$$i(v)(dZ_{I,I'} \cdot Z_{I,I'}^{-1}) = (i(v)\,dZ_{I,I'}) \cdot Z_{I,I'}^{-1} = (\mathfrak{Z}N)_{I,I'} \cdot Z_{I,I'}^{-1}. \quad (12.18)$$

Hence

$$i(v)(dZ_{I,I'} \cdot Z_{I,I'}^{-1})|_{e_{I,I'}} = \begin{pmatrix} \lambda_{i_1} & & & & & \\ & \ddots & & & & \\ & & \lambda_{i_k} & & & \\ & & & -\lambda_{j_1} & & \\ & & & & \ddots & \\ & & & & & -\lambda_{j_{n-k}} \end{pmatrix}. \quad (12.19)$$

Now assume that

$$H^0(G_{II}(2n, n), \theta_Z) = F_{(n(n+1)/2} \supset F_{n(n+1)/2-1} \supset \cdots \supset F_0 = \mathbb{C} \quad (12.20)$$

is the filtration of the Carrell–Liebermann theorem. Let $\sigma_j(x_1, \ldots, x_n)$ indicate the j-degree elementary symmetry homogeneous form for n variables. We define a function $\tilde{\sigma}_j \in H^0(G_{II}(2n, n), \theta_Z)$ on Z as follows:

$$\tilde{\sigma}_j(e_{I,I'}) = \sigma_j(\lambda_{i_1}, \ldots, \lambda_{i_k}, -\lambda_{j_1}, \ldots, -\lambda_{j_{n-k}}), \qquad 1 \leq j \leq n. \quad (12.21)$$

By (12.12) and (12.13), there exists $\tilde{\sigma}_j \in F_j/F_{j-1}$. From the isomorphism φ, we find

$$\varphi(\tilde{\sigma}_j) = \sigma_j(C(E)) \in H^j(G_{II}(2n, n), \Omega^j) \cong H^{j,j}(G_{II}(2n, n), \mathbb{C}).$$

The key point here is to find all the bases for F_j/F_{j-1} $(j = 1, \ldots, \tfrac{1}{2}n(n + 1))$.

In this connection, we define a map π from the ring of symmetric polynomials to the ring of functions on Z:

$$\pi: p(x_1, \ldots, x_n) \to \tilde{p} \in H^0(G_{II}(2n, n), \theta_Z),$$

$$\tilde{p}(e_{I,I'}) = p(\lambda_{i_1}, \ldots, \lambda_{i_k}, -\lambda_{j_1+n}, \ldots, -\lambda_{j_{n-k}+n}). \quad (12.22)$$

It is well known that the ring of symmetric polynomials is generated by $1, \sigma_1, \ldots, \sigma_n$, as an algebra, i.e., $C[\sigma_1, \ldots, \sigma_n]$. Now let H be the linear subspace generated by $\sigma_{i_1} \cdots \sigma_{i_k}$, $i_1 < i_2 < \cdots < i_k$, $1 \le k \le n$,

$$H = \{\sigma_{i_1} \cdots \sigma_{i_k} \mid 1 \le i_1 < \cdots < i_k \le n, 1 \le k \le n\}. \qquad (12.23)$$

Obviously dim $H = 2^n$. Then the following result is valid.

LEMMA 12.3. $\pi|_H \; H \to H^0(G_{11}(2n, n), \theta_Z)$ is an isomorphism.

PROOF: Since H and $H^0(G_{11}(2n, n), \theta_Z)$ have the same dimension, it suffices to show π is surjective. For this reason, we need to prove that for each $e_{I,I'} \in Z$ the character function

$$\delta_{I,I'}(Z) = \begin{cases} 1, & e_{I,I'}, \\ 0, & \text{otherwise} \end{cases}$$

is in the image of π. Without loss of generality, we may assume that $I = (1, \ldots, k)$, $I' = (k+1, \ldots, n)$. Write

$$f_i(x) = (x_i + \lambda_i) \cdots (x_n + \lambda_i), \qquad i = 1, \ldots, k,$$

$$g_j(x) = (x_1 - \lambda_j) \cdots (x_n - \lambda_j), \qquad j = k+1, \ldots, n;$$

that is,

$$f_i(x) = \lambda_i^n + \sigma_1(x_1, \ldots, x_n)\lambda_i^{n-1} + \cdots + \sigma_n(x_1, \ldots, x_n), \qquad i = 1, \ldots, k,$$

$$g_i(x) = (-1)^n\lambda_j^n + (-1)^{n-1}\sigma_1(x_1, \ldots, x_n)\lambda_j^{n-1} + \cdots + \sigma_n(x_1, \ldots, x_n),$$

$$j = k+1, \ldots, n.$$

Define

$$f_{I,I'}(x) = f_1(x) \cdots f_k(x)g_{k+1}(x) \cdots g_n(x).$$

Since

$$I = (1, \ldots, k), I' = (k+1, \ldots, n),$$

$$\tilde{\sigma}_i(e_{I,I'}) = \sigma_i(\lambda_1, \ldots, \lambda_k, -\lambda_{k+1}, \ldots, -\lambda_n),$$

it is easily seen that

$$\pi(f_{I,I'})(e_{J,J'}) = \tilde{f}_{I,I'}(e_{J,J'})$$

$$= \begin{cases} 0, & J \neq I, \\ (-1)^{n-k}2^n\lambda_1 \cdots \lambda_k\lambda_{k+1} \cdots \lambda_n, & J = I. \end{cases}$$

Hence, $\pi(f_{I,I'}) = (-1)^{n-k}2^n\lambda_1 \cdots \lambda_k\lambda_{k+1} \cdots \lambda_n\delta_{I,I'}.$ $\qquad\square$

Notice that $\lambda_i \neq 0$, $\forall i$. Clearly, if we are able to justify $\pi(f_{I,I'}) \in \pi(H)$, then Lemma 12.3 will be established. But $\pi(f_{I,I'}) \in \pi(H)$ is ensured by the following lemma.

LEMMA 12.4. *For any $k \leq n$, $1 \leq i_1 \leq \cdots \leq i_k \leq n$, we have*

$$\pi(\sigma_{i_1} \cdots \sigma_{i_k}) = \tilde{\sigma}_{i_1} \cdots \tilde{\sigma}_{i_k} \in \pi(H).$$

PROOF: By the definition of H, if $i_1 < \cdots < i_k$, then $\tilde{\sigma}_{i_1} \cdots \tilde{\sigma}_{i_k} \in \pi(H)$, so we may assume that at least two subscripts are equal.

Denote by $1(Z)$ the constant function 1 on Z. Obviously, for any symmetric function p of $x_1^2, \ldots, x_n^2$, we have $\tilde{p} = \pi(p) \in C \cdot 1(Z) \in \pi(H)$, especially, $\tilde{\sigma}_n^2 = \lambda_1^2 \cdots \lambda_n^2 \cdot 1(Z) \in \pi(H)$.

Assuming $j < n$, we consider

$$\sigma_j^2 = \left(\sum_{i_1 < \cdots < i_j} x_{i_1} \cdots x_{i_j} \right)^2 = \sum x_{i_1}^2 \cdots x_{i_j}^2 + \sum_{\substack{i_1 < \cdots < i_j \\ k_1 < \cdots < k_j \\ (i) \neq (k)}} x_{i_1} \cdots x_{i_j} x_{k_1} \cdots x_{k_j}.$$

Let $x_1 \cdots x_j x_{k_1} \cdots x_{k_j}$ be a representation monomial of the second summation with $(1, \ldots, j) \neq (k_1, \ldots, k_j)$. If r $(0 \leq r \leq j)$ of the indices, for example $x_1 \cdots x_r$, are the same, then by symmetry the sums of all the terms are exactly

$$(x_1 \cdots x_j + \cdots)(x_1 \cdots x_r x_{k_{r+1}} \cdots x_{k_j} + \cdots)$$

$$= (x_1 \cdots x_r + \cdots)(x_1 \cdots x_r x_{r+1} \cdots x_j x_{k_{r+1}} \cdots x_{k_j} + \cdots) = \sigma_r \cdot \sigma_{2j-r}.$$

$r = 0$ being admitted in the above formula, we only need to define $\sigma_0 = 1$. Therefore we obtain $\sigma_j^2 =$ symmetric polynomial of

$$(x_1^2, \ldots, x_n^2) + C \sum_{i<j} \sigma_i \cdot \sigma_{2j-i},$$

hence

$$\pi(\sigma_j^2 \sigma_{i_1} \cdots \sigma_{i_{k-1}}) \in C_\pi(\sigma_{i_1} \cdots \sigma_{i_{k-1}}) + \sum C_\pi(\sigma_i \sigma_{2j-i} \sigma_{i_1} \cdots \sigma_{i_{k-1}}).$$

Now resorting to the induction method on k, we infer

$$\pi(\sigma_j^2 \sigma_{i_1} \cdots \sigma_{i_{k-1}}) \equiv \sum_{i<j} \pi(\sigma_i \sigma_{2j-i} \sigma_{i_1} \cdots \sigma_{i_{k-1}}) \qquad (\mathrm{mod}\ \pi(H)).$$

If all subscripts of $\sigma_i \sigma_{2j-i} \sigma_{i_1} \cdots \sigma_{i_{k-1}}$ are different, then they belong to $\pi(H)$ already; otherwise, we can continue these processes. In short, we can decrease the least subscript i_1 in $\sigma_{i_1} \cdots \sigma_{i_k}$ and finally obtain

$$\pi(\sigma_1^2 \sigma_{j_1} \cdots \sigma_{j_{k-1}}) = \tilde{\sigma}_1^2 \tilde{\sigma}_{j_1} \cdots \tilde{\sigma}_{j_{k-1}}.$$

However,

$$\tilde{\sigma}_1^2 = (x_1^2 + \cdots + x_n^2) - 2\tilde{\sigma}_2 = (\lambda_1^2 + \cdots + \lambda_n^2)1(Z) - 2\tilde{\sigma}_2,$$

So

$$\tilde{\sigma}_1^2 \tilde{\sigma}_{j_1} \cdots \tilde{\sigma}_{j_{k-2}} \equiv 2\tilde{\sigma}_2 \tilde{\sigma}_{j_1} \cdots \tilde{\sigma}_{j_{k-2}} \equiv 0 \qquad (\mathrm{mod}\ \pi(H)).$$

The last step comes from induction. The lemma is proved.   □

So far we have found a set of bases of $H^0(G_{II}(2n, n), \theta_Z)$:

$$\{1, \pi(\sigma_{i_1} \cdots \sigma_{i_k}) | 1 \le i_1 < \cdots < i_k \le n, 1 \le k \le n\}$$

$$= \{1, \tilde{\sigma}_{i_1} \cdots \tilde{\sigma}_{i_k} | 1 \le i_1 < \cdots < i_k \le n, 1 \le k \le n\}. \qquad (12.24)$$

From (12.12) and (12.13) we find $\tilde{\sigma}_j \in F_j/F_{j-1}$. By the Carrell–Lieberman theorem $F_i \cdot F_j \subseteq F_{i+j}$, we get

$$\tilde{\sigma}_{i_1} \cdots \tilde{\sigma}_{i_k} \in F_{i_1+\cdots+i_k}.$$

In what follows we claim

$$\tilde{\sigma}_{i_1} \cdots \tilde{\sigma}_{i_k} \in \frac{F_{i_1+\cdots+i_k}}{F_{i_1+\cdots+i_{k-1}}}. \qquad (12.25)$$

To prove this, let

$$b_{2p} = \dim \frac{F_p}{F_{p-1}} = \dim H^{p,p}(G_{II}(2n, n), C), \qquad p = 0, 1, \ldots, \frac{n(n+1)}{2}$$

$$C_{2p} = \left\{ \text{number of } (i_1, \ldots, i_k) \left| \begin{matrix} i_1 + \cdots + i_k = p \\ 1 \leq i_1 < \cdots < i_k \leq n \end{matrix} \right. \right\}$$

$$= \left\{ \text{number of } \tilde{\sigma}_{i_1} \cdots \tilde{\sigma}_{i_k} \left| \begin{matrix} i_1 + \cdots + i_k = p \\ 1 \leq i_1 < \cdots < i_k \leq n \end{matrix} \right. \right\},$$

$$p = 0, \ldots, \frac{n(n+1)}{2}.$$

For $G_{II}(2n, n)$, obviously $b_0 = b_{2 \cdot n(n+1)/2} = 1$, and by definition we infer $C_0 = C_{2 \cdot n(n+1)/2} = 1$.

Furthermore, by using the Lefschetz theorem, when M is a compact Kähler manifold the relation $H^{2p}(M, C) \cong H^{2n-2p}(M, C)$ holds. Moreover, by the Carrell–Liebermann theorem, $H^{p,q}(G_{II}(2n, n), C) = 0$ for $p \neq q$. We get $b_{2p} = b_{2(n(n+1)/2-p)}$. On the other hand, from the definition of C_{2p} it is easy to see that

$$\sum_{p=0}^{n(n+1)/2} C_{2p} t^{2p} = (1 + t^2)(1 + t^4) \cdots (1 + t^{2n}).$$

Therefore we also have $C_{2p} = C_{2(n(n+1)/2-p)}$.

Now we need a lemma.

LEMMA 12.5. $b_{2p} = C_{2p}$; $p = 0, 1, \ldots, n(n+1)/2$.

PROOF. Since $H^0(G_{II}(2n, n)C) = F_{n(n+1)/2} \supset F_{n(n+1)/2-1} \supset \cdots \supset F_0 = C$, $b_{2p} = \dim(F_p/F_{p-1})$, we have

$$\dim \frac{H^0(G_{II}(2n, n), C)}{F_{p-1}} = b_{2 \cdot n(n+1)/2} + \cdots + b_{2p}.$$

On the other hand, $\tilde{\sigma}_{i_1} \cdots \tilde{\sigma}_{i_k} \in F_{i_1 + \cdots + i_k}$, and so if $i_1 + \cdots + i_k < p$, then $\tilde{\sigma}_{i_1} \cdots \tilde{\sigma}_{i_k} \in F_{p-1}$. Therefore only when $i_1 + \cdots + i_k \geq p$, may we have $\tilde{\sigma}_{i_1} \cdots \tilde{\sigma}_{i_k} \in H^0(G_{II}(2n, n), C)/F_{p-1}$. The following inequality is then obtained:

$$b_{2 \cdot n(n+1)/2} + \cdots + b_{2p} \geq C_{2 \cdot n(n+1)/2} + \cdots + C_{2p}, \quad \forall p \qquad (12.26)$$

But $b_{2j} = b_{2(n(n+1)/2-j)}$ and $C_{2j} = C_{2(n(n+1)/2-j)}$, so (12.26) is equivalent to

$$b_0 + b_2 + \cdots + b_{2p} \geq C_0 + C_2 + \cdots + C_{2p}. \qquad (12.27)$$

Adding p ($p = 0, \ldots, n(n + 1)/2$) to both sides of (12.26) and (12.27) yields

$$b_0 + b_2 + \cdots + b_{2(n(n+1)/2)} \geq C_0 + C_2 + \cdots + C_{2(n(n+1)/2)}. \quad (12.28)$$

However, $\Sigma b_{2j} = \Sigma C_{2j} = 2^n$, so for each p, (12.26) must be an equality. Then from $b_{2(n(n+1)/2)} = C_{2(n(n+1)/2} = 1$, we have $b_{2p} = C_{2p}$, $\forall 0 \leq p \leq \frac{1}{2}n(n + 1)$. This completes the proof of the lemma. $\square$

To sum up our discussion, we have proved the following theorem.

THEOREM 12.6. *For* $G_{\mathrm{II}}(2n, n)$, *and under the isomorphism* φ *of the Carrell-Liebermann theorem,* $\{\varphi(\tilde{\sigma}_{i_1} \cdots \tilde{\sigma}_{i_k}) | i_1 + \cdots + i_k = p\}$ *constitutes a basis for* $H^{p,p}(G_{\mathrm{II}}(2n, n), C)$. *In particular,*

$$\sum_{p=0}^{n(n+1)/2} b_{2p} t^{2p} = \prod_{i=1}^{n} (1 + t^{2j})$$

is the Poincaré polynomial of $G_{\mathrm{II}}(2n, n)$.

12.3. The Case of $G_{\mathrm{III}}(2n, n)$

In this section, we shall deal with the cohomology of the extension space of the third classical domain

$$G_{\mathrm{III}}(2n, n) = \left\{ \mathfrak{Z} \in G(2n, n) \, \middle| \, \mathfrak{Z} \begin{pmatrix} 0 & I \\ I & 0 \end{pmatrix} \mathfrak{Z}' = 0 \right\}.$$

As proved in Ref. 4, it is a homogeneous Kähler manifold of dimension $\frac{1}{2}n(n - 1)$, but differs from $G_{\mathrm{II}}(2n, n)$. The difference lies in the fact that $G_{\mathrm{III}}(2n, n)$ is not connected but is divided into two connected components. We propose the next lemma.

LEMMA 12.7. *The number of connected components of* $G_{\mathrm{III}}(2n, n)$ *is* 2.

PROOF. As proved in Ref. 7, $G_{\mathrm{III}}(2n, n)$ is transitive with respect to the following transformation group:

$$\mathfrak{Z} \to \mathfrak{Z}Q, \qquad QJQ' = J, \quad J = \begin{pmatrix} 0 & I \\ I & 0 \end{pmatrix}.$$

Since J is symmetric, $\Gamma_J = \{Q \,|\, QJQ' = J\} \cong O(2n, C)$, where $O(2n, C)$ is the orthogonal group of rank $2n$.

Now let $(I, 0) \in G_{\mathrm{III}}(2n, n)$. We want to find the isotropy group at $(I, 0)$,

$$\{Q \in \Gamma_J \,|\, (I, 0)Q = (*, 0)\}.$$

Writing $Q = \begin{pmatrix} A & B \\ C & D \end{pmatrix}$, we can readily see that B should be zero. Simultaneously from $QJQ' = J$, we obtain the isotropy group at $(I, 0)$ as

$$\left\{ Q = \begin{pmatrix} A & 0 \\ C & A'^{-1} \end{pmatrix} \middle| CA^{-1} \text{ is antisymmetric} \right\} \det \Gamma_J^0(2n).$$

Hence $G_{\mathrm{III}}(2n, n) = \Gamma_J(2n)/\Gamma_J^0(2n)$.

If $Q \in \Gamma_J(2n)$, then $(\det Q)^2 = 1$. Thus, analogous to $O(2n, C)$, $\Gamma_J(2n)$ is divided into two connected components:

$$\Gamma_J^+(2n) = \{Q \in \Gamma_J(2n) | \det Q = 1\},$$

$$\Gamma_J^-(2n) = \{Q \in \Gamma_J(2n) | \det Q = -1\}.$$

Obviously,

$$\Gamma_J^0(2n) \subset \Gamma_J^+(2n).$$

Then $G_{\mathrm{III}}(2n, n)$ is divided into two connected components analytically homeomorphic to each other.

$$G_{\mathrm{III}}(2n, n) = M_1 \cup M_2, \qquad M_1 = \frac{\Gamma_J^+(2n)}{\Gamma_J^0(2n)}, \qquad M_2 = \frac{\Gamma_J^-(2n)}{\Gamma_J^0(2n)}.$$

This concludes the proof of Lemma 12.7.   □

As a consequence, we get $H^0(G_{\mathrm{III}}(2n, n), C) = 2H^0(M_1, C)$.

Similarly to the preceding section, let $I = \{1 \le i_1 < i_2 < \cdots < i_n \le 2n\}$, and let Z_I be the submatrix formed by the i_1-,..., i_n-th columns of $\mathfrak{Z}$. Then $U_I = \{\mathfrak{Z} \in G_{\mathrm{III}}(2n, n) | \det Z_I \ne 0\}$ is an open set of $G_{\mathrm{III}}(2n, n)$, and $\{U_I\}$ constitutes an open cover of $G_{\mathrm{III}}(2n, n)$.

Let $\tilde{E}$ be the bundle of $G_{\mathrm{III}}(2n, n)$ such that $\{f_{IJ} = Z_I \cdot Z_J^{-1}, U_I \cap U_J\}$ are the transition functions of E. Similar to Section 12.2, the Atiyah–Chern class of E is written as

$$C(E) = \{df_{IJ} \cdot f_{IJ}^{-1}\}. \tag{12.29}$$

Now let v be a holomorphic vector field, and let Z be the set of zero points of v. Under the isomorphism φ (see (12.12) and (12.13)), before we compute $H^0(G_{III}(2n, n), C)$, we have first of all to find

$$i(v)(dZ_I \cdot Z_I^{-1})|_{Z \cap U_I}. \tag{12.30}$$

It is natural that the choice of the holomorphic vector field v be determined by a one-parameter subgroup of the analytic automorphic group of $G_{III}(2n, n)$, where the one-parameter subgroup is chosen to be

$$\mathfrak{Z} \to \mathfrak{Z}e^{tN}, \qquad NJ + JN' = 0, \qquad J = \begin{pmatrix} 0 & I \\ I & 0 \end{pmatrix}. \tag{12.31}$$

For better determination, we choose N to be

$$N = \begin{pmatrix} \lambda_1 & & & & & \\ & \ddots & & & & \\ & & \lambda_n & & & \\ & & & -\lambda_1 & & \\ & & & & \ddots & \\ & & & & & -\lambda_n \end{pmatrix}, \qquad \lambda_i \neq 0, \quad \forall i.$$

By direct computation, the vector field determined by N is

$$v = \sum_{\substack{i \leq i \leq n \\ 1 \leq j \leq n}} \lambda_j Z_{ij} \frac{\partial}{\partial Z_{ij}} - \sum_{\substack{i \leq i \leq n \\ 1 \leq j \leq n}} \lambda_j Z_{i,n+j} \frac{\partial}{\partial Z_{i,n+j}}. \tag{12.32}$$

The $2n$-dimensional vector

$$e_i = (0, \ldots, \overset{i}{1}, 0, \ldots, 0), \qquad 1 \leq i \leq n,$$

corresponds to an eigenvector of λ_i, and

$$e_{n+i} = (0, \ldots, \overset{n+i}{1}, \ldots, 0)$$

corresponds to an eigenvector of $-\lambda_i$. Arbitrarily choose n vectors among $\{e_1, \ldots, e_{2n}\}$, $1 \le i_1 < \cdots < i_n \le 2n$, to form a point

$$\beta = \begin{pmatrix} e_{i_1} \\ \vdots \\ e_{i_n} \end{pmatrix}.$$

Then it is easy to verify that $\beta \in G_{\text{III}}(2n, n)$ if and only if no pairs of the subscripts $(1, n+1), \ldots, (n, 2n)$ appear. Following Section 12.2, such points can be indexed as follows: let $I = (i_1, \ldots, i_k) \subset (1, 2, \ldots, n)$, and suppose $I' = (j_1, \ldots, j_{n-k})$ is the complement of I. Then

$$Z = \left\{ e_{I,I'} = \begin{pmatrix} e_{i_1} \\ \vdots \\ e_{i_k} \\ e_{n+j_1} \\ \vdots \\ e_{n+j_{n-k}} \end{pmatrix} \right\}$$

constitutes the set of all zero points of v. Obviously, $\deg Z = 2^n$.

Z is divided into two parts: $Z = Z_1 \cup Z_2$,

$$Z_1 = \{ e_{I,I'} \mid |I'| = n - k \text{ is even} \},$$
$$Z_2 = \{ e_{I,I'} \mid |I'| = n - k \text{ is odd} \}, \tag{12.33}$$

and $\deg Z_1 = \deg Z_2 = \frac{1}{2} \deg Z$.

LEMMA 12.8. $Z_1 \subset M_1$, $Z_2 \subset M_2$.

PROOF. From the definition of M_1 and M_2, we need only show that for any $e_{I,I'} \in Z_1$, there exists $Q \in \Gamma_j^+(2n)$ such that $e_{I,I'}Q = (I, 0)$; and for any $e_{I,I'} \in Z_2$, there exists $Q \in \Gamma_j^-(2n)$ such that $e_{I,I'}Q = (I, 0)$.

Now choose arbitrarily $e_{I,I'} \in Z$, where $I = (i_1, \ldots, i_k)$, $I' = (j_1, \ldots, j_{n-k})$. Then clearly, there exists a permutation matrix p so as to transform $(i_1, \ldots, i_k, j_1, \ldots, j_{n-k})$ to $(1, \ldots, k, k+1, \ldots, n)$. Thus

$$e_{I,I'}\begin{pmatrix} p & 0 \\ 0 & p \end{pmatrix} = \begin{pmatrix} I & 0 & 0 & 0 \\ 0 & 0 & 0 & I \end{pmatrix} \begin{matrix} k \\ n-k \end{matrix}$$
$$\phantom{e_{I,I'}\begin{pmatrix} p & 0 \\ 0 & p \end{pmatrix} = } \begin{matrix} k & n-k & k & n-k \end{matrix}$$

and

$$\begin{pmatrix} p & 0 \\ 0 & p \end{pmatrix} \begin{pmatrix} 0 & I \\ I & 0 \end{pmatrix} \begin{pmatrix} p' & 0 \\ 0 & p' \end{pmatrix} = \begin{pmatrix} 0 & I \\ I & 0 \end{pmatrix},$$

which means $\begin{pmatrix} p & 0 \\ 0 & p \end{pmatrix} \in \Gamma_J(2n)$. Again choose Q to be

$$Q = \begin{pmatrix} I^{(k)} & 0 & 0 & 0 \\ 0 & 0 & 0 & I^{(k)} \\ 0 & 0 & I^{(k)} & 0 \\ 0 & I^{(n-k)} & 0 & 0 \end{pmatrix}.$$

We can easily check that

$$Q \begin{pmatrix} 0 & I \\ I & 0 \end{pmatrix} Q' = \begin{pmatrix} 0 & I \\ I & 0 \end{pmatrix}, \qquad \begin{pmatrix} I & 0 & 0 & 0 \\ 0 & 0 & 0 & I \end{pmatrix} Q = \begin{pmatrix} I & 0 & 0 & 0 \\ 0 & I & 0 & 0 \end{pmatrix}.$$

$$k \quad n-k \quad k \quad n-k$$

Therefore

$$e_{I,I'} \begin{pmatrix} p & 0 \\ 0 & p \end{pmatrix} Q = (I, 0) \qquad \text{and} \qquad \begin{pmatrix} p & 0 \\ 0 & p \end{pmatrix} Q \in \Gamma_J(2n),$$

$$n \quad n$$

$$\det \begin{pmatrix} p & 0 \\ 0 & p \end{pmatrix} Q = \det Q = \det \begin{pmatrix} 0 & I^{(n-k)} \\ I^{(n-k)} & 0 \end{pmatrix} = (-1)^{n-k}.$$

This completes the proof.   □

Now we compute (12.30). Let $I = (i_1, \ldots, i_k)$, and let $I' = (j_1, \ldots, j_{n-k})$ be the complement of I in $(1, \ldots, n)$. Suppose $U_{I,I'}$ is an open set of nonsingular submatrices of the $(i_1, \ldots, i_k, n+j_1, \ldots, n+j_{n-k})$th columns. Then $e_{I,I'} \in U_{I,I'}$ and corresponds to the zero point of the local coordinates in $U_{I,I'}$. (As usual, if $\mathfrak{Z} \in U_{I,I'}$, the $(i_1, \ldots, i_k, n+j_1, \ldots, n+j_{n-k})$th columns in $\mathfrak{Z}$ are written as $Z_{I,I'}$; the rest of the submatrices are written as $\check{Z}_{I,I'}$. Then the local coordinates of $U_{I,I'}$ can be chosen to be $w = Z_{I,I'}^{-1} \check{Z}_{I,I'}$.) By (12.32), we have

$$i(v)(dZ_{I,I'} \cdot Z_{I,I'}^{-1}) = (\mathfrak{Z}N)_{I,I'} \cdot Z_{I,I'}^{-1},$$

$$i(v)(dZ_{I,I'} \cdot Z_{I,I'}^{-1})|_{e_{I,I'}} = \begin{pmatrix} \lambda_{i_1} & & & & & & \\ & \ddots & & & & & \\ & & \lambda_{i_k} & & & & \\ & & & -\lambda_{j_1} & & & \\ & & & & \ddots & & \\ & & & & & -\lambda_{j_{n-k}} \end{pmatrix}.$$

As mentioned above, the one-parameter group of N is

$$e^{tN} = \begin{pmatrix} e^{t\lambda_1} & & & & & \\ & \ddots & & & & \\ & & e^{t\lambda_n} & & & \\ & & & e^{-t\lambda_1} & & \\ & & & & \ddots & \\ & & & & & e^{-t\lambda_n} \end{pmatrix} \in \Gamma_J^+(2n),$$

which maps $M_1 \rightarrow M_1$, $M_2 \rightarrow M_2$, and determines the vector field v on M_1 and M_2, respectively. In the following, we shall confine ourselves to M_1.

Assume that

$$H^0(M_1, \theta_{Z_1}) = F_{n(n-1)/2} \supset F_{n(n-1)/2-1} \supset \cdots \supset F_0 = C, \qquad F_i \cdot F_j \subseteq F_{i+j}$$

is the filtration of the Carrell–Liebermann theorem. Let $\sigma_j(x_1, \ldots, x_n)$ be the j-degree elementary symmetric homogeneous polynomial of n variables. Define a function $\tilde{\sigma}_j \in H^0(M_1, \theta_{Z_1})$,

$$\tilde{\sigma}_j(e_{I,I'}) = \sigma_j(\lambda_{i_1}, \ldots, \lambda_{i_k}, -\lambda_{j_1}, \ldots, -\lambda_{j_{n-k}}),$$

thus having $\tilde{\sigma}_j \in F_j/F_{j-1}$ $(j \leq n)$.

But here we should note that when $j = n$, it follows that

$$\tilde{\sigma}_n(e_{I,I'}) = \lambda_1 \cdots \lambda_n (-1)^{|I'|} = \lambda_1 \cdots \lambda_n \cdot 1(e_{I,I'}) \in F, \qquad (12.34)$$

where $1(Z_1)$ is taken identically as the constant function 1 on Z_1. By using (12.34) and applying Lemmas 12.3–12.5 to the rest, we finally get the following theorem.

THEOREM 12.9. *For* $G_{III}(2n, n) = M_1 \cup M_2$ *and under the isomorphism* ϕ *of the Carrell–Liebermann theorem,* $\{\varphi(\tilde{\sigma}_{i_1} \cdots \tilde{\sigma}_{i_k}) | i_1 + i_2 + \cdots + i_k = p, 1 \leq i_1 < \cdots < i_k \leq n - 1\}$ *furnishes a basis of* $H^{p,p}(M_1, C)$. *In particular, the Poincaré polynomial of* $G_{III}(2n, n)$ *is* $2(1 + t^2)(1 + t^4) \cdots (1 + t^{2n-2})$.

References

1. M. F. Atiyah, Complex analytic connections in fiber bundles, *Trans. Amer. Math. Soc.* **85**, 181–207 (1957).
2. J. Carrell, Chern classes of the Grassmannians and Schubert calculus, *Topology* **17** (2), 177–183 (1978).
3. J. Carrell and D. Lieberman, Holomorphic vector fields and Kähler manifolds, *Invent. Math.* **21**, 303–309 (1973).
4. J. Carrell and D. Lieberman, Vector fields and Chern numbers, *Math. Ann.* **225**, 263–273 (1977).
5. P. Griffiths and J. Harris, *Principles of Algebraic Geometry*, Wiley, New York (1978).
6. L. K. Hua, *Harmonic Analysis of Functions of Several Complex Variables in Classical Domains*, Vol. 6 of *Transl. Math. Monographs*, Amer. Math. Soc., Providence, RI (1963).
7. Lu Qi-Keng (K.-H. Look), *Classical Domains and Classical Manifolds*, Scientific Publishers, Shanghai (1963) (in Chinese).

13

The Degree of Strong Nondegeneracy of the Bisectional Curvature of Exceptional Bounded Symmetric Domains

In Refs. 3 and 4 Siu discovered the complex analyticity of harmonic maps between two Kähler manifolds under some conditions and proved the strong rigidity of compact quotients of irreducible bounded symmetric domains of dimension at least 2. Furthermore, he proposed the following conjecture.

Suppose $f: N \to M$ is a harmonic map between compact Kähler manifolds and M is a compact quotient of an irreducible symmetric domain. Let r be the maximal rank of df (over $\mathbb{R}$). Then f is either holomorphic or antiholomorphic provided that r is appropriately large.

In Ref. 5 Siu confirmed this conjecture for the four classical domains and indicated that the confirmation for the two exceptional domains depends on the computation of the degree of the strong nondegeneracy of the bisectional curvature in the two exceptional cases.

This chapter gives the computation of the degrees of strong nondegeneracy for the two exceptional cases needed for the confirmation of Siu's conjecture. The main result is the following.

THEOREM 13.1. *For the exceptional bounded symmetric domains D_V and D_{VI}, the degrees of strong nondegeneracy of bisectional curvature are 6 and 11, respectively.*

This work was finished during my visit at Stanford University in 1981. I am greatly indebted to Professor Siu for his help and encouragement during that time.

Let M be a compact Kähler manifold and $x \in M$.

DEFINITION 13.2 (*Siu* [5]). The bisectional curvature of M is said to be strongly s-nondegenerate at x if the following holds: If k and l are positive integers and if $\xi_{(1)}, \xi_{(2)}, \ldots, \xi_{(k)}$ (respectively, $\eta_{(1)}, \eta_{(2)}, \ldots, \eta_{(l)}$) are $\mathbb{C}$-linearly independent tangent vectors of type $(1, 0)$ at x, such that

$$R_{\alpha\bar{\beta}\gamma\bar{\delta}} \xi^{\alpha}_{(u)} \bar{\xi}^{\beta}_{(u)} \eta^{\gamma}_{(\nu)} \bar{\eta}^{\delta}_{(\nu)} = 0, \qquad \begin{array}{l} 1 \le u \le k, \\ 1 \le \nu \le l, \end{array}$$

then $k + l \le s$, where $R_{\alpha\bar{\beta}\gamma\bar{\delta}}$ is the curvature tensor of M.

We first state briefly the facts about the curvature tensor of Hermitian symmetric domains [2].

Let G be a real, simple, compact, and simply connected Lie group with trivial center and K the identity component of the fixed point of an involutive automorphism of G such that the center of K is one-dimensional. Let g, k be the Lie algebras of G, K, and $g_{\mathbb{C}}$, $k_{\mathbb{C}}$ their complexifications. Let $G_{\mathbb{C}}$ be the complex Lie group with Lie algebra $g_{\mathbb{C}}$. Let p be the orthogonal complement of k in g with respect to the Killing form of g. It is well known that $g = k + p$ is the Cartan decomposition of g. Let $g_0 = k + \sqrt{-1}p$, and let G_0 be the analytic subgroup of $G_{\mathbb{C}}$ corresponding to the $\mathbb{R}$-Lie algebra g_0. G_0 is a noncompact Lie group, and K is its maximal compact subgroup. From the theory of symmetric spaces we know that the quotient G/K is a simply connected irreducible compact Hermitian manifold, and its "dual" G_0/K is an irreducible bounded symmetric domain. Moreover, the curvature tensor of G_0/K is the curvature tensor of G/K with opposite sign:

$$\text{curvature of } G/K = (-1) \text{ curvature of } G_0/K.$$

Therefore instead of treating the case G_0/K we can deal with the curvature of G/K.

Let t be the Cartan subalgebra of k (and g) so that its complexification $t_{\mathbb{C}}$ is the Cartan subalgebra of $g_{\mathbb{C}}$. Let Δ denote the root system of g with respect to t. One can choose a partial order in Δ such that the center of k corresponds to a simple root.

Take the root space decomposition of $g_{\mathbb{C}}$:

$$g_{\mathbb{C}} = t_{\mathbb{C}} + \sum_{\varphi \in \Delta} g^{\varphi}.$$

We know that the center of k (which we denote by Z) is one-dimensional, and $J = \text{ad } Z$ is the complex structure of $M = G/K$.

For $\varphi \in \Delta$, let e_φ denote the root vector corresponding to φ. If $\varphi \in \Delta^+$, then $-\varphi \in \Delta^-$ and $e_{-\varphi} = \bar{e}_\varphi$. Take $M = G/K$. Let $0 = \{K\}$. Then we have

$$T_0 M = p, \qquad T_0^{(1,0)} M = \bigoplus_{\varphi \in \psi} \mathbb{C} e_\varphi,$$

where ψ is the set of noncompact roots. Recall that a root φ is a noncompact root iff $e_\varphi \in p_\mathbb{C}$, where $p_\mathbb{C}$ is the complexification of p. We know that $\forall \alpha, \beta \in \psi$, $[e_\alpha, e_\beta] = 0$.

Because $J = \mathrm{ad}\, Z|_{p_\mathbb{C}}$ is the complex structure, we have $J^2 = -1$. So $p_\mathbb{C}$ can be decomposed into eigenspace according to eigenvalues $\pm i$. Denote by $p^\pm$ the eigenspaces with respect to $\pm i$. Then we have

$$g_\mathbb{C} = k_\mathbb{C} + p^+ + p^-.$$

From the theory of symmetric spaces we know that p^+ is spanned by all noncompact positive root vectors and p^- is spanned by all noncompact negative root vectors, and the $\mathbb{R}$-basis of $T_0 M$ can be taken as $\{\mathrm{Re}\, e_\alpha, \mathrm{Im}\, e_\alpha \mid \alpha \in \psi\}$. Therefore $T_0^{(1,0)} M = \{\sum_{\alpha \in \psi} \xi_\alpha e_\alpha \mid \xi_\alpha \in \mathbb{C}\}$.

The Killing form of g is negative definite since G is compact. The negative of the Killing form gives the invariant metric on M, which we denote by $\langle \cdot, \cdot \rangle_\mathbb{R}$. Extend it by $\mathbb{C}$-bilinearity to $g_\mathbb{C}$ and denote this extension also by $\langle \cdot, \cdot \rangle_\mathbb{R}$. Then the Hermitian metric on $T_0^{(1,0)} M$ is given by

$$\langle e_\alpha, e_\beta \rangle = \langle e_\alpha, \bar{e}_\beta \rangle_\mathbb{R}, \qquad \alpha, \beta \in \psi.$$

The curvature tensor for the metric $\langle \cdot, \cdot \rangle_\mathbb{R}$ on M is

$$\langle R(X, Y)Y, X \rangle = \| [X, Y] \|^2, \qquad X, Y \in T_0 M \cong p,$$

where $\| \cdot \|^2 = \langle \cdot, \cdot \rangle_\mathbb{R}$ (see Ref. 1, p. 76).

Take $\xi^\alpha \in \mathbb{C}$. Let $X = 2 \mathrm{Re} \sum \xi^\alpha e_\alpha$, $JX = 2 \mathrm{Im} \sum \xi^\alpha e_\alpha$. Then it is easy to verify that

$$\sum R_{\alpha \bar{\beta} \gamma \bar{\delta}} \xi^\alpha \bar{\xi}^\beta \xi^\gamma \bar{\xi}^\delta = -\| [\sum \xi^\alpha e_\alpha, \overline{\sum \xi^\alpha e_\alpha}] \|^2,$$

$$R_{\alpha \bar{\beta} \gamma \bar{\delta}} = -\langle [e_\alpha, e_{-\beta}], [e_\delta, e_{-\gamma}] \rangle,$$

and the bisectional curvature for $\sum \xi^\alpha e_\alpha, \sum \eta^\alpha e_\alpha$ is given by

$$\sum R_{\alpha \bar{\beta} \gamma \bar{\delta}} \xi^\alpha \bar{\xi}^\beta \eta^\gamma \bar{\eta}^\delta = -\| [\sum \xi^\alpha e_\alpha, \overline{\sum \eta^\beta e_\beta}] \|^2.$$

From the theory of symmetric spaces we know that one can find a maximal strongly orthogonal basis Λ in the set ψ of noncompact positive roots so that $\alpha, \beta \in \Lambda \subset \psi \Leftrightarrow \alpha \pm \beta \notin \Delta$. If $\alpha, \beta \in \Lambda$, it follows that $[e_\alpha, e_{\pm\beta}] = 0$ and $\langle \alpha, \beta \rangle = 0$. The subalgebra $\sum_{\varphi \in \Lambda} \mathbb{C} e_\varphi$ is in fact a maximal abelian Lie subalgebra in $p_\mathbb{C}$. The dimension of $\sum_{\varphi \in \Lambda} \mathbb{C} e_\varphi$ is called the rank of symmetric space M.

Set $a^+ = \sum_{\varphi \in \Lambda} \mathbb{R} e_\varphi$ and $a = \sum_{\varphi \in \Lambda} \mathbb{R}(e_\varphi + e_{-\varphi})$. a^+ is a real abelian Lie algebra.

PROPOSITION 13.3.

$$p^+ = (I - iJ)p, \qquad a^+ = (I - iJ)a. \tag{13.1}$$

PROOF. For any $\varphi \in \Lambda$,

$$(I - iJ)e_\varphi = e_\varphi - iJe_\varphi = e_\varphi - i(ie_\varphi) = 2e_\varphi,$$

$$(I - iJ)e_{-\varphi} = e_{-\varphi} - iJe_{-\varphi} = e_{-\varphi} - i(-ie_\varphi) = 0,$$

$$(I - iJ)(e_\varphi + e_{-\varphi}) = 2e_\varphi. \qquad \square$$

PROPOSITION 13.4.

$$p^+ = \bigcup_{k \in K} \text{Ad}(k)a^+. \tag{13.2}$$

PROOF. First we prove $p = \bigcup_{k \in K} \text{Ad}(k)a$. Since Ad k is the adjoint representation and its Lie algebra is ad k (where k is the Lie algebra of K), it suffices to prove $[k, a] = p$. It is trivial to see that $[k, a] \subset [k, p] \subset p$. If $[k, a] \subsetneq p$, then there exists $x \neq 0$, $x \in p$, such that $\langle [k, a], x \rangle = 0$. But since a is a maximal strongly orthogonal set in p, $0 = \langle [k, a], x \rangle = -\langle a, [k, x] \rangle$ implies $[k, x] = 0$ and $x = 0$, which is a contradiction. The proposition now follows from

$$p^+ = (I - iJ)p = (I - iJ) \bigcup_{k \in K} \text{Ad}(k)a$$

$$= \bigcup_{k \in K} \text{Ad}\, k(I - iJ)a = \bigcup_{k \in K} \text{Ad}(k)a^+,$$

where we have used Proposition 13.3 and the fact that Ad k and J commute.

Now we are in a position to prove the main theorem. Because the process of proof is similar for D_V and D_{VI}, for simplicity we give the proof of the theorem in detail for D_V and only sketch it for D_{VI}.

It is well known that the dual compact Hermitian symmetric manifold of D_V is $M = E_6/\mathrm{Spin}(10) \times T^1$. Let x_i $(i = 1, \ldots, 6)$ denote the coordinates of $\mathbb{R}$. We regard the root systems of E_6 as the linear forms on $\mathbb{R}^6$. We list the simple roots, the positive roots Δ^+, the positive noncompact roots ψ, the set of maximal strongly orthogonal roots Λ, and the rank as follows:

simple roots: $x_i - x_{i+1}$ $(1 \le i \le 5)$, $x_4 + x_5 + x_6$.

$$\Delta^+: x_i - x_j \ (1 \le i < j \le 6), \quad x_i + x_j + x_k \ (1 \le i < j < k \le 6).$$

$$\sum_1^6 x_i$$

$$\psi: x_1 - x_i \ (2 \le i \le 6), \quad x_1 + x_i + x_j \ (2 \le i < j \le 6), \quad \sum_1^6 x_i.$$

$$\Lambda: x_1 - x_2, \quad x_1 + x_2 + x_3.$$

rank: 2.

In view of the above discussions, the theorem we want to prove can be stated as follows.

THEOREM 13.5. *For* $M = E_6/\mathrm{Spin}(10) \times T^1$, *if* $\xi_1, \ldots, \xi_k \in p^+$, $\eta_1, \ldots, \eta_l \in p^-$, ξ_i *(respectively,* η_j *) are* $\mathbb{C}$-*independent such that*

$$[\xi_i, \eta_j] = 0, \qquad \forall 1 \le i \le k, 1 \le j \le l,$$

then $k + l \le 6$.

The proof is a consequence of the following steps.

LEMMA 13.6. *If* $\xi \ne 0$, $\eta \ne 0$, $\xi \in p^+$, $\eta \in p^-$ *such that* $[\xi, \eta] = 0$, *then* ξ *can be either transformed into* $\{e_{x_1-x_2}\}$ *or* $\{e_{x_1+x_2+x_3}\}$ *by* $\mathrm{ad}\, K$, *but it can never be transformed into* $ae_{x_1-x_2} + be_{x_1+x_2+x_3}$, $a, b \ne 0$.

PROOF. By Proposition 13.4 it suffices to prove that, if $a, b \ne 0$, $[ae_{x_1-x_2} + be_{x_1+x_2+x_3}, \eta] = 0$, $\eta \in p^-$, then $\eta = 0$. Set

$$\eta = \sum_{i=2}^6 a_i e_{-(x_1-x_i)} + \sum_{2 \le i < j \le 6} b_{ij} e_{-(x_1+x_i+x_j)} + c e_{-\sum_1^6 x_i}.$$

$$0 = [ae_{x_1-x_2} + be_{x_1+x_2+x_3}, \eta]$$

$$= \sum_{i \ge 2} aa_i \{e_{x_i-x_2}\} + \sum_{2 \le i < j \le 6} ab_{ij} \{e_{-(x_2+x_i+x_j)}\} + \sum_{2 \le i < j \le 6} bb_{ij} \{e_{x_2+x_3-x_i-x_j}\}$$

$$+ \sum_{i \ge 2} ba_{ij} \{e_{x_2+x_3+x_i}\} + bc \{e_{-(x_4+x_5+x_6)}\}. \tag{13.3}$$

So $c = 0$ and $a_i = 0$ $(i = 4, 5, 6)$. From the term $\sum ab_{ij}\{e_{-(x_2+x_i+x_j)}\}$, we have $b_{ij} = 0$ $(j > i > 2)$. It follows that

$$\eta = a_2 e_{-(x_1-x_2)} + a_3 e_{-(x_1-x_3)} + \sum_{j \geq 3}^{6} b_{2j} e_{-(x_1+x_2+x_j)}.$$

Thus (13.3) can be written as

$$0 = aa_2(H_{x_1-x_2}) + aa_3\{e_{-(x_2-x_3)}\} + bb_{23}(H_{x_1+x_2+x_3}) + \sum_{j \geq 4}^{6} bb_{2j}\{e_{x_3-x_j}\},$$

from which we conclude that $a_3 = b_{2j} = 0$ $(4 \leq j \leq 6)$. Furthermore, $H_{x_1-x_2}$ and $H_{x_1+x_2+x_3}$ are independent, because $x_1 - x_2$ is strongly orthogonal to $x_1 + x_2 + x_3$. Hence $a_2 = b_{23} = 0$, $\eta = 0$, and the lemma is proved. $\qquad \square$

For the subspace A of p^+, we set

$$A^{\perp} = \{\eta \mid \eta \in p^-, [\xi, \eta] = 0, \qquad \forall \xi \in A\}.$$

LEMMA 13.7.

$$(e_{x_1-x_2})^{\perp} = \{e_{-(x_1+x_2+x_i)}, \qquad i = 3, 4, 5, 6; e_{-\sum x_i}\}. \qquad (13.4)$$

$$\{(e_{x_1-x_2})^{\perp}\}^{\perp} = \{e_{x_1-x_2}\}. \qquad (13.5)$$

$$(e_{x_1+x_2+x_3})^{\perp} = \{e_{-(x_1-x_2)}, e_{-(x_1-x_3)}, e_{-(x_1+x_4+x_5)}, e_{-(x_1+x_4+x_6)}, e_{-(x_1+x_5+x_6)}\}. \qquad (13.6)$$

$$\{(e_{x_1+x_2+x_3})^{\perp}\}^{\perp} = \{e_{x_1+x_2+x_3}\}. \qquad (13.7)$$

PROOF. The verification of (13.4) and (13.6) is a direct and easy computation. Formula (13.5) follows from

$$\{e_{-(x_1+x_2+x_i)}, 3 \leq i \leq 6, e_{-\sum x_j}\}^{\perp} = \bigcap_{i=3}^{6} (e_{-(x_1+x_2+x_i)})^{\perp} \cap (e_{-\sum x_j})^{\perp}$$

and

$$\{e_{-(x_1+x_2+x_i)}\}^{\perp} = \{e_{x_1-x_2}, e_{x_1-x_i}, e_{x_1+x_k+x_l}(k, l \neq 1, 2, i)\} \qquad (3 \leq i \leq 6),$$

$$\{e_{-\sum x_j}\}^{\perp} = \{e_{x_1-x_2}, e_{x_1-x_i} \ (i = 3, 4, 5, 6)\}.$$

The proof of (13.7) is similar.   □

COROLLARY 13.8. *For any $\xi \in p^+$, $\xi \neq 0$, $\dim \{\xi\}^\perp \leq 5$.*

PROOF: By Proposition 13.4, $\exists k \in K$ such that $(\mathrm{ad}\, k)\xi \in a^+$. If $\mathrm{ad}\, k \xi = c e_{x_1-x_2} + d e_{x_1+x_2+x_3}$, $c, d \neq 0$, then $[(\mathrm{ad}\, k)\xi]^\perp = 0$, and it follows from Lemma 13.7 that either $\{\xi\}^\perp = 0$ or $\dim \{\xi\}^\perp = 5$.

LEMMA 13.9. *If ξ_1, ξ_2 are independent vectors in p^+, then $\dim \{\xi_1, \xi_2\}^\perp \leq 4$.*

PROOF. By Proposition 13.4, without loss of generality we can assume $\xi_1 = \mathrm{Ad}\, k\, e_{x_1-x_2}$. Set $F = \{e_{x_1-x_2}\}^\perp$. Because $[\mathrm{Ad}\, k\, \xi, \mathrm{Ad}\, k\, \eta] = \mathrm{Ad}\, k\, [\xi, \eta]$, we have $\mathrm{Ad}\, k^{-1}\{\xi\}^\perp = \{\mathrm{Ad}\, k^{-1}\xi\}^\perp$. So

$$\{\xi_1\}^\perp = \{\mathrm{Ad}\, k\, e_{x_1-x_2}\}^\perp = \mathrm{Ad}\, k\{e_{x_1-x_2}\}^\perp = \mathrm{Ad}\, k\, F.$$

But it is clear that $\{\xi_1, \xi_2\}^\perp \subseteq \{\xi_1\}^\perp = \mathrm{Ad}\, k\, F$. If $\{\xi_1, \xi_2\}^\perp = \mathrm{Ad}\, k\, F$, then we have

$$[\xi_1, \mathrm{Ad}\, k\, F] = [\xi_2, \mathrm{Ad}\, k\, F] = 0,$$

and $\mathrm{Ad}\, k^{-1}\xi_i \in F^\perp$ $(i = 1, 2)$. By Lemma 13.7, $\dim F = 1$. So $\mathrm{Ad}\, k^{-1}\xi_1 = \lambda \, \mathrm{Ad}\, k^{-1}\xi_2$, $\xi_1 = \lambda \xi_2$, contradicting the assumptions. Hence, $\{\xi_1, \xi_2\}^\perp \subsetneq \mathrm{Ad}\, k\, F$, and it follows that

$$\dim \{\xi_1, \xi_2\}^\perp < \dim \mathrm{Ad}\, k\, F = \dim F = 5.\qquad □$$

LEMMA 13.10. *If ξ_1, ξ_2, ξ_3 are independent vectors in p^+, then $\dim \{\xi_1, \xi_2, \xi_3\}^\perp \leq 3$.*

PROOF. Clearly $\{\xi_1, \xi_2, \xi_3\}^\perp = \{\xi_1\}^\perp \cap \{\xi_2\}^\perp \cap \{\xi_3\}^\perp$. From Lemma 13.9, if $\xi_1 \notin \{\xi_2\}$, then $\{\xi_1\}^\perp \neq \{\xi_2\}^\perp$. So $\dim \{\xi\}^\perp \cap \{\xi\}^\perp \leq 4$. Furthermore, from $\xi_3 \notin \{\xi_1, \xi_2\}$ it follows that $\{\xi_1\}^\perp \cap \{\xi_2\}^\perp \not\subset \{\xi_3\}^\perp$. Therefore $\dim \{\xi_1, \xi_2, \xi_3\}^\perp = \dim \{\xi_1\}^\perp \cap \{\xi_2\}^\perp \cap \{\xi_3\}^\perp \leq 3$.   □

PROOF OF THEOREM 13.5. From $[\xi_i, \eta_j] = 0$, $1 \leq i \leq k$, $1 \leq j \leq l$, and $k \geq 1$, it follows that $l \leq 5$ (Lemma 13.7). Because of the symmetry of the ξ_i, η_j, the only cases needed to be verified are $k = 1$, $l \leq 5$ (or $l = 1$, $k \leq 5$); $k = 2$, $l \leq 4$ (or $l = 2$, $k \leq 4$); $k = 3$, $l \leq 3$ (or $l = 3$, $k \leq 3$).

For $M = E_7/E_6 \times T^1$ (the compact dual of D_{VI}), we have the following list.

Simple roots: $\quad x_i - x_{i+1}$ $(1 \le i \le 6)$, $\quad x_5 + x_6 + x_7$,

Δ^+: $\quad x_i - x_j$ $(1 \le i < j \le 7)$, $\quad x_i + x_j + x_k$ $(1 \le i < j < k \le 7)$,

$$d - x_i \ (1 \le i \le 7), \quad \text{where } d = \sum_{i=1}^{7} x_i,$$

ψ: $\quad x_i - x_i$ $(2 \le i \le 7)$, $\quad x_1 + x_i + x_j$ $(2 \le i < j \le 7)$,

$$d - x_i \ (2 \le i \le 7),$$

Λ: $\quad x_1 - x_2$, $\quad x_1 + x_2 + x_3$, $\quad d - x_3$,

rank: $\quad$ 3.

By direction computation we have the next lemma.

LEMMA 13.11.

$$\dim \{e_{x_1-x_2}\}^\perp = \dim \{e_{x_1+x_2+x_3}\}^\perp = \dim \{e_{d-x_3}\}^\perp = 10. \tag{13.8}$$

For example,

$$\{e_{x_1-x_2}\}^\perp = \{e_{-(d-x_i)} \ (3 \le i \le 7), \ e_{-(x_1+x_2+x_j)} \ (3 \le j \le 7)\}.$$

LEMMA 13.12. *For any* $\xi \in p^+$, $\dim \{\xi\}^\perp \le 10$.

PROOF. Without loss of generality we can assume

$$\xi = ae_{x_1-x_2} + be_{x_1+x_2+x_3} + ce_{d-x_3}.$$

Take

$$\eta = \sum_{i \ge 2}^{7} a_i e_{-(x_1-x_i)} + \sum_{2 \le i < j \le 7} b_{ij} e_{-(x_1+x_i+x_j)} + \sum_{i \ge 2}^{7} c_i e_{-(d-x_i)}.$$

From $[\xi, \eta] = 0$ we get

$$cc_i = 0 \ (i = 2, 4, 5, 6, 7), \qquad ca_i = 0 \ (2 \le i \le 7), \qquad cb_{ij} = 0 \ (j > i \ge 4),$$

$$bb_{2j} = 0 \ (j \ge 4), \qquad bb_{3j} = 0 \ (j \ge 4), \qquad ba_i = 0 \ (i \ge 4),$$

$$bc_j = 0 \ (j \ge 4), \qquad ab_{ij} = 0 \ (j > i \ge 4), \tag{13.9}$$

$$bb_{23} + cc_3 = 0.$$

From (13.9) we conclude the following.

If $a, c \neq 0$, then dim $\{\xi\}^\perp \leq 5$,

$a, b \neq 0$ then dim $\{\xi\}^\perp \leq 4$,

$b, c \neq 0$, then dim $\{\xi\}^\perp \leq 3$.

In view of Lemma 13.11, we have in general dim $\{\xi\}^\perp \leq 10$. $\Box$

LEMMA 13.13. *If $\xi_1, \xi_2 \in p^+$, then* dim $\{\xi_1, \xi_2\}^\perp \leq 9$.

PROOF. Transform ξ_1 (or ξ_2) by Ad k into $ae_{x_1-x_2} + be_{x_1+x_2+x_3} + ce_{d-x_3}$ in the case when two of the $\{a, b, c\} \neq 0$. From Lemma 13.12 and $\{\xi_1, \xi_2\}^\perp \subset \{\xi_1\}^\perp$, we have

$$\dim \{\xi_1, \xi_2\}^\perp \leq \dim \{\xi_1\}^\perp \leq 5.$$

For the other cases, by the same reasoning as in the proof of Lemma 13.9 we have dim $\{\xi_1, \xi_2\}^\perp \leq 9$. $\Box$

The remainder of the proof of the theorem for D_{VI} is similar to that of Theorem 13.5.

References

1. J. Cheeger and D. Ebin, *Comparison Theorems in Riemannian Geometry*, North-Holland, Amsterdam (1975).
2. S. Helgason, *Differential Geometry and Symmetric Spaces*, Academic Press, New York (1962).
3. Yum-Tong Siu, The complex-analyticity of harmonic maps and the strong rigidity of compact Kahler manifolds, *Ann. of Math.* **112**, 73–111 (1980).
4. Yum-Tong Siu, Strong rigidity of compact quotients of exceptional domains, *Duke Math. J.* **48**, 857–871 (1981).
5. Yum-Tong Siu, Complex-analyticity of harmonic maps, vanishing and Lefschetz theorems, *J. Diff. Geom.* **17**, 55–138 (1982).

14

The Estimate of the First Eigenvalue of a Compact Riemannian Manifold

Abstract

The main theorem proved in this chapter is: Let M be a compact Riemannian manifold with nonnegative Ricci curvature. Then the first eigenvalue $-\lambda_1$ of the Laplace operator of M satisfies $\lambda_1 \geq \pi^2/d^2$, where d denotes the diameter of M. This estimate improves the recent results due to S. T. Yau and P. Li [1, 2] and gives the best estimate for this kind of manifold.

14.1. Introduction

There have been many works discussing the estimates of the eigenvalues of compact Riemannian manifolds. In this field, the latest results are due to P. Li and S. T. Yau [1, 2]. One of the main results in Ref. 2 is the following.

THEOREM 14.1 (*P. Li and S. T. Yau*). *Let M be a compact Riemannian manifold with nonnegative Ricci curvature. Then the first eigenvalue $-\lambda_1$ of the Laplace operator of M satisfies*

$$\lambda_1 \geq \frac{1}{4}\frac{\pi^2}{d^2}, \tag{14.1}$$

where d denotes the diameter of M.

Joint work with Yang Hong-Cang.

In Ref. 1 the above result was improved to

$$\lambda_1 \geq \frac{1}{2} \frac{\pi^2}{d^2}. \tag{14.2}$$

It is reasonable to think that the best estimate of the lower bound of λ_1 should be

$$\lambda_1 \geq \frac{\pi^2}{d^2}. \tag{14.3}$$

This chapter aims at proving this conjecture, the main result of which is the following.

THEOREM 14.2. *Let M be a compact Riemannian manifold with nonnegative Ricci curvature and let d be the diameter of M. Then,*

$$\lambda_1 \geq \frac{\pi^2}{d^2}. \tag{14.3'}$$

14.2. Notations and Formulas

Let M be a smooth orientable compact Riemannian manifold of dimension m. Take a local orthonormal frame field $\{e_i\}$ and denote its coframe fields as $\{\theta^i\}$. The Riemann metric of M is $ds^2 = \sum_i (\theta^i)^2$. It is well known that there exists a Riemann connection $\{\theta^i_j\}$ and the structure equations:

$$d\theta^i + \Sigma \theta^i_j \wedge \theta^j = 0,$$

$$\theta^i_j + \theta^j_i = 0,$$

$$d\theta^i_j + \Sigma \theta^i_k \wedge \theta^k_j = \tfrac{1}{2}\Sigma R^i_{jkl}\theta^k \wedge \theta^l, \tag{14.4}$$

$$R^i_{jkl} = R_{ijkl} = -R_{jikl} = -R_{ijlk},$$

where R^i_{jkl} denotes the Riemann curvature tensor.

Suppose $u: M \to \mathbf{R}$ is a differentiable function. Define the covariant differentials u_i, u_{ij}, u_{ijk} successively by

$$Du = du \equiv \Sigma u_i \theta^i,$$

$$Du_i = du_i - \Sigma u_j \theta^j_i \equiv \Sigma u_{ij}\theta^j, \tag{14.5}$$

$$Du_{ij} = du_{ij} - \Sigma u_{ik}\theta^k_j - \Sigma u_{kj}\theta^k_i \equiv \Sigma u_{ijk}\theta^k.$$

Then, from the structure equations (14.4) it is easy to deduce

$$u_{ij} = u_{ji}, \qquad u_{ijk} - u_{ikj} = \Sigma u_l R^l_{ijk}. \tag{14.6}$$

Equation (14.6) is the Bianchi identity.

The Laplace operator Δ on M is defined by

$$\Delta u = \sum_i u_{ii}. \tag{14.7}$$

Now suppose that $u: M \to \mathbf{R}$ is a normal eigenfunction which corresponds to the first eigenvalue $-\lambda_1$ of Δ. Here, "normal eigenfunction" means

$$\Delta u = -\lambda_1 u,$$

$$\max u = 1, \tag{14.8}$$

$$\min u = -k, \qquad 0 < k \leq 1.$$

Set

$$\tilde{u} = \left(u - \frac{1-k}{2}\right) \Big/ \frac{1+k}{2},$$

$$\tag{14.9}$$

$$a = \frac{1-k}{1+k}, \qquad 0 \leq a < 1.$$

Then (14.8) can be rewritten as

$$\Delta \tilde{u} = -\lambda_1(\tilde{u} + a),$$

$$\max \tilde{u} = 1, \tag{14.10}$$

$$\min \tilde{u} = -1.$$

For simplicity, we will still use u to denote $\tilde{u}$ below. Consider the function $f: M \to \mathbf{R}$,

$$f(x) = \frac{|\nabla u(x)|^2}{1 - u^2} = \frac{\Sigma u_i^2(x)}{1 - u^2}. \tag{14.11}$$

Note. Strictly speaking, we use $u/(1 + \varepsilon)$ in place of u in (14.11), where $\varepsilon > 0$ is a sufficient small constant. At the end, we let $\varepsilon \to 0$ to get our conclusions.

Set $u = \sin \theta$, $-\pi/2 \le \theta \le \pi/2$; then $\theta: M \to \mathbf{R}$ is a differentiable function and

$$f(x) = |\nabla \theta(x)|^2. \tag{14.12}$$

Define $F: (-\pi/2, \pi/2) \to \mathbf{R}$ by

$$F(\theta_0) = \max_{x \in M, \theta(x) = \theta_0} f(x) \qquad \forall \theta_0 \in \left(-\frac{\pi}{2}, \frac{\pi}{2}\right). \tag{14.13}$$

It is not difficult to verify, in view of (14.12) that the function $F(\theta)$ is continuous in $(-\pi/2, \pi/2)$. Moreover, because of max $u = 1$, min $u = -1$, and the remark below (14.11), we can consider

$$F\left(-\frac{\pi}{2}\right) = F\left(-\frac{\pi}{2} + 0\right) = 0,$$

$$F\left(\frac{\pi}{2}\right) = F\left(\frac{\pi}{2} + 0\right) = 0.$$

So, $F(\theta)$ can be viewed as a continuous function on $[-\pi/2, \pi/2]$. The following fact is obviously true: $\forall \theta_0 \in (-\pi/2, \pi/2)$, $\exists x_0 \in M$ s.t. $\theta(x_0) = \theta_0, f(x_0) = F(\theta_0)$.

14.3. The Estimation of $F(\theta)$

The estimation of the upper bound of $F(\theta)$ plays an important role in the estimation of the lower bound for λ_1. To explain this, we first establish a rough estimation for $F(\theta)$.

LEMMA 14.3. *The following estimate holds*:

$$F(\theta) \le \lambda_1(1 + a). \tag{14.14}$$

PROOF. Assume that the function $F(\theta)$ attains its maximum at θ_0. So, there exists $x_0 \in M$ such that $\theta(x_0) = \theta_0$, $f(x_0) = F(\theta_0)$. Let $\phi: M \to \mathbf{R}$ be the function

$$\phi(x) = [f(x) - F(\theta_0)] \cos^2 \theta(x).$$

Then ϕ attains a maximum at x_0. Applying the maximum principle to the differentiable function ϕ, we know that, at point x_0,

$$\nabla \phi = 0, \qquad \Delta \phi \le 0, \tag{14.15}$$

i.e.,

$$2 \sum_i u_i u_{ij} - F(\theta_0) \cos \theta_0 (-2 \sin \theta_0) \theta_j|_{x_0} = 0,$$

$$2 \sum_{i,j} u_{ij}^2 + 2 \sum_{i,j} u_i u_{ijj} - F(\theta_0)\Delta \cos \theta|_{x_0} \leq 0. \tag{14.16}$$

Since $u = \sin \theta$, we have (note formula (14.10)),

$$u_j = \cos \theta \cdot \theta_j,$$

$$\Delta \theta = \frac{\Delta u}{\cos \theta} + \frac{\sin \theta}{\cos^2 \theta} \nabla u \cdot \nabla \theta = \frac{-\lambda(\sin \theta + a)}{\cos \theta} + \frac{\sin \theta}{\cos \theta} |\nabla \theta|^2, \tag{14.17}$$

$$\Delta \cos^2 \theta = 2(\sin^2 \theta - \cos^2 \theta)|\nabla \theta|^2 - 2 \sin \theta \cos \theta \cdot \Delta \theta$$

$$= 2\lambda_1 \sin \theta(\sin \theta + a) - 2|\Delta \theta|^2 \cos^2 \theta. \tag{14.17'}$$

On the other hand, we have

$$\sum_{i,j} u_{ij}^2 \geq \left(\sum_{i,j} u_{ij} u_u u_j \right)^2 \Big/ (\Sigma u_i^2)^2. \tag{14.18}$$

From (14.6),

$$\sum_{i,j} u_i u_{ijj} = \sum_i u_i \sum_j u_{jij}$$

$$= \sum_i u_i \sum_j \left(u_{jji} + \sum_l R_{jij}^l u_l \right)$$

$$= \sum_i u_i \left(-\lambda_1 u_i + \sum_l R_{li} u_l \right)$$

$$= [-\lambda_1 + R(\nabla u)]|\nabla \theta|^2 \cos^2 \theta, \tag{14.19}$$

where $r_{li} = \sum_j R_{lij}$ and $R(X)$ denotes the Ricci curvature corresponding to vector X. When the Ricci curvature of M is nonnegative, we obtain

$$\sum_{i,j} u_i u_{ijj} \geq -\lambda_1 \cos^2 \theta|\nabla \theta|^2. \tag{14.20}$$

By virtue of (14.17′), (14.18), and (14.20), it is easy to deduce from (14.16) that

$$2 \sin^2 \theta_0 \cdot F^2(\theta_0) - 2\lambda_1 \cos^2 \theta_0 \cdot F(\theta_0)$$

$$- 2F(\theta_0)[\lambda_1 \sin \theta_0(\sin \theta_0 + a) - F(\theta_0) \cos^2 \theta_0] \le 0.$$

By straightening out the above expression, we obtain

$$2F^2(\theta_0) - 2\lambda_1(1 + a \sin \theta_0)F(\theta_0) \le 0,$$

$$F(\theta_0) \le \lambda_1(1 + a \sin \theta_0) \le \lambda_1(1 + a).$$

The proof is complete. □

Applying Lemma 14.3, we obtain a simple estimate of λ_1 as follows.

LEMMA 14.4. *If M is a compact Riemannian manifold, and its Ricci curvature is nonnegative with diameter d, then*

$$\lambda_1 \ge \frac{1}{1+a} \frac{\pi^2}{d^2}. \tag{14.21}$$

PROOF. Take $x_1, x_2 \in M$ such that $\theta(x_1) = \pi/2$, $\theta(x_2) = -\pi/2$. The length d' of the shortest curve r on M connecting x_1 and x_2 is obviously less than d (i.e., $d' \le d$). From (14.14),

$$\lambda_1^{1/2} \ge \frac{1}{\sqrt{1+a}} |\nabla \theta|.$$

Hence,

$$\int_r \lambda_1^{1/2} \, ds \ge \frac{1}{\sqrt{1+a}} \int_r |\nabla \theta| \, ds$$

$$\ge \frac{1}{\sqrt{1+a}} \int_r d\theta$$

$$= \frac{1}{\sqrt{1+a}} \int_{-\pi/2}^{\pi/2} d\theta = \frac{\pi}{\sqrt{1+a}}.$$

This completes the proof. □

Since $a < 1$, the inequality (14.21) is equivalent to or stronger in some sense than the result (14.2), cited in the introduction. From (14.21) it follows that if we can prove $a = 0$ (or equivalently, $k = 1$, i.e., max $u = -$min u, which we call "symmetric"), we shall obtain $\lambda_1 \geq \pi^2/d^2$. But we do not know whether all eigenfunctions are symmetric. Even if they are symmetric, perhaps it is more difficult to prove their symmetry than to prove (14.3). So, below we will prove (14.3) under the assumption $a > 0$.

Introduce the function $\varphi(\theta)$ such that

$$F(\theta) \equiv \lambda_1(1 + a\varphi(\theta)). \tag{14.22}$$

By Lemma 14.3, we have $\varphi \leq 1$. We are going to estimate $\varphi(\theta)$. First of all, we prove the next lemma.

LEMMA 14.5. *Assume that a C^2 function $y: [-\pi/2, \pi/2] \to \mathbf{R}$ satisfies*

1. $y(\theta) \geq \varphi(\theta)$.
2. *There exists some $\theta_0 \in (-\pi/2, \pi/2)$, $y(\theta_0) = \varphi(\theta_0) \geq -1$.*
3. $y'(\theta_0) \geq 0$.

Then the following estimate holds at θ_0:

$$\varphi(\theta_0) \leq \sin \theta_0 - \sin \theta_0 \cos \theta_0 \cdot y'(\theta_0) + \tfrac{1}{2} \cos^2 \theta_0 \cdot y''(\theta_0). \tag{14.23}$$

PROOF. Take $x_0 \in M$ such that $\theta(x_0) = \theta_0, f(x_0) = F(\theta_0)$. Consider the function $\Phi: M \to \mathbf{R}$:

$$\Phi(x) = [f(x) - \lambda_1(1 + ay(\theta(x)))] \cos^2 \theta(x).$$

It is easy to verify that $\Phi(x)$ attains its maximum at x_0. By the maximum principle (14.15), we obtain at x_0,

$$2 \sum_i u_i u_{ij} = \lambda_1[(1 + ay)(-2 \cos \theta \sin \theta) + ay' \cos^2 \theta]\theta_j,$$

$$2 \sum_{i,j} u_{ij}^2 + 2\Sigma u_i u_{ijj} - \lambda_1 a(y''|\nabla \theta|^2 + y'\Delta\theta) \cos \theta$$

$$- 2\lambda_1 ay'(-2 \cos \theta \sin \theta)|\nabla \theta|^2 - \lambda_1(1 + ay)\Delta \cos^2 \theta \leq 0. \tag{14.24}$$

By (14.17'), (14.17), (14.18), (14.20), it follows from (14.24) at x_0 that

$$\tfrac{1}{2}\lambda_1^2[(1 + ay)(-2 \sin \theta) + ay' \cos \theta]^2 - 2\lambda_1^2 \cos^2 \theta(1 + ay)$$

$$- \lambda_1^2 a(1 + ay)y'' \cos^2 \theta - \lambda_1^2 a[-\cos \theta(\sin \theta + a)$$

$$+ \sin \theta \cos \theta(1 + ay)]y' + 4\lambda_1^2 a \cos \theta \sin \theta(1 + ay)y'$$

$$- 2\lambda_1^2(1 + ay)[\sin \theta(\sin \theta + a) - (1 + ay) \cos^2 \theta] \le 0. \qquad (14.25)$$

Rearranging and eliminating $\lambda_1^2 a$, we have

$$\frac{1}{a} 2(1 + ay)[1 + ay - (1 + a \sin \theta)] + \frac{1}{2} a \cos^2 \theta \cdot y'^2$$

$$+ \cos \theta \sin \theta(1 + ay)y' + \cos \theta(\sin \theta + a)y'$$

$$- (1 + ay)y'' \cos^2 \theta \le 0. \qquad (14.26)$$

Hence, we have at x_0,

$$y - \sin \theta \le -\left(\frac{1}{2} \cos \theta \sin \theta + \frac{1}{2} \frac{a + \sin \theta}{1 + ay} \cos \theta\right)y' + \frac{1}{2} \cos^2 \theta \, y''. \quad (14.27)$$

Since $|y| \le 1$ (at θ_0), $0 < a < 1$, from $y \sin \theta \le 1$, it follows that $a \ge ay \sin \theta$, $1 + ay > 0$,

$$a + \sin \theta \ge ay \sin \theta + \sin \theta = \sin \theta(1 + ay),$$

so

$$\frac{a + \sin \theta}{1 + ay} \ge \sin \theta.$$

Therefore, from (14.27), we obtain

$$\varphi(\theta_0) = y(\theta_0) \le \sin \theta_0 - \cos \theta_0 \sin \theta_0 \cdot y'(\theta_0) + \tfrac{1}{2} \cos^2 \theta_0 \cdot y''(\theta_0).$$

The conclusion of the lemma follows.   □

LEMMA 14.6. *Define a function* $\psi(\theta)$,

$$\psi(\theta) = \frac{(4/\pi)(\theta + \cos \theta \sin \theta) - 2 \sin \theta}{\cos^2 \theta}, \quad \theta \in \left(-\frac{\pi}{2}, \frac{\pi}{2}\right), \quad (14.28)$$

$$\psi\left(\frac{\pi}{2}\right) = 1, \quad \psi\left(-\frac{\pi}{2}\right) = -1.$$

Then $\psi \in C^0[-\pi/2, \pi/2] \cap C^2(-\pi/2, \pi/2)$. *Moreover,* $y = \psi(\theta)$ *satisfies the differential equation*

$$y - \sin \theta + \sin \theta \cos \theta \cdot y' - \tfrac{1}{2} \cos^2 \theta \cdot y'' = 0 \qquad (14.29)$$

and

$$\psi'(\theta) \geq 0. \qquad (14.29')$$

PROOF. Let us solve the differential equation (14.29). Set $y = Z/(\cos^2 \theta)$; (14.29) is reduced to

$$\frac{Z}{\cos^2 \theta} - \sin \theta + \sin \theta \cos \theta \left(\frac{Z'}{\cos^2 \theta} + \frac{2 \sin \theta}{\cos^3 \theta} Z\right)$$

$$- \frac{1}{2} \cos^2 \theta \left(\frac{Z''}{\cos^2 \theta} + \frac{4 \sin \theta}{\cos^3 \theta} Z' + \frac{6 \sin^2 \theta + 2 \cos \theta}{\cos^4 \theta} Z\right) = 0.$$

Arranging it in order, we obtain

$$-\frac{1}{2} Z'' - \frac{\sin \theta}{\cos \theta} Z' - \sin \theta = 0, \qquad (14.30)$$

which is a differential equation of order 1 for Z'. Solve it successively to get

$$Z' = \cos^2 \theta \left(2A - \frac{2}{\cos \theta}\right), \qquad (14.31)$$

$$Z = B + A(\theta + \sin \theta \cos \theta) - 2 \sin \theta. \qquad (14.32)$$

Finally we get the general solution of (14.29),

$$y = \frac{B + A(\theta + \sin \theta \cos \theta) - 2 \sin \theta}{\cos^2 \theta}, \qquad (14.33)$$

where A, B are constants. Taking $B = 0$, $A = 4/\pi$ in (14.33), we obtain the $\psi(\theta)$ expressed in (14.28). It follows from L'Hospital's rule that

$$\psi\left(\frac{\pi}{2} - 0\right) = \lim_{\theta \to \pi/2} \frac{(8/\pi)\cos^2\theta - 2\cos\theta}{-2\cos\theta\sin\theta} = 1$$

and $\psi(-\pi/2 + 0) = -1$. So $\psi \in C[-\pi/2, \pi/2]$. By direct computation, we know that

$$\psi'(\theta) = 2\frac{g(\theta)}{\cos^3\theta}, \tag{14.34}$$

$$g(\theta) = \frac{4}{\pi}\theta\sin\theta + \frac{4}{\pi}\cos\theta - 1 - \sin^2\theta, \tag{14.35}$$

where $g(\theta)$ is a differential function on $[-\pi/2, \pi/2]$. It attains its minimum at some critical points or at boundary points. The critical points satisfy

$$g'(\theta) = \left(\frac{4}{\pi}\theta - 2\sin\theta\right)\cos\theta = 0. \tag{14.36}$$

They are either $\theta = 0$ or $\theta = \pm\pi/2$. But $g(\pm\pi/2) = 0$, $g(0) = (4/\pi) - 1 > 0$. Consequently $g(\theta) \geq 0$ and so $\psi'(\theta) \geq 0$. Lemma 14.6 is proved.   □

LEMMA 14.7. *Assume that $\psi(\theta)$ is the function defined by* (14.28) *and $\varphi(\theta)$ is the function defined in* (14.22). *Then*

$$\varphi(\theta) \leq \psi(\theta). \tag{14.37}$$

PROOF. By reduction to absurdity, if (14.37) is not true, in view of

$$\varphi\left(-\frac{\pi}{2}\right) < -1, \qquad \varphi\left(\frac{\pi}{2}\right) < -1,$$

there exists $\theta_0 \in (-\pi/2, \pi/2)$ and some constant $b > 0$ such that

$$\varphi(\theta_0) - \psi(\theta_0) = \max_{\theta \in (-\pi/2, \pi/2)} (\varphi(\theta) - \psi(\theta)) = b. \tag{14.38}$$

From this we can verify easily that $y(\theta) = \psi(\theta) + b$ satisfies the hypothesis in Lemma 14.5 at θ_0. By Lemma 14.5, it implies

$$\varphi(\theta_0) \leq \sin\theta_0 - \sin\theta_0\cos\theta_0 \cdot \psi'(\theta_0) + \tfrac{1}{2}\cos^2\theta_0 \cdot \psi''(\theta_0). \tag{14.39}$$

From the definition of $\psi(\theta)$ and (14.29) in Lemma 14.6, the right side of (14.39) equals $\psi(\theta_0)$. By Lemma 14.7, it follows that

$$\varphi(\theta_0) \leq \psi(\theta_0). \tag{14.40}$$

But this contradicts (14.38), so the lemma is proved. $\Box$

COROLLARY 14.8.

$$F(\theta) \leq \lambda_1(1 + a\psi(\theta)), \tag{14.41}$$

where $F(\theta)$ and $\psi(\theta)$ are functions defined respectively in (14.13) and (14.28).

Inequality (14.41) is the estimate of the upper bound of $F(\theta)$ required.

14.4. Estimation of λ_1

By means of the preceding estimate for $F(\theta)$, we are now in a position to give the proof of the main theorem. Equation (14.41) means

$$\lambda_1^{1/2} \geq \frac{|F(\theta)|^{1/2}}{\sqrt{1 + a\psi(\theta)}} \geq \frac{|\nabla \theta|}{\sqrt{1 + a\psi(\theta)}}, \tag{14.42}$$

where $\psi(\theta)$ is the function defined in (14.28). It is easy to verify from (14.28) that

$$\psi(0) = 0, \qquad \psi(-\theta) = -\psi(\theta). \tag{14.43}$$

Using the same method as we used in the proof of Lemma 14.4, we obtain

$$\lambda_1^{1/2}d \geq \lambda_1^{1/2}d' \geq \int_{-\pi/2}^{\pi/2} \frac{d\theta}{\sqrt{1 + a\psi(\theta)}}$$

$$= \int_0^{\pi/2} \left(\frac{1}{\sqrt{1 + a\psi(\theta)}} + \frac{1}{\sqrt{1 - a\psi(\theta)}} \right) d\theta. \tag{14.44}$$

Noting $|a\psi| < 1$ and applying the Taylor's formula, we have

$$\frac{1}{\sqrt{1 + a\psi}} + \frac{1}{\sqrt{1 - a\psi}} = 2\left[1 + \sum_{k=1}^{\infty} \frac{1 \cdot 3 \cdots (4k - 1)}{2 \cdot 4 \cdots (4k)} a^{2k}\psi^{2k} \right]. \tag{14.45}$$

Set $C_k = (2/\pi) \int_0^{\pi/2} \psi^{2k}(\theta)\, d\theta$, which are positive constants. Then (14.45) can be rewritten as

$$\int_0^{\pi/2} \left(\frac{1}{\sqrt{1+a\psi}} + \frac{1}{\sqrt{1-a\psi}} \right) d\theta$$

$$= \pi \left[1 + \sum_{k=1}^{\infty} \frac{1 \cdot 3 \cdots (4k-1)}{2 \cdot 4 \cdots (4k)} C_k a^{2k} \right]. \tag{14.46}$$

Thus, we have proved the following theorem.   □

THEOREM 14.9. *If M is a compact Riemannian manifold whose Ricci curvature is nonnegative with diameter d, then the first eigenvalue $-\lambda_1$ of M satisfies*

$$\lambda_1 \geq \frac{\pi^2}{d^2} \left[1 + \sum_{k=1}^{\infty} \frac{1 \cdot 3 \cdots (4k-1)}{2 \cdot 4 \cdots (4k)} C_k a^{2k} \right], \tag{14.47}$$

where $C_k = (2/\pi) \int_0^{\pi/2} \psi^{2k}(\theta)\, d\theta$, $\psi(\theta)$ is the function defined in (14.28), $0 \leq a < 1$.

Consequently, when $a > 0$, it shows $\lambda_1 > \pi^2/d^2$. This combined with Lemma 14.4 completely proves Theorem 14.2.   □

References

1. P. Li, *Ann. of Math. Stud.*, Vol. 102, pp. 73–85. Princeton University Press (1982).
2. P. Li and S. T. Yau, *Proc. Symp. Pure Math.*, Vol. 36, 1980.

15

Curvature Characterization of Compact Hermitian Symmetric Spaces

In the study of complex manifolds the following conjecture is a well-known and natural analog of the elliptic case of the uniformization theorem.

CONJECTURE 15.1. *Suppose X is a compact Kähler manifold of nonnegative holomorphic bisectional curvature and positive Ricci curvature. Then X is biholomorphic to a compact Hermitian symmetric space.*

The special case, when X is of positive bisectional curvature and conjectured to be $\mathbf{P}^n$, is the Frankel conjecture, resolved simultaneously and independently by Mori [19] and Siu and Yau [22] in 1979 using very different methods. The general case of Conjecture 15.1 is at present still open. A related conjecture in case X is assumed to be Kähler–Einstein is the following.

CONJECTURE 15.2. *Suppose X is a compact Kähler–Einstein manifold of nonnegative holomorphic bisectional curvature and positive Ricci curvature. Then X is isometric to a compact Hermitian symmetric space.*

The first efforts to resolve Conjecture 15.2 were due to Berger [3], who showed in 1966 that a compact Kähler–Einstein manifold of positive sectional curvature is isometric. to $\mathbf{P}^n$ and equipped with the Fubini-Study metric (up to a scalar factor). This was reformulated by Goldberg and Kobayashi to the case of positive holomorphic bisectional curvature. Later, Gray [8] proved Conjecture 15.2 in 1973 under the stronger assumption of nonnegative Riemannian sectional curvature. He introduced on the unit

---

Joint work with Ngaiming Mok.

sphere bundle of X a (degenerate) elliptic operator D and developed a Bochner–Kodaira formula for DR, R denoting the curvature tensor, to prove the vanishing of ∇R on X. The last property is the simplest characterization of locally symmetric spaces in terms of the curvature tensor. Apparently, there are serious difficulties in modifying Gray's argument to the general case of nonnegative holomorphic bisectional curvature since D will in general not be (degenerate) elliptic. This has left Conjecture 15.2 open for a long time. It was one of the open questions in Kähler geometry raised by Siu [21] in his address in 1983 to the International Congress of Mathematics at Warsaw.

One connection between Conjectures 15.1 and 15.2 is inspired by the work of Hamilton [10] on deforming Riemannian metrics of positive curvature on a compact 3-manifold to an Einstein metric. It is hoped that such an approach can be applied to compact Kähler manifolds of nonnegative holomorphic bisectional curvature. In this connection we refer to the reader to a recent article of Bando [2], who used the evolution equation of Hamilton [10] and results of Siu [20] on characterizing hyperquadrics to obtain an affirmative answer to Conjecture 15.1 in the case of dimension 3. (The cases of dimensions 1 and 2 are well known.)

In this chapter we resolve Conjecture 15.2 in the affirmative. Our starting point is the method of Berger [3] on characterizing $\mathbf{P}^n$ with the Fubini–Study metric. He did this by showing that the Kähler manifold X under consideration has constant holomorphic sectional curvature. To do this, he considered a point x_0 on X and a unit tangent vector α of type $(1, 0)$ at x_0, where the global maximum of holomorphic sectional curvatures is attained, and applied the maximum principle to $\Delta R_{\alpha \bar{\alpha} \alpha \bar{\alpha}}(x_0)$. For Conjectire 15.2, we used the characterization of Hermitian symmetric spaces by the vanishing of ∇R, a property not verifiable by a direct application of Berger's method. In a similar setting as above, assuming X Kähler–Einstein of nonnegative bisectional curvature at $\alpha \in T^{1,0}_{x_0}(X)$, one can show that relative to the Hermitian bilinear form $H_\alpha(\xi, \xi') = R_{\alpha \bar{\alpha} \xi \bar{\xi}}(x_0)$, $T^{1,0}_{x_0}(X)$ decomposes into the orthogonal direct sum of eigenspaces $\mathbf{C}\alpha \oplus \mathcal{H}_\alpha \oplus \mathcal{N}_\alpha$, where $R_{\alpha \bar{\alpha} \xi \bar{\xi}}(x_0) = \frac{1}{2} R_{\alpha \bar{\alpha} \alpha \bar{\alpha}}(x_0)$ for $\xi \in \mathcal{H}_\alpha$, $= 0$ for $\xi \in \mathcal{N}_\alpha$, and moreover $\Delta R_{\alpha \bar{\alpha} \alpha \bar{\alpha}}(x_0) = 0$.

Our idea is to prove first of all the invariance of $R_{\alpha \bar{\alpha} \alpha \bar{\alpha}}$ under parallel transport of α along certain curves emanating from x_0. To start with we prove, using the maximum principle, that the global maximum of holomorphic sectional curvatures is attained at every point $x \in X$. Let γ be an integral curve of any vector field of "maximal directions" $\alpha(x)$; we prove the stronger fact that the curvature tensor R is invariant under parallel transport along the curve γ. Using an orthonormal basis $\{e_i\}$ at $x \subset \gamma$ consisting of eigenvectors of the Hermitian form $H_\alpha(\xi, \xi') = R_{\alpha \bar{\alpha} \xi \bar{\xi}}(x)$, $\alpha = \alpha(x)$, we shall actually prove the vanishing of all terms $\nabla_\alpha R_{i \bar{j} k \bar{l}}(x)$.

The proof of the vanishing of such covariant derivatives will occupy the bulk of this chapter.

Our original aim was to prove that $\nabla^i_\eta R_{\alpha\bar\alpha\alpha\bar\alpha}(x) = 0$ at a global maximum direction α for all real tangent vectors η at x and all positive integers i. Since the Kähler–Einstein metric is real analytic, this would allow us to conclude the invariance of $R_{\alpha\bar\alpha\alpha\bar\alpha}$ under parallel transport along geodesics. Although this scheme is too involved for higher-order radial derivatives, it will be enough to show $\nabla^i_\eta R_{\alpha\bar\alpha\alpha\bar\alpha}(x) = 0$ for $1 \le i \le 7$, which is sufficient to imply the invariance of R under parallel transport along integral curves of maximal directions. The point of departure is the observation that Berger's formula implies $\Delta R_{\alpha\bar\alpha\alpha\bar\alpha}(x) = 0$ and hence $\nabla^i_\eta R_{\alpha\bar\alpha\alpha\bar\alpha}(x) = 0$ for $1 \le i \le 3$ in view of the global maximality of $R_{\alpha\bar\alpha\alpha\bar\alpha}$. It follows that $\nabla^4_\eta R_{\alpha\bar\alpha\alpha\bar\alpha}(x) \ge 0$. Define a $(2k)$th-order elliptic operator $S^{(2k)}$ on smooth tensors T by taking, at each point y where T is defined, $S^{(2k)}T$ to be the average, suitably normalized, of $\nabla^{2k}_\eta T$ over all $\eta \in T_y(X)$ of unit length. Clearly, $S^{(4)}R_{\alpha\bar\alpha\alpha\bar\alpha}(x) \le 0$. On the other hand, we show that $\Delta^2 R_{\alpha\bar\alpha\alpha\bar\alpha}(x) \ge 0$. For the Euclidean case, $S^{(4)}$ agrees with Δ^2. However, for Kähler manifolds in general $S^{(4)}$ differs from Δ^2 by some zero-order terms. Such zero-order terms are obtained by a number of commutations. At x we have sufficient knowledge of zero-order terms to conclude that $S^{(4)}R_{\alpha\bar\alpha\alpha\bar\alpha}(x) = \Delta^2 R_{\alpha\bar\alpha\alpha\bar\alpha}(x)$, implying both are zero and that $\nabla^i_\eta R_{\alpha\bar\alpha\alpha\bar\alpha}(x) = 0$ for $1 \le i \le 5$.

To proceed further one can consider similarly $S^{(2k)}R_{\alpha\bar\alpha\alpha\bar\alpha}(x)$ and $S^{(2k-2)}\Delta R_{\alpha\bar\alpha\alpha\bar\alpha}(x)$. In general the difference between $S^{(2k)}$ and $S^{(2k-2)}\Delta$ is a differential operator of order $(2k - 4)$. We are able to prove in a way similar to the above that $S^{(6)}R_{\alpha\bar\alpha\alpha\bar\alpha}(x) = S^{(4)}\Delta R_{\alpha\bar\alpha\alpha\bar\alpha}(x) = 0$, implying $\nabla^i_\eta R_{\alpha\bar\alpha\alpha\bar\alpha}(x)$ for $1 \le i \le 7$. This involves proving the vanishing of commutation terms which are second-order covariant derivatives of terms of the type $R_{\alpha\bar\alpha\alpha\bar\alpha}$ or $R_{\alpha\bar\alpha k\bar k}$. These are obtained from variation equalities or Taylor series expansions of curvature functions along geodesics issuing from x.

In order to prove the vanishing of $\nabla_\alpha R_{i\bar j k \bar l}(x)$ we make full use of gradient terms arising in formulas $\Delta^2 R_{\alpha\bar\alpha\alpha\bar\alpha}(x)$ and $S^{(4)}\Delta R_{\alpha\bar\alpha\alpha\bar\alpha}(x)$. To prove $\nabla_\alpha R_{i\bar j k \bar l}(x) = 0$ it will actually be necessary also to show $\Delta^3 R_{\alpha\bar\alpha\alpha\bar\alpha}(x) = 0$ and to make use of gradient terms arising from $\Delta^3 R_{\alpha\bar\alpha\alpha\bar\alpha}$. One surprising thing in this scheme of proof is that, under our special choice of basis at x, $\alpha \in T^{1,0}_x(X)$ a fixed maximal direction, we show that there are only a few types of nonvanishing curvature terms. Such information is also used in the proof of $\nabla_\alpha R_{i\bar j k \bar l}(x) = 0$.

At each $x \in X$ let V_x be the real linear subspace of $T^{1,0}_x(X)$ generated by the nonempty set of maximal directions $\alpha \in T_x(X)$. We can use the invariance of R under parallel transport along integral curves of vector fields of maximal directions to show that at some point x, the vector subspaces $\mathrm{Re}\, V_y \subset T_y(X)$ for adjacent points y constitute an integrable distribution. The integral submanifolds are moreover complex, totally geodesic and locally symmetric. Then, we use the theorem of Bonnet–Myers to show that these integral submanifolds extend to complex submanifolds of X for a suitable choice of x, and that they are mutually nonintersecting. We use this to show that the curvature tensor is reducible at each point, that the vector subspaces V_x, $x \in X$, constitute a differentiable vector bundle invariant under parallel transport and that the foliation of X by integral submanifolds of the distribution $x \mapsto \mathrm{Re}\, V_x$ actually corresponds to a global decomposition of X up to a finite covering. This allows us to prove Conjecture 15.2 inductively.

We believe that our analysis of the curvature tensor should also be useful in other problems in Kähler geometry related to locally symmetric Hermitian manifolds.

The main results of the present article, together with a sketch of the methods of proof, has appeared in Mok and Zhong [18].

We would like to thank Professor Siu and Professor Yau for their interest in the research project. The research was carried out while the second author was a visiting member at the Institute for Advanced Study. He would like to thank the Institute for its support and hospitality.

15.1. Statement of Results

Our main theorem is the following confirmation of Conjecture 15.2 (in the introduction).

THEOREM 15.3. *Let X be a compact Kähler–Einstein manifold of nonnegative holomorphic bisectional curvature and positive Ricci curvature. Then X is isometric to a compact Hermitian symmetric space.*

We remark that the nonnegativity of holomorphic bisectional curvatures is strictly weaker than the nonnegativity of Riemannian sectional curvatures and that the former concept is more natural in the context of complex differential geometry. For a compact Kähler manifold, holomorphic bisectional curvatures are nonnegative if and only if the unit ball bundle of the dual tangent bundle is weakly pseudoconvex.

Our method of proof yields the following generalizations of Theorem 15.3.

COROLLARY 15.4. *Let X be a compact Kähler manifold of nonnegative holomorphic bisectional curvature and constant scalar curvature. Then X is isometric to a compact Hermitian locally symmetric space.*

COROLLARY 15.5. *Let X be a complete Kähler manifold of nonnegative holomorphic bisectional curvature such that the Ricci tensor is parallel. Then X is isometric to a complete Hermitian locally symmetric space.*

From the proof given for Theorem 15.3, it is immediate to generalize to the case when X is assumed to be of nonnegative holomorphic bisectional curvature and the Ricci tensor is parallel and positive. The proofs of the corollaries involve essentially a splitting of the flat directions of the Ricci tensor. Corollary 15.4 follows from Corollary 15.5 and results of Bishop and Goldberg [4-6] which assert that under the hypothesis of Corollary 15.4, the Ricci tensor is automatically parallel.

15.2. Background Material

15.2.1. The Curvature Tensor on Kähler Manifolds and Commutation Formulas

Let X be a Kähler manifold. Denote by $R = R\langle\cdot,\cdot,\cdot,\cdot\rangle$ the Riemannian curvature tensor on the underlying Riemannian manifold. By complexification, R acts on $(CT(X))^4$, $CT(X)$ denoting the complexified tangent bundle. On X we have a decomposition of $CT(X)$ into the orthogonal direction sum $T^{1,0}(X) \oplus T^{0,1}(X)$. If we choose a system of holomorphic coordinates $(z_1, \ldots, z_n)$ at $x \in X$, then $\{\partial/\partial z_1, \ldots, \partial/\partial z_n\}$ and $\{\partial/\partial\bar{z}_1, \ldots, \partial/\partial\bar{z}_n\}$ constitute bases of $T_x^{1,0}(X)$ and $T_x^{0,1}(X)$ respectively. In terms of the corresponding decomposition of tensors into (p, q)-types on a Kähler manifold, R is of type $(2, 2)$. In terms of the basis $\{\partial/\partial z_1, \ldots, \partial/\partial z_n\,;$ $\partial/\partial\bar{z}_1, \ldots, \partial/\partial\bar{z}_n\}$ of $CT_{x_0}(X)$ and writing $R_{i\bar{j}k\bar{l}} = R\langle\partial/\partial z_i, \partial/\partial\bar{z}_j,$ $\partial/\partial z_k, \partial/\partial\bar{z}_l\rangle$, etc., the only possible nonzero terms of R_{****} (indices with or without bars) are given by $R_{i\bar{j}k\bar{l}}$ and accompanying terms obtained by permutation of indices. We write $R_{i\bar{j}}$ for the Ricci curvature tensor in terms of coordinates. Our convention on R is such that $R_{1\bar{1}1\bar{1}} > 0$ for the Riemann sphere with the standard Hermitian metric of constant positive curvature.

We say that X is of nonnegative holomorphic bisectional curvature if $R\langle\xi, \bar{\xi}; \zeta, \bar{\zeta}\rangle \geq 0$ for all $x \in S$ and $\xi, \zeta \in T_x^{(1,0)}(X)$. In terms of indices this means that $\sum R_{i\bar{j}k\bar{l}} a_i\bar{a}_j b_k\bar{b}_l \geq 0$ for all n-tuples $(a_1, \ldots, a_n)$, $(b_1, \ldots, b_n)$ of complex numbers. Every Hermitian (globally) symmetric space carries an invariant Kähler-Einstein metric of nonnegative holomorphic bisectional curvature. In terms of the curvature tensor we have the following characterization of locally symmetric Riemannian manifolds.

PROPOSITION 15.6 (cf. Kobayashi and Nomizu [12]). A Riemannian manifold X is locally symmetric at $x \in X$ if and only if in a neighborhood of x, $\nabla R \equiv 0$ for the Riemannian curvature tensor R.

The curvature tensor measures analytically the commutation of covariant differentiations. For example, for a covariant tensor of type

$T_{i\bar{j}k\bar{l}}$, we have, denoting $\nabla_t\nabla_s T_{i\bar{j}k\bar{l}} = T_{i\bar{j}k\bar{l},st}$ etc.,

$$T_{i\bar{j}k\bar{l},st} = T_{i\bar{j}k\bar{l},ts}, \qquad T_{i\bar{j}k\bar{l},\bar{s}\bar{t}} = T_{i\bar{j}k\bar{l},\bar{t}\bar{s}},$$

$$T_{i\bar{j}k\bar{l},s\bar{t}} = T_{i\bar{j}k\bar{l},\bar{t}s} + R_{i\bar{\mu}s\bar{t}}T_{\mu\bar{j}k\bar{l}} - R_{\mu\bar{j}s\bar{t}}T_{i\bar{\mu}k\bar{l}} + R_{k\bar{\mu}s\bar{t}}T_{i\bar{j}\mu\bar{l}} - R_{\mu\bar{l}s\bar{t}}T_{i\bar{j}k\bar{\mu}}.$$

All three equations follow from the definition of R in terms of commutation of covariant differentiations. The first two are consequences of the fact that R is of type $(2, 2)$.

In general, for any covariant tensor field $T_{**...**}$, commutation for second-order covariant differentiation occurs only if we commute two indices of opposite type (on barred, one unbarred), in which case there are as many commutation terms as there are indices in T, the sign attached to a commutation term in $T_{**...**,s\bar{t}} - T_{**...,\bar{t}s}$ is positive if it corresponds to a substitution of an unbarred index in $T_{**...**}$, and negative otherwise. This is simply because $-R_{\mu\bar{j}s\bar{t}} = R_{\bar{j}\mu s\bar{t}}$.

Finally we recall that the Bianchi identity implies the equality $R_{i\bar{j}k\bar{l},m} = R_{i\bar{j}m\bar{l},k}$ in the case of Kähler manifolds.

15.2.2. Computation of $\Delta R_{\alpha\bar{\alpha}\alpha\bar{\alpha}}$

At $x \in X$ fixed let $(z_1,\ldots,z_n)$ be a system of local holomorphic coordinates. For any smooth tensor T we shall denote by ΔT the operator $\sum_{i,j} g^{i\bar{j}}(\nabla_i\nabla_{\bar{j}}T + \nabla_{\bar{j}}\nabla_i T)$, where $g^{i\bar{j}}$ is the contravariant metric tensor. (See Section 15.2.3 for the meaning of ΔT and other averaging differential operators.) We recall here the computation of $\Delta R_{\alpha\bar{\alpha}\alpha\bar{\alpha}}$ in Berger [3] for any tangent vector α of type $(1, 0)$.

PROPOSITION 15.7. *Let* $(z_1,\ldots,z_n)$ *be a system of local holomorphic coordinates at* $x \in X$ *such that* $g_{ij}(x) = \delta_{ij}$ *for the Kähler metric tensor* (g_{ij}). *Then, denoting by* ρ *the Einstein constant, i.e., Ricci form* $= \rho(K\ddot{a}hler form)$, *we have*

$$\tfrac{1}{2}\Delta R_{\alpha\bar{\alpha}\alpha\bar{\alpha}} = \sum_{i,j}|R_{\alpha\bar{i}\alpha\bar{j}}|^2 + \rho R_{\alpha\bar{\alpha}\alpha\bar{\alpha}} - 2\sum_{i,j}|R_{\alpha\bar{\alpha}i\bar{j}}|^2.$$

PROOF. Obviously the right-hand side is independent of the choice of holomorphic coordinates at x as long as they are unitary at x, and it suffices to prove the proposition for e_α of unit length. We may therefore choose $(z_1,\ldots,z_n)$ so that $\partial/\partial z_1 = \alpha$. Then, from

$$R_{\alpha\bar{\alpha}\alpha\bar{\alpha},i\bar{i}} = R_{\alpha\bar{\alpha}\alpha\bar{\alpha},\bar{i}i} + 2\sum_\mu R_{\alpha\bar{\mu}i\bar{i}}R_{\mu m\bar{\alpha}\alpha\bar{\alpha}} - 2\sum_\mu R_{\mu\bar{\alpha}i\bar{i}}R_{\alpha\bar{\mu}\alpha\bar{\alpha}}$$

we have, using $\sum_i R_{\alpha\bar{\mu}i\bar{i}} = \rho\delta_{\alpha\mu}$,

$$\sum_i R_{\alpha\bar{\alpha}\alpha\bar{\alpha},i\bar{i}} = \sum_i R_{\alpha\bar{\alpha}\alpha\bar{\alpha},\bar{i}i} + 2\rho R_{\alpha\bar{\alpha}\alpha\bar{\alpha}} - 2\rho R_{\alpha\bar{\alpha}\alpha\bar{\alpha}} = \sum_i R_{\alpha\bar{\alpha}\alpha\bar{\alpha},\bar{i}i}.$$

Hence

$$\tfrac{1}{2}\Delta R_{\alpha\bar{\alpha}\alpha\bar{\alpha}} = \sum_i R_{\alpha\bar{\alpha}\alpha\bar{\alpha},i\bar{i}} = \sum_i R_{\alpha\bar{\alpha}i\bar{\alpha},\alpha\bar{i}}$$

$$= \sum_i R_{\alpha\bar{\alpha}i\bar{\alpha},\bar{i}\alpha} + \sum_{i,\mu} R_{\alpha\bar{\mu}\alpha\bar{i}} R_{\mu\bar{\alpha}i\bar{\alpha}} - \sum_{i,\mu} R_{\mu\bar{\alpha}\alpha\bar{i}} R_{\alpha\bar{\mu}i\bar{\alpha}}$$

$$+ \sum_{i,\mu} R_{i\bar{\mu}\alpha\bar{i}} R_{\alpha\bar{\alpha}\mu\bar{\alpha}} - \sum_{i,\mu} R_{\mu\bar{\alpha}\alpha\bar{i}} R_{\alpha\bar{\alpha}i\bar{\mu}}$$

$$= \sum_i R_{\alpha\bar{\alpha}i\bar{i},\bar{\alpha}\alpha} + \sum_{i,j} |R_{\alpha\bar{i}\alpha\bar{j}}|^2 + \rho R_{\alpha\bar{\alpha}\alpha\bar{\alpha}} - 2\sum_{i,j} |R_{\alpha\bar{\alpha}i\bar{j}}|^2$$

by the Bianchi identity.

Since the metric on X is Kähler–Einstein, we have

$$\sum_i R_{\alpha\bar{\alpha}i\bar{i},\bar{\alpha}\alpha} = R_{\alpha\bar{\alpha},\bar{\alpha}\alpha} = 0,$$

proving the proposition.   □

15.2.3. Averaging Operators of Radial Derivatives

Let T be a covariant smooth tensor of order m defined on an open subset U of X. At $x \in U$ let η be a real tangent vector of unit length. Let γ be a geodesic passing through x and within the cut locus of x with η tangent to γ at x. Let $v_1, \ldots, v_k$ be complexified tangent vectors at x and denote by the same symbols the vector fields defined on γ obtained by parallel transport. Then

$$\nabla^i_\eta T\langle v_1, \ldots, v_m\rangle(x) = (\nabla^i_\eta T)\langle v_1, \ldots, v_m\rangle(x)$$

since $\nabla_\eta v_j = 0$ for $1 \le j \le k$ along γ. Letting k be a positive integer, we define the operator $S^{(2k)}T$ by setting

$$S^{(2k)}T\langle v_1, \ldots, v_m\rangle(x) = c_{2k}\int_\eta \nabla^{2k}_\eta T\langle v_1, \ldots, v_m\rangle(x)$$

at each x where T is defined, where the integration is over the unit sphere of the real tangent space at x, endowed with the unique rotation-invariant metric of unit mass, and where c_{2k} is a constant to be determined. In case $k = 1$ we have

$$S^{(2)}T_{**...**}(x) = c_2 \int_\eta \nabla_\eta^2 T_{**...**}(x).$$

Let $(z_1, \ldots, z_n)$ be a local holomorphic coordinate system unitary at x. Then $\eta = \xi + \bar{\xi}$ for some $\xi \in T_x^{1,0}(X)$ of length $1/\sqrt{2}$. Write $\xi = \sum_i a_i(\eta)\partial/\partial z_i$,

$$S^{(2)}T_{**...**}(x) = c_2 \int_\eta (\nabla_{\sum_i a_i(\eta)\partial/\partial z_i + \overline{\sum_i a_i(\eta)\partial/\partial z_i}})^2 T_{**...**}(x)$$

$$= 2 \operatorname{Re} c_2 \int_\eta \sum_{i,j} a_i(\eta)a_j(\eta)\nabla_i\nabla_j T_{**...**}(x)$$

$$+ c_2 \int \sum_{i,j} a_i(\eta)\overline{a_j(\eta)}(\nabla_i\bar{\nabla}_j + \bar{\nabla}_j\nabla_i)T_{**...**}(x).$$

Denote

$$b_{ij} = \int_\eta a_i(\eta)a_j(\eta), \qquad b_{i\bar{j}} = \int_\eta a_i(\eta)\overline{a_j(\eta)}.$$

Consider on $T_x^{1,0}(X) \approx \mathbf{C}^n$ the transformation

$$(a_1, \ldots, a_i, \ldots, a_j, \ldots, a_n) \to (a_1, \ldots, e^{i\theta_i}a_i, \ldots, e^{i\theta_j}a_j, \ldots, a_n),$$

where the left-hand side stands for $\sum a_i\partial/\partial z_i$. This induces an orthogonal transformation $\eta \mapsto \eta'$ on $T_x(X) = \{\xi + \bar{\xi}: \xi \in T_x^{(1,0)}(X)\}$. It follows that

$$b_{ij} = \int_\eta a_i(\eta')a_j(\eta') = e^{i(\theta_i+\theta_j)}b_{ij},$$

$$b_{i\bar{j}} = \int_\eta a_i(\eta')\overline{a_j(\eta')} = e^{i(\theta_i-\theta_j)}b_{i\bar{j}}.$$

Choosing θ_i, θ_j suitably we see that $b_{ij} = 0$ for all i, j and that $b_{i\bar{j}} = 0$ unless $i = j$. By symmetry, clearly $b_{1\bar{1}} = \cdots = b_{n\bar{n}}$. These constants can be computed by taking T to be the function $\sum_i |z_i|^2$ and comparing coefficients. In any case $b_{i\bar{i}} > 0$ and we choose $c_2 b_{1\bar{1}} = 1$, giving

$$S^{(2)}T_{**...**} = \sum_i \nabla_i\nabla_{\bar{i}} T_{**...**} + \sum_i \nabla_{\bar{i}}\nabla_i T_{**...**};$$

i.e., $S^{(2)}T = \sum_i (\nabla_i \nabla_{\bar\imath} + \nabla_{\bar\imath} \nabla_i) T$. We use ΔT to denote $S^{(2)}T$. There are two related fourth order averaging differential operators, namely Δ^2 and $S^{(4)}$. We have

$$\Delta^2 T = \sum_{i,j} (\nabla_i \nabla_{\bar\imath} + \nabla_{\bar\imath} \nabla_i)(\nabla_j \nabla_{\bar\jmath} + \nabla_{\bar\jmath} \nabla_j) T$$

$$= \sum_{i,j} T_{,i\bar\imath j\bar\jmath} + T_{,i\bar\imath \bar\jmath j} + T_{,\bar\imath i j \bar\jmath} + T_{,\bar\imath i \bar\jmath j}.$$

By the same argument as above we have

$$S^{(4)}T = c_4 \sum_i b_{i\bar\imath i\bar\imath} (T_{,(i\bar\imath + \bar\imath i)(i\bar\imath + \bar\imath i)} + T_{,ii\bar\imath\bar\imath} + T_{,\bar\imath\bar\imath ii})$$

$$+ c_4 \sum_{i<j} b_{i\bar\imath j\bar\jmath}(T_{,(i\bar\imath + \bar\imath i)(j\bar\jmath + \bar\jmath j)} + T_{,(j\bar\jmath + \bar\jmath j)(i\bar\imath + \bar\imath i)} + T_{,(ij+ji)(\bar\imath\bar\jmath + \bar\jmath\bar\imath)}$$

$$+ T_{,(\bar\imath\bar\jmath + \bar\jmath\bar\imath)(ij+ji)} + T_{,(i\bar\jmath + \bar\jmath i)(j\bar\imath + \bar\imath j)} + T_{,(j\bar\imath + \bar\imath j)(i\bar\jmath + \bar\jmath i)}).$$

Here we are adopting the convention

$$T_{,(i\bar\imath + \bar\imath i)(j\bar\jmath + \bar\jmath j)} = T_{,i\bar\imath j\bar\jmath} + T_{,i\bar\imath \bar\jmath j} + T_{,\bar\imath i j\bar\jmath} + T_{,\bar\imath i \bar\jmath j},$$

etc. The equality above is obtained by noting that the only nonzero terms in b_{****} must come from two pairs of conjugates, e.g. $b_{i\bar\imath j\bar\jmath}$, $b_{ij\bar\jmath\bar\imath}$, etc., which follows from using the transformation

$$(a_1, \ldots, a_i, \ldots, a_j, \ldots, a_k, \ldots, a_l, \ldots, a_n)$$

$$\rightarrow (a_1, \ldots, e^{i\theta_i} a_i, \ldots, e^{i\theta_j} a_j, \ldots, e^{i\theta_k} a_k, \ldots, e^{i\theta_l} a_l, \ldots, a_n).$$

Obviously $b_{i\bar\imath j\bar\jmath}$ remains unchanged when indices are permuted, but the corresponding covariant differentiation may differ because of the curvature. We write the expansion for $S^{(4)}T$ in a more uniform manner. Denote by S_4 the permutation group of four elements. For any $\sigma \in S_4$ and any fourth-order covariant differentiation $T_{,\alpha\beta\gamma\delta}$ (indices with or without bars), we denote by $T^\sigma_{,\alpha\beta\gamma\delta}$ the fourth-order covariant derivative obtained by formally permuting the four indices using $\sigma \in S_4$. In this notation we can write

$$S^{(4)}T = c_4 \sum_i \frac{b_{i\bar\imath i\bar\imath}}{4} \sum_{\sigma \in S_4} T^\sigma_{,i\bar\imath i\bar\imath} + c_4 \sum_{i<j} b_{i\bar\imath j\bar\jmath} \sum_{\sigma \in S_4} T^\sigma_{,i\bar\imath j\bar\jmath}.$$

Note that in the original expression that there are 6 terms attached to $b_{i\bar{i}i\bar{i}}$ and 24 terms attached to $b_{i\bar{i}j\bar{j}}$, $i < j$. This accounts for the factor of 1/4 in the first term of the new expression. Our main result in this section is the following:

PROPOSITION 15.8. *For a suitable choice of positive constant c_2, we have the expansion*

$$6S^{(4)}T = \sum_{i,j} \sum_{\sigma \in S_4} T^\sigma_{,i\bar{i}j\bar{j}}.$$

Similarly, for any positive integer k

$$\frac{(2k)!}{2^k} S^{(2k)}T = \sum_{i_1,\ldots,i_k} \sum_{\sigma \in S_{2k}} T^\sigma_{,i_1\bar{i}_1\ldots i_k\bar{i}_k}.$$

PROOF. We will only prove the special case $k = 2$ since the proof of the general case is exactly the same. Recall that $b_{i\bar{i}j\bar{j}}$ is defined by

$$b_{i\bar{i}j\bar{j}} = \int_\eta a_i(\eta)\overline{a_i(\eta)}a_j(\eta)\overline{a_j(\eta)}.$$

Obviously $b_{i\bar{i}i\bar{i}} = b_{i'\bar{i}'i'\bar{i}'}$ for all i and i', $1 \le i, i' \le n$, and $b_{i\bar{i}j\bar{j}} = b_{i'\bar{i}'j'\bar{j}'}$ for $i \ne j$ and $i' \ne j'$. It follows that $S^{(4)}T$ must be of the form

$$S^{(4)}T = c\sum_i \sum_{\sigma \in S_4} T^\sigma_{,i\bar{i}i\bar{i}} + c' \sum_{i \ne j} \sum_{\sigma \in S_4} T^\sigma_{,i\bar{i}j\bar{j}}. \tag{15.1}$$

We claim that in the Euclidean case $S^{(4)}T$ agrees with $\Delta^2 T$ for a suitable choice of c_{2k}. We denote the operators $S^{(4)}$ and Δ^2 in the Euclidean case by $S_0^{(4)}$ and Δ_0^2 respectively. Then, the symbol of the fourth-order operator $S_0^{(4)}$ with constant coefficients is a fourth-order polynomial on $\mathbf{C}^n$ (with coordinates ξ) in $\xi_1, \ldots, \xi_n; \bar{\xi}_1, \ldots, \bar{\xi}_n$ invariant under rotations, so that it must be a (positive) multiple of $(\sum_i |\xi_i|^2)^2$, the symbol of Δ_0^2, hence proving the claim. From now on we will choose the constant $c_{2k} > 0$ such that $S_0^{(4)} = \Delta_0^2$. Now from (15.1) we have

$$c\sum_i \sum_{\sigma \in S_4} T^\sigma_{,i\bar{i}i\bar{i}} + c' \sum_{i \ne j} \sum_{\sigma \in S_4} T^\sigma_{,i\bar{i}j\bar{j}} = \sum_{i,j} T_{,(i\bar{i}+\bar{i}i)(j\bar{j}+\bar{j}j)}$$

in the Euclidean case. But in this case $T^\sigma_{,\alpha\beta\gamma\delta} = T_{,\alpha\beta\gamma\delta}$. By comparing coefficients this yields that $c' = c$, so that in general

$$S^{(4)}T = c\sum_{i,j} \sum_{\sigma \in S_4} T^\sigma_{,i\bar{i}j\bar{j}}.$$

The constant c can be obtained by setting the right-hand side equal to $\Delta_0^2 T = 4 \sum_l i, j T_{,i\bar{l}j\bar{j}}$ in the Euclidean case, yielding the special case of $k = 2$ and in an analogous manner the general case of Proposition 15.8. □

15.2.4. Conversion of Radial Derivatives to Mixed Covariant Derivatives

In the argument of showing that certain components of covariant derivatives of the curvature tensor vanish at a given point x, it will be a typical situation first to show radial derivatives of certain orders along geodesics γ through the point x vanishing and then to show similar vanishing phenomena for mixed covariant derivatives. Suppose for some fixed positive integer k we have $\nabla_\eta^k R_{i\bar{j}k\bar{l}}(x) = 0$ for all real tangent vectors η at x. In the Euclidean case this would mean that for $R_{i\bar{j}k\bar{l}}$ all covariant derivatives of degree k at x, symbolically $\nabla^k R_{i\bar{j}k\bar{l}}(x)$, would vanish at x. However, for a general Kähler manifold this is not the case. We have the following proposition in the general case of Riemannian manifolds.

PROPOSITION 15.9. *Let M be an m-dimensional Riemannian manifold and let $x \in M$ such that for the given smooth tensor T, the covariant of order k, $\nabla_\eta^k T_{i_1 i_2 \cdots i_l}(x) = 0$ for some specific indices $i_1, \ldots, i_s$ and for all real tangent vectors η at x. For any $\sigma \in S_k$, we denote by $(\nabla_1^{k_1} \nabla_2^{k_2} \cdots \nabla_m^{k_m})^\sigma T_{i_1 i_2 \cdots i_l}(x)$ the components of covariant derivatives obtained by permuting the order of the $k = k_1 + \cdots + k_m$ derivatives using σ. Then we have*

$$\sum_{\sigma \in S_k} (\nabla_1^{k_1} \nabla_2^{k_2} \cdots \nabla_m^{k_m})^\sigma T_{i_1 \cdots i_l}(x) = 0$$

for any set of nonnegative integers $k_1, \ldots, k_m$ such that $k_1 + \cdots + k_m = k$.

PROOF. Let $x_1, \ldots, x_m$ be real normal geodesic coordinates at the point $x \in M$. The point x is then the origin in this coordinate system. For any real tangent vector $\eta = a_1 \partial/\partial x_1 + \cdots + a_m \partial/\partial x_m$ of unit length we have

$$T_{i_1 i_2 \cdots i_l}(t\eta) = \sum_{s \geq 0} \frac{t^s}{s!} \nabla_\eta^s T_{i_1 i_2 \cdots i_l} = \sum_{s \geq 0} \frac{1}{s!} \nabla_{t\eta}^s T_{i_1 i_2 \cdots i_l}.$$

Writing $t\eta = (x_1, x_2, \ldots, x_m)$ we have

$$T_{i_1 i_2 \cdots i_l}(x_1, x_2, \ldots, x_m)$$

$$= \sum_{s \geq 0} \sum_{\substack{s_1 + \cdots + s_m = s \\ s_i \geq 0}} \sum_{\sigma \in S_s} (\nabla_1^{s_1} \nabla_2^{s_2} \cdots \nabla_m^{s_m})^\sigma T_{i_1 i_2 \cdots i_l}(0) x_1^{s_1} x_2^{s_2} \cdots x_m^{s_m},$$

where S_s is the group of formal permutations of the s indices involved in the covariant differentiation. The proposition follows immediately by setting equal to zero all the coefficients of kth-order monomials $x_1^{k_1} x_2^{k_2} \cdots x_m^{k_m}$, $k_1 + k_2 + \cdots + k_m = k$, which must be the case when $\nabla_\eta^k T_{i_1 i_2 \cdots i_l}(0) = 0$ for all real tangent vectors η at x. $\qquad\square$

REMARKS 15.10. In the complex case we can rewrite the formula in Proposition 15.9 by allowing the differentiations to be against barred or unbarred indices.

15.2.5. Second Variation Inequalities Associated with the Curvature Tensor

Let X be a Kähler manifold and $x \in X$ be a point where holomorphic bisectional curvatures are nonnegative. There are two important and well-known inequalities associated to $R_{i\bar{j}k\bar{l}}(x)$. They are respectively related to maximal directions of holomorphic sectional curvatures and flat (zero) directions of holomorphic bisectional curvatures. We formulate them in the form of two lemmata.

LEMMA 15.11. *Let X be a Kähler manifold and $x \in X$ be a point where holomorphic bisectional curvatures are nonnegative. Suppose $\alpha \in T_x^{1,0}(X)$ of unit length is the direction attaining the maximum of all holomorphic sectional curvatures at x. Then, for any $\xi \in T_x^{1,0}(X)$ of unit length which is perpendicular to α, we have*

$$0 \leq 2R_{\alpha\bar{\alpha}\xi\bar{\xi}}(x) + |R_{\alpha\bar{\xi}\alpha\bar{\xi}}(x)| \leq R_{\alpha\bar{\alpha}\alpha\bar{\alpha}}(x).$$

PROOF. We shall henceforth call α a maximal direction of holomorphic sectional curvatures at x or simply a maximal direction at x. We have, for any $\varepsilon > 0$ and any real θ,

$$R\langle \alpha + \varepsilon e^{i\theta}\xi, \overline{\alpha + \varepsilon e^{i\theta}\xi}; \alpha + \varepsilon e^{i\theta}\xi, \overline{\alpha + \varepsilon e^{i\theta}\xi}\rangle \leq (1 + \varepsilon^2)^2 R_{\alpha\bar{\alpha}\alpha\bar{\alpha}}.$$

Since ξ is orthogonal to α, $\|\alpha\| = \|\xi\| = 1$ and α is a maximal direction at x. Here and henceforth in the article we will sometimes drop the reference to the point x when there is no danger of confusion. Comparing the coefficients of ε immediately yields $R_{\alpha\bar{\alpha}\alpha\bar{\xi}} = 0$. Comparing the coefficients of ε^2 then yields

$$4R_{\alpha\bar{\alpha}\xi\bar{\xi}} + 2 \operatorname{Re}(e^{2i\theta} R_{\alpha\bar{\xi}\alpha\bar{\xi}}) \leq 2R_{\alpha\bar{\alpha}\alpha\bar{\alpha}}.$$

We can always choose the angle θ so that $2 \operatorname{Re}(e^{2i\theta} R_{\alpha\bar{\xi}\alpha\bar{\xi}}) \geq 0$, yielding

$$0 \leq 2R_{\alpha\bar{\alpha}\xi\bar{\xi}} + |R_{\alpha\bar{\xi}\alpha\bar{\xi}}| \leq R_{\alpha\bar{\alpha}\alpha\bar{\alpha}},$$

the desired inequality. $\qquad\square$

LEMMA 15.12. *Let X be a Kähler manifold and $x \in X$ be a point where holomorphic bisectional curvatures are nonnegative. Suppose $\alpha, \beta \in T_x^{1,0}(X)$ are such that $R_{\alpha\bar{\alpha}\beta\bar{\beta}}(x) = 0$. Let $\xi, \zeta \in T_x^{1,0}(X)$ be arbitrary. Then*

$$|R_{\alpha\bar{\xi}\beta\bar{\zeta}}|^2 + |R_{\alpha\bar{\beta}\xi\bar{\zeta}}|^2 \le R_{\alpha\bar{\alpha}\xi\bar{\xi}} R_{\beta\bar{\beta}\zeta\bar{\zeta}}.$$

PROOF. Since holomorphic bisectional curvatures are nonnegative at x, we have for all $\delta, \varepsilon > 0$ and θ, ϕ real

$$R\langle \alpha + \delta e^{i\theta}\zeta, \overline{\alpha + \delta e^{i\theta}\zeta}; \beta + \varepsilon e^{i\phi}\xi, \overline{\beta + \varepsilon e^{i\phi}\xi} \rangle \ge 0.$$

Expanding in terms of δ, z and writing out the coefficients of δ, ε we obtain

$$R_{\alpha\bar{\alpha}\beta\bar{\xi}} = R_{\beta\bar{\beta}\alpha\bar{\zeta}} = 0.$$

From the second-order terms we obtain

$$\varepsilon^2 R_{\alpha\bar{\alpha}\xi\bar{\xi}} + \delta^2 R_{\beta\bar{\beta}\zeta\bar{\zeta}} + 2 \operatorname{Re} \varepsilon\delta(e^{-i(\theta+\phi)}R_{\alpha\bar{\xi}\beta\bar{\zeta}}) + 2 \operatorname{Re} \varepsilon\delta(e^{i(\phi-\theta)}R_{\alpha\bar{\beta}\xi\bar{\zeta}}) \ge 0$$

for δ, ε sufficiently small and hence for all $\delta, \varepsilon > 0$. By making the transformations $\xi \mapsto e^{i\theta_0}\xi$ and $\zeta \mapsto e^{i\phi_0}\zeta$ so that $R_{\alpha\bar{\xi}\beta\bar{\zeta}}$ is changed to $e^{-i(\theta_0+\phi_0)}R_{\alpha\bar{\xi}\beta\bar{\zeta}}$ and $R_{\alpha\bar{\beta}\xi\bar{\zeta}}$ is changed to $e^{i(\theta_0-\phi_0)}R_{\alpha\bar{\beta}\xi\bar{\zeta}}$ we may without loss of generality assume that to start with both $R_{\alpha\bar{\xi}\beta\bar{\eta}}$ and $R_{\alpha\bar{\beta}\xi\bar{\zeta}}$ are real. Then by choosing $\theta = \phi = 0$, we obtain from the discrimnant

$$|R_{\alpha\bar{\xi}\beta\bar{\zeta}} + R_{\alpha\bar{\beta}\xi\bar{\zeta}}|^2 \le R_{\alpha\bar{\alpha}\xi\bar{\xi}} R_{\beta\bar{\beta}\zeta\bar{\zeta}}.$$

By choosing $\phi = \pi/2, \theta = -\pi/2$, we obtain, on the other hand,

$$|R_{\alpha\bar{\xi}\beta\bar{\zeta}} - R_{\alpha\bar{\beta}\xi\bar{\zeta}}|^2 \le R_{\alpha\bar{\alpha}\xi\bar{\xi}} R_{\beta\bar{\beta}\zeta\bar{\zeta}}.$$

Summing up the two inequalities and dividing by 2, we obtain the desired inequality

$$|R_{\alpha\bar{\xi}\beta\bar{\zeta}}|^2 + |R_{\alpha\bar{\beta}\xi\bar{\zeta}}|^2 \le R_{\alpha\bar{\alpha}\xi\bar{\xi}} R_{\beta\bar{\beta}\zeta\bar{\zeta}}. \qquad \square$$

15.3. Zero-Order Information on Curvature Terms Associated with a Global Maximal Direction

15.3.1. Berger's Computation

By using the computation of Section 15.2.2 and Lemma 15.11, Berger [3] obtained, in the case of positive sectional curvature, that for a unit vector $\alpha \in T_x^{1,0}(X)$ attaining the global maximum of all holomorphic sectional curvatures, $R_{\alpha\bar{\alpha}\xi\bar{\xi}} = \frac{1}{2}R_{\alpha\bar{\alpha}\alpha\bar{\alpha}}$ for any $\xi \in T_x^{1,0}(X)$ of unit length and orthogonal to α. In the case of nonnegative holomorphic bisectional curvature, his computation yields immediately.

PROPOSITION 15.13. $T_x^{1,0}(X)$ *splits into the orthogonal direct sum* $\mathbf{C}\alpha \oplus \mathcal{H} \oplus \mathcal{N}$, *where* $\mathcal{H}$ *consists of all* $\xi \in T_x^{1,0}(X)$ *such that* $R_{\alpha\bar{\alpha}\xi\bar{\xi}} > 0$ *and* $\mathcal{N}$ *consists of all* $\zeta \in T_x^{1,0}(X)$ *such that* $R_{\alpha\bar{\alpha}\xi\bar{\xi}} = 0$.

PROOF. Since $R_{\alpha\bar{\alpha}\alpha\bar{\alpha}}$ is a global maximum of all holomorphic sectional curvatures, we have

$$\Delta R_{\alpha\bar{\alpha}\alpha\bar{\alpha}}(x) = \text{const} \int_{\eta} \nabla_{\eta}^2 R\langle \alpha(y), \bar{\alpha}(y), \alpha(y), \bar{\alpha}(y)\rangle(x) \le 0$$

as in Section 15.2.3 where η ranges over all real tangent vectors of unit length, $\alpha(y)$ denotes on a neighborhood of x the vector field obtained from $\alpha(x) = \alpha$ by parallel transport along geodesics from x and the integration is with respect to the rotation-symmetric measure of the unit sphere of $T_x(X)$. On the other hand, from Section 15.2.2 we have

$$\tfrac{1}{2}\Delta R_{\alpha\bar{\alpha}\alpha\bar{\alpha}} = \sum_{\substack{i\neq 1 \\ \text{or } j\neq 1}} |R_{\alpha\bar{i}\alpha\bar{j}}|^2 + \sum_{i\neq 1} (R_{\alpha\bar{\alpha}\alpha\bar{\alpha}} - 2R_{\alpha\bar{\alpha}i\bar{i}})R_{\alpha\bar{\alpha}i\bar{i}}$$

for an orthonormal basis $\{e_i\}$ of $T_x^{1,0}(X)$ such that $e_1 = \alpha$ and $R_{\alpha\bar{\alpha}i\bar{j}} = 0$ for $i \neq j$. From Lemma 15.11 we have $R_{\alpha\bar{\alpha}i\bar{i}} \le \frac{1}{2}R_{\alpha\bar{\alpha}\alpha\bar{\alpha}}$ yielding

$$\Delta R_{\alpha\bar{\alpha}\alpha\bar{\alpha}} \ge 0.$$

Equality holds if and only if $R_{\alpha\bar{\alpha}i\bar{i}} = \frac{1}{2}R_{\alpha\bar{\alpha}\alpha\bar{\alpha}}$ or 0 for $i > 1$ and that $R_{\alpha\bar{i}\alpha\bar{j}} = 0$ for all i, j except $i = j = 1$. In particular we have the orthogonal decomposition of $T_x^{1,0}(X)$ into eigenspaces of the Hermitian bilinear form $H_\alpha(\xi, \xi) = R_{\alpha\bar{\alpha}\xi\bar{\xi}}$. $\square$

From now on we shall fix an α and call $\mathcal{H}_\alpha = \mathcal{H}$ the half-space and $\mathcal{N}_\alpha = \mathcal{N}$ the null space associated to α.

Also from here on we shall fix an $\alpha \in T^{1,0}(X)$ of unit length such that $R_{\alpha\bar{\alpha}\alpha\bar{\alpha}} = \sup_{\xi \in T^{1,0}(X), \|\xi\|=1} R_{\xi\bar{\xi}\xi\bar{\xi}}$, where $\alpha \in T_x^{1,0}(X)$ shall be termed a global maximal direction (of holomorphic sectional curvatures). We shall also fix an orthonormal basis of $T_x^{1,0}(X)$, according to the orthogonal decomposition $T_x^{1,0}(X) = \mathbf{C}\alpha \oplus \mathcal{H} \oplus \mathcal{N}$, consisting of $\{e_1, \ldots, e_n\}$ such that $e_1 = \alpha$, $e_p \in \mathcal{H}$ for $2 \le p \le m$ and $e_q \in \mathcal{N}$ for $m + 1 \le q \le n$. Write $H = \{2, \ldots, m\}$ and $N = \{m + 1, \ldots, n\}$. For $\alpha \in T_x^{1,0}(X)$ fixed, we shall typically use the indices given by the above choice of bases. Since the choice of $\{e_p : p \in H\}$ and $\{e_q : q \in N\}$ is arbitrary within $\mathcal{H}$ and $\mathcal{N}$ as long as they form orthonormal bases of $\mathcal{H}$ and $\mathcal{N}$ respectively, we shall also use the notations e_p and e_q for general elements of $\mathcal{H}$ and $\mathcal{N}$ of unit length. Any orthonormal basis $\{e_1\} \cup \{e_p : p \in H\} \cup \{e_q : q \in N\}$ associated to $\alpha \in T_{x_0}^{1,0}(X)$ will be called a privileged orthonormal basis associated to α.

To systematize once and for all the choice of notations, we use, as has been the case, ξ, ζ to denote the complexified tangent vectors of type $(1, 0)$ and η to denote the general real tangent vectors. The new indices arising from substitution in commutation formulas will be denoted by μ, ν.

15.3.2. Equations Satisfied by Curvature Terms Associated to a Global Maximal Direction

We collect here the necessary information on $R_{i\bar{j}k\bar{l}}$ associated to a global maximal direction of holomorphic sectional curvatures $\alpha = e_1 \in T_x^{1,0}(X)$. Some of these equations are already contained in Proposition 15.13 and its proof.

PROPOSITION 15.14. *Let* $\alpha \in T_{x_0}^{1,0}(X)$ *of unit length be a global maximal direction of holomorphic sectional curvatures and let* $\{e_1\} \cup \{e_p : p \in H\} \cup \{e_q : q \in N\}$ *be a privileged orthonormal basis of* $T_{x_0}^{1,0}(X)$ *associated to* α. *In terms of this basis, we have*

(a) $R_{1\bar{1}p\bar{p}} = \frac{1}{2}R_{1\bar{1}1\bar{1}} > 0$ *for* $p \in H$.
(b) $R_{1\bar{i}1\bar{j}} = 0$ *for* $i \neq 1$ *or* $j \neq 1$.
(c) $R_{1\bar{q}i\bar{j}} = 0$ *for* $q \in N$ *and* $1 \le i, j \le n$. (*In particular* $R_{1\bar{1}q\bar{q}} = R_{1\bar{1}p\bar{q}} = 0$ *for* $p \in H$ *and* $q \in N$.)
(d) $R_{1\bar{p}p\bar{p}} = 0$ *for* $p \in H$.
(e) $\sum_{r \in H} |R_{1\bar{\xi}q\bar{r}}|^2 = R_{1\bar{1}\xi\bar{\xi}} R_{\xi\bar{\xi}q\bar{q}}$ *for* $\xi \in \mathcal{H}$ *and* $q \in N$.
(e)' $R_{p\bar{p}'q\bar{q}'} = (2/R_{1\bar{1}1\bar{1}}) \sum_{r \in H} R_{1\bar{p}'q\bar{r}} R_{p\bar{1}r\bar{q}'}$.

Before proving Proposition 15.14 we give a few remarks on our formulation of the equations. Since the equations are stated for arbitrary choices of privileged orthonormal basis associated to α, the equations are satisfied for e_p an arbitrary unit vector of the half-space $\mathcal{H}$ and e_q an arbitrary unit vector of the null-space $\mathcal{N}$. For example, the equation $R_{1\bar{1}q\bar{q}} = 0$ implies by polarization $R_{1\bar{1}q\bar{q}'} = 0$ for $q, q' \in N$, and the equation $R_{1\bar{p}p\bar{p}} = 0$ for $p \in H$ implies by polarization $R_{1\bar{p}p'\bar{p}''} = 0$ for $p, p', p'' \in H$. Equation (e) implies by polarization the representation $(e)'$ of curvature terms $R_{p\bar{p}'q\bar{q}'}$ in terms of $R_{1\bar{i}jk}$.

PROOF OF PROPOSITION 15.14. (a), (b), and the special case of (c) in parentheses are included either in the statement or the proof of Proposition 15.13. We first prove (d) by a third-order variation equality at $\alpha = e_1$. Consider the function $F(\varepsilon)$ in the real variable ε defined by

$$F(\varepsilon) = \frac{1}{(1 + \varepsilon^2)^2} R\langle e_1 + \varepsilon e_p, \overline{e_1 + \varepsilon e_p}, e_1 + \varepsilon e_p, \overline{e_1 + \varepsilon e_p}\rangle.$$

Proposition 15.14 implies the vanishing of both the first and the second variation of $F(\varepsilon)$ at $\varepsilon = 0$. In fact

$$F(\varepsilon) = \frac{1}{1 + 2\varepsilon^2 + \varepsilon^4} (R_{1\bar{1}1\bar{1}} + 4\varepsilon^2 R_{1\bar{1}p\bar{p}} + 4\varepsilon^3 \operatorname{Re} R_{1\bar{p}p\bar{p}} + \varepsilon^4 R_{p\bar{p}p\bar{p}})$$

since $R_{1\bar{1}1\bar{p}} = R_{1\bar{p}1\bar{p}} = 0$.

Since $R_{1\bar{1}p\bar{p}} = \frac{1}{2}R_{1\bar{1}1\bar{1}}$ we obtain

$$F(\varepsilon) = R\left\langle \frac{e_1 + \varepsilon e_p}{\sqrt{1 + \varepsilon^2}}, \frac{\overline{e_1 + \varepsilon e_p}}{\sqrt{1 + \varepsilon^2}}, \frac{e_1 + \varepsilon e_p}{\sqrt{1 + \varepsilon^2}}, \frac{\overline{e_1 + \varepsilon e_p}}{\sqrt{1 + \varepsilon^2}}\right\rangle$$

$$= \frac{1}{(1 + 2\varepsilon^2 + \varepsilon^4)} (R_{1\bar{1}1\bar{1}} + 2\varepsilon^2 R_{1\bar{1}1\bar{1}} + 4\varepsilon^3 \operatorname{Re} R_{1\bar{p}p\bar{p}} + \varepsilon^4 R_{p\bar{p}p\bar{p}}).$$

Since $F(0) = R_{1\bar{1}1\bar{1}}$, by comparing coefficients of Taylor expansions of the denominator and the numerator, we have immediately $dF(0)/d\varepsilon = d^2F(0)/d\varepsilon^2 = 0$. Since $F(\varepsilon) \le R_{1\bar{1}1\bar{1}}$ the third-order variation equality yields

$$\frac{d^3F}{d\varepsilon^3}(0) = 4 \operatorname{Re} R_{1\bar{p}p\bar{p}} \le 0.$$

Since the same equality holds with e_p replaced by $e^{i\theta}e_p$, we conclude the equation (d)

$$R_{1\bar{p}p\bar{p}} = 0 \quad \forall p \in H.$$

(Recall that $e_p \in \mathcal{H}$, $\|e_p\| = 1$, is arbitrary so that by polarization $R_{1\bar{p}p'\bar{p}''} = 0$ for $p, p', p'' \in H$.)

To finish the proof of Proposition 15.14 we shall need the following analysis at the zero directions $R_{1\bar{1}q\bar{q}}(x_0)$ of holomorphic bisectional curvatures.   □

PROPOSITION 15.15. *Let* $q \in N$ *and* $\alpha = e_1 \in T_{x_0}^{1,0}(X)$ *be a global maximal direction of holomorphic sectional curvatures. Then,* $\Delta R_{1\bar{1}q\bar{q}}(x_0) = 0$. *Hence,* $\nabla_\eta^i R_{1\bar{1}q\bar{q}}(x_0) = 0$ *for* $0 \le i \le 3$.

PROOF. First, we compute $\Delta R_{1\bar{1}q\bar{q}}$. It is immediate that $R_{1\bar{1}q\bar{q},i\bar{i}} = R_{1\bar{1}i\bar{q},q\bar{i}}$ by a commutation formula. Hence

$$\tfrac{1}{2}\Delta R_{1\bar{1}q\bar{q}} = \sum_i R_{11q\bar{q},i\bar{i}} \quad \text{(summations over } i \text{ with unspecified}$$
$$\text{ranges will henceforth mean } 1 \le i \le n)$$

$$= \sum_i R_{1\bar{1}i\bar{q},q\bar{i}}$$

$$= \sum_i R_{1\bar{1}i\bar{q},\bar{i}q}$$

$$+ \sum_{i,\mu} (R_{1\bar{\mu}q\bar{i}}R_{\mu\bar{1}i\bar{q}} - R_{\mu\bar{1}q\bar{i}}R_{1\bar{\mu}i\bar{q}} + R_{i\bar{\mu}q\bar{i}}R_{1\bar{1}\mu\bar{q}} - R_{\mu\bar{q}q\bar{i}}R_{1\bar{1}i\bar{\mu}}).$$

Since

$$\sum_i R_{1\bar{1}i\bar{q},\bar{i}q} = \sum_i R_{1\bar{1}i\bar{i},\bar{q}q} = R_{1\bar{1},\bar{q}q} = 0,$$

we obtain at x_0

$$\tfrac{1}{2}\Delta R_{1\bar{1}q\bar{q}} = \sum_{i,\mu} |R_{1\bar{\mu}q\bar{i}}|^2 + R_{1\bar{1}q\bar{q}} - \sum_{i,\mu} |R_{1\bar{q}i\bar{\mu}}|^2 - \sum_{i,\mu} R_{1\bar{1}i\bar{\mu}}R_{\mu\bar{i}q\bar{q}}.$$

From $R_{1\bar{1}q\bar{q}} = 0$ and first variation equalities, noting that bisectional curvatures on X are nonnegative, we obtain immediately

$$R_{1\bar{i}q\bar{q}} = R_{1\bar{1}q\bar{i}} = 0 \quad \text{for } 1 \le i \le n, \, e_q \in \mathcal{N}, \|e_q\| = 1 \text{ arbitrary.}$$

This yields at x_0

$$\tfrac{1}{2}\Delta R_{1\bar{1}q\bar{q}} = \sum_{p,r \in H} |R_{q\bar{p}q\bar{r}}|^2 - \sum_{i,j} |R_{1\bar{q}i\bar{j}}|^2 - \sum_{p \in H} R_{1\bar{1}p\bar{p}}R_{p\bar{p}q\bar{q}}.$$

We claim that from the second-variation inequality Lemma 15.12

$$\Delta R_{1\bar{1}q\bar{q}}(x_0) \le 0.$$

Since X carries nonnegative holomorphic bisectional curvatures

$$\Delta R_{1\bar{1}q\bar{q}}(x_0) \geq 0,$$

yielding Proposition 15.15. To prove the inequality $\Delta R_{1\bar{1}q\bar{q}}(x_0) \leq 0$ we give two different approaches. First, we recall the following lemma in linear algebra.

LEMMA 15.16. *Let $S(z; z')$ be a complex symmetric bilinear form on a complex vector space $\mathbf{C}^n$ represented by the matrix S with respect to the canonical coordinates of $\mathbf{C}^n$. Then there exists a unitary transformation U of $\mathbf{C}^n$ (relative to the Euclidean Hermitian structure) such that $U^t S U$ is a diagonal matrix.*

Using Lemma 15.16, we diagonalize the complex symmetric bilinear form $S(\xi; \xi) = R_{1\bar{\xi}q\bar{\xi}}$ on $\mathcal{H}$ yielding $R_{1\bar{p}q\bar{p}'} = 0$ for $p \neq p$ for some choice of orthonormal basis $\{e_p\}$ of $\mathcal{H}$. In this coordinate system we have

$$\tfrac{1}{2}\Delta R_{1\bar{1}q\bar{q}} = \sum_{p \in H} |R_{1\bar{p}q\bar{p}}|^2 - \sum_{i,j} |R_{1\bar{q}i\bar{j}}|^2 - \sum_{p \in H} R_{1\bar{1}p\bar{p}} R_{p\bar{p}q\bar{q}}.$$

Note that $\sum_{p \in H} |R_{1\bar{p}q\bar{r}}|^2$ is invariant under the unitary transformations on $\mathcal{H}$ since $\mathrm{tr}\,(U^t S \bar{S} \bar{U}) = \mathrm{tr}\,(U^t S \overline{SU}) = \mathrm{tr}\,(S\bar{S})$. From the second-variation inequality in Lemma 15.12 we have

$$|R_{1\bar{p}q\bar{p}}|^2 + |R_{1\bar{q}p\bar{p}}|^2 \leq R_{1\bar{1}p\bar{p}} R_{p\bar{p}q\bar{q}}.$$

This yields

$$\tfrac{1}{2}\Delta R_{1\bar{1}q\bar{q}} \leq -\sum_{i,j} |R_{i\bar{q}i\bar{j}}|^2 \leq 0.$$

This yields a proof of Proposition 15.15 and with it also the equation (d)

$$R_{i\bar{q}i\bar{j}} = 0 \quad \forall i, j, 1 \leq i, j \leq n.$$

Since Lemma 15.16 is proved entirely by algebraic means, it would be desirable to give a geometric proof of

$$\sum_{p,r} |R_{1\bar{p}q\bar{r}}|^2 \leq \sum_{p} R_{1\bar{1}p\bar{p}} R_{p\bar{p}q\bar{q}} \tag{15.2}$$

in our situation without a special choice of coordinates. We claim that for any $\xi \in \mathcal{H}$ and any orthonormal basis $\{e_r\}$ of $\mathcal{H}$,

$$\sum_{r \in H} |R_{1\bar{\xi}q\bar{r}}|^2 + \sum_{r \in H} |R_{1\bar{q}\xi\bar{r}}|^2 \leq \tfrac{1}{2} R_{1\bar{1}1\bar{1}} R_{\xi\bar{\xi}q\bar{q}}. \tag{15.3}$$

Then, integrating (15.3) over $\xi \in \mathcal{H}$ of unit length using a rotation-symmetric metric on the unit sphere yields immediately (15.2) and hence another proof of Proposition 15.15.

To prove (15.3) observe first of all that Lemma 15.12 yields only

$$|R_{1\bar{p}q\bar{r}}|^2 \leq \tfrac{1}{2} R_{1\bar{1}1\bar{1}} R_{\xi\bar{\xi}q\bar{q}}.$$

We shall now prove (15.3) by making a better use of the argument of the second-variation used in Lemma 15.12. Consider the Taylor expansion of

$$G(\varepsilon) = R\left(e_1 + \varepsilon\xi, \overline{e_1 + \varepsilon\xi}, e_q + \varepsilon \sum_{r \in H} C_r e_r, \overline{e_q + \varepsilon \sum_{r \in H} C_r e_r}\right).$$

We then have

$$G(\varepsilon) = \varepsilon^2\left(R_{\xi\bar{\xi}q\bar{q}} + \sum_{r \in H} |C_r|^2 R_{1\bar{1}r\bar{r}} + \sum_{r \in H} 2\,\mathrm{Re}\,\bar{C}_r R_{1\bar{\xi}q\bar{r}} + \sum_{r \in H} 2\,\mathrm{Re}\,C_r R_{1\bar{q}r\bar{\xi}}\right).$$

Since X carries nonnegative holomorphic bisectional curvatures, $\partial^2 G(0)/\partial\varepsilon^2$ is always nonnegative for any choice of complex numbers C_r. It follows that the quadratic form Q in $(z_1, z_2, \ldots, z_m)$, defined by

$$Q((z_1, \ldots, z_m); (z_1, \ldots, z_m)) = |z_1|^2 R_{\xi\bar{\xi}q\bar{q}} + \sum_{2 \leq r \leq m} |z_r|^2 \frac{R_{1\bar{1}1\bar{1}}}{2}$$

$$+ \sum_{2 \leq r \leq m} 2\,\mathrm{Re}\,(z_1 \bar{z}_r R_{1\bar{\xi}q\bar{r}} + \bar{z}_1 z_r R_{1\bar{q}r\bar{\xi}}),$$

is positive semidefinite. Now take z_r to be of the form $x_r\,e^{i\theta_r}$, x_r, θ_r real, for $2 \leq r \leq m$ and take $z_1 = x_1$ real and positive. Choose θ_r and replace e_q by $e^{i\alpha} e_q$ for an appropriate real α so that $e^{-i\theta_r} R_{1\bar{\xi}q\bar{r}}$ is real and ≥ 0 while $e^{i\theta_r} R_{1\bar{q}\xi\bar{r}}$ is ≤ 0. By computing the determinant of the symmetric matrix representing the real symmetric bilinear form Q_θ given by

$$Q_\theta((x_1, \ldots, x_m); (x_1, \ldots, x_m))$$

$$= Q((x_1, x^2\,e^{i\theta_2}, \ldots, x_m\,e^{i\theta_m}); (x_1, x\,e^{i\theta_2}, \ldots, x\,e^{i\theta_m})),$$

we conclude immediately that

$$\sum_{r \in H} ||R_{1\bar{\xi}q\bar{r}}| - |R_{1\bar{q}\xi\bar{r}}||^2 \leq \frac{R_{1\bar{1}1\bar{1}}}{2} R_{\xi\bar{\xi}q\bar{q}}.$$

By a similar argument we have

$$\sum_{r \in H} ||R_{1\bar{\xi}q\bar{r}}| + |R_{1\bar{q}\bar{\xi}\bar{r}}||^2 \le \frac{R_{1\bar{1}1\bar{1}}}{2} R_{\xi\bar{\xi}q\bar{q}}.$$

Adding the two equations and dividing by 2, we obtain immediately (15.3).

END OF PROOF OF PROPOSITION 15.14. The second proof of Proposition 15.15 yields immediately from the equality $\Delta R_{1\bar{1}q\bar{q}}(x_0) = 0$ the equality

$$\sum_{r \in H} |R_{1\bar{\xi}q\bar{r}}|^2 = \frac{R_{1111}}{2} R_{\xi\bar{\xi}q\bar{q}}. \tag{15.4}$$

This is (e) of Proposition 15.14. (e') of this proposition is obtained easily from (e) by polarization. To see this define a tensor $T_{p\bar{p}'q\bar{q}'}$ of type $(2, 2)$ for $p, p' \in H$ and $q, q' \in N$ by

$$T_{p\bar{p}'q\bar{q}'} = \frac{2}{R_{1\bar{1}1\bar{1}}} \sum_{r \in H} R_{1\bar{p}'q\bar{r}} R_{p\bar{1}r\bar{q}'}.$$

Clearly $T_{\xi\bar{\xi}\zeta\bar{\zeta}} = R_{\xi\bar{\xi}\zeta\bar{\zeta}}$ for $\xi \in \mathcal{H}$ and $\zeta \in \mathcal{N}$. It suffices therefore to show that from $S_{\xi\bar{\xi}\zeta\bar{\zeta}} = T_{\xi\bar{\xi}\zeta\bar{\zeta}} - R_{\xi\bar{\xi}\zeta\bar{\zeta}} = 0$, one can prove $S_{p\bar{p}'q\bar{q}'} = 0$, proving $T_{p\bar{p}'q\bar{q}'} = R_{p\bar{p}'q\bar{q}'}$, i.e., (15.4)'. But now for ε, δ real and θ, ϕ real angles

$$S(e_p + \varepsilon\, e^{i\theta} e'_p, \overline{e_p + \varepsilon\, e^{i\theta} e'_p}; e_q + \delta\, e^{i\phi} e'_q, \overline{e_q + \delta\, e^{i\phi} e'_q}) \equiv 0$$

giving by computing the coefficient of $\varepsilon\delta$,

$$2\, \mathrm{Re}\, e^{-i(\theta+\phi)} R_{p\bar{p}'q\bar{q}'} + 2\, \mathrm{Re}\, e^{i(\theta-\phi)} R_{p'\bar{p}q\bar{q}'} = 0,$$

for all real θ and ϕ. This implies $R_{p\bar{p}'q\bar{q}'} = R_{p'\bar{p}q\bar{q}'} = 0$. $\Box$

15.4. Structure of the Space of Maximal Directions on $T^{1,0}(X)$

15.4.1. Everywhere Existence of Global Maxima of Holomorphic Sectional Curvature

From now on X will stand for a compact Kähler–Einstein manifold of nonnegative holomorphic bisectional curvatures. We denote by $S_x^{1,0}(X)$ the unit sphere of the Hermitian vector space $T_x^{1,0}(X)$ of complexified tangent vectors of type $(1, 0)$. $S^{1,0}(X)$ will denote the sphere bundle thus obtained. Define the function f on X by $f(x) = \sup_{\xi \in S_x^{1,0}(X)} R_{\xi\bar{\xi}\xi\bar{\xi}}$. If $\alpha \in S_x^{1,0}(X)$ and $R_{\alpha\bar{\alpha}\alpha\bar{\alpha}} = f(x)$ we shall call α a maximal direction (of holomorphic sectional curvature at x). Clearly f is a continuous function. Our main result here is the following proposition:

PROPOSITION 15.17. *The function f is constant on X. In other words, the global maximum $\sup_{\xi \in S^{1,0}(X)} R_{\xi\bar{\xi}\xi\bar{\xi}}$ of all holomorphic sectional curvatures is attained at every single point of X.*

PROOF. We prove the proposition by using the maximum principle. It suffices to show that f is subharmonic in the generalized sense. The starting point is the following consequence of Berger's computation.

Let $\alpha \in S_x^{1,0}(X)$ be a maximal direction of holomorphic sectional curvatures at x. Then $\Delta R_{\alpha\bar{\alpha}\alpha\bar{\alpha}}(x) \geq 0$. (15.5)

The proof was given in Proposition 15.14. There, for the verification $\Delta R_{\alpha\bar{\alpha}\alpha\bar{\alpha}}(x) \geq 0$, it suffices to assume that $R_{\alpha\bar{\alpha}\alpha\bar{\alpha}}(x) = \sup_{\xi \in S_x^{1,0}(X)} R_{\xi\bar{\xi}\xi\bar{\xi}}$. We note that since $\Delta R_{\alpha\bar{\alpha}\alpha\bar{\alpha}}$ is the Laplacian of the tensor $R_{\alpha\bar{\alpha}\alpha\bar{\alpha}}$ evaluated at $\langle \alpha, \bar{\alpha}; \alpha, \bar{\alpha} \rangle$ we cannot apply the maximum principle directly. Instead we claim that at each $x \in X$, and for any $\alpha \in S_x^{1,0}(X)$ such that $R_{\alpha\bar{\alpha}\alpha\bar{\alpha}} = f(x)$,

$$\Delta f(x) \geq \Delta R_{\alpha\bar{\alpha}\alpha\bar{\alpha}}(x) \geq 0 \qquad (15.6)$$

in the generalized sense. To prove (15.6) we construct local barrier functions for f, denoted by g_x, as follows. Fix $x \in X$ and $\alpha \in S_x^{1,0}(X)$ with $R_{\alpha\bar{\alpha}\alpha\bar{\alpha}} = f(x) = \sup_{\xi \in S_x^{1,0}(X)} R_{\xi\bar{\xi}\xi\bar{\xi}}$. In an open neighborhood U_x of x within the cut-locus of x we shall denote by $\alpha(y)$ the complexified tangent vector at y of type $(1, 0)$ obtained by parallel transport of $\alpha = \alpha(x)$ along the unique geodesic joining x to y within the cut-locus of x. Define $g_x(y) = R\langle \alpha(y), \overline{\alpha(y)}, \alpha(y), \overline{\alpha(y)} \rangle$ for $y \in U_x$. From the discussion in Section 15.2.3 of averaging operators of radial derivatives we know that

$$\Delta g_x(x) = \Delta R_{\alpha\bar{\alpha}\alpha\bar{\alpha}}(x).$$

We know that on U_x, $g_x \leq f$ and that $g_x(x) = f(x)$. For the Laplacian of continuous functions, we have the generalized definition (following Oka)

$$\Delta f(x) = c_{2n} \lim_{r \to 0} \frac{1}{\gamma^2} \left(\frac{\int_{B(x;r)} f}{\int_{B(x;r)} 1} - f(x) \right).$$

With this definition f is subharmonic on X if and only if $\Delta f(x) \geq 0$ at each point $x \in X$. Obviously $\Delta f(x) \geq \Delta g_x(x)$ since $g_x(x) = f(x)$ and $g_x \leq f$ on $B(x; r)$. It follows that

$$\Delta f(x) \geq \Delta R_{\alpha\bar{\alpha}\alpha\bar{\alpha}}(x) \geq 0$$

in the generalized sense. Thus, f is subharmonic and hence constant on X.   □

15.4.2. Structure of the Bundle of "Maximal Subspaces"

On the unit sphere $S_x^{1,0}(X)$ of $T_x^{1,0}(X)$, we shall denote by $\mathcal{M}_x$ the set of all $\alpha \in S_x^{1,0}(X)$ attaining the global maximum of holomorphic sectional curvatures. By Proposition 15.17, $\mathcal{M}_x$ is nonempty for any $x \in X$. We denote by V_x the complex linear span of $\mathcal{M}_x$ and call it the "maximal subspace" at x. We call $V = \bigcup_{x \in X} V_x \subset T^{1,0}(X)$ the bundle of maximal subspaces. Note that we do not know at this point that V is a differentiable vector subbundle of $T^{1,0}(X)$. Denoting by $\pi: T^{1,0}(X) \to X$ the canonical projection, we shall write $V|_U = V \cap \pi^{-1}(U)$ for the restriction of V to the open set U. We claim that

PROPOSITION 15.18. *There exists a point $x \in X$ such that in some open neighborhood U_x of x, $V|_{U_x}$ is a differentiable complex vector subbundle of $T^{1,0}(U_x)$.*

PROOF. Denote by $\mathcal{M} = \bigcup_{x \in X} \mathcal{M}_x$ the bundle of maximal direction $\mathcal{M} \subset S^{1,0}(X)$, $\mathcal{M}$ is defined by the real-analytic equation $R_{\alpha\bar{\alpha}\alpha\bar{\alpha}} = \sup_{\xi \in S^{1,0}(X)} R_{\xi\bar{\xi}\xi\bar{\xi}}$, so that it is a real-analytic subvariety of the compact real-analytic manifold $S^{1,0}(X)$. Let $\mathcal{M} = \mathcal{M}_1 \cup \cdots \cup \mathcal{M}_l$ be the decomposition of $\mathcal{M}$ into irreducible components. Since the global maximum of holomorphic sectional curvatures is attained at each $x \in X$, we have $\bigcup_{1 \le i \le l} \pi(\mathcal{M}_i) = X$. We arrange the $\mathcal{M}_i$ such that π is a submersion at some regular point of $\mathcal{M}_i$ if and only if $1 \le i \le k$, and denote by $\mathcal{M}'$ the union $\bigcup_{1 \le i \le k} \mathcal{M}_i$. Denote by U the nonempty open set $X - \pi(\bigcup_{i > k} \mathcal{M}_i)$. Then for each $x \in U$, $\mathcal{M}_x \subset \mathcal{M}'$. Let E be the union of singular points of $\mathcal{M}'$ and regular points of $\mathcal{M}'$ at which $\pi: \text{Reg}(\mathcal{M}') \to X$ fails to be a submersion.

We claim the following lemma is true.

LEMMA 15.19. *E is a real-analytic subvariety of $\mathcal{M}'$.*

PROOF. The problem being local, it suffices to prove the following

> Let W be an irreducible real-analytic subvariety of some open subset G of $\mathbf{R}^N$ such that the projection $\rho_w(x_1, \ldots, x_N) = (x_1, \ldots, x_n)$ onto the first n coordinates is of rank n at some point of W. Let S be the union of the singular points of W and the regular points of W at which ρ fails to be a submersion. Then S is a real-analytic subvariety of W. (15.7)

To prove (15.7) assume $0 \in W$ and let $f_1, \ldots, f_k$ be generators of the reduced ideal sheaf $\mathscr{F}_W \cap G'$ for some open neighborhood G' of 0, $G' \Subset G$. We can regard $\mathbf{R}^N$ as the real part of $\mathbf{C}^N$ and extend (cf. Gunning and Rossi [9]) $f_1, \ldots, f_k$ to holomorphic functions $F_1, \ldots, F_k$ on a Stein neighborhood D of 0 in $\mathbf{C}^N$ such that $D \cap \mathbf{R}^N = G'$. Then the common zero set of $F_1, \ldots, F_k$ is a complex-analytic subvariety C of D such that $C \cap G' = W \cap G'$. We can assume, by shrinking D if necessary, that C is an irreducible complex-analytic variety such that the complex dimension of C equals the real dimension of W. Moreover, C is smooth at smooth points of $W \cap G'$. Now $\rho_W : W \to \mathbf{R}^n$ is a submersion at $x \in W$ if and only if the real n form $\rho_W^*(dx_1 \wedge \cdots \wedge dx_n)$ vanishes at x. Consider the projection $\rho_C : C \to \mathbf{C}^n$ defined by $\rho_C(z_1, \ldots, z_N) = (z_1, \ldots, z_N)$ extending $\rho_W|_{W \cap G'}$. Then clearly $\rho_W^*(dx_1 \wedge \cdots \wedge dx_n) = 0$ if and only if $\rho_C^*(dz_1 \wedge \cdots \wedge dz_n)(x) = 0$. Let H be the union of the singular set of C and regular points of C at which ρ_C fails to be a submersion. Then, $S \cap G' = (H \cap G') \cup \operatorname{Sing}(W \cap G')$ which clearly yields (15.7) if we know that H is a complex-analytic subvariety of C.

To show that H is a complex-analytic subvariety of C, we resort to the coherence theorem of Oka and Theorem A of Cartan and Oka (cf. Gunning and Rossi [9]). Consider the sheaf mapping $\phi : \mathcal{O}_C^N \to \mathcal{O}_C^k$, $\mathcal{O}_C$ denoting the reduced structure sheaf of C, defined by

$$\phi(g_1, \ldots, g_N) = \left(\left\langle dF_1, g_1 \frac{\partial}{\partial z_1} + \cdots + g_N \frac{\partial}{\partial z_N} \right\rangle, \right.$$
$$\left. \ldots, \left\langle dF_k, g_1 \frac{\partial}{\partial z_1} + \cdots + g_N \frac{\partial}{\partial z_N} \right\rangle \right),$$

where the pairing between 1-forms and vector fields is defined by

$$\left\langle \omega_1 \, dz_1 + \cdots + \omega_N \, dz_n, g_1 \frac{\partial}{\partial z_1} + \cdots + g_N \frac{\partial}{\partial z_n} \right\rangle = \omega_1 g_1 + \cdots + \omega_N g_N.$$

Clearly ϕ is a morphism between coherent sheaves. Denote the kernel by $\mathcal{F}$. $\mathcal{F}$ can be regarded as a coherent sheaf of restrictions of local holomorphic vector fields on D to C which are tangent to regular points of C. Since $\mathcal{F}$ is a coherent subsheaf of $\mathcal{O}_C^N$ we can define the coherent subsheaf $\wedge^n \mathcal{F}$ of $\wedge^n \mathcal{O}_C^N$. At a regular point z of C, $\rho_C : C \to \mathbf{C}^n$ fails to be a submersion if and only if, under the natural pairing of n-forms and n-vector fields, we have

$$\langle dz_1 \wedge \cdots \wedge dz_n, v_1 \wedge \cdots \wedge v_n \rangle(z) = 0$$

for all holomorphic tangent vectors $v_1, \ldots, v_n$ of C at z. By Theorem A of Cartan and Oka, $(\wedge^n \mathcal{F})_z$ is generated by $\Gamma(D, \wedge^n \mathcal{F})$ since D is Stein. It follows that H is the set of common zeros on C of $\langle dz_1 \cdots dz_n, h \rangle$, where h runs over all holomorphic sections of $\bigwedge^n \mathcal{F}$ on D. Hence, H is a complex analytic subvariety of C. This finishes the proof of the lemma. $\square$

CONTINUATION OF PROOF OF PROPOSITION 15.18. Recall that U is an open subset of X, $\pi^{-1}(U) \cap \mathcal{M} = \mathcal{M}' = \bigcup_{1 \le i \le k} \mathcal{M}_i$, $\pi|_{\mathcal{M}_i}$ is a submersion at some regular point, and E is the union of singular points of $\mathcal{M}'$ and regular points of $\mathcal{M}'$ at which π fails to be a submersion. By the preceding lemma, E is a real-analytic subvariety which obviously does not contain any component of $\mathcal{M}'$. Recall also that the bundle $V = \bigcup_{x \in X} V_x$ is obtained by taking $V_x = $ C-linear span of $\mathcal{M}_x$. We assert

> There exists a nonempty open subset U' of U and a finite
> number of subsets $S_1, \ldots, S_m$ of $\mathcal{M}' \cap \pi^{-1}(U')$ such that   (15.8)
>
> 1. each S_i is a locally closed real-analytic submanifold (possibly discon-
>   nected) of $\pi^{-1}(U')$,
> 2. $\pi|_{S_i}$ is everywhere a submersion,
> 3. $\mathcal{M}' \cap \pi^{-1}(U') = S_1 \cup \cdots \cup S_m$.

We now set forth to prove (15.8). Let $E = \bigcup_{1 \le i \le l} E_i$ be a decomposition of E into irreducible components and assume $E' = \bigcup_{1 \le i \le k} E_i$ is the union of irreducible components containing the branches of $E \cap \mathrm{Reg}\,(\mathcal{M}')$. By Sard's theorem $\pi|_{E_i \cap \mathrm{Reg}(\mathcal{M}')}$ is not a submersion at any point. Hence $\pi(E_i)$ is a closed semianalytic subset (in the sense of Łojasiewicz [17]) of U of measure zero. Define $U_1 = U - \pi(E')$. Then $\pi|_{\mathrm{Reg}(\mathcal{M}') \cap \pi^{-1}(U_1)}$ is everywhere a submersion. We shall choose some $U_1' \subset U_1$, to be determined later, and define S_1 by $S_1 = \mathrm{Reg}\,(\mathcal{M}') \cap \pi^{-1}(U')$. On U_1, let $\mathrm{Sing}\,(\mathcal{M}') \cap \pi^{-1}(U_1) = \bigcup_{1 \le i \le p} T_i$ be a decomposition of $\mathrm{Sing}\,(\mathcal{M}') \cap \pi^{-1}(U_1)$ into irreducible components. For each T_i either $\pi(T_i)$ is a closed semianalytic subset of U_1 or π is a submersion at some regular point of T_i. We arrange T_i such that π is a submersion at some regular point of T_i if and only if $1 \le i \le q$. Now let $1 \le i \le q$. $\pi|_{T_i} : T_i \to U_1$ is not necessarily surjective. Let T_i' be the union of the singular set of T_i and regular points of T_i at which $\pi|_{T_i}$ fails to be a submersion. $\pi(T_i)$ is a closed semianalytic subset of U_1 (because of properness) and $\pi(T_i - T_i')$ is an open subset of U_1 dense in $\pi(T_i)$. Define $F_i = \pi(T_i) - \pi(T_i - T_i')$. Then, on each connected component Ω of $U_1 - F_i$, either $\pi|_{T_i \cap \pi^{-1}(\Omega)}$ maps $T_i \cap \pi^{-1}(\Omega)$ properly and surjectively onto Ω or $T_i \cap \pi^{-1}(\Omega) = \varnothing$. Applying Sard's theorem to the mapping $\pi|_{\mathrm{Reg}(T_i)} : \mathrm{Reg}\,(T_i) \to u_1$ we obtain then a closed subset $\tilde{F}_i$ of measure zero of $U_1 - F_i$ such that $\pi|_{\mathrm{Reg}(T_i) \cap \pi^{-1}(U_1 - F_i - \tilde{F}_i)}$ is everywhere a submersion. Since the boundary of each Ω in U_1 is contained in F_i, clearly $F_i \cup \tilde{F}_i$ is a closed subset (of measure zero) of U_1. We now define $U_2 = U_1 - \bigcup_{1 \le i \le q} (F_i \cup \tilde{F}_i) - \bigcup_{q \le i \le p} \pi(T_i)$. For $U' \subset U_2$ to be determined we define $S_2 = \text{Reg}\,(T_1) \cap \pi^{-1}(U')$, etc. It is now clear that one can go on by removing step by step the singular set of irreducible components of the preceding singular set in order to obtain open sets $U, U_1, U_2, \ldots, U_s$ all derived from the preceding set by removing a closed (semianalytic) subset of measure zero until we obtain the last open set $U' = U_s$ and the closed real-analytic submanifolds $S_i, 1 \le i \le m$, of $\pi^{-1}(U')$ on which $\pi|_{S_i}$ is everywhere a submersion. Obviously $\mathcal{M}' \cap \pi^{-1}(U') = S_1 \cup \cdots \cup S_m$.

Proposition 15.18 will now be proved by picking some point $x \in U'$ and some open neighborhood U_x of x contained in U'. Let $x \in U'$ be a point such that V_x is of maximum dimension among points on U'. Suppose $\{v_1, \ldots, v_s\}$ is a basis of V_s with $v_i \in \mathcal{M}'_x$. Each v_i is contained in one of the pieces $S_j, 1 \le j \le m$. Since $\pi|_{S_j} : S_j \to U'$ is a submersion it follows that there exist vector fields $v_i(y)$ defined for y sufficiently close to x such that $v_i(x) = v_i, 1 \le i \le s$. For y sufficiently close to x, say $y \in U_x$, $\{v_1(y), \ldots, v_s(y)\}$ are linearly independent. But since $\dim_{\mathbb{C}} V_y \le \dim V_x$ it follows that $V|_{U_x}$ is an s-dimensional complex vector bundle generated at each point by $\{v_1(y), \ldots, v_s(y)\}$.

15.5. The Maximum Principle for Fourth-Order Radial Derivatives

15.5.1. The Equality $\Delta^2 R_{\alpha\bar{\alpha}\alpha\bar{\alpha}} \ge 0$ at Maximal Directions α

The main objective of this section is to prove the vanishing of fourth-order radial derivatives $\nabla^4_\eta R_{\alpha\bar{\alpha}\alpha\bar{\alpha}}$ for any maximal direction α of holomorphic sectional curvatures and any real tangent vector η at $x = \pi(\alpha)$. As was explained in the introduction, we know $\nabla^i_\eta R_{\alpha\bar{\alpha}\alpha\bar{\alpha}}$ for $1 \le i \le 3$. We will first prove $\Delta^2 R_{\alpha\bar{\alpha}\alpha\bar{\alpha}} \ge 0$ and then compute the difference between $\Delta^2 R_{\alpha\bar{\alpha}\alpha\bar{\alpha}}$ and $S^{(4)} R_{\alpha\bar{\alpha}\alpha\bar{\alpha}}$, the averaging operator of fourth-order radial derivatives introduced in Proposition 15.8. Then we will conclude $S^{(4)} R_{\alpha\bar{\alpha}\alpha\bar{\alpha}} \ge 0$, implying $\nabla^4_\eta R_{\alpha\bar{\alpha}\alpha\bar{\alpha}} = 0$.

PROPOSITION 15.20. *Let α be a maximal direction of holomorphic sectional curvatures, $\pi(\alpha) = x$. Then $\Delta^2 R_{\alpha\bar{\alpha}\alpha\bar{\alpha}}(x) \ge 0$.*

PROOF. (I) Without loss of generality we may assume that α is a unit vector. Let $\{e_1, \ldots, e_n\}$ be a privileged basis of $T^{1,0}_x(X)$, $x = \pi(\alpha)$, associated with the unit maximal direction α. From Proposition 15.7 we have

$$\tfrac{1}{2}\Delta R_{1\bar{1}1\bar{1}} = \sum_{i,j} |R_{1\bar{i}1\bar{j}}|^2 + \rho R_{1\bar{1}1\bar{1}} - 2\sum_{i,j} |R_{1\bar{1}i\bar{j}}|^2,$$

where ρ is the Einstein constant and the equality holds in a neighborhood of x when $R_{i\bar{j}k\bar{l}}(y)$, for y sufficiently close to x, is interpreted as $R\langle e_i(y), \overline{e_j(y)}, e_k(y), \overline{e_l(y)}\rangle$ with $e_i(y)$ obtained from $e_i = e_i(x)$ by parallel transport along geodesics emanating from x. Letting η be a real tangent vector at x, we have, at x

$$\tfrac{1}{2}\Delta R_{1\bar{1}1\bar{1},\eta\eta} = 2\sum_{i,j}|R_{1\bar{i}1\bar{j},\eta}|^2 - 4\sum_{p\in H}R_{1\bar{1}p\bar{p}}R_{1\bar{1}p\bar{p},\eta\eta} - 4\sum_{i,j}|R_{1\bar{1}i\bar{j},\eta}|^2.$$

Here we have used the equalities of Proposition 15.14 at x and the fact that $R_{1\bar{1}1\bar{1},\eta\eta}(x) = 0$. It follows that at x,

$$\tfrac{1}{8}\Delta R_{1\bar{1}1\bar{1},\eta\eta} \geq -\sum_{p\in H}R_{1\bar{1}p\bar{p}}R_{1\bar{1}p\bar{p},\eta\eta} - \sum_{i,j}|R_{1\bar{1}i\bar{j},\eta}|^2$$

$$= -\frac{R_{1\bar{1}1\bar{1}}}{2}\sum_{p\in H}R_{1\bar{1}p\bar{p},\eta\eta} - \sum_{i,j}|R_{1\bar{1}i\bar{j},\eta}|^2.$$

Since X is Kähler-Einstein, we have $\sum_i R_{1\bar{1}i\bar{i},\eta\eta} = 0$, giving

$$\sum_{p\in H}R_{1\bar{1}p\bar{p},\eta\eta} = -R_{1\bar{1}1\bar{1},\eta\eta} - \sum_{q\in N}R_{1\bar{1}q\bar{q},\eta\eta} = -\sum_{q\in N}R_{1\bar{1}q\bar{q},\eta\eta} \quad \text{at } x.$$

Hence, we obtain the inequality

$$\tfrac{1}{8}\Delta R_{1\bar{1}q\bar{q},\eta\eta} \geq \frac{R_{1\bar{1}1\bar{1}}}{2}\sum_{q\in N}R_{1\bar{1}q\bar{q},\eta\eta} - \sum_{i,j}|R_{1\bar{1}i\bar{j},\eta}|^2 \quad \text{at } x. \qquad (15.9)$$

(II) We claim that the only possible nonzero terms in the summation $\sum_{i,j}|R_{1\bar{1}i\bar{j},\eta}|^2$ are of the type $|R_{1\bar{1}p\bar{q},\eta}|^2$. In other words, we have

LEMMA 15.21. *For any real tangent vector η at x, we have*

1. $R_{1\bar{1}1\bar{j},\eta} = 0$ *for* $1 \leq j \leq n$,
2. $R_{1\bar{1}p\bar{p},\eta} = 0$ *for* $p \in H$,
3. $R_{1\bar{1}q\bar{q},\eta} = 0$ *for* $q \in N$.

PROOF. To prove (1) and (2) we consider the Taylor series expansion of $R_{1\bar{1}1\bar{1}}$ along geodesics issuing from x. Let $\gamma(t)$, $-\delta < t < \delta$, be a geodesic parametrized by arc length such that $\gamma(0) = x$, $\dot{\gamma}(0) = \eta$. We know that

$$R_{1\bar{1}1\bar{1}}(\gamma(t)) = R_{1\bar{1}1\bar{1}}(x) + \frac{1}{4!}R_{1\bar{1}1\bar{1},\eta\eta\eta\eta}(x)t^4 + \cdots \quad \text{with } R_{1\bar{1}1\bar{1},\eta\eta\eta\eta} \leq 0$$

by the maximality of $R_{1\bar{1}1\bar{1}}(x)$. Recall that $\nabla_\eta^i R_{1\bar{1}1\bar{1}}(x) = 0$ for $1 \leq i \leq 3$. Consider the expansion

$$R\langle e_1 + \varepsilon e_j, \overline{e_1 + \varepsilon e_j}; e_1 + \varepsilon e_j, \overline{e_1 + \varepsilon e_j}\rangle(\gamma(t))$$

$$= R_{1111}(\gamma(t)) + 2\varepsilon \,\mathrm{Re}\, R_{1\bar{1}1\bar{j}}(\gamma(t))$$

$$+ \varepsilon^2(4R_{1\bar{1}j\bar{j}} + 2\,\mathrm{Re}\, R_{1\bar{j}1\bar{j}})(\gamma(t)) + \cdots. \qquad (15.10)$$

Now choose $\varepsilon = t^2$ and change e_j to $e^{i\theta} e_j$ for some real θ so that $R_{1\bar{1}1\bar{j},\eta}(x)$ is real and ≥ 0. We have

$$\frac{1}{(1+\varepsilon)^2} R\langle e_1 + \varepsilon e_j, \overline{e_1 + \varepsilon e_j}; e_1 + \varepsilon e_j, \overline{e_1 + \varepsilon e_j}\rangle(\gamma(t))$$

$$= \frac{1}{(1+t^4)^2} (R_{1\bar{1}1\bar{1}}(x) + O(t^4) + (2R_{1\bar{1}1\bar{j},\eta})(t^3) + O(t^4)).$$

From the fact that $R_{1\bar{1}1\bar{1}}(x) = \sup_{\xi \in T^{1,0}(x)} R_{\xi\bar{\xi}\xi\bar{\xi}}$ and comparing the Taylor expansions of the denominator and the numerator, we obtain immediately

$$R_{1\bar{1}1\bar{j},\eta}(x) = 0 \quad \text{for } j \geq 1 \tag{15.11}$$

To prove 3, $R_{1\bar{1}p\bar{p},\eta} = 0$, we also use (15.10). Choosing $j = p \in H$ in the expansion (15.10) and setting $\varepsilon = t$ for $t > 0$ we define

$$F(t) = R\langle e_1 + t^\sigma e_p, e_1 + t^\sigma e_p; e_1 + t^\sigma e_p, \overline{e_1 + t^\sigma e_p}\rangle(\gamma(t))$$

$$= R_{1\bar{1}1\bar{1}}(\gamma(t)) + 2 \operatorname{Re} R_{1\bar{1}1\bar{p}}(\gamma(t))t^\sigma$$

$$+ (4R_{1\bar{1}p\bar{p}} + 2 \operatorname{Re} R_{1\bar{p}1\bar{p}})(\gamma(t)) \cdot t^{2\sigma} + 4 \operatorname{Re} R_{1\bar{p}p\bar{p}}(\gamma(t)) \cdot t^{3\sigma}$$

$$+ R_{p\bar{p}p\bar{p}}(\gamma(t)) \cdot t^{4\sigma}.$$

We have $R_{1\bar{1}1\bar{p},\eta}(x) = 0$ and $R_{1\bar{p}1\bar{p}}(x) = R_{1\bar{p}p\bar{p}}(x) = 0$ (Proposition 15.14), so that

$$F(t) = R_{1\bar{1}1\bar{1}}(x) + O(t^4) + O(t^{\sigma+2}) + 2R_{1\bar{1}1\bar{1}}(x) \cdot t^{2\sigma}$$

$$+ (4R_{1\bar{1}p\bar{p},\eta}(x) + 2 \operatorname{Re} R_{1\bar{p}1\bar{p},\eta}(x))t^{2\sigma+1} + O(t^{3\sigma+1}) + O(t^{4\sigma}).$$

Now choose $\sigma = 0.9$. We get

$$F(t) = R_{1\bar{1}1\bar{1}}(x)(1 + 2t^{1.8}) + (4R_{1\bar{1}p\bar{p},\eta}(x) + 2 \operatorname{Re} R_{1\bar{p}1\bar{p},\eta}(x))t_{2.8} + O(t^{2.9}).$$

By comparing the Taylor expansion of $(1 + \varepsilon^2)^2$, $\varepsilon = t^{0.9}$, and that of $x(t)$, we obtain from $F(t)/(1 + t^{1.8})^2 \geq R_{1\bar{1}1\bar{1}}(x)$ the inequality

$$4R_{1\bar{1}p\bar{p}}(x) + 2 \operatorname{Re} R_{1\bar{p}1\bar{p},\eta}(x) \leq 0.$$

Without loss of generality we may assume Re $R_{1\bar{p}1\bar{p},\eta}(x) \geq 0$ (by some change $e_p \mapsto e^{i\theta} e_p$) so that

$$R_{11pp,\eta}(x) \leq 0.$$

Since the same inequality applies to the geodesic γ with orientation reversed we have also $R_{1\bar{1}p\bar{p},-\eta}(x) \leq 0$, giving

$$R_{1\bar{1}p\bar{p},\eta}(x) = 0. \tag{15.12}$$

Finally,

$$R_{1\bar{1}q\bar{q},\eta}(x) = 0 \tag{15.13}$$

follows immediately from $R_{1\bar{1}q\bar{q}}(x) = 0$ and the fact that X carries nonnegative holomorphic bisectional curvature, so that $R_{1\bar{1}q\bar{q}}(x)$ is a minimum $R_{\xi\bar{\xi}\xi'\bar{\xi}'}$, $\xi, \xi' \in T^{1,0}(X)$.

(III) The equations $R_{1\bar{1}1\bar{1},\eta}(x) = 0$, (i) $R_{1\bar{1}1\bar{j},\eta}(x) = 0$ for $j > 1$, (ii) $R_{1\bar{1}p\bar{p},\eta}(x) = 0$ and (iii) $R_{1\bar{1}q\bar{q},\eta}(x) = 0$ can now be used to yield the estimate from (15.9)

$$\tfrac{1}{8}\Delta R_{1\bar{1}1\bar{1},\eta\eta} \geq \frac{R_{1\bar{1}1\bar{1}}}{2} \sum_{q \in N} R_{1\bar{1}q\bar{q},\eta\eta} - \sum_{\substack{p \in N \\ q \in N}} |R_{1\bar{1}p\bar{q},\eta}|^2 \quad \text{at } x. \tag{15.14}$$

In order to finish the proof of Proposition 15.20 it suffices to prove the inequality, for each $q \in N$,

$$\frac{R_{1\bar{1}1\bar{1}}}{2} R_{1\bar{1}q\bar{q},\eta\eta} \geq \sum_{p \in H} |R_{1\bar{1}p\bar{q},\eta}|^2 \quad \text{at } x. \tag{15.15}$$

In fact, from the discussion of Section 15.2.3, $\Delta^2 R_{1\bar{1}1\bar{1}}(x)$ is the average of $\Delta R_{1\bar{1}1\bar{1},\eta\eta}$ over the unit sphere $S_x^{1,0}(X)$ of $T_x^{1,0}(X)$, up to a multiplicative constant.

(IV) To prove (15.15) we apply the Schwarz inequality. Let $\gamma(t)$, $-\delta < t < \delta$, denote the same geodesic as above. Then

$$R_{1\bar{1}q\bar{q}}(\gamma(t)) = \tfrac{1}{2} R_{1\bar{1}q\bar{q},\eta\eta}(x) t^2 + \cdots.$$

On the other hand, for any $p \in H$

$$R_{1\bar{1}p\bar{p}}(\gamma(t)) = \frac{R_{1\bar{1}1\bar{1}}}{2}(x) + \tfrac{1}{2} R_{1\bar{1}p\bar{p},\eta\eta}(x) t^2 + \cdots.$$

By the Schwarz inequality applied to the semidefinite Hermitian form $H_t(\xi, \zeta) = R_{1\bar{1}\xi\bar{\zeta}}(\gamma(t))$ at $\gamma(t)$

$$|R_{1\bar{1}p\bar{q}}(\gamma(t))|^2 \le R_{1\bar{1}p\bar{p}}(\gamma(t))R_{1\bar{1}q\bar{q}}(\gamma(t)),$$

yielding

$$|R_{1\bar{1}p\bar{q}}(\gamma(t))|^2 \le \frac{R_{1\bar{1}1\bar{1}}}{2}(x)R_{1\bar{1}q\bar{q},\eta\eta}(x)t^2.$$

But $R_{1\bar{1}p\bar{q}}(x) = 0$ and $R_{1\bar{1}p\bar{q}}(\gamma(t)) = R_{1\bar{1}p\bar{q},\eta}(x)t + \cdots$, so that by comparing the Taylor expansion we have immediately

$$|R_{1\bar{1}p\bar{q},\eta}(x)|^2 \le \frac{R_{1\bar{1}1\bar{1}}}{4}(x)R_{1\bar{1}q\bar{q},\eta\eta}(x). \tag{15.16}$$

(V) At first glance (15.16) is not strong enough to yield (15.15) unless $\mathcal{H} = \mathcal{H}_\alpha$ is at most two-dimensional. However, the estimate (15.16), for $e_q \in \mathcal{N}_\alpha$ fixed, $\|e_q\| = 1$, is true for any $e_p \in \mathcal{H}_\alpha$ of unit length. This means that, if we fix one privileged orthonormal basis of $T_x^{1,0}(X)$ associated to α, (15.16) can be applied to $\sum_{p \in H} a_p e_p$ in place of e_p for any $(q_1, \ldots, a_p)$ such that $\sum_{p \in H} |a_p|^2 = 1$. In general,

$$\left| \sum_{p \in H} a_p R_{1\bar{1}p\bar{q},\eta}(x) \right|^2 \le \left(\frac{R_{1\bar{1}1\bar{1}}}{4}(x)R_{1\bar{1}q\bar{q},\eta\eta}(x) \right) \left(\sum_{p \in H} |a_p|^2 \right).$$

In particular, if we choose $a_p = \overline{R_{1\bar{1}p\bar{q},\eta}}$, then

$$\left(\sum_{p \in H} |R_{11pq,\eta}(x)|^2 \right)^2 \le \left(\frac{R_{1\bar{1}1\bar{1}}}{4}(x)R_{1\bar{1}q\bar{q},\eta\eta}(x) \right) \left(\sum_{p \in H} |R_{1\bar{1}p\bar{q},\eta}(x)|^2 \right),$$

yielding

$$\sum_{p \in H} |R_{1\bar{1}p\bar{q},\eta}(x)|^2 \le \frac{R_{1\bar{1}1\bar{1}}}{4}(x)R_{1\bar{1}q\bar{q},\eta\eta}(x),$$

an equality even sharper than the required inequality (15.15), proving Proposition 15.20. $\qquad\qquad\qquad\qquad\qquad\qquad\qquad\qquad\qquad\qquad\qquad\square$

15.5.2. Comparing $S^{(4)}R_{\alpha\bar{\alpha}\alpha\bar{\alpha}}$ and $\Delta^2 R_{\alpha\bar{\alpha}\alpha\bar{\alpha}}$

Recall that from Section 15.2.3 $S^{(4)}R_{\alpha\bar{\alpha}\alpha\bar{\alpha}}(x)$ was defined by

$$S^{(4)}R_{\alpha\bar{\alpha}\alpha\bar{\alpha}}(x) = c_4 \int_\eta \nabla_\eta^4 R_{\alpha\bar{\alpha}\alpha\bar{\alpha}}(x),$$

where c_4 is a positive constant, η a real tangent vector of unit length, and the integral is over the unit tangent sphere with the canonical metric.

Recall that from Proposition 15.8 we have the formula

$$6S^{(4)}R_{\alpha\bar{\alpha}\alpha\bar{\alpha}} = \sum_{i,j} \sum_{\sigma \in S_4} R^\sigma_{\alpha\bar{\alpha}\alpha\bar{\alpha},i\bar{i}j\bar{j}}, \qquad (15.17)$$

where S_4 denotes the symmetry group of order 4, and $R^\sigma_{\alpha\bar{\alpha}\alpha\bar{\alpha},i\bar{i}j\bar{j}}$ is the fourth-order covariant derivative of $R_{\alpha\bar{\alpha}\alpha\bar{\alpha}}$ obtained by formally permuting the last four elements using σ. Our main result in this section is the following proposition.

PROPOSITION 15.22. *Let α be a maximal direction of holomorphic sectional curvatures, $\pi(\alpha) = x$. Then $S^{(4)}R_{\alpha\bar{\alpha}\alpha\bar{\alpha}}(x) = \Delta^2 R_{\alpha\bar{\alpha}\alpha\bar{\alpha}}(x) = 0$. Hence $\nabla_\eta^i R_{\alpha\bar{\alpha}\alpha\bar{\alpha}}(x) = 0$ for $1 \le i \le 5$.*

PROOF. (I) To prove Proposition 15.22 it suffices to show that

$$S^{(4)}R_{\alpha\bar{\alpha}\alpha\bar{\alpha}}(x) = \Delta^2 R_{\alpha\bar{\alpha}\alpha\bar{\alpha}}(x).$$

In fact, since $\nabla_\eta^i R_{\alpha\bar{\alpha}\alpha\bar{\alpha}}(x) = 0$, $1 \le i \le 3$, and is a global maximal direction of holomorphic sectional curvatures, we have

$$\nabla_\eta^4 R_{\alpha\bar{\alpha}\alpha\bar{\alpha}}(x) \le 0.$$

Integrating over η of unit length, we have

$$S^{(4)}R_{\alpha\bar{\alpha}\alpha\bar{\alpha}}(x) \le 0.$$

From Proposition 15.20 the equality $S^{(4)}R_{\alpha\bar{\alpha}\alpha\bar{\alpha}}(x) = \Delta^2 R_{\alpha\bar{\alpha}\alpha\bar{\alpha}}(x)$ would imply $S^{(4)}R_{\alpha\bar{\alpha}\alpha\bar{\alpha}}(x) = 0$ and hence $\nabla_\eta^i R_{\alpha\bar{\alpha}\alpha\bar{\alpha}}(x) = 0$ for $1 \le i \le 5$ and for all $\eta \in T_x(X)$.

(II) From (15.14) we have

$$6S^{(4)}R_{\alpha\bar{\alpha}\alpha\bar{\alpha}} = 4 \operatorname{Re} \sum_{i,j} R_{\alpha\bar{\alpha}\alpha\bar{\alpha},i\bar{i}j\bar{j}+i\bar{i}\bar{j}j+ij\bar{i}\bar{j}+ij\bar{j}\bar{i}+i\bar{j}\bar{i}j+ij\bar{j}\bar{i}}. \qquad (15.18)$$

To see this, there are 24 terms on the right-hand side of (15.18) of the form i_{***}, $\bar{i}_{***}$, j_{***} or $\bar{j}_{***}$ in the order of differentiation. By interchanging the roles of i and j in the same terms of the expansion of (15.18) we obtain (15.14). Furthermore we have the equalities

$$R_{\alpha\bar{\alpha}\alpha\bar{\alpha},ij\bar{i}\bar{j}} = R_{\alpha\bar{\alpha}\alpha\bar{\alpha},ij\bar{j}\bar{i}}, \qquad R_{\alpha\bar{\alpha}\alpha\bar{\alpha},i\bar{i}\bar{j}j} = R_{\alpha\bar{\alpha}\alpha\bar{\alpha},i\bar{j}\bar{i}j}.$$

Recall that by our definition of Δ^2 we have

$$\Delta^2 R_{\alpha\bar{\alpha}\alpha\bar{\alpha}} = \sum_{i,j} R_{\alpha\bar{\alpha}\alpha\bar{\alpha},i\bar{i}j\bar{j}+i\bar{i}\bar{j}j+\bar{i}ij\bar{j}+\bar{i}i\bar{j}j}.$$

Our approach of computing $S^{(4)}R_{\alpha\bar{\alpha}\alpha\bar{\alpha}} - \Delta^2 R_{\alpha\bar{\alpha}\alpha\bar{\alpha}}$ at x is by converting all terms to $\sum_{i,j} R_{\alpha\bar{\alpha}\alpha\bar{\alpha},i\bar{i}j\bar{j}}$. To start with we fix at x a privileged system of orthonormal basis of $T_x^{1,0}(X)$ associated with the maximal direction α. Then, at x,

$$R_{1\bar{1}1\bar{1},i\bar{i}j\bar{j}} = R_{1\bar{1}1\bar{1},i\bar{i}\bar{j}j} + 2\sum_{\mu} R_{\mu\bar{1}1\bar{1},i\bar{i}}R_{1\bar{\mu}j\bar{j}} - 2\sum_{\mu} R_{1\bar{\mu}1\bar{1},i\bar{i}}R_{\mu\bar{1}j\bar{j}}$$

$$+ \sum_{\mu} R_{1\bar{1}1\bar{1},\mu\bar{i}}R_{i\bar{\mu}j\bar{j}} - \sum_{\mu} R_{1\bar{1}1\bar{1},i\bar{\mu}}R_{\mu\bar{i}j\bar{j}}.$$

Summing up over j we immediately have, using the Einstein condition,

$$\sum_{j} R_{1\bar{1}1\bar{1},i\bar{i}j\bar{j}} = \sum_{j} R_{1\bar{1}1\bar{1},i\bar{i}\bar{j}j}.$$

In particular, we have

$$\sum_{i,j} R_{1\bar{1}1\bar{1},i\bar{i}j\bar{j}} = \sum_{i,j} R_{1\bar{1}1\bar{1},i\bar{i}\bar{j}j}, \qquad \sum_{i,j} R_{1\bar{1}1\bar{1},\bar{i}ij\bar{j}} = \sum_{i,j} R_{1\bar{1}1\bar{1},\bar{i}ij\bar{j}},$$

where the second equation is obtained from the first by conjugation. Furthermore,

$$R_{1\bar{1}1\bar{1},i\bar{i}j\bar{j}} = \left(R_{1\bar{1}1\bar{1},\bar{i}i} + 2\sum_{\mu} R_{\mu\bar{1}1\bar{1}}R_{1\bar{\mu}i\bar{i}} - 2\sum_{\mu} R_{1\bar{\mu}1\bar{1}}R_{\mu\bar{1}i\bar{i}} \right)_{j\bar{j}}$$

$$= R_{1\bar{1}1\bar{1},\bar{i}ij\bar{j}} + 2\left(\sum_{\mu} R_{\mu\bar{1}1\bar{1}}R_{1\bar{\mu}i\bar{i}} \right)_{j\bar{j}} - 2\left(\sum_{\mu} R_{1\bar{\mu}1\bar{1}}R_{\mu\bar{1}i\bar{i}} \right)_{j\bar{j}}.$$

Summing up over i, we have, using the Einstein condition,

$$\sum_{i} R_{1\bar{1}1\bar{1},i\bar{i}j\bar{j}} = \sum_{i} R_{1\bar{1}1\bar{1},\bar{i}ij\bar{j}}.$$

In particular, combined with equalities above

$$\sum_{i,j} R_{1\bar{1}1\bar{1},i\bar{i}j\bar{j}} = \sum_{i,j} R_{1\bar{1}1\bar{1},i\bar{i}\bar{j}j} = \sum_{i,j} R_{1\bar{1}1\bar{1},\bar{i}ij\bar{j}} = \sum_{i,j} R_{1\bar{1}1\bar{1},\bar{i}i\bar{j}j}$$

so that

$$\Delta^2 R_{1\bar{1}1\bar{1}} = 4\sum_{i,j} R_{1\bar{1}1\bar{1},i\bar{i}j\bar{j}}.$$

(Hence the last term is real.)

(III) From (15.18) for $S^{(4)}R_{\alpha\bar{\alpha}\bar{\alpha}} = S^{(4)}R_{1\bar{1}1\bar{1}}$, we now have

$$6S^{(4)}R_{1\bar{1}1\bar{1}} = 4\left(3\sum_{i,j} R_{1\bar{1}1\bar{1},i\bar{i}j\bar{j}} + 2\operatorname{Re}\sum_{i,j} R_{1\bar{1}1\bar{1},ij\bar{i}\bar{j}} + \operatorname{Re}\sum_{i,j} R_{1\bar{1}1\bar{1},ij\bar{j}\bar{i}}\right).$$

Now we convert the last two terms to $\sum_{i,j} R_{1\bar{1}1\bar{1},i\bar{i}j\bar{j}}$:

$$R_{1\bar{1}1\bar{1},ij\bar{i}\bar{j}} = \left(R_{1\bar{1}1\bar{1},i\bar{i}j} + 2\sum_{\mu} R_{\mu\bar{1}1\bar{1},i}R_{1\bar{\mu}j\bar{i}}\right.$$

$$\left. - 2\sum_{\mu} R_{1\bar{\mu}1\bar{1},i}R_{\mu\bar{1}j\bar{i}} + \sum_{\mu} R_{1\bar{1}1\bar{1},\mu}R_{i\bar{\mu}j\bar{i}}\right)_{\bar{j}}$$

$$= R_{1\bar{1}1\bar{1},i\bar{i}j\bar{j}} + 2\sum_{\mu} R_{\mu\bar{\omega}1\bar{1},i\bar{j}}R_{1\bar{\mu}j\bar{i}} + 2\sum_{\mu} R_{\mu\bar{1}1\bar{1},i}R_{1\bar{\mu}j\bar{i},\bar{j}}$$

$$- 2\sum_{\mu} R_{1\bar{\mu}1\bar{1},i\bar{j}}R_{\mu\bar{1}j\bar{i}} - 2\sum_{\mu} R_{1\bar{\mu}1\bar{1},i}R_{\mu\bar{1}j\bar{i},\bar{j}}$$

$$+ \sum_{\mu} R_{1\bar{1}1\bar{1},\mu\bar{j}}R_{i\bar{\mu}j\bar{i}} + \sum_{\mu} R_{1\bar{1}1\bar{1},\mu}R_{i\bar{\mu}j\bar{i},\bar{j}}.$$

Summing up over i, j, and applying the Bianchi identity and the Einstein condition, we obtain

$$\sum_{i,j} R_{1\bar{1}1\bar{1},ij\bar{i}\bar{j}} = \sum_{i,j} R_{1\bar{1}1\bar{1},i\bar{i}j\bar{j}} + 2\sum_{i,j,\mu} R_{\mu\bar{1}1\bar{1},i\bar{j}}R_{1\bar{\mu}j\bar{i}}$$

$$- 2\sum_{i,j,\mu} R_{1\bar{\mu}1\bar{1},i\bar{j}}R_{\mu\bar{1}j\bar{i}} + \Delta R_{1\bar{1}1\bar{1}}.$$

From previous information we know the first term is real and equal to $\frac{1}{4}\Delta^2 R_{1\bar{1}1\bar{1}}$, so that at x

$$2\operatorname{Re}\sum_{i,j} R_{1\bar{1}1\bar{1},ij\bar{i}\bar{j}} = 2\sum_{i,j} R_{1\bar{1}1\bar{1},i\bar{i}j\bar{j}} + \operatorname{Re}\sum_{i,j,\mu} R_{\mu\bar{1}1\bar{1},i\bar{j}}R_{1\bar{\mu}j\bar{i}}$$

$$- 4\operatorname{Re}\sum_{i,j,\mu} R_{1\bar{\mu}1\bar{1},i\bar{j}}R_{\mu\bar{1}j\bar{i}}.$$

Regrouping the terms we have, at x,

$$2 \operatorname{Re} \sum_{i,j} R_{1\bar{1}1\bar{1},ij\bar{j}} = 2 \sum_{i,j} R_{1\bar{1}1\bar{1},i\bar{i}j\bar{j}} + 2 \sum_{i,j,\mu} (R_{\mu\bar{1}1\bar{1},ij} - R_{\mu\bar{1}1\bar{1},\bar{j}i}) R_{1\bar{\mu}j\bar{i}}$$

$$+ 2 \sum_{i,j,\mu} (R_{1\bar{\mu}1\bar{1},\bar{j}i} - R_{1\bar{\mu}1\bar{1},ij}) R_{\mu\bar{1}j\bar{i}}.$$

We compute the commutation terms inside the parentheses to obtain

$$R_{\mu\bar{1}1\bar{1},ij} - R_{\mu\bar{1}1\bar{1},\bar{j}i} = \sum_\nu R_{\nu\bar{1}1\bar{1}} R_{\mu\bar{\nu}i\bar{j}} - 2 \sum_\nu R_{\mu\bar{\nu}1\bar{1}} R_{\nu\bar{1}i\bar{j}} + \sum_\nu R_{\mu\bar{1}\nu\bar{1}} R_{1\bar{\nu}i\bar{j}}.$$

The other commutation can be computed by conjugation, yielding at x

$$2 \operatorname{Re} \sum_{i,j} R_{1\bar{1}1\bar{1},\bar{i}j\bar{i}\bar{j}} = 2 \sum_{i,j} R_{1\bar{1}1\bar{1},i\bar{i}j\bar{j}} + 4 \operatorname{Re} \sum_{i,j,\mu} R_{1\bar{1}1\bar{1}} |R_{\mu\bar{1}i\bar{j}}|^2$$

$$- 8 \operatorname{Re} \sum_{i,j,\mu} R_{1\bar{1}\mu\bar{\mu}} |R_{\mu\bar{1}i\bar{j}}|^2 + 4 \operatorname{Re} \sum_{i,j} R_{1\bar{1}1\bar{1}} |R_{1\bar{1}i\bar{j}}|^2.$$

Here we have used equation (b) of Proposition 15.14, i.e., $R_{1\bar{1}i\bar{j}} = 0$ unless $i = j = 1$. We can furthermore regroup the commutation terms according to whether $\mu = 1$, $\mu \in H$ or $\mu \in N$, yielding

$$2 \operatorname{Re} \sum_{i,j} R_{1\bar{1}1\bar{1},\bar{i}j\bar{i}\bar{j}} - 2 \sum_{i,j} R_{1\bar{1}1\bar{1},i\bar{i}j\bar{j}}$$

$$= \sum_{i,j} \sum_{p \in H} (4R_{1\bar{1}1\bar{1}} - 8R_{1\bar{1}p\bar{p}}) |R_{p\bar{1}i\bar{j}}|^2 + 4R_{1\bar{1}1\bar{1}} \sum_{q \in N} \sum_{i,j} |R_{q\bar{1}i\bar{j}}|^2.$$

Since $R_{1\bar{1}p\bar{p}} = \frac{1}{2} R_{1\bar{1}1\bar{1}}$ and $R_{q\bar{1}i\bar{j}} = 0$ for all $i, j, 1 \le i, j \le n$, we have obtained

$$2 \operatorname{Re} \sum_{i,j} R_{1\bar{1}1\bar{1},\bar{i}j\bar{i}\bar{j}} = 2 \sum_{i,j} R_{1\bar{1}1\bar{1},i\bar{i}j\bar{j}}.$$

(IV) Similarly, we compute

$$\operatorname{Re} \sum_{i,j} R_{1\bar{1}1\bar{1},ij\bar{j}\bar{i}} - \sum_{i,j} R_{1\bar{1}1\bar{1},i\bar{i}j\bar{j}}$$

by commutation at x,

$$R_{1\bar{1}1\bar{1},ij\bar{j}\bar{i}} = R_{1\bar{1}1\bar{1},ij\bar{i}\bar{j}} + 2 \sum_\mu R_{\mu\bar{1}1\bar{1},ij} R_{1\bar{\mu}j\bar{i}} - 2 \sum_\mu R_{1\bar{\mu}1\bar{1},ij} R_{\mu\bar{1}j\bar{i}}$$

$$+ \sum_\mu R_{1\bar{1}1\bar{1},\mu\bar{j}} R_{i\bar{\mu}j\bar{i}} - \sum_\mu R_{1\bar{1}1\bar{1},i\bar{\mu}} R_{\mu\bar{j}j\bar{i}}.$$

Summing over i, j and using the Einstein condition we have

$$2 \operatorname{Re} \sum_{i,j} R_{1\bar{1}1\bar{1},ij\bar{j}\bar{i}} - 2 \sum_{i,j} R_{1\bar{1}1\bar{1},i\bar{i}j\bar{j}} = 4 \operatorname{Re} \sum_{i,j,\mu} R_{\mu\bar{1}1\bar{1},ij} R_{1\bar{\mu}j\bar{i}}$$

$$- 4 \operatorname{Re} \sum_{i,j,\mu} R_{1\bar{\mu}1\bar{1},ij} R_{\mu\bar{1}j\bar{i}}.$$

The same computation as in (III) yields the equality

$$\operatorname{Re} \sum_{i,j} R_{1\bar{1}1\bar{1},ij\bar{j}\bar{i}} = \sum_{i,j} R_{1\bar{1}1\bar{1},i\bar{i}j\bar{j}} = \sum_{i,j} R_{1\bar{1}1\bar{1},i\bar{i}j\bar{j}}.$$

From this and equalities in (II) we obtain at x

$$6 S^{(4)} R_{1\bar{1}1\bar{1}} = 24 \sum_{i,j} R_{1\bar{1}1\bar{1},i\bar{i}j\bar{j}} = 6 \Delta^2 R_{1\bar{1}1\bar{1}},$$

proving Proposition 15.22. $\square$

15.6. The Maximum Principle for Sixth-Order Radial Derivatives and Computation of $\Delta^3 R_{\alpha\bar{\alpha}\alpha\bar{\alpha}}$

15.6.1. Further Zero-Order Information on the Curvature

The major objective of this section is to extract further zero-order information on the curvature tensor with respect to a privileged orthonormal basis relative to any maximal direction α at any point $x \in X$. (Results of this section will be used in Section 15.7 to prove the crucial fact $\nabla_\alpha R = 0$ for most $\alpha \in \mathcal{M}$.) In order to do this it will be necessary to make use of gradient terms arising in the expressions of $S^{(4)} \Delta R_{\alpha\bar{\alpha}\alpha\bar{\alpha}}(x)$ and $\Delta^3 R_{\alpha\bar{\alpha}\alpha\bar{\alpha}}(x)$. Since the computation of these two quantities resemble the computation of $S^{(4)} R_{\alpha\bar{\alpha}\alpha\bar{\alpha}}(x)$ and $\Delta^2 R_{\alpha\bar{\alpha}\alpha\bar{\alpha}}(x)$, which were carried out in the last section, we will be contented with sketching the steps of such computation, and indicating only the necessary modifications and new methods of applying the computation.

Keeping notations as before, we will fix some $x \in X$, some maximal direction α at x and use a fixed privileged orthonormal basis of $T_x^{1,0}(X)$ adapted to α. Recall that $T_x^{1,0}(x) = \mathbf{C}\alpha \oplus \mathcal{H}_\alpha \oplus \mathcal{N}_\alpha$; the index set of the basis for $\mathcal{H} = \mathcal{H}_\alpha$ is denoted by H and that $\mathcal{N} = \mathcal{N}_\alpha$ is denoted by N. The indices will be denoted respectively by $p, p', \ldots$ and $q, q', \ldots$. For the sake of simplicity we shall say that a curvature from $R_{1\bar{p}p'\bar{p}''}$, for example, is of type $R_{1\bar{p}p\bar{p}}$, etc., meaning that the indices p, q appearing in terms of type R_{****} can take arbitrary values in H and N respectively. We can therefore group the curvature terms into those of types $R_{1\bar{1}1\bar{1}}$, $R_{1\bar{1}1\bar{p}}$, $R_{1\bar{1}p\bar{p}}$, ... We shall say that a curvature term $R_{ij k\bar{l}}$ is of type R_{****} up to conjugation and permutation if $R_{ij k\bar{l}}$ can be obtained from R_{****} by conjugation, the allowable permutations of indices due to symmetry and by substituting any p and q indices by arbitrary indices in H and N respectively. Our major objective here is the following result on the structure of R.

PROPOSITION 15.23. *With $x \in X$ and $\alpha \in \mathcal{M}_x$ fixed as above, the only possible nonvanishing terms of $R_{ij k\bar{l}}$ are those of the following types up to conjugation and permutation:*

$$R_{1\bar{1}1\bar{1}}, R_{1\bar{1}p\bar{p}}, R_{p\bar{p}q\bar{q}}, R_{p\bar{p}p\bar{p}}, R_{q\bar{q}q\bar{q}} \text{ and } R_{1\bar{p}q\bar{p}}.$$

Proposition 15.23 says that almost all nonvanishing curvature terms are of bisectional type $R_{k\bar{k}l\bar{l}}$, $k, l = 1, p, q$, with the possible exception of $R_{1\bar{p}q\bar{p}}$. It contains, in addition to results of Proposition 15.14 the fact that all curvature terms of types $R_{p\bar{p}p\bar{q}}$, $R_{p\bar{q}q\bar{q}}$ and $R_{p\bar{q}p\bar{q}}$ are zero. It is somewhat surprising that the vanishing of such terms can be derived from computations related to $R_{1\bar{1}1\bar{1}}$ since all the information in Proposition 15.14 obtained from variational inequalities are on terms associated to the maximal direction $e_1 = \alpha$. For the derivation of Proposition 15.23 we need the following lemma.

LEMMA 15.24. *Suppose second order covariant derivatives of $R_{1\bar{1}p\bar{q}}$ vanish at x. Then, all curvature terms of types $R_{p\bar{p}p\bar{q}}$, $R_{p\bar{q}q\bar{q}}$, and $R_{p\bar{q}p\bar{q}}$ vanish.*

PROOF. By polarization it suffices to prove the vanishing of the given terms; i.e., the indices p and q can be assumed to carry the same meaning. Under the hypothesis of the lemma, we have, at x,

1. $R_{1\bar{1}p\bar{q}, p\bar{p}} - R_{1\bar{1}p\bar{q}, \bar{p}p} = 0.$
2. $R_{1\bar{1}p\bar{q}, q\bar{q}} - R_{1\bar{1}p\bar{q}, \bar{q}q} = 0.$
3. $R_{1\bar{1}p\bar{q}, p\bar{q}} - R_{1\bar{1}p\bar{q}, \bar{q}p} = 0.$

We compute these differences by commutation separately.

1. $\qquad R_{1\bar{1}p\bar{q}, p\bar{p}} - R_{1\bar{1}p\bar{q}, \bar{p}p} = \sum_{\mu} R_{\mu\bar{1}p\bar{q}} R_{1\bar{\mu}p\bar{p}} - \sum_{\mu} R_{1\bar{\mu}p\bar{q}} R_{\mu\bar{1}p\bar{p}}$

$$+ \sum_{\mu} R_{1\bar{1}\mu\bar{q}} R_{p\bar{\mu}p\bar{p}} - \sum_{\mu} R_{1\bar{1}p\bar{\mu}} R_{\mu\bar{q}p\bar{p}}.$$

From Proposition 15.14, statements (c) and (d), we have the vanishing of curvature terms of types $R_{1\bar{q}i\bar{j}}$ and $R_{1\bar{p}p\bar{p}}$, so that

$$0 = R_{1\bar{1}p\bar{q},\,p\bar{p}} - R_{1\bar{1}p\bar{q},\,\bar{p}p} = -R_{1\bar{1}p\bar{p}}\,R_{p\bar{p}p\bar{q}}.$$

Since $R_{1\bar{1}p\bar{p}} = \frac{1}{2}R_{1\bar{1}1\bar{1}} \neq 0$, we obtain immediately $R_{p\bar{p}p\bar{q}} = 0$. Similarly we have

2. $R_{1\bar{1}p\bar{q},\,q\bar{q}} - R_{1\bar{1}p\bar{q},\,\bar{q}q} = -R_{1\bar{1}p\bar{p}}\,R_{p\bar{q}q\bar{q}}$,

3. $R_{1\bar{1}p\bar{q},\,p\bar{q}} - R_{1\bar{1}p\bar{q},\,\bar{q}p} = -R_{1\bar{1}p\bar{p}}\,R_{p\bar{q}p\bar{q}}$.

It follows therefore, under the hypothesis of Lemma 15.24, that

$$R_{p\bar{p}p\bar{q}} = R_{p\bar{q}q\bar{q}} = R_{p\bar{q}p\bar{q}} = 0.$$

Our next step is therefore to prove the vanishing of second-order covariant derivatives of $R_{1\bar{1}p\bar{q}}$. Recall that from the computation of $S^{(4)}R_{1\bar{1}1\bar{1}} = \Delta^2 R_{1\bar{1}1\bar{1}} = 0$ we obtain at the same time the vanishing of $R_{1\bar{1}p\bar{q},\eta}$ for any real tangent vector η at x. The term $R_{1\bar{1}p\bar{q},\eta}$ appears in the expression

$$\Delta R_{1\bar{1}1\bar{1},\eta\eta} = \sum_{i,j} |R_{1\bar{1}1\bar{j},\eta}|^2 + R_{1\bar{1}1\bar{1},\eta\eta}$$

$$- 2\sum_{i,j} |R_{1\bar{1}i\bar{j},\eta}|^2 - 2\,\mathrm{Re}\sum_{i,j} R_{1\bar{1}i\bar{j}}\,\overline{R_{1\bar{1}i\bar{j},\eta\eta}}.$$

Similarly

$$\Delta R_{1\bar{1}1\bar{1},\eta\eta\eta\eta} = \sum_{i,j} |R_{1\bar{1}1\bar{j},\eta\eta}|^2 + R_{1\bar{1}1\bar{1},\eta\eta\eta\eta}$$

$$- 2\sum_{i,j} |R_{1\bar{1}i\bar{j},\eta\eta}|^2 - 2\,\mathrm{Re}\sum_{i,j} R_{1\bar{1}i\bar{j}}\,\overline{R_{1\bar{1}i\bar{j},\eta\eta\eta\eta}}.$$

Here we have already used the facts $R_{1\bar{1}1\bar{j},\eta} = R_{1\bar{1}i\bar{j},\eta} = 0$ derived together with the vanishing of $\Delta R_{1\bar{1}1\bar{1},\eta\eta}$. It is plausible from the preceding expression that the vanishing of $\Delta R_{1\bar{1}1\bar{1},\eta\eta\eta\eta}$ can be used to derive the vanishing of second-order *radial* derivatives of $R_{1\bar{1}p\bar{q}}$. This is in fact the case. For this purpose we are going to compute the sixth-order term $S^{(4)}\Delta R_{1\bar{1}1\bar{1}}$ in the same spirit as in Section 15.5 for $S^{(2)}\Delta R_{1\bar{1}1\bar{1}} = \Delta^2 R_{1\bar{1}1\bar{1}}$. Notice that the equation $R_{1\bar{1}p\bar{q},\eta\eta} = 0$ does not imply the vanishing of second-order covariant derivatives. However, if instead we compute the expression

$$\Delta^3 R_{1\bar{1}1\bar{1}} = \sum_{\alpha,\beta} \Delta R_{1\bar{1}1\bar{1},(\alpha\bar{\alpha}+\bar{\alpha}\alpha)(\beta\bar{\beta}+\bar{\beta}\beta)},$$

then the gradient terms attached to $R_{1\bar{1}i\bar{j}}$ will be of the form $|R_{1\bar{1}i\bar{j},\alpha\beta}|^2$, $|R_{1\bar{1}i\bar{j},\alpha\bar{\beta}}|^2$, etc. Therefore we will further compute the commutation from $S^{(4)}\Delta R_{1\bar{1}1\bar{1}}$ to $\Delta^3 R_{1\bar{1}1\bar{1}} = S^{(2)}\Delta^2 R_{1\bar{1}1\bar{1}}$.

We will now collect the computational results into the following two propositions.

PROPOSITION 15.25. *At an arbitrary $x \in X$ and for any $\alpha \in \mathcal{M}_x$, we have, in terms of notations used before, $S^{(4)}\Delta R_{1\bar{1}1\bar{1}} = S^{(6)}R_{1\bar{1}1\bar{1}} = 0$, so that $\nabla_\eta^i R_{1\bar{1}1\bar{1}} = 0$ for $1 \le i \le 7$.*

PROPOSITION 15.26. *At an arbitrary $x \in X$ and for any $\alpha \in \mathcal{M}_x$, in terms of notations used before,*

$$\Delta^3 R_{1\bar{1}1\bar{1}} = S^{(4)}\Delta R_{1\bar{1}1\bar{1}} = 0.$$

This implies in particular that for any real tangent vector η at x

$$R_{1\bar{1}p\bar{q},\alpha\beta} = R_{1\bar{1}p\bar{q},\alpha\bar{\beta}} = R_{1\bar{1}p\bar{q},\bar{\alpha}\beta} = R_{1\bar{1}p\bar{q},\bar{\alpha}\bar{\beta}} = 0$$

for $p \in H$, $q \in N$, and $1 \le \alpha, \beta \le n$.

15.6.2. Sketch of the Proof of Proposition 15.25— $S^{(4)}\Delta R_{1\bar{1}1\bar{1}} = S^{(6)}R_{1\bar{1}1\bar{1}} = 0$

By the same argument as step (I) of Proposition 15.22 the equality $S^{(6)}R_{1\bar{1}1\bar{1}} = 0$ will imply $\nabla_\eta^i R_{1\bar{1}1\bar{1}} = 0$ for $1 \le i \le 7$. The proof of Proposition 15.25 therefore reduces to the following two statements.

PROPOSITION 15.25, PART I. *Notations as above, for any real tangent vector η at x we have $\Delta R_{1\bar{1}1\bar{1},\eta\eta\eta\eta} \ge 0$ so that in particular $S^{(4)}\Delta R_{1\bar{1}1\bar{1}} \ge 0$.*

PROPOSITION 15.25, PART II. *Notations as above, we have at x $S^{(6)}R_{1\bar{1}1\bar{1}} \ge S^{(4)}\Delta R_{1\bar{1}1\bar{1}} \ge 0$.*

For the derivation of the first inequality we need the following lemma for which we include a proof.

LEMMA 15.27. *At x, for any i, j, $1 \le i, j \le n$, for any $p \in H$, $q \in N$, and η any real tangent vector at x, we have*

1. $R_{1\bar{i}1\bar{j},\eta} = R_{1\bar{1}i\bar{j},\eta} = 0.$
2. $R_{1\bar{1}1\bar{j},\eta\eta} = 0.$
3. $R_{1\bar{1}q\bar{q},\eta\eta} = 0.$
4. $R_{1\bar{1}p\bar{p},\eta\eta} = R_{1\bar{p}1\bar{p},\eta\eta} = 0.$

PROOF. From the proof of Proposition 15.20 [see (15.14) where we dropped the first term and obtained an inequality] we have

$$\tfrac{1}{8}\Delta R_{1\bar{1}1\bar{1},\eta\eta} = \tfrac{1}{2}\sum_{i,j}|R_{1\bar{i}1\bar{j},\eta}|^2 + \frac{R_{1\bar{1}1\bar{1}}}{2}\sum_{q\in N}R_{1\bar{1}q\bar{q},\eta\eta} - \sum_{\substack{p\in H \\ q\in N}}|R_{1\bar{1}p\bar{q},\eta\eta}|^2. \qquad (15.19)$$

From the last formula of step (V) we have at x,

$$\frac{R_{1111}}{4}R_{11qq,\eta\eta} \geq \sum_{p\in H}|R_{1\bar{1}p\bar{q},\eta}|^2.$$

From these we derive

$$\tfrac{1}{8}R_{1\bar{1}1\bar{1},\eta\eta} \geq \tfrac{1}{2}\sum_{i,j}|R_{1\bar{i}1\bar{j},\eta}|^2 + \sum_{\substack{p\in H \\ q\in N}}|R_{1\bar{1}p\bar{q},\eta}|^2.$$

Recalling Lemma 15.21, we have

$$R_{1\bar{1}1\bar{j},\eta} = R_{1\bar{1}p\bar{p},\eta} = R_{1\bar{1}q\bar{q},\eta} = 0 \quad \text{for } 1 \leq i, j \leq n.$$

Since $\Delta R_{1\bar{1}1\bar{1},\eta\eta} = 0$ by Proposition (15.22) we thus obtain

$$R_{1\bar{i}1\bar{j},\eta} = R_{1\bar{1}i\bar{j},\eta} = 0,$$

proving statement 1 of Lemma 15.27.

To prove statement 2, we need only to consider the case $j > 1$. Let $\gamma(t), -\delta < t < \delta$, be a geodesic parametrized by arc length such that $\gamma(0) = x$ and $\gamma(0) = \eta$. Since $\nabla_\eta^i R_{1\bar{1}1\bar{1}} = 0$ at x for $1 \leq i \leq 5$, we have

$$R_{1111}(\gamma(t)) = R_{1111}(x) + \frac{1}{6!}\nabla^6 R_{1111}(x)t^6 + \cdots = R_{1111}(x) + O(t^6).$$

On the other hand, we have

$$R\langle e_1 + \varepsilon e_j, \overline{e_1 + \varepsilon e_j}; e_1 + \varepsilon e_j, \overline{e_1 + \varepsilon e_j}\rangle(\gamma(t))$$

$$= R_{1\bar{1}1\bar{1}}(\gamma(t)) + 2\varepsilon \operatorname{Re} R_{1\bar{1}1\bar{j}}(\gamma(t))$$

$$+ \varepsilon^2(4R_{1\bar{1}j\bar{j}} + 2\operatorname{Re} R_{1\bar{j}1\bar{j}})(\gamma(t)) \qquad (15.20)$$

$$+ \varepsilon^2(4\operatorname{Re} R_{1\bar{j}j\bar{j}}(\gamma(t)))$$

$$+ \varepsilon^4 R_{j\bar{j}j\bar{j}}(\gamma(t)).$$

Substituting $\varepsilon = t^3$ and changing e_j to $e^{i\theta}e_j$ for some real θ so that $R_{1\bar{1}1\bar{j},\eta\eta}(x)$ is real and ≥ 0 (recall that $R_{1\bar{1}1\bar{j},\eta} = 0$ at x for $j > 1$), we have, for $j > 1$,

$$\frac{1}{(1+\varepsilon^2)^2} R\langle e_1 + \varepsilon e_j, \overline{e_1 + \varepsilon e_j}; e_1 + \varepsilon e_j, \overline{e_1 + \varepsilon e_j}\rangle(\gamma(t))$$

$$= \frac{1}{(1+t^6)^2} (R_{1\bar{1}1\bar{1}}(x) + O(t^6) + 2R_{1\bar{1}1\bar{j},\eta\eta}(x)t^5 + O(t^6)).$$

But since the holomorphic sectional curvature of $(e_1 + \varepsilon e_j)/\sqrt{1+\varepsilon^2}$ at $\gamma(t)$ is smaller than $R_{1\bar{1}1\bar{1}}(x)$, we see immediately that $R_{1\bar{1}1\bar{j},\eta\eta}(x) = 0$.

Equation 3, $R_{1\bar{1}q\bar{q},\eta\eta} = 0$ for all $q \in N$, has already been proved in Proposition 15.15. To prove 4 we use (15.20). We choose $j = p \in H$ and set $\varepsilon = t^\sigma$ for $t > 0$. By taking $\sigma = 1.5$ and comparing expansions in terms of t we obtain

$$4R_{1\bar{1}p\bar{p},\eta\eta}(x) + 2\,\text{Re}\,R_{1\bar{p}1\bar{p},\eta\eta}(x) \leq 0.$$

By replacing e_p by $e^{i\theta}e_p$ for a suitable real θ we may assume that $\text{Re}\,R_{1\bar{p}1\bar{p},\eta\eta}(x) \geq 0$, so that

$$R_{1\bar{1}p\bar{p},\eta\eta}(x) \leq 0.$$

Since $R_{1\bar{1}1\bar{1},\eta\eta}(x) = R_{1\bar{1}q\bar{q},\eta\eta}(x) = 0$ for all $q \in N$, from the Einstein condition we obtain $\sum_{p \in H} R_{1\bar{1}p\bar{p},\eta\eta}(x) = 0$, giving

$$R_{1\bar{1}p\bar{p},\eta\eta}(x) = R_{1\bar{p}1\bar{p},\eta\eta}(x) = 0$$

and completing the proof of Lemma 15.27. □

Using Lemma 15.27, one obtains immediately from the formula of Berger,

$$\tfrac{1}{8}\Delta R_{1\bar{1}1\bar{1},\eta\eta\eta\eta} = \tfrac{3}{2}\sum_{i,j}|R_{1\bar{i}1\bar{j},\eta\eta}|^2 + 3\sum_{\substack{p \in H \\ q \in N}}|R_{1\bar{1}p\bar{q},\eta\eta}|^2$$

$$\times \sum_{q \in N}\left(\frac{R_{1\bar{1}1\bar{1}}}{2}R_{1\bar{1}q\bar{q},\eta\eta\eta\eta} - 3\sum_{p \in H}|R_{1\bar{1}p\bar{q},\eta\eta}|^2\right). \qquad (15.21)$$

By the same application of the Schwarz inequality as in Proposition 15.22 we can show that the term inside the bracket is always nonnegative, proving $\frac{1}{8}\Delta R_{1\bar{1}1\bar{1},\eta\eta\eta\eta} \geq 0$ and hence the integrated form $S^{(4)}\Delta R_{1\bar{1}1\bar{1}} \geq 0$, proving Part I of Proposition 15.25.   □

To prove Part II we need only to show $S^{(6)}R_{1\bar{1}1\bar{1}} \geq S^{(4)}\Delta R_{1\bar{1}1\bar{1}}$. This is done by the same commutation technique as in Proposition 15.22. Since covariant derivatives with both barred and unbarred indices are involved, we will need a conversion of our knowledge of radical derivatives into that of general covariant derivatives. We shall only indicate the procedure by an example. It will be necessary, for example, to use the fact that all second order covariant derivatives of $R_{1\bar{1}p\bar{p}}$, $p \in H$, vanish. By Lemma 15.27 we know that all second-order radial derivatives of $R_{1\bar{1}p\bar{p}}$, $p \in H$, vanish at z, i.e., $R_{1\bar{1}p\bar{p},\eta\eta} = 0$ for all $\eta \in T_x(X)$. By polarization (Proposition 15.9) we obtain, for $1 \leq \alpha, \beta \leq n$,

$$R_{11p\bar{p},\alpha\beta} = R_{1\bar{1}p\bar{p},\bar{\alpha}\bar{\beta}} = R_{1\bar{1}p\bar{p},\alpha\bar{\beta}} + R_{1\bar{1}p\bar{p},\bar{\beta}\alpha} = 0.$$

To prove $\nabla^2 R_{1\bar{1}p\bar{p}} = 0$ it suffices therefore to show $R_{1\bar{1}p\bar{p},\alpha\bar{\beta}} - R_{1\bar{1}p\bar{p},\bar{\beta}\alpha} = 0$, which can be obtained by the formula for commutation and our knowledge of zero order information on the curvature tensor.

The rest of the proof of $S^{(6)}R_{1\bar{1}1\bar{1}} \geq S^{(4)}\Delta R_{1\bar{1}1\bar{1}} \geq 0$ follows the same line of thought as in Proposition 15.22 and will be omitted.   □

15.6.3. Sketch of Proof of Proposition 15.26—$\Delta^3 R_{1\bar{1}1\bar{1}} = S^{(4)}\Delta R_{1\bar{1}1\bar{1}} = 0$

From Proposition 15.25 one can derive the vanishing of a number of second-order radial derivatives of the curvature tensor. In addition to the list given in the lemma, we obtain from the actual expression (15.21) of $\Delta R_{1\bar{1}1\bar{1},\eta\eta\eta\eta}$ (recall $\Delta R_{1\bar{1}1\bar{1},\eta\eta\eta\eta} \geq 0$ for $\eta \in T_x(X)$) the vanishing of $R_{1\bar{i}1\bar{j},\eta\eta}$ and $R_{1\bar{1}p\bar{q},\eta\eta}$ for $1 \leq i,j \leq n$, $p \in H$ and $q \in N$. Recall that for the derivation we needed the vanishing of $\nabla^2 R_{1\bar{1}p\bar{q}}$. For this purpose we need Proposition 15.26. First we write down the following simplified formula of $\Delta^3 R_{1\bar{1}1\bar{1}}$ using our knowledge of certain vanishing covariant derivatives as indicated at the end of Section 15.6.2:

$$\frac{1}{2}\Delta^3 R_{1\bar{1}1\bar{1}} = 2R_{1\bar{1}1\bar{1}} \sum_{q \in N} \Delta^2 R_{1\bar{1}q\bar{q}} - \sum_{k,l} \sum_{\substack{p \in H \\ q \in N}} (|R_{1\bar{1}p\bar{q},k\bar{l}}|^2 + |R_{1\bar{1}p\bar{q},\bar{k}l}|^2). \quad (15.22)$$

The derivation of this formula is very much the same as the formula for $\Delta R_{1\bar{1}1\bar{1},\eta\eta\eta\eta}$ in Section 15.6.2. Here the covariant derivatives associated to $R_{1\bar{i}1\bar{j}}$ are discarded because one can derive the equality $\nabla^2 R_{1\bar{i}1\bar{j}} = 0$ from

$R_{1\bar{1}j,\eta\eta} = 0$ for all $\eta \in T_x(X)$, by the argument of the last paragraph of Section 15.6.2. To prove Proposition 15.26 it suffices therefore to show $\Delta^3 R_{1\bar{1}1\bar{1}} = 0$ and $\Delta^2 R_{1\bar{1}q\bar{q}} \leq 0$. The derivation of $\Delta^3 R_{1\bar{1}1\bar{1}} = 0$ follows the same pattern as the derivation of the inequality $S^{(6)} R_{1\bar{1}1\bar{1}} \geq S^{(4)} \Delta R_{1\bar{1}1\bar{1}} \geq 0$. Namely, we compare $\Delta^3 R_{1\bar{1}1\bar{1}} = S^{(2)} \Delta^2 R_{1\bar{1}1\bar{1}}$ against $S^{(4)} \Delta R_{1\bar{1}1\bar{1}}$ by the formula for commutation. (See the proof of Proposition 15.22. In the present situation we actually obtain $\Delta^3 R_{1\bar{1}1\bar{1}} = S^{(4)} \Delta R_{1\bar{1}1\bar{1}} = 0$ directly. The derivation of $\Delta^2 R_{1\bar{1}q\bar{q}} \leq 0$ is more involved conceptually. It suffices to show $\Delta R_{1\bar{1}q\bar{q},\eta\eta} \leq 0$ for all $\eta \in T_x(X)$. We derive from Proposition 15.15 (which gives $\Delta R_{1\bar{1}q\bar{q}} = 0$) the expression

$$\Delta R_{1\bar{1}q\bar{q},\eta\eta} = \sum_{p,r \in H} |R_{1\bar{p}p\bar{r}}|^2_{,\eta\eta} - \frac{R_{1\bar{1}1\bar{1}}}{2} \sum_{p \in H} R_{p\bar{p}q\bar{q},\eta\eta}$$

$$= \int_{\|\xi\|=1,\xi \in H} \left[\sum_{r \in H} |R_{1\bar{\xi}q\bar{r}}|^2 - R_{1\bar{1}\xi\bar{\xi}} R_{\xi\bar{\xi}q\bar{q}} \right]_{\eta\eta}.$$

Note that the term inside the bracket in the integrand vanishes at x by formula (e) of Proposition 15.14. Call this expression $\delta_q(\xi)$ in a neighborhood of x. (As usual the vectors e_q and ξ in a neighborhood of x are understood to be obtained by parallel transport from x of $e_q(x)$ and $\xi(x)$ along geodesics.) Recall that in Proposition 15.15 $\delta_q(\xi)(x)$ was interpreted as the discriminant of some quadratic polynomial associated with $R_{1\bar{1}q\bar{q}}$. In fact, we defined at x,

$$G(\varepsilon) = R\left(e_1 + \varepsilon\xi, \overline{e_1 + \varepsilon\xi}, e_q + \varepsilon \sum_{r \in H} C_r e_r, \overline{e_q + \varepsilon \sum_{r \in H} C_r e_r}\right),$$

and $\delta_q(\xi)$ is the discriminant of the coefficient of ε^2 in the Taylor expansion of $G(\varepsilon)$ in ε, regarded as a quadratic polynomial in the variables C_r, $r \in H$. This quadratic polynomial is positive definite (since $R_{1\bar{1}q\bar{q}} = 0$ and $G(\varepsilon) \geq 0$ because X carried semipositive bisectional curvature). The vanishing of the discriminant then implies the existence of a nonzero set of coefficients $(C_r)_{r \in H}$ such that the coefficient of ε^2 in $G(\varepsilon)$ vanishes. In fact, this is given by the formula $C_r = -(R_{1\bar{\xi}q\bar{r}} / R_{1\bar{1}r\bar{r}})(x)$. Fix a geodesic $\gamma(t)$ passing through x with $\gamma(0) = x$ and $\dot{\gamma}(0) = \eta$ and define now $C_r(t) = -(R_{1\bar{\xi}q\bar{r}} / R_{1\bar{1}r\bar{r}})(\gamma(t))$ obtained by parallel transport. Consider the function for $t \geq 0$,

$$F_\sigma(t) = R\left(e + t^\sigma\xi, \overline{e_1 + t^\sigma\xi}, e_q + t^\sigma \sum_{r \in H} C_r(t)e_r,\right.$$

$$\left. e_q + t^\sigma \sum_{r \in H} C_r(t)e_r\right)(\gamma(t)) \geq 0.$$

Writing $R_{1\bar{\xi}q\bar{r}(t)}$ for $R_{1\bar{\xi}q\bar{r}}(\gamma(t))$, etc., the coefficient of $t^{2\sigma}$ in the expansion of $F_\sigma(t)$ in t is given by

$$K(t) = R_{\xi\bar{\xi}q\bar{q}}(t) - \sum_{r\in H} \frac{|R_{1\bar{\xi}q\bar{r}}|^2}{R_{1\bar{1}r\bar{r}}}(t).$$

We note that $K(0) = -4/R_{1\bar{1}1\bar{1}}(0)$, $\delta_q(\xi)(0) = 0$ and that $K''(0) = (-4/R_{1\bar{1}1\bar{1}}(0))\nabla_\eta^2\delta_q(\xi)(0)$. To finish the proof of $\Delta^2 R_{1\bar{1}q\bar{q}}(x) \leq 0$, it suffices to show $K''(0) \geq 0$. The proof of this follows the same line of argument as in Lemma 15.27. Namely, by choosing appropriate σ, we conclude successively the vanishing of certain coefficients of powers of t. The starting point of this algorithm is the estimate $R_{1\bar{1}q\bar{q}}(t) = O(t^6)$. To see this from (15.21) of $\Delta R_{1\bar{1}1\bar{1},\eta\eta\eta\eta}$ and its vanishing at x we have $\nabla_\eta^i R_{1\bar{1}q\bar{q}}(x) = 0$ for $0 \leq i \leq 4$, yielding immediately $R_{1\bar{1}q\bar{q}}(t) = O(t^6)$ since $R_{1\bar{1}q\bar{q}}(x) = 0$ is a minimum of bisectional curvatures. The rest of the argument is routine and will be omitted.

The vanishing of $\Delta^3 R_{1\bar{1}1\bar{1}}$ and $\Delta^2 R_{1\bar{1}q\bar{q}} \leq 0$ imply the vanishing of $\nabla^2 R_{1\bar{1}p\bar{q}}$ by the expansion (15.22) of $\Delta^3 R_{1\bar{1}1\bar{1}}$, which in turn implies the main result Proposition 15.23 of this section, as was indicated in Section 15.6.1. □

15.7. Invariance of R along Integral Curves of Vector Fields of Maximal Directions

We will make use of our preceding knowledge of the curvature tensor and first-order covariant derivatives to show that there is a nonempty open set U such that for any $x \in U$ and any $\alpha \in \mathcal{M}_x$, $\nabla_\alpha R(x) = 0$. It follows immediately that if $\gamma(t)$ is a curve in U, $\gamma(t)$ is a multiple of some $\alpha \in \mathcal{M}_{\gamma(t)}$, then the curvature tensor is invariant under parallel transport along γ. In order to prove $\nabla_\alpha R(x) = 0$ we first collect all information about first-order covariant derivatives at x. As usual we will fix $x \in X$, $\alpha \in \mathcal{M}_x$ and use a privileged orthonormal basis $\{e_1, \ldots, e_n\}$ of $T_x^{1,0}(X)$ adapted to $\alpha = e_1$.

LEMMA 15.28. At x, $\nabla R_{1\bar{1}1\bar{j}} = \nabla R_{1\bar{1}i\bar{j}} = \nabla R_{1\bar{q}i\bar{j}} = \nabla R_{1\bar{p}p\bar{p}} = 0$ for all $p \in H$, $q \in N$ and for $1 \leq i, j \leq n$.

PROOF. The only thing that was not already contained in Lemma 15.27 is the equation $\nabla R_{1\bar{p}p\bar{p}} = 0$. To prove this consider the expansion along any geodesic $\gamma(t)$, $-\delta < t < \delta$, passing through x with $\gamma(0) = x$ and $\dot\gamma(0) = \eta$. Writing $R_{1\bar 1 1\bar 1}(\gamma(t)) = R_{1\bar 1 1\bar 1}(t)$, etc., we define for $\sigma > 0, 0 \le t < \delta$,

$$F_\sigma(t) = R(e_1 + t^\sigma e_p, \overline{e_1 + t^\sigma e_p}, e_1 + t^\sigma e_p, \overline{e_1 + t^\sigma e_p})(\gamma(t))$$

$$= R_{1\bar 1 1\bar 1}(t) + 4t^\sigma \operatorname{Re} R_{1\bar 1 1\bar p}(t) + t^{2\sigma}(4R_{1\bar 1 p\bar p}(t) + 2 \operatorname{Re} R_{1\bar p 1\bar p}(t))$$

$$+ 4t^{3\sigma} \operatorname{Re} R_{1\bar p p\bar p}(t) + t^{4\sigma} R_{p\bar p p\bar p}(t).$$

Recall that from Proposition 15.25 and Lemma 15.27 we have

$$R_{1\bar 1 1\bar 1}(t) = R_{1\bar 1 1\bar 1}(0) + O(t^8), \qquad R_{1\bar 1 1\bar p}(t) = O(t^3),$$

$$R_{1\bar p 1\bar p}(t) = O(t^3), \qquad R_{1\bar 1 p\bar p}(t) = \tfrac{1}{2}R_{1\bar 1 1\bar 1}(0) + O(t^3).$$

From the maximality of $R_{1\bar 1 1\bar 1}(0)$ we have

$$F_\sigma(t) \le R_{1111}(0) \cdot (1 + t^{2\sigma})^2.$$

Take any $\sigma > 0$ and comparing the coefficients on both sides of the inequality, we have

$$1 + O(t^8) + O(t^{\sigma+3}) + 2t^{2\sigma} + O(t^{2\sigma+3}) + 4t^{3\sigma+1} R_{1\bar p p\bar p, \eta} + O(t^{4\sigma})$$

$$\le 1 + 2t^{2\sigma} + t^{4\sigma},$$

noting that $R_{1\bar p p\bar p} = 0$ by Proposition 15.15. Substituting $\sigma = 0.5$ we immediately obtain $R_{1\bar p p\bar p, \eta} \le 0$. Applying the inequality to the geodesic $\gamma^-(t) = \gamma(-t)$ we obtain $-R_{1\bar p p\bar p, \eta} \le 0$ and hence $R_{1\bar p p\bar p, \eta} = 0$, proving the lemma. $\qquad\square$

The main result of Section 15.7 is the following proposition:

PROPOSITION 15.29. *In the notation of Lemma* 15.28, *there exists a nonempty dense open set* U *such that we have, at any point* $x \in U$ *and for any* $e_1 = \alpha \in \mathcal{M}_x$,

$$\nabla_1 R_{i\bar j k\bar l} = 0 \quad for \ 1 \le i, j, k, l \le n.$$

PROOF. By means of polarization it suffices to prove $\nabla_1 R_{i\bar{j}k\bar{l}} = 0$ for $i, j, k, l = 1, p$ or q, where p and q represent typical elements of H and N respectively. We will first prove this for all types with one exception by using Lemma 15.28 and the Bianchi identity. First, we can classify curvature terms into groups of types up to conjugation and permutation of indices:

1. $R_{1\bar{i}j\bar{k}}$ for $1 \leq i, j, k \leq n$.
2. $R_{p\bar{p}p\bar{p}}$ for $p \in H$.
3. $R_{q\bar{q}q\bar{q}}$ for $q \in N$.
4. $R_{p\bar{p}q\bar{q}}$ for $p \in H, q \in N$.
5. $R_{p\bar{p}p\bar{q}}$ for $p \in H, q \in N$.
6. $R_{p\bar{q}q\bar{q}}$ for $p \in H, q \in N$.
7. $R_{p\bar{q}p\bar{q}}$ for $p \in H, q \in N$.

Since this division is up to conjugation (and permutation of indices), it is necessary to prove $R_{i\bar{j}k\bar{l},1} = R_{i\bar{j}k\bar{l},\bar{1}} = 0$ for all terms given in the list. We have

1. $R_{1\bar{i}j\bar{k},1} = R_{1\bar{i}1\bar{k},j} = 0$, $R_{1\bar{i}j\bar{k},\bar{1}} = R_{1\bar{1}j\bar{k},i} = 0$.
2. $\overline{R_{p\bar{p}p\bar{p},\bar{1}}} = R_{p\bar{p}p\bar{p},1} = R_{1\bar{p}p\bar{p},p} = 0$.
3. $\overline{R_{q\bar{q}q\bar{q},\bar{1}}} = R_{1\bar{q}q\bar{q},1} = R_{1\bar{q}q\bar{q},q} = 0$.
4. $\overline{R_{p\bar{p}q\bar{q},\bar{1}}} = R_{p\bar{p}q\bar{q},1} = R_{1\bar{q}p\bar{p},q} = 0$.
5. $R_{p\bar{p}p\bar{q},1} = R_{1\bar{q}p\bar{p},p} = 0$, $R_{p\bar{p}p\bar{q},\bar{1}} = R_{p\bar{p}p\bar{1},\bar{q}} = 0$.
6. $R_{p\bar{q}q\bar{q},1} = R_{1\bar{q}p\bar{q},q} = 0$, $R_{p\bar{q}q\bar{q},\bar{1}} = R_{q\bar{1}p\bar{q},\bar{q}} = 0$.
7. $R_{p\bar{q}p\bar{q},1} = R_{1\bar{q}p\bar{q},p} = 0$, $R_{p\bar{q}p\bar{q},\bar{1}} = 0$.

Everything is proved except for $R_{p\bar{q}p\bar{q},\bar{1}}$, because the only possible application of Bianchi identity $R_{p\bar{q}p\bar{q},\bar{1}} = R_{p\bar{1}p\bar{q},\bar{q}}$ does not yield a curvature term for which one can apply Lemma 15.28. To complete the proof of Proposition 15.29 it suffices therefore to prove: □

LEMMA 15.30. *There exists a dense open set U of X such that for any $\alpha \in \mathcal{M}_x$ and for any $\eta \in T_x(X)$, we have, in terms of a privileged basis at x adapted to α, $R_{p\bar{q}p\bar{q},\eta} = 0$.*

PROOF OF LEMMA 15.30. In Proposition 15.23 we used the vanishing of $\nabla^2 R_{1\bar{1}p\bar{q}}$ to conclude that $R_{p\bar{q}p\bar{q}} = R_{p\bar{p}p\bar{q}} = R_{p\bar{q}q\bar{q}} = 0$ at x. By the same argument it would be possible to deduce $\nabla R_{p\bar{q}p\bar{q}} = 0$, etc., at x if we know $\nabla^3 R_{1\bar{1}p\bar{q}} = 0$. But this would necessitate the computation of $\Delta^4 R_{1\bar{1}1\bar{1}}$. Instead we will show that our knowledge of the structure of the curvature tensor (Proposition 15.23) and additional knowledge on covariant derivatives is sufficient for proving $\nabla R_{p\bar{q}p\bar{q}} = 0$ at x wherever $\alpha \in \mathcal{M}'$, where $\mathcal{M}'$ as defined in Section 15.4.1 is the union of components $\mathcal{M}_i$, $1 \leq i \leq i$, such that $\pi|_{\mathcal{M}_i}$

is a submersion at some smooth point. This contains in particular Lemma 15.30. To do this it suffices by continuity to prove $\nabla R_{p\bar{q}p\bar{q}} = 0$ at x for $\alpha \in \text{Reg } \mathcal{M}_i - \{\alpha \in \mathcal{M}_i \text{ where } \pi \text{ fails to be a submersion at } \alpha\}$ for $1 \le i \le k$. For any such $\alpha \in \mathcal{M}_x$ there exists a smooth vector field $\tilde{\alpha}(y)$ defined on a neighborhood W of x such that $\tilde{\alpha}(x) = \alpha$ and $\tilde{\alpha}(y) \in \mathcal{M}_y$ for each $y \in W$. At $y \in W$ are we have the orthogonal decomposition

$$T_y^{1,0} = \mathbf{C}\tilde{\alpha}(y) \oplus \mathcal{H}_{\tilde{\alpha}(y)} \oplus \mathcal{N}_{\tilde{\alpha}(y)}.$$

Since the dimension of $\mathcal{H}_{\tilde{\alpha}(y)}$ and $\mathcal{N}_{\tilde{\alpha}(y)}$ are both independent of y, as a consequence of the Einstein condition or simply of the fact that the Ricci tensor is continuous, the splitting given above for each $y \in W$ actually yields an orthogonal splitting of the smooth vector bundle $T^{1,0}(W)$ as

$$T^{1,0}(W) = \mathcal{A}(W) \oplus \mathcal{H}(W) \oplus (W),$$

where by definition $\mathcal{A}(W) = \bigcup_{y \in W} \mathbf{C}\tilde{\alpha}(y)$, etc. Fix a geodesic $\gamma(t)$, $-\delta < t < \delta$, $\gamma(0) = x$ through x lying in W and denote by $\alpha(y) \in T_y^{1,0}(X)$ obtained by parallel transport of $\alpha(x) = \alpha$ along γ. Denote by η the tangent vector $\gamma'(0)$ at x. Fix some $e_q \in \mathcal{N}_\alpha$, $\|e_q\| = 1$, and denote by $e_q(y) \in T_y^{1,0}(X)$, $y \in \gamma$, the corresponding vector similarly obtained by parallel transport. We write $\alpha(t)$ for $\alpha(y)$, etc. for $y = \gamma(t)$. We have the orthogonal decomposition

$$\alpha(t) = a(t)\tilde{\alpha}(t) + \xi(t) + \zeta(t), \quad \text{with } \xi(t) \in \mathcal{H}_{\tilde{\alpha}(t)}, \zeta(t) \in \mathcal{N}_{\tilde{\alpha}(t)},$$

$$e_q(t) = b(t)\tilde{\alpha}(t) + \xi'(t) + \zeta'(t), \quad \text{with } \xi'(t) \in \mathcal{H}_{\tilde{\alpha}(t)}, \zeta'(t) \in \mathcal{N}_{\tilde{\alpha}(t)}.$$

Here obviously $\xi(0) = \zeta(0) = \xi'(0) = b(0)\tilde{\alpha}(0) = 0$. We assert that

$$b(t) = O(t^2). \tag{15.23}$$

The estimate (15.23) will be used to study the behavior of $R_{p\bar{q}p\bar{q}}(t)$ for $p \in H$ and $q \in N$ in order to conclude $R_{p\bar{q}p\bar{q},\eta}(0) = 0$. To prove (15.23) recall that we have

$$R_{1\bar{1}1\bar{q}}(0) = R_{1\bar{1}1\bar{q},\eta} = 0$$

which means that

$$R(\alpha(t), \overline{\alpha(t)}, \alpha(t), \overline{e_q(t)}) = O(t^2).$$

Substituting the decompositions of $\alpha(t)$ and $e_q(t)$ into the preceding equation and using the fact that

$$R(\tilde{\alpha}(t), \overline{\tilde{\alpha}(t)}, \tilde{\alpha}(t), \overline{\xi'(t) + \zeta'(t)}) \quad \text{(type } R_{1\bar{1}1\bar{j}} \text{ for } j > 1)$$

$$= R(\tilde{\alpha}(t), \overline{\tilde{\alpha}(t)}, \alpha(t), \overline{\zeta'(t)}) \quad \text{(type } R_{1\bar{q}i\bar{j}})$$

$$= R(\tilde{\alpha}(t), \overline{\xi(t) + \zeta(t)}, \tilde{\alpha}(t), \overline{\zeta'(t) + \xi'(t)}) \quad \text{(type } R_{1\bar{i}1\bar{j}}, i, j \neq 1)$$

$$= 0,$$

as could be read off from Proposition 15.14 on the structure of R_{1***}, we see immediately that

$$R(\alpha(t), \overline{\alpha(t)}, \alpha(t), \overline{e_q(t)})$$

$$= a^3(t)b(t)R(\tilde{\alpha}(t), \tilde{\alpha}(t), \tilde{\alpha}(t), \tilde{\alpha}(t)) + O(t^2).$$

Since $R(\tilde{\alpha}(t), \overline{\tilde{\alpha}(t)}, \tilde{\alpha}(t), \overline{\tilde{\alpha}(t)}) = R_{\alpha\bar{\alpha}\alpha\bar{\alpha}} > 0$ for all t, $-\delta < t < \delta$, we have established the estimate $(*)$.

To make use of (15.23), let e_p be a fixed unit vector in $\mathcal{H}$ and consider the decomposition

$$e_p(t) = c(t)\tilde{\alpha}(t) + \xi''(t) + \zeta''(t), \quad \text{with } \xi''(t) \in \mathcal{H}_{\tilde{\alpha}(t)}, \zeta''(t) \in \mathcal{N}_{\tilde{\alpha}(t)}.$$

Clearly, $c(0)\tilde{\alpha}(0) = \zeta''(0) = 0$. Then,

$$R(e_p(t), \overline{e_q(t)}, e_p(t), \overline{e_q(t)})$$

$$= R(\xi''(t), \overline{\zeta'(t)}, \xi''(t), \overline{\zeta'(t)})$$

$$+ 2R(c(t)\tilde{\alpha}(t) + \zeta''(t), \overline{\zeta'(t)}, \xi''(t), \overline{\zeta'(t)})$$

$$+ 2R(\xi''(t), \overline{b(t)\tilde{\alpha}(t) + \xi'(t)}, \xi''(t), \overline{\zeta'(t)}) + O(t^2).$$

From Proposition 15.23 on the structure of the curvature tensor we obtain

$$R(\xi''(t), \overline{\zeta'(t)}, \xi''(t), \overline{\zeta'(t)}) \quad \text{(type } R_{p\bar{q}p\bar{q}})$$

$$= R(\tilde{\alpha}(t), \overline{\zeta'(t)}, \xi''(t), \overline{\zeta''(t)}) \quad \text{(type } R_{1\bar{q}i\bar{j}})$$

$$= R(\zeta''(t), \overline{\zeta'(t)}, \xi''(t), \overline{\zeta'(t)}) \quad \text{(type } R_{p\bar{q}q\bar{q}})$$

$$= R(\xi''(t), \overline{\xi'(t)}, \xi''(t), \overline{\zeta'(t)}) \quad \text{(type } R_{p\bar{p}p\bar{q}})$$

$$= 0$$

so that

$$R(e_p(t), \overline{e_q(t)}, e_p(t), \overline{e_q(t)})$$

$$= 2b(t)R(\xi''(t), \overline{\alpha(t)}, \xi''(t), \overline{\zeta'(t)}) + O(t^2).$$

Notice that curvature terms of type $R_{p\bar{1}p\bar{q}}$ may be nonzero. However, by (15.23) we have $b(t) = O(t^2)$, yielding

$$R(e_p(t), \overline{e_q(t)}, e_p(t), \overline{e_q(t)}) = O(t^2),$$

hence, $R_{p\bar{q}p\bar{q},\eta}(0) = 0$, proving Lemma 15.30 and thus establishing Proposition 15.29.   □

15.8. Totally Geodesic Hermitian Symmetric Integral Submanifolds and Isometric Decomposition of X

15.8.1. The Integrability of V

Recall that by Proposition 15.18 there exists a nonempty open subset U of X such that the bundle of maximal subspaces $V = \bigcup_{x \in X} V_x$, $V_x = $ C-linear span of $\mathcal{M}_x$, is a differentiable vector bundle on U. By Proposition 15.29, by shrinking U if necessary, we can assume that for any $\alpha \in \mathcal{M}_x$, $x \in U$, we have $\nabla_\alpha R_{ij\bar{k}\bar{l}} = 0$ for $1 \le i, j, k, l \le n$, so that $\nabla_\xi R_{ij\bar{k}\bar{l}}$ for all $\xi \in V_x$. On U we now consider the distribution Re $V|_U = \{\xi + \bar{\xi} \colon \xi \in V|_U\}$ of vector subspaces of the tangent spaces. Our main result in this section is the following.

PROPOSITION 15.31. *The distribution* Re $V|_U$ *of vector subspaces of* $T_x(X)$, $x \in U$, *is integrable. Moreover, the integral submanifolds are complex, totally geodesic, and locally symmetric.*

PROOF. By the theorem of Frobenius, to prove Re $V|_U$ is integrable, all we need to show is that it is closed under taking Lie brackets. Since the metric on X is Riemannian, for any smooth tangent vector fields Y, Z on any open set,

$$\nabla_Y Z - \nabla_Y Z - [Y, Z] = 0.$$

It suffices for the proof of the integrability of Re $V|_U$ to show that $\nabla_\eta \eta'$ takes values in Re $V|_U$ for any tangent vector fields η, η' on an open subset of U with values in Re $V|_U$. Fix $x \in U$ and let $\alpha_1, \ldots, \alpha_m$ be a basis of V_x consisting of maximal directions. We may further assume, as in Proposition 15.29, that there exist smooth vector fields $\alpha_1(y), \ldots, \alpha_m(y)$ in a neighborhood of x such that $\alpha_i(y) \in \mathcal{M}_y$ and $\alpha_i(x) = \alpha_i$. Let $\gamma = \gamma(t)$, $-\delta < t < \delta$, $\gamma(0) = x$, be any integral curve of Re $V|_U$, i.e., $\dot{\gamma}(t) \in$ Re $V_{\gamma(t)}$ for each t. Then the curvature tensor is invariant under parallel transport along γ. In particular, if $\beta_i(t)$ is the parallel transport of α_i along γ to $\gamma(t)$, then

$$\frac{d}{dt} R(\beta(t), \overline{\beta(t)}, \beta(t), \overline{\beta(t)}) \equiv 0$$

since $\nabla_{\dot{\gamma}(t)} R \equiv 0$ and $\nabla_{\dot{\gamma}(t)} \beta(t) = 0$. It follows that $\beta(t)$ is also a maximal direction. In particular, $\beta(t) \in V_{\gamma(t)}$. Write

$$\beta_i(t) = \sum_j a_{ij}(t) \alpha_j(t), \qquad \alpha_j(t) = \alpha_j(\gamma(t)).$$

Write $\eta = \dot{\gamma}(0)$. Then, at x, $\nabla_\eta \beta_i(t) = 0$; i.e.,

$$\sum a_{ij}(0) \nabla_\eta \alpha_j(0) + \sum_j a_{ij}'(0) \alpha_j(0) = 0.$$

From the definition of β_i it is clear that $a_{ij}(0) = \delta_{ij}$, so that

$$\nabla_\eta \alpha_i(0) = -\sum_j a_{ij}'(0) \alpha_j(0),$$

proving that $\nabla_\eta \alpha_i(0) \in V_x$. Since η is real, we obviously have

$$\nabla_\eta (\text{Re } \alpha_i)(0) \in \text{Re } V_x.$$

But this applies to any $\eta = \dot{\gamma}(0)$ and γ an integral curve of Re $V|_U$. It follows therefore that for any open $U' \subset U$ and any real tangent vector fields η, η' on U' such that $\eta(x)$, $\eta'(x) \in$ Re V_x, we have

$$\nabla_\eta \eta'(x) \in \text{Re } V_x \quad \text{for all } x \in U', \tag{15.24}$$

proving in particular the integrability of Re $V|_U$. Obviously the integral submanifolds are complex because $\eta \in V_x$ implies $J_\eta \in V_x$ for the J-operator on the complex manifold X. Finally (15.24) implies that $\nabla'_\eta \eta'(x) = \nabla_\eta \eta'(x)$ for the Riemannian connection ∇' on Z, from which it follows that Z is totally geodesic, proving Proposition 15.31. $\square$

15.8.2. The Global Structure of the Foliation

The local foliation on U by locally symmetric complex totally geodesic submanifolds Z_x is a strong indication that X is itself Hermitian symmetric. In this subsection our contention is that each Z_x is contained in a compact locally Hermitian symmetric submanifold $\tilde{Z}_x$. To be precise, we have

PROPOSITION 15.32. *For each $x \in U$ there exists a totally geodesic, compact, locally Hermitian symmetric submanifold $\tilde{Z}_x$ containing x such that $\tilde{Z}_x \cap U = Z_x$ is the integral submanifold of* Re $V|_U$ *passing through x.*

PROOF. We will prove Proposition 15.32 using the theorem of Bonnet and Meyers, which asserts that every complete Riemannian manifold of Ricci curvature bounded from below by a positive constant is necessarily compact. Let $r > 0$ be less than the injectivity radius of X so that for any $y \in X$, the exponential map at y is a diffeomorphism on the Euclidean ball $\overline{B(r)} = \overline{B(0; r)}$ on the tangent space $T_y(X)$, equipped with the obvious Euclidean metric. Without loss of generality we may let U be the open geodesic ball $B(x; r)$ so that Z_x is nothing other than $\exp_x (B(r) \cap \text{Re } V_x)$. We can step by step enlarge the piece $Z = Z_x$ as follows. Define $Z_0 = Z$. We will define Z_i in general as a locally closed extendable submanifold of X, in the sense that there exists some locally closed submanifold Z_i' of X such that $Z_i \Subset Z_i'$. Suppose Z_i is defined. Fix $\varepsilon > 0$ such that $r + \varepsilon <$ injectivity radius. Choose a finite subset S_i of Z_i such that for each $y_0 \in Z_i$ there exists $y \in Z_i$ such that $d(y_0, y) < \varepsilon$. This can be done because Z_i is extendible. Define

$$Z_{i+1} = \bigcup_{y \in S_i} A_{i+1}(y),$$

where

$$A_{i+1}(y) = \exp_y (B(r) \cap T_y(Z_i)).$$

We have chosen S_i so that $Z_i \Subset Z_{i+1}$. We claim that Z_{i+1} is a locally closed extendible submanifold. By definition Z_{i+1} is locally closed. To show that Z_{i+1} is a submanifold locally, it suffices to show that for $y, y' \in S_i$, either $A_{i+1}(y) \cap A_{i+1}(y') = \varnothing$ or $A_{i+1}(y) \cup A_{i+1}(y')$ is a locally closed connected submanifold extending both $A_{i+1}(y)$ and $A_{i+1}(y')$. To prove this one has to rule out the possibility that they intersect each other in a subset of smaller dimension. If $A_{i+1}(y) \cup A_{i+1}(y')$ is not smooth, we would have either

1. $A_{i+1}(y)$ intersects $A_{i+1}(y')$ tangentially at some y'', or
2. there exists $y'' \in A_{i+1}(y) \cap A_{i+1}(y')$ such that $T_{y''}(A_{i+1}(y)) \cup T_{y''}(A_{i+1}(y'))$ span a real linear subspace of $T_{y''}(X)$ of dimension larger than $2 \dim_{\mathbb{C}} V_x = $ real dimension of Z_x.

Possibility 1 cannot happen because both $A_{i+1}(y)$ and $A_{i+1}(y')$ must be totally geodesic at y'' (by the identity theorem for real analytic functions), so that they are determined by their tangent planes at y''. To rule out possibility 2 observe that both $T_{y''}(A_{i+1}(y))$ and $T_{y''}(A_{i+1}(y'))$ are generated by real parts of maximal directions at y'' (obtained by parallel transport from y and y' respectively). Then translating them back from y'' to the point x along broken geodesics on $Z_i \cup A_{i+1}(y)$ will yield more than $\dim_\mathbb{C} V_x$ $\mathbb{C}$-linearly independent maximal directions at x, contradicting with the definition of V_x. This establishes our claim that $Z_{i+1} = \bigcup_{y \in S_i} A_{i+1}(y)$ is a locally closed submanifold. That Z_{i+1} is extendible follows easily by taking

$$Z'_{i+1} = \bigcup_{y \in S_i} \exp_y (B(r + \varepsilon) \cap T_y(Z_i)) \quad \text{for some } \varepsilon > 0 \text{ sufficiently small.}$$

For $r + \varepsilon <$ injectivity radius of X, clearly Z'_{i+1} is also a locally closed submanifold such that $Z_{i+1} \Subset Z'_{i+1}$.

We now have a sequence of real-analytic manifolds Z_i such that

$$Z_1 \Subset Z_2 \Subset \cdots \Subset Z_k \Subset Z_{k+1} \Subset \cdots,$$

where Z_k, equipped with the restriction of the Kähler metric on X, is necessarily locally symmetric by the identity theorem of real-analytic functions. Moreover, if we define $\tilde{Z}$ to be the union $\bigcup_{k \geq 1} Z_k$ equipped with the induced metric, $\tilde{Z}$ is necessarily a complete Kähler manifold. In fact, at each $z \in Z_k$ there exists some $y \in S_k$, $d(y, z) < \varepsilon$ so that Z_{k+1} contains $B(z; r - \varepsilon)$. This implies that for each $z \in Z$ we have $B(z; r - \varepsilon) \Subset Z$, which in turn implies the completeness of $\tilde{Z}$. Recall that at each $z \in \tilde{Z}$, $T_z(\tilde{Z})$ is generated by the real parts of maximal directions at z. Because of local symmetry, $\tilde{Z}$ with the induced metric splits locally into products of Hermitian symmetric spaces and flat tori. If there is a flat torus as a local factor, it would not be possible for $T_z(\tilde{Z})$ to be generated by real parts of maximal directions. Hence $\tilde{Z}$ is a complete Kähler manifold with positive Ricci curvature bounded from below by some $c > 0$. By the theorem of Bonnet and Myers, $\tilde{Z}$ must be compact, proving Proposition 15.32. $\qquad\square$

15.8.3. Pointwise Reducibility of Bisectional Curvatures

Let $x \in U$. Denote by ξ a typical element of $T_x^{1,0}(Z_x) = V_x$ and by ζ a typical element of $V_x^\perp$, the orthogonal complement of V_x in $T_x^{1,0}(X)$. To prove that X is Hermitian symmetric it suffices to show that

$$\nabla_\xi R_{i\bar{j}k\bar{l}} = \nabla_\zeta R_{i\bar{j}k\bar{l}} = 0 \quad \text{for } 1 \leq i, j, k, l \leq n.$$

What remains to be proved is the vanishing of $\nabla_\zeta R_{\zeta\bar\zeta\bar\zeta}$. In fact, any terms of the form $\nabla_\zeta R_{\xi***}$ or $\nabla_\zeta R_{*\bar\xi**}$ would also be zero because of Proposition 15.29 and the Bianchi identity. We may assume without loss of generality that $V|_U \neq T^{1,0}(U)$. In order to prove $\nabla_\zeta R_{\zeta\bar\zeta\bar\zeta} = 0$ we will first show that Re $V^\perp|_U$ is an integrable distribution, where $V_x^\perp$ denotes the orthogonal complement of V_x in $T_x^{1,0}(X)$. From this and the arguments of Proposition 15.32 we will be able to obtain integral submanifolds $Z^\perp$ of Re $V^\perp|_U$ which extend to totally geodesic compact complex submanifolds $\tilde Z^\perp$ of X. Moreover, in the process of proof we will also show that $R_{\xi\bar\xi\zeta\bar\zeta} = 0$ for all $\xi \in V_x$ and $\zeta \in V_x^\perp, x \in U$. This allows us to conclude that each such $\tilde Z^\perp$ is Kähler–Einstein. Moreover, holomorphic bisectional curvatures are non-negative on $\tilde Z^\perp$ (because $\tilde Z^\perp$ is totally geodesic). To prove Theorem 15.3 by induction on dimension we can assume that $\tilde Z^\perp$ is isometric to a Hermitian symmetric space, so that $\nabla_\zeta R_{\zeta\bar\zeta\bar\zeta} = 0$ for all $\zeta \in V_x^\perp$, proving $\nabla R \equiv 0$ on X by the identity theorem for real-analytic functions, thus establishing Theorem 15.3.

In order to show that Re $V^\perp|_U$ is integrable we will first of all show that $V|_U$ is invariant under parallel transport along all curves on U. For the proof of this we will need the reducibility of bisectional curvatures as stated above.

PROPOSITION 15.33. *For each $x \in U$ and for all $\xi \in V_x$, $\zeta \in V_x^\perp$, we have* $R_{\xi\bar\xi\zeta\bar\zeta} = 0$.

PROOF. For each $x \in U$ there exists a totally geodesic compact complex submanifold $\tilde Z_x$ of X such that $Z_x = \tilde Z_x \cap U$ is an integral submanifold of the distribution Re $V|_U$. Suppose $y \in U, y \notin Z_x$; we assert that $\tilde Z_x \cap \tilde Z_y = \varnothing$. In fact, the proof of this is exactly as in Proposition 15.32, where it was shown that the $\tilde Z_x$ cannot have self-intersections. Since adjacent extended integral submanifolds are mutually nonintersecting, the normal bundle $N_\mathbf{R}$ of the real manifold $\tilde Z_x$ in X must be trivial as a differentiable vector bundle. As a differentiable bundle, $N_\mathbf{R}$ is simply isomorphic to the bundle Re $V^\perp|_{\tilde Z_x}$, where obviously we can assume that the open set U contains $\tilde Z_x$. Since $N_\mathbf{R}$ is differentiable trivial, the complex bundle $N_\mathbf{C} = N_\mathbf{R} \otimes_\mathbf{R} \mathbf{C}$ is a differentiably trivial complex vector bundle. We have the decomposition

$$N_\mathbf{C} = N^{1,0} \oplus N^{0,1},$$

where $N_x^{1,0}$ is the eigenspace of J on $N_{\mathbf{C},x} = N_{\mathbf{R},x} \otimes_\mathbf{R} \mathbf{C}$ corresponding to the eigenvalue i, and $N_x^{0,1}$ is that corresponding to the eigenvalue $-i$. Note that this decomposition is possible because $V_x^\perp$ is closed under the J-operator. Let N be the holomorphic normal bundle on $\tilde Z_x$, i.e., $N = T^{1,0}(X)|_{\tilde Z_x} / T^{1,0}(\tilde Z_x)$. As a differentiable $\mathbf{C}$-vector bundle, N is isomorphic to $N^{1,0} \cong V^{\perp}|_{\tilde{Z}_x}$. It is well known that any Hermitian holomorphic quotient bundle of a Hermitian holomorphic vector bundle of semipositive curvature remains semipositive, so that N, with the induced metric, is semipositive on $\tilde{Z}_x$. It follows that the first Chern class of N is represented by a semipositive closed $(1, 1)$ form. Now, $N_C = N^{1,0} \oplus N^{0,1} \cong N \oplus \bar{N}$ as differentiable C-vector bundles, where $\bar{N}$ is the antiholomorphic vector bundle obtained by taking conjugates of transition functions of N. By defining the length of $\bar{v}$ to be that of v for $v \in N_x$, $\bar{v} \in \bar{N}_x$, we see that $c_1(\bar{N}) = c_1(N)$. It follows that

$$c_1(N_C) = c_1(N^{1,0}) + c_1(N^{0,1}) = c_1(N) + c_1(\bar{N}) = 2c_1(N)$$

is represented by a semipositive closed $(1, 1)$ form. Hence, the triviality of N_C as a differentiable C-vector bundle implies that $c_1(N) = 0$ and that the curvature form of N is identically zero on $\tilde{Z}_x$. We assert that the flatness of the Hermitian holomorphic vector bundle N implies the proposition, i.e., $R_{\xi\bar{\xi}\zeta\bar{\zeta}} = 0$ for $\xi \in V_x$, $\zeta \in V_x^{\perp}$ and $x \in U$. To see this we examine the curvature form of N more closely. Consider the exact sequence

$$0 \to N^* \to T^{1,0}(X)^*|_{\tilde{Z}_x} \to T^{1,0}(\tilde{Z}_x)^* \to 0.$$

The flatness of N implies that of the dual bundle N^*. By the curvature decreasing property of Hermitian holomorphic vector subbundles, we have, denoting by $\Theta' = \Theta_{N^*}$ and $\Theta = \Theta_{T^{1,0}(X)^*}$ the curvature forms of N^* and $T^{1,0}(X)^*$ with the induced metrics respectively,

$$\Theta'(\xi, \bar{\xi}; \zeta^*, \bar{\zeta}^*) \leq \Theta(\xi, \bar{\xi}; \zeta^*, \bar{\zeta}^*)$$

for $\xi \in T_x^{1,0}(\tilde{Z}_x)$ and $\zeta^* \in N^*$. Now let $\{e_1, \ldots, e_n\}$ be a basis of $T_x^{1,0}(X)$ and $\{e_1^*, \ldots, e_n^*\}$ be the dual basis. Then, for $\zeta^* = \sum a_i e_i^*$ belonging to N_x^*, we have

$$\Theta'(\xi, \bar{\xi}, \zeta^*, \bar{\zeta}^*) \leq \Theta(\xi, \bar{\xi}, \zeta^*, \bar{\zeta}^*) = \sum a_i a_j \Theta(\xi, \bar{\xi}, e_i^*, \overline{e_j^*})$$

$$= -\sum a_i a_j \Theta_{T^{1,0}(x)}(\xi, \bar{\xi}; e_j, \overline{e_i}) = -\sum a_i \bar{a}_j R_{\xi\bar{\xi}i\bar{j}}.$$

Here the next to last inequality follows from the standard relation between curvatures of dual bundles and the last equality comes from the definition of $\Theta_{T^{1,0}(x)}$, the curvature tensor of $T^{1,0}(X)$ with the induced Hermitian matrix. It follows now that $\Theta'(\xi, \bar{\xi}, \zeta^*, \bar{\zeta}^*) \leq 0$ and that equality hold for all $\zeta^* \in N_x^*$ only if $R_{\xi\bar{\xi}\zeta\bar{\zeta}} = 0$ for all $\zeta = \sum a_i e_i$ such that $\sum a_i e_i^* \in N_x^*$. But $\sum a_i e_i^* \in N_x^*$ if and only if $(\sum a_i e_i^*)(\xi) = 0$ for all $\xi \in T_x^{1,0}(\tilde{Z}_x)$, i.e., if and only if $\langle \sum a_i e_i, \xi \rangle = 0$, so that

$$R_{\xi\bar{\xi}\zeta\bar{\zeta}} = 0 \quad \text{for all } \zeta \in V_x^{\perp},$$

proving Proposition 15.33. $\square$

15.8.4. Invariance of V under Parallel Transport

Recall that $\mathcal{M}|_U$ is invariant under parallel transport along curves on $\tilde{Z}_x$, which implies in particular that $V|_U$ is invariant under parallel transport along any curves (not necessarily geodesics) $\gamma(t)$ such that $\dot{\gamma}(t) \in V_{\gamma(t)}$. In this subsection we assert the stronger statement:

PROPOSITION 15.34. *The bundle V of maximal subspaces, defined by $V = \bigcup_{x \in X} V_x$ and $V_x = $ C-linear span of $\mathcal{M}_x$, is invariant under parallel transport along any smooth curve l on X. In particular, V is a bona fide differentiable vector bundle on X.*

PROOF. (I) First we assert that it suffices to prove that $V|_U$ is invariant under parallel transport along geodesics. First of all, we contend that the latter statement would imply that maximal directions remain maximal directions when translated by parallel transport along any geodesic passing through U. At each $x' \in X$ there exists a geodesic γ joining x' to some point $x \in U$. Adjacent geodesics emanating from x will also intersect U so that maximal directions at x' remain maximal when translated by parallel transport along some open cone of geodesics emanating from x' and hence along all geodesics emanating from x', by the identity theorem for real-analytic functions. It follows that the bundle V of maximal subspaces is invariant under translation by parallel transport along any geodesic. In particular, V is a differentiable vector subbundle of $T^{1,0}(X)$. Let $l(t)$, $-\delta < t < \delta$, be any smooth curve. Suppose $\chi \in V_{l(0)}$ and

$$\chi(t) = \xi(t) + \zeta(t), \qquad \xi(t) \in V_{l(t)}, \zeta(t) \in V_{l(t)}^{\perp},$$

is the decomposition of the parallel transport $\chi(t) \in T^{1,0}_{l(t)}$ according to the orthogonal decomposition $T^{1,0}(X) = V \oplus V^{\perp}$. From the invariance of V under parallel transport along geodesics it follows readily that $\|\zeta(t)\| = O(t^2)$, $\|\cdot\|$ denoting the length. To show that $\chi(t) \in V_{l(t)}$ it suffices to show that $d\|\zeta(t)\|^2/dt \equiv 0$. Let $\xi_t(s)$ and $\zeta_t(s)$ denote the translation of $\xi(t)$ and $\zeta(t)$ to $l(s)$, for s sufficiently close to t, by parallel transport. Obviously, $\xi_t(t) = \xi(t)$, $\zeta_t(t) = \zeta(t)$ and $\xi_t(s) + \zeta_t(s) = \chi(s)$. Let $\xi_t(s) = \xi'_t(s) + \zeta'_t(s)$ and $\zeta_t(s) = \xi''_t(s) + \zeta''_t(s)$ denote the decompositions of ξ_t and ζ_t according to the decomposition $T^{1,0}(X) = V \oplus V^{\perp}$. Then

$$\zeta(s) = \zeta'_{t_0}(s) + \zeta''_{t_0}(s).$$

We have

$$\frac{d}{dt}\|\zeta\|^2(t_0) = 2\,\mathrm{Re}\left\langle\frac{\nabla\zeta}{ds},\zeta\right\rangle(t_0)$$

$$= 2\,\mathrm{Re}\left\langle\frac{\nabla}{ds}\zeta'_{t_0},\zeta\right\rangle(t_0) + 2\,\mathrm{Re}\left\langle\frac{\nabla}{ds}\zeta''_{t_0},\zeta\right\rangle(t_0).$$

Just as $\|\zeta(t)\| = O(t^2)$ we also have $\|\zeta'_t(s)\| = O((t-s)^2)$, so that $(\Delta/ds)\zeta'_{t_0}(t_0) = 0$. To estimate $(\nabla/ds)\zeta''_{t_0}(t_0)$ we observe first of all the following fact.

LEMMA 15.35. *For any t such that $-\delta < t_0 < \delta$ and any $\mu_t \in T^{1,0}_{l(t)}(X)$, let*

$$\mu_t(s) = \tilde{\xi}(s) + \tilde{\zeta}(s)$$

be the decomposition of the translation $\mu(s)$ of $\mu(t) = \mu$ by parallel transport along l. Then

$$\left\|\frac{\nabla}{ds}\tilde{\zeta}(t)\right\| \le K\|\mu_t\|$$

with a positive constant K independent of t_0.

PROOF. Suppose $\mu' = c\mu$, and $\mu'_t(s) = \tilde{\xi}'(s) + \tilde{\zeta}'(s)$ is the corresponding decomposition of μ'. Then, obviously $\tilde{\zeta}'(s) = c\tilde{\zeta}(s)$ so that

$$\frac{\nabla}{ds}\tilde{\zeta}'(t) = c\frac{\nabla}{ds}\tilde{\zeta}(t).$$

Now let K be the supremum of all $\|(\nabla/ds)\tilde{\zeta}(t)\|$ obtained from all possible t with $-\delta < t < \delta$ and from all possible $\mu_t \in T^{1,0}_{l(t)}(X)$ of unit length. K is clearly finite by the real-analyticity of $\mu_t(s)$ jointly in μ_t and s, when $\mu_t(s)$ is defined on $T^{1,0}(X)|_{\hat{l}} \times (-2\delta, 2\delta)$, where $\hat{l}$ is an extension of l to $(-2\delta, 2\delta)$, assumed to lie within the cut-locus of $x \in X$. Lemma 15.35 is obviously valid with this constant K.

Given the lemma, we can now estimate

$$\left|\frac{d}{dt}\|\zeta\|^2(t_0)\right| = 2\left|\mathrm{Re}\left\langle\frac{\nabla}{ds}\zeta''_{t_0},\zeta\right\rangle(t_0)\right|$$

$$\le K\|\zeta_0(t_0)\|\,\|\zeta(t_0)\| = K\|\zeta(t_0)\|^2.$$

To show that $d\|\zeta(t)\|^2/dt \equiv 0$ and hence that V is invariant under parallel transport along any curve it suffices therefore to show that any real-analytic function $f(t)$ defined on $(-\delta, \delta)$ satisfying $|df/dt| \le K|f|$, $f(0) = 0$, must necessarily be identically zero. (Observe that $d\|\zeta\|^2(0)/dt = \langle\zeta(0), (\nabla/dt)\zeta(0)\rangle = 0$.) In fact, if $f = c_m t^m + O(t^{m+1})$, $c_m \ne 0$,

$$\left|\frac{df}{dt}\right| = |mc_m t^{m-1} + O(t^m)| = m|c_m|t^{m-1},$$

which clearly dominates $K|f|$ in a neighborhood of 0 for any constant K, proving the assertion that Proposition 15.34 can be reduced to the corresponding statement on the open set U for geodesics γ.

(II) The proof of the reduction on (I) implies that to prove Proposition 15.34 it suffices to show that if $\alpha \in \mathcal{M}_x$, $x \in U$, and $\gamma(t)$, $-\delta < t < \delta$, is any geodesic on U with $\gamma(0) = x$, then, for the decomposition $\alpha(t) = \xi(t) + \zeta(t)$ of the translation $\alpha(t)$ of $\alpha = \alpha(0)$ by parallel transport along γ according to the decomposition $T^{1,0}(U) = V|_U \oplus V^\perp|_U$, we have

$$\|\zeta(t)\| = O(t^2).$$

In fact, this would imply that $\alpha(t) \in V_{\gamma(t)}$, so that $V_{\gamma(0)}$ is translated to $V_{\gamma(t)}$ since $V_{\gamma(0)}$ is generated as a C-linear space by the space of maximal directions $\mathcal{M}_{\gamma(0)} = \mathcal{M}_x$.

Suppose now $\zeta(t) = ct\tilde{\zeta}(t) + O(t^2)$ with $\|\tilde{\zeta}(t)\| = 1$, where $O(t^2)$ stands for a vector-valued function of length of order $O(t^2)$. Then,

$R(\alpha(t), \overline{\alpha(t)}, \alpha(t), \overline{\alpha(t)})$

$\quad = R(\xi(t) + \zeta(t), \overline{\xi(t) + \zeta(t)}, \xi(t) + \zeta(t), \overline{\xi(t) + \zeta(t)})$

$\quad = R(\xi(t), \overline{\xi(t)}, \xi(t), \overline{\xi(t)}) + 4\,\mathrm{Re}\,R(\xi(t), \overline{\xi(t)}, \xi(t), \overline{\zeta(t)})$

$\qquad + 4R(\xi(t), \overline{\xi(t)}, \zeta(t), \overline{\zeta(t)}) + 2\,\mathrm{Re}\,R(\xi(t), \overline{\zeta(t)}, \xi(t), \overline{\zeta(t)})$

$\qquad + 4\,\mathrm{Re}\,(\xi(t), \overline{\zeta(t)}, \zeta(t), \overline{\zeta(t)}) + O(t^4).$

By Proposition 15.33 $R(\xi(t), \overline{\xi(t)}, \zeta(t), \overline{\zeta(t)}) = 0$. We claim that actually for any $x \in U$, any $\xi \in V_x$ and any $\zeta \in V_x^\perp$ we have

$$R_{\xi\bar{\zeta}x\bar{x}} = 0 \quad \text{for all } \chi \in T_x^{1,0}(X). \tag{15.25}$$

To see this, suppose $\xi = \zeta \in \mathcal{M}_x$. From Proposition 15.33 we have $R_{\alpha\bar{\alpha}\zeta\bar{\zeta}} = 0$ for all $\zeta \in V_x^{\perp}$, so that $\zeta \in \mathcal{N}_\alpha$. However, by Proposition 15.14 now we have

$$R_{\alpha\bar{\xi}\chi\bar{\chi}} = 0 \quad \text{for all } \chi \in T_x^{1,0}(X),$$

which yields (15.25) since V_x is the linear span of $\mathcal{M}_x$. It follows now from (15.25) that

$$R(\alpha(t), \overline{\alpha(t)}, \alpha(t), \overline{\alpha(t)}) = R(\xi(t), \overline{\xi(t)}, \xi(t), \overline{\xi(t)}) + O(t^4).$$

On the other hand, if $\zeta(t) = ct\tilde{\zeta}(t) + O(t^2)$, $\|\tilde{\zeta}(t)\| = 1$, we have

$$\|\xi(t)\|^2 = 1 - c^2 t^2 + O(t^4),$$

so that

$$R(\alpha(t), \overline{\alpha(t)}, \alpha(t), \overline{\alpha(t)}) \leq (1 - c^2 t^2 + O(t^4))^2 R_{\alpha\bar{\alpha}\alpha\bar{\alpha}}(0) + O(t^4)$$

$$= (1 - 2c^2 t^2) R_{\alpha\bar{\alpha}\alpha\bar{\alpha}}(0) + O(t^4).$$

If $c \neq 0$ we would have $R_{\alpha\bar{\alpha}\alpha\bar{\alpha},\eta}(0) = 0$ and $R_{\alpha\bar{\alpha}\alpha\bar{\alpha},\eta\eta}(0) \leq -4c^2 R_{\alpha\bar{\alpha}\alpha\bar{\alpha}}(0)$ for $\eta = \gamma(0)$. But we have simply, from Berger's formula, $R_{\alpha\bar{\alpha}\alpha\bar{\alpha},\eta\eta}(0) = 0$. This proves $c = 0$, so that in the decomposition $\alpha(t) = \xi(t) + \zeta(t)$ we have

$$\|\zeta(t)\| = O(t^2)$$

implying by the reduction method of (I) that V is a differentiable C-vector bundle invariant under parallel transport along any smooth curve on X, proving Proposition 15.34. $\square$

15.8.5. Integral Submanifolds of Re $V^{\perp}$

Recall that V is a distribution of tangent vectors of type $(1, 0)$ invariant under parallel transport and for each $x \in X$ there is a compact totally geodesic complex submanifold $\tilde{Z}_x$ which is locally symmetric. (Such a Z_x exists now for each $x \in X$ because one can take the open set U to be X, since we know now that $V = \bigcup_{x \in X} V_x$ is a differentiable C-vector bundle.) The $\tilde{Z}_x$ are integral submanifolds of Re V. According to the orthogonal decomposition $T^{1,0}(X) = V \oplus V^{\perp}$ we can divide vectors of $T^{1,0}(X)$ into types $\xi \in V$ and $\zeta \in V^{\perp}$. In Section 15.8.2 we deduced that X is locally symmetric if $\nabla_{\zeta} R_{\xi \bar{\zeta} \zeta \bar{\zeta}} = 0$ for all $\zeta \in V^{\perp}$.

Since V is invariant under parallel transport, the same applies to $V^{\perp}$. The arguments in Proposition 15.31 for V now apply to $V^{\perp}$ to show that Re $V^{\perp}$ is an integrable distribution of tangent vectors. For each $x \in X$, let $Z_x^{\perp}$ be the leaf passing through x of the foliation defined by the distribution Re $V^{\perp}$. The arguments of Proposition 15.31 imply that $Z_x^{\perp}$ is totally geodesic and complex analytic. From Proposition 15.33 we know that $R_{\xi \bar{\xi} \zeta \bar{\zeta}} = 0$ whenever $\xi \in V_x$ and $\zeta \in V_x^{\perp}$. It follows by the hypothesis of Theorem 15.3 that $Z_x^{\perp}$ can be regarded as a complete Kähler–Einstein manifold of positive Ricci curvature. By the theorem of Bonnet and Meyers each $Z_x^{\perp}$ is compact. Now it is obvious how one can prove Theorem 15.3 by induction on the complex dimension of X. In fact, $Z_x^{\perp}$ satisfies the hypothesis on X in Theorem 15.3. We can therefore assume as an induction hypothesis that $Z_x^{\perp}$ is a Hermitian symmetric manifold unless $V = T^{1,0}(X)$, in which case there is nothing to prove. Hence, $\nabla_{\zeta} R_{\xi \bar{\zeta} \zeta \bar{\zeta}} = 0$ whenever $\zeta \in V^{\perp}$, proving $\nabla R \equiv 0$ on X for the curvature tensor R, completing the proof of Theorem 15.3. □

REMARKS 15.36. 1. By a result of Kobayashi [11], all locally symmetric compact complex manifolds of positive Ricci curvature must be simply connected. Hence, the manifold X in Theorem 15.3 is globally symmetric.

2. It is clear that the proof of Theorem 15.3 implies immediately the more general case when the Ricci tensor of X is only assumed to be parallel and positive. Positivity of the Ricci tensor is only used in proving compactness of certain integral submanifolds, using the theorem of Bonnet and Meyers.

15.9. A Generalization of Theorem 15.3

Recall Corollaries 15.4 and 15.5, which assert that Theorem 15.3 can be generalized to the case when the Ricci tensor is parallel and the Kähler manifold X is complete, possibly noncompact. By results of Bishop and Goldberg [4–6], for a compact Kähler manifold X of nonnegative holomorphic bisectional curvature, the Ricci tensor of X is parallel if and only if X has constant scalar curvature. To complete the Chapter it suffices to prove Corollary 15.5, where X is only assumed to be complete and the Ricci tensor of X is assumed parallel, in place of being Kähler–Einstein of positive Ricci curvature.

The only places where the positivity of the Ricci curvature is used are Sections 15.8.2 and 15.8.5, where we applied the theorem of Bonnet and Meyers. The Kähler–Einstein condition was only used in obtaining formulas for computing $\nabla R_{i\bar{j}k\bar{l}}$, etc. But it is clear that these formulas (obtained by commutation) would still hold if $\nabla(\text{Ric}) \equiv 0$; i.e., the Ricci tensor is parallel.

PROOF OF COROLLARY 15.5 (AND HENCE COROLLARY 15.4). The point of the proof is simply to split off the directions where the Ricci tensor vanishes. Define $W_x \subset T_x^{1,0}(X)$ to be the subspace of all $\chi \in T_x^{1,0}(X)$ such that Ric $(\chi, \chi) = 0$. Clearly W_x is a C-vector subspace of $T_x^{1,0}(X)$ since the Ricci form is a Hermitian symmetric bilinear form on X. Since the Ricci tensor $W = \bigcup_{x \in X} W_x$ is a differentiable vector bundle on X invariant under parallel transport, by the arguments of Proposition 15.32. Re W as an integral distribution of real tangent vectors. The leaves L_x of the foliation defined by Re W are flat since they are totally geodesic, the Ricci tensor on L is everywhere zero and holomorphic bisectional curvatures of L_x are nonnegative. Let $W_x^\perp$ be the orthogonal complement of W_x in $T_x^{1,0}(X)$. Then $W = \bigcup_{x \in X} W_x^\perp$ is invariant under parallel transport. Denote by Z_x a leaf of the foliation defined by Re $W^\perp$. Then Z_x carries positive Ricci curvature by the definition of $W_x^\perp$. Moreover, the Ricci tensor of Z_x is parallel since $R_{\xi\bar{\xi}\chi\bar{\chi}} = 0$ for all $\xi \in W_x^\perp$ and $\chi \in W_x$. By Remark 15.36(2) we conclude that each Z_x is a (global) Hermitian symmetric space, so that $\nabla_\xi R_{\xi\bar{\xi}\xi\bar{\xi}} = 0$ for all $\xi \in W^\perp$. Since W is invariant under parallel transport and $R_{\chi\bar{\chi}\chi\bar{\chi}} = 0$ for all $\chi \in W$, it follows that $\nabla_\xi R_{\chi\bar{\chi}\chi\bar{\chi}} = \nabla_\chi R_{\chi\bar{\chi}\chi\bar{\chi}} = 0$ for $\xi \in W_x^\perp$ and $\chi \in W_x$. The only other terms of ∇R, up to conjugation and permutation of indices, are of the types $\nabla_\mu R_{\xi\bar{\chi}\mu\bar{\mu}}$, where $\mu \in T_x^{1,0}(X)$ is arbitrary. It suffices therefore to show that

$$R_{\xi\bar{\chi}\mu\bar{\mu}} = 0 \quad \text{for all } \mu \in T_x^{1,0}(X). \tag{15.26}$$

To prove (15.26) it is equivalent to prove $R_{\chi\bar{\mu}\mu\bar{\mu}} = 0$ for all $\mu \in T_x^{1,0}(X)$ since $R_{\chi\bar{\chi}\mu\bar{\mu}} = 0$ for all $\chi \in W_x$ and $\mu \in T_x^{1,0}(X)$. Let $\mu \in T_x^{1,0}(X)$ and consider the function

$$F(\varepsilon) = R(\chi + \varepsilon\mu, \overline{\chi + \varepsilon\mu}, \chi + \varepsilon\mu, \overline{\chi + \varepsilon\mu})$$

defined for ε real. Then, from $F(\varepsilon) \geq 0$ we obtain by variation formulas that $R_{\chi\bar{\chi}\chi\bar{\mu}} = 0$ and $4R_{\chi\bar{\chi}\mu\bar{\mu}} + 2 \text{Re } R_{\chi\bar{\mu}\chi\bar{\mu}} \geq 0$. But since $R_{\chi\bar{\chi}\mu\bar{\mu}} = 0$ and we can always assume Re $R_{\chi\bar{\mu}\chi\bar{\mu}} \leq 0$ if μ is replaced by $e^{i\theta}\mu$ for an appropriate real angle θ, it follows that $R_{\chi\bar{\mu}\chi\bar{\mu}} = 0$. Computing now the third variation of F against ε at 0, we obtain $R_{\chi\bar{\mu}\mu\bar{\mu}} = 0$, proving (15.26), thus showing that $\nabla R \equiv 0$ on X and proving Corollary 15.5 (and hence Corollary 15.4). $\square$

CONCLUDING REMARKS 1. By Koszul [13] and Lichnerowicz [16] every compact homogeneous Kähler manifold X carries a Kähler metric with parallel and semipositive Ricci tensor. Analogous to the situation of Gray [18] our theorem shows that such a Kähler metric on X cannot have nonnegative holomorphic bisectional curvature everywhere unless X is Hermitian symmetric (cf. Lichnerowicz [14, 15]).

2. As was indicated in Gray [8], every compact homogeneous space X of the form G/T, where G is a compact Lie group and T is a maximal torus of G, admits an Einstein and bi-invariant metric of nonnegative sectional (and hence nonnegative bisectional) curvature. This metric is in general not Kählerian.

3. See Auslander [1] for an example of compact flat Kähler manifolds which are not homogeneous. Hence, in the formulation of Corollaries 15.4 and 15.5, we can only conclude that X is locally symmetric.

References

1. L. Auslander, Four dimensional compact locally Hermitian manifolds, *Trans. Amer. Math. Soc.* **84**, 379–391 (1957).
2. S. Bando, On three dimensional compact Kähler manifolds of nonnegative bisectional curvature, *J. Diff. Geom.* **19**, 283–297 (1984).
3. M. Berger, Sur les variétés d'Einstein compactes, C. R. IIIᵉ Réunion Math Expression Latine, Namur, 35–55 (1965).
4. R. L. Bishop and S. I. Goldberg, On the second cohomology group of a Kähler manifold of positive curvature, *Proc. Amer. Math. Soc.* **16**, 119–122 (1965).
5. R. L. Bishop and S. I. Goldberg, Rigidity of positively curved Kähler manifolds, *Proc. Nat. Acad. Sci. U.S.A.* **54**, 1037–1041 (1965).
6. R. L. Bishop and S. I. Goldberg, On the topology of positively curved Kähler manifolds. II, *Tôhoku Math. J.* **11**, 310–318 (1965).
7. S. I. Goldberg and S. Kobayashi, On holomorphic bisectional curvature, *J. Differential Geometry* **1**, 225–233 (1967).
8. A. Gray, Compact Kähler manifolds with nonnegative sectional curvature, *Invent. Math.* **41**, 33–43 (1977).
9. R. Gunning and H. Rossi, *Analytic Functions of Several Complex Variables*, Prentice-Hall, Englewood Cliffs, NJ (1965).
10. R. S. Hamilton, Three-manifolds with positive Ricci curvature, *J. Differential Geometry* **17**, 255–306 (1982).
11. S. Kobayashi, On compact Kähler manifolds with positive Ricci tensor, *Ann. of Math.* (2) **74**, 570–575 (1961).
12. S. Kobayashi and K. Nomizu, *Foundations of Differential Geometry*, Wiley, New York (1969).
13. J. L. Koszul, Sur la forme hermitienne canonique des espaces homogènes complexes, *Canad. J. Math.* **7**, 562–576 (1955).

14. A. Lichnerowicz, Variétés pseudokählériennes à courbure de Ricci non nulle; applications aux domaines bornés homogènes de c″, *C.R. Acad. Sci. Paris* **237**, 695-697 (1953).
15. A. Lichnerowicz, Espaces homogènes kähleriens, *C.R. Acad. Sci. Paris* **237**, 695-697 (1953).
16. A. Lichnerowicz, *Some Aspects of the Theory of Homogeneous Kählerian Spaces*, Lecture Notes, Princeton (1958).
17. S. Łojasiewicz, Triangulation of semi-analytic sets, *Ann. Scuola Norm. Sup. Pisa* **18**, 449-474 (1964).
18. N. Mok and J.-Q. Zhong, Variétés compactes kählériennes d'Einstein de courbure bisectionnelle semipositive, *C.R. Acad. Sci. Paris, Sér. I* **296**, 473-475 (1985).
19. S. Mori, Projective manifolds with ample tangent bundles, *Ann. of Math.* **110**, 593-606 (1979).
20. Y.-T. Siu, Curvature characterization of hyperquadratics, *Duke Math. J.* **47**, 641-654 (1980).
21. Y.-T. Siu, Some recent developments in complex differential geometry, preprint, 1983.
22. Y.-T. Siu and S.-T. Yau, Compact Kähler manifolds of positive bisectional curvature, *Invent Math.* **59**, 189-204 (1980).

16

Schubert Calculus and Schur Functions

Abstract

Given two Schubert cycles of a Grassmann manifold, σ_a and σ_b, we have the product formula

$$\sigma_a \cdot \sigma_b = \sum_c \delta(a, b, c)\sigma_c.$$

The method for calculating $\delta(a, b, c)$ has already been given in a previous work [5] with the aid of Schur functions in the representation theory of the unitary group. On this basis of [5], making use of the similarity between the Schubert calculus and Schur functions and the branching formula of group representation theory, we investigate the question as to which of those can make σ_c take $\delta(a, b, c)$ as coefficients in $\sigma_a\sigma_b$ for given σ_c, σ_b.

16.1. Introduction

The so-called Schubert calculus is the calculation of the multiplicative structure of the homology ring of a Grassmann manifold. Grassmann manifold $G(n, d)$ is a manifold composed of all d-dimensional linear subspaces in $\mathbb{C}^n$, and its complex dimension is $N = d(n - d)$. For $G(n, d)$, a cell decomposition may be made as follows: take an arbitrary flag, $V = \{V_1 \subset V_2 \subset \cdots \subset V_n = \mathbb{C}^n, \dim V_k = k\}$ and a d-tuple

$$a = (a_1, a_2, \ldots, a_d), \quad n - d \geq a_1 \geq \cdots \geq a_d \geq 0$$

of nonnegative integers. The point set of $G(n, d)$,

$$\sigma_a(V) = \sigma_{(a_1,...,a_d)}(V) = \{\Lambda| \dim (\Lambda \cap V_{n-d+i-a_i}) \geq i\},$$

is called a Schubert cycle with subscript a. Generally speaking, $\sigma_a(V)$ is an analytical subvariety with (real) codimension $2 \sum a_i$. The set of all $\sigma_a(V)$, denoted by $\{\sigma_a(V)\}$, constitutes a cell decomposition of $G(n, d)$. Because the homology class to which the $\sigma_a(V)$ belongs is independent of the choice of V, in terms of homology, $\sigma_a(V)$ can be simply denoted by $\sigma(a_1, \ldots, a_d)$. Therefore,

$$H_*(G(n, d), \mathbb{Z}) = \{\sigma_{(a_1,...,a_d)}| n - d \geq a_1 \geq \cdots \geq a_d \geq 0\}.$$

As is well known, for a general homology group $H_*(G(n, d), \mathbb{Z})$, its multiplication may be defined by "intersection number,"

$$H_{2N-2m_1}(G(n, d), \mathbb{Z}) \times H_{2N-2m_2}(G(n, d), \mathbb{Z}) \to H_{2N-2(m_1+m_2)}(G(n, d), \mathbb{Z}).$$

Such multiplication, if expressed by its basis, Schubert cycles, is

$$\sigma_a \cdot \sigma_b = \sum \delta(a, b; c)\sigma_c, \qquad (16.1)$$

where $a = (a_1, \ldots, a_{k_1})$, $b = (b_1, \ldots, b_{k_2})$, $c = (c_1, \ldots, c_l)$, $l = k_1 + k_2$, $\sum c_i = \sum a_i + \sum b_i$. Note that (16.1) is a universal product formula and is applicable to all $G(n, d)$, e.g.,

$$\sigma_1 \cdot \sigma_{(2,1)} = \sigma_{(3,1)} + \sigma_{(2,2)} + \sigma_{(2,1,1)} \qquad (\text{in } G(n, 3)),$$

$$\sigma_1 \cdot \sigma_{(2,1)} = \sigma_{(3,1)} + \sigma_{(2,2)} \qquad (\text{in } G(n, 2))$$

because in $G(n, 2)$, $(2, 1, 1)$ should be considered as zero.

The coefficients $\delta(a, b; c)$ of the general term in formula (16.1) express the intersection number, $\delta(a, b; c) = \#(\sigma_a \cdot \sigma_b \cdot \sigma_c)$. The central topic of the Schubert calculus is how to give a method for calculating $\delta_{(a,b;c)}$. The following are the basic relevant results.

1. $$\#(\sigma_a \cdot \sigma_b) = \delta_{a_1}^{n-d-b_d} \cdots \delta_{a_d}^{n-d-b_1}. \qquad (16.2)$$

In general, in $G(n, d)$, $\sigma_{(a_1,...,a_d)}$ and $\sigma_{(b_1,...,b_d)}$ are said to be complementary, if $a = \tilde{b} \triangleq (n - d - b_d, \ldots, n - d - b_1)$. Formula (16.2) means that the intersection number of two Schubert cycles is 1 if and only if they are complementary.

2. Pieri formula. If $\sigma_a = \sigma_{(a,0,\ldots,0)}$ (called special Schubert cycle), $\sigma_b = \sigma_{(b_1,\ldots,b_d)}$, then

$$\sigma_a \cdot \sigma_b = \sum_{\substack{b_i \le c_i \le b_{i-1} \\ \sum c_i = a + \sum b_i}} \sigma_c. \tag{16.3}$$

3. Giambelli formula

$$\sigma_{(a_1,\ldots,a_d)} = \begin{vmatrix} \sigma_{a_1} & \sigma_{a_1+1} & \cdots & \sigma_{d_1+d-1} \\ \sigma_{a_2-1} & \sigma_{a_2} & \cdots & \sigma_{a_2+d-2} \\ & & \cdots & \\ \sigma_{a_d-d+1} & & \cdots & \sigma_{a_d} \end{vmatrix}. \tag{16.4}$$

In the above determinant all σ_{a_i} are special Schubert cycles (cf. Ref. 2).

In the history of its development similarities have been repeatedly found between Schubert calculus and the theory of Schur functions in the representation theory of symmetric groups (or equivalently the representation theory of unitary groups), (for general theory of the Schur function, refer to Ref. 4). As we know, the Schur function $S_f(x_1,\ldots,x_n)$ with subscript $f = (f_1,\ldots,f_n)$, $f_1 \ge \cdots \ge f_n \ge 0$, is the characteristic of the irreducible representation of the unitary group $U(n)$ (or general linear group $GL(n,\mathbb{C})$) with signature $f = (f_1,\ldots,f_n)$, $f_1 \ge \cdots \ge f_n \ge 0$,

$$S_f(x_1,\ldots,x_n) = \frac{M_f(x_1,\ldots,x_n)}{D(x_1,\ldots,x_n)},$$

where $M_f(x_1,\ldots,x_n) = \det(x_i^{\tilde{f}_j})$, $D(x_1,\ldots,x_n) = \prod_{i<j}(x_i - x_j)$, $\tilde{f}_1 = f_1 + n - 1, \ldots, \tilde{f}_n = f_n$. Similar to the case of (16.3) and (16.4) for Schur functions, we have [4]

$$S_{(a,0,\ldots,0)}(x_1,\ldots,x_n)S_{(b_1,\ldots,b_n)}(x_1,\ldots,x_n) = \sum_{\substack{b_{i-1} \ge c_i \ge b_i \\ \sum c_i = a + \sum b_i}} S_{(c_1,\ldots,c_n)}(x_1,\ldots,x_n),$$

$$\tag{16.3$'$}$$

$$S_{(a_1,\ldots,a_n)}(x_1,\ldots,x_n) = \det(e_{a_i+j-1}(x_1,\ldots,x_n)), \tag{16.4$'$}$$

where $e_{a_i+j-1}(x_1,\ldots,x_n) = S_{(a_i+j-1,0,\ldots,0)}(x_1,\ldots,x_n)$ is a special Schur function. Equation (16.4)$'$ is also called the Jacobi identity. From the similarity between (16.3), (16.4), and (16.3)$'$, (16.4)$'$, and considering that (16.1) may be obtained by combining (16.3) and (16.4) in finite steps, it is known that if treating (16.3)' and (16.4)' in the same way, for Schur functions (when n is fixed) we can get

$$S_{(a_1,\ldots,a_n)}(x_1,\ldots,x_n)S_{(b_1,\ldots,b_n)}(x_1,\ldots,x_n) = \sum \delta(a,b;c)S_{(c_1,\ldots,c_n)}(x_1,\ldots,x_n).$$
$$(16.5)$$

That is, the Schubert coefficient $\delta(a,b;c)$ can also be obtained by the coefficient of the product of Schur functions being expanded into Schur functions.

Another formula relating Schubert calculus with Schur function is the following theorem established by the author in Ref. 5.

THEOREM 16.1 [5]. *Suppose that H is a commutative ring with unit. Let $\varphi(t)$ be the formal power series*

$$\varphi(t) = \sum_{j=0}^{\infty} b_j t^j, \qquad b_j \in H, \quad b_0 = 1.$$

Then

$$\varphi(t_1) \cdots \varphi(t_n) = \sum_{a_1 \geq \cdots \geq a_n \geq 0} S_{(a_1,\ldots,a_n)}(t_1,\ldots,t_n)g_{(a_1,\ldots,a_n)}, \qquad (16.6)$$

where $g_{(a_1,\ldots,a_n)} = \det(b_{a_i+j-i})$ is a determinant of Giambelli type that takes values in the ring H.

For example, taking $\varphi(t) = \sum_{a=0}^{\infty} \sigma_a t^a$, $\sigma_a = \sigma_{(a,0,\ldots,0)}$ and considering (16.4), we have the next result.

COROLLARY 16.2. *If $\varphi(t) = \sum_{a=0}^{\infty} \sigma_a t^a$, $\sigma_a = \sigma_{(a,0,\ldots,0)}$, then*

$$\varphi(t_1) \cdots \varphi(t_n) = \sum_{a_1 \geq \cdots \geq a_n \geq 0} \sigma_{(a_1,\ldots,a_n)}S_{(a_1,\ldots,a_n)}(t_1,\ldots,t_n). \qquad (16.7)$$

Applying (16.7) to the unitary group and making use of the orthogonality of the irreducible characters, we can get the coefficient of the general product formula (16.1).

THEOREM 16.3 [5].

$$\delta(a,b;c)$$

$$= \int_{U(k_1)} \int_{U(k_2)} S_c(t_1,\ldots,t_l)\overline{S_a(t_1,\ldots,t_{k_1})S_b(t_{k+1},\ldots,t_l)}\, \dot{U}(k_1)\dot{U}(k_2).$$
$$(16.8)$$

16.2. Involution

Denote the commutative ring of symmetric polynomials of $x_1, \ldots, x_n$ by $S[x_1, \ldots, x_n]$. According to the basic theorems of symmetric polynomials, the ring can be generated by elementary symmetric polynomials. The basis of the polynomial ring can take either form of the following two elementary symmetric polynomials:

$$e_j = S_{(1,1,\ldots,1,0,\ldots,0)}(x) = \sum_{i_1 < \cdots < i_j} x_{i_1} \cdots x_{i_j}, \qquad 1 \le j \le n,$$
$$\underbrace{}_{j}$$

$$h_j = S_{(j,0,\ldots,0)}(x) = \sum_{i_1 \le \cdots \le i_j} x_{i_1} \cdots x_{i_j}, \qquad 1 \le j \le n.$$

According to Ref. 4, in the symmetric polynomial ring $S[x_1, \ldots, x_n] = \mathbb{C}[e_1, \ldots, e_n] = \mathbb{C}[h_1, \ldots, h_n]$, an involutive holomorphism T can be defined as

$$Te_j = h_j, \qquad T^2 = I. \tag{16.9}$$

Hence $Th_j = e_j$. Under the involution T, the Schur function $S_\lambda = S_{(\lambda_1, \ldots, \lambda_n)}(x)$ becomes

$$TS_\lambda = S_{\lambda^*}, \tag{16.10}$$

λ^* being the conjugate of λ. Make $\lambda = (\lambda_1, \ldots, \lambda_n)$ correspond to a Young tableau (λ). The Young tableau (λ^*) is the transpose of (λ), and then λ^* is obtained. For example, $(3, 2, 2, 1)^* = (4, 3, 1, 0)$.

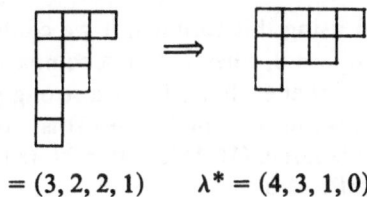

$$\lambda = (3, 2, 2, 1) \qquad \lambda^* = (4, 3, 1, 0)$$

Thus we can see that the entire basis for the Schubert calculus is the Pieri formula (16.3) and the Giambelli formula (16.4) as well as formula (16.1) derived from them. Considering the linear space that takes the finite formal sum of σ_a, say $\sum c_a \sigma_a$, as its elements. Formula (16.1), being the product in the space, becomes a commutative ring with unit $\sigma_0 = 1$, which we call the Schubert ring S. The Giambelli formula shows that S is the ring of finite polynomials that takes σ_k ($k = 1, 2, \ldots$) as its basis. Write

$$\sigma_k = \sigma_{(k,0,\ldots,0)}, \qquad \sigma_{[k]} = \sigma_{(\underbrace{1,1,\ldots,1}_{k},0,\ldots,0)}, \tag{16.11}$$

and we can define an involutive holomorphism T in S as follows: If $a = (a_1, \ldots, a_n)$, we let

$$T\sigma_a = \sigma_{a^*}, \tag{16.12}$$

where a^* is the conjugate of a (see Ref. 2, p. 200).

PROPOSITION 16.4. *T is indeed an involution.*

PROOF. It is evident that $T^2 = I$. We need only prove $T(\sigma_a \cdot \sigma_b) = T\sigma_a \cdot T\sigma_b$. Formula (16.1) gives

$$\sigma_a \cdot \sigma_b = \sum \delta(a, b; c)\sigma_c$$

$$T(\sigma_a \cdot \sigma_b) = \sum \delta(a, b; c)T\sigma_c = \sum \delta(a, b; c)\sigma_{c^*}.$$

Reference 2 indicates that $\delta(a^*, b^*; c^*) = \delta(a, b; c)$. Therefore

$$T(\sigma_a \cdot \sigma_b) = \sum \delta(a, b; c)\sigma_{c^*} = \sum \delta(a^*, b^*; c^*)\sigma_{c^*}$$

$$= \sigma_{a^*} \cdot \sigma_{b^*} = (T\sigma_a)(T\sigma_b). \qquad \square$$

A direct corollary of this theorem is (using involution T to formula (16.4)).

COROLLARY 16.5 (*Dual Giambelli formula*). *If $a = (a_1, \ldots, a_n)$, then*

$$\sigma_{a^*} = \begin{vmatrix} \sigma_{[a_1]} & \sigma_{[a_1+1]} & \cdots & \sigma_{[a_1+n-1]} \\ \sigma_{[a_2-1]} & \sigma_{[a_2]} & \cdots & \sigma_{[a_2+n-2]} \\ \sigma_{[a_n-n+1]} & & \cdots & \sigma_{[a_n]} \end{vmatrix}. \tag{16.13}$$

Notice that, when using this formula, if we confine our discussion to $H_*(G(m, n), \mathbb{Z})$, then $\sigma_{[k]} = 0$, when $k > n$. A comparison between formula (16.13) and (16.4) shows that each has its own strong points and shortcomings. When a_1 is large, while n is small, (16.13) is convenient; conversely, when a_1 is small and n is large, (16.4) is better. Note that it is a convention that $\sigma_{[0]} = 1$, $\sigma_{[k]} = 0$, if $k < 0$.

COROLLARY 16.6 (*Dual Pieri formula*).

$$\sigma_{[k]} \cdot \sigma_{(b_1,\ldots,b_n)} = \sum_{i_1 < \cdots < i_k}^{\wedge} \sigma_{(b_1,\ldots,b_{i_1}+1,\ldots,b_{i_k}+1,\ldots,b_n)}, \tag{16.14}$$

where $\sum^{\wedge}$ stands for summing over subscripts that satisfy the descending relation. The meaning that subscripts satisfy descending relations is that $a_1 \geq \cdots \geq a_n \geq 0$.

PROOF. We may make an involution of the Pieri formula to prove it, but it is rather intricate. A simple method is to use the similarity between the Schubert calculus and the product of Schur functions ((16.1) and (16.5)). Formula (16.5) in group representation theory expresses the decomposition of the tensor product of two irreducible representations (into the sum of an irreducible representation). For the irreducible representation of $U(n)$, Ref. 4 shows that

$$S_{(1,1,...1,0,...0)}(x_1, \ldots, x_n) S_{(b_1,...,b_n)}(x_1, \ldots, x_n) = \sum_{i_1 < \cdots < i_k}^{\wedge} S_{(b_1,...,b_{i_1}+1,...,b_{i_k}+1,...,b_n)}.$$

Hence, (16.14) holds.   □

16.3. Branching Formula and Schubert Calculus

The connection between the Schubert calculus and Schur functions is not only reflected by (16.1) and (16.5) but also by the branching formula. According to group representation theory, any irreducible representation $\Pi_{(f_1,...,f_n)}$ of $U(n)$ of order n, $((f_1, \ldots, f_n)$ being its young signature) if confined to a unitary group of order $n - 1$ is no longer irreducible in general, and it may be decomposed into the sum of irreducible representations of $U(n - 1)$. Such a decomposition is given by the so-called branching formula

$$\Pi_{(f_1,...,f_n)} = \sum_{f_1 \geq f_1' \geq f_2 \geq f_2' \geq \cdots \geq f_{n-1}' \geq f_n} \Pi_{(f_1',...,f_{n-1}')}. \tag{16.15}$$

In the following, for the sake of convenience, for two descending integer sequences, $f_1 \geq \cdots \geq f_n$ and $f_1' \geq \cdots \geq f_{n-1}'$, we set

$$f = (f_1, \ldots, f_n) > f' = (f_1', \ldots, f_{n-1}') \Leftrightarrow f_1 \geq f_1' \geq f_2 \geq f_2' \geq \cdots \geq f_{n-1}' \geq f_n.$$

Thus, (16.15), if written in the form of irreducible characters (i.e., Schur functions), becomes

$$S_f(x_1, \ldots, x_n) = \sum_{f' < f} x_n^{\Sigma f_i - \Sigma f_i'} S_{f'}(x_1, \ldots, x_{n-1})$$

$$= \sum_{\substack{f' < f \\ m+|f'|=|f|}} S_m(x_n) S_{f'}(x_1, \ldots, x_{n-1}), \tag{16.16}$$

where $|f| = \sum_{i=1}^{n} f_i$, $|f'| = \sum_{i=1}^{n-1} f_i'$.

We will show that the branching formula (16.16) corresponds to the Pieri formula through (16.8). In fact, for any $g = (g_1, \ldots, g_n)$, (16.8) gives

$$\delta(\sigma_{m_1}, \sigma_{(f_1, \ldots f_{n-1})}; \sigma_{(g_1, \ldots, g_n)})$$

$$= \int_{U(1) \times U(n-1)} S_m(x_1) S_{(f_1, \ldots, f_{n-1})}(x_2, \ldots, x_n) \overline{S_g(x_1, \ldots, x_n)} \, \dot{U}(1) \dot{U}(n-1).$$

From (16.16), it follows that

$$S_g(x_1, \ldots, x_n) = \sum_{\substack{g' < g \\ k + |g'| = |g|}} S_k(x_1) S_{(g'_1, \ldots, g'_{n-1})}(x_2, \ldots, x_n)$$

such that

$$\delta(\sigma_m, \sigma_{(f_1, \ldots, f_{n-1})}; \sigma_g)$$

$$= \sum_{\substack{g' < g \\ k + |g'| = |g|}} \int_{U(1) \times U(n-1)} S_m(x_1) \overline{S_k(x_1)} \, S_{(f_1, \ldots, f_{n-1})}(x_2, \ldots, x_n)$$

$$\times S_{(g'_1, \ldots, g'_{n-1})}(x_2, \ldots, x_n) \dot{U}(1) \dot{U}(n+1)$$

$$= \begin{cases} 1, & \text{if } k = m, (g'_1, \ldots, g'_{n-1}) = (f_1, \ldots, f_{n-1}), \\ 0, & \text{otherwise,} \end{cases}$$

i.e., in $\sigma_m \cdot \sigma_{(f_1, \ldots, f_{n-1})}$ there can only be σ_g's that satisfy $g > (f_1, \ldots, f_{n-1})$, $|g| = |f| + m$, whose corresponding coefficient is 1. This is the Pieri formula

$$\sigma_m \cdot \sigma_{(f_1, \ldots, f_{n-1})} = \sum_{\substack{g > f \\ |g| = |f| + m}} \sigma_{(g_1, \ldots, g_n)}.$$

Now we prove a generalization of branching formula (16.16).

THEOREM 16.7. *Let* $n \geq 2k$. *We have the identity*

$$S_{(l, \ldots, l, 0, \ldots, 0)}(x_1, \ldots, x_k, y_1, \ldots, y_{n-k})$$

$$= \sum_{l \geq m_1 \geq \cdots \geq m_k \geq 0} S_{(m_1, \ldots, m_k)}(x_1, \ldots, x_k) S_{\underbrace{(l - m_k, \ldots, l - m, 0, \ldots, 0)}_{n-k}}(y_1, \ldots, y_{n-k}).$$

$$\tag{16.17}$$

PROOF. From formula (16.16),

$$S_{(l,\ldots,l,0,\ldots,0)}(x_1,\ldots,x_k,y_1,\ldots,y_{n-k})$$

$$= \sum_{l \geq i \geq 0} y_1^{l-i} S_{(l,\ldots,l,i,0,\ldots,0)}(x_1,\ldots,x_k,y_2,\ldots,y_{n-k})$$

$$= \sum_{l \geq m_1 \geq 0} S_{m_1}(y_1) S_{(l,\ldots,l,l-m,0,\ldots,0)}(x_1,\ldots,x_k,y_2,\ldots,y_{n-k})$$

$$= \sum_{l \geq m_1 \geq 0} S_{m_1}(y_1) \sum_{l \geq i_2 \geq m_1 \geq i_1 \geq 0} y_2^{i_1+i_2-m_1}$$

$$\times S_{(l,\ldots,l,l-i_1,l-i_2,0,\ldots,0)}(x_1,\ldots,x_k,y_3,\ldots,y_{n-k})$$

$$= \sum_{l \geq i_2 \geq i_1 \geq 0} S_{(l,\ldots,l,l-i_1,l-i_2,0,\ldots,0)}(x_1,\ldots,x_k,y_3,\ldots,y_{n-k})$$

$$\times \sum_{i_2 \geq m_1 \geq i_1} y_2^{i_1+i_2-m_1} S_{m_1}(y_1)$$

$$= \sum_{l \geq i_2 \geq i_1 \geq 0} S_{(i_1,i_2)}(y_1,y_2) S_{(l,\ldots,l,l-i_1,l-i_2,0,\ldots,0)}(x_1,\ldots,x_k,y_3,\ldots,y_{n-k}).$$

The inner summation of the above formula is also carried out by branching formula (16.16). Continuing the above process, we see that the above formula

$$= \sum S_{(i_1,i_2)}(y_1,y_2) y_3^{p_1+p_2+p_3-i_1-i_2}$$

$$\times S_{(l,\ldots,l,l-p_1,l-p_2,l-p_3,0,\ldots,0)}(x_1,\ldots,x_k,y_4,\ldots,y_{n-k})$$

$$= \sum_{l \geq p_3 \geq p_2 \geq p_1 \geq 0} \left(\sum_{p_3 \geq i_1 \geq p_2 \geq i_2 \geq p_1} y_3^{p_1+p_2+p_3-i_1-i_2} S_{(i_1,i_2)}(y_1,y_2) \right)$$

$$\times S_{(l,\ldots,l,l-p_1,l-p_2,l-p_3,0,\ldots,0)}(x_1,\ldots,x_k,y_4,\ldots,y_{n-k})$$

$$= \sum_{l \geq p_3 \geq p_2 \geq p_1 \geq 0} S_{(p_3,p_2,p_1)}(y_1,y_2,y_3)$$

$$\times S_{(l,\ldots,l,l-p_1,l-p_2,l-p_3,0,\ldots,0)}(x_1,\ldots,x_k,y_4,\ldots,y_{n-k}).$$

After k steps, the desired conclusion is reached. $\quad\square$

Now we ask the question: For fixed σ_c, σ_b, and arbitrary σ_a, what is the value of $\delta(a,b;c)$? We can express the question by formula (16.18):

$$\sigma_c \cdot \sigma_b = \sum_a \delta(a,b;c)\sigma_a. \tag{16.18}$$

When $c = (c_1, \ldots, c_{n+1})$, $b = (b_1, \ldots, b_n)$, the Pieri formula has already answered this question, because the Pieri formula can be written as

$$\sigma_c \cdot \sigma_b = \begin{cases} \sum_{a+|b|=|c|} \sigma_a & \text{if } b < c, \\ 0 & \text{otherwise.} \end{cases} \tag{16.19}$$

It is hard to solve (16.18) in a general way. But when $c = (c_1, \ldots, c_{n+2})$ and $b = (b_1, \ldots, b_n)$, we may make full use of the branching formula so as to deduce a simple calculating method. For this purpose, we put $c = (c_1, \ldots, c_{n+2})$, $b = (b_1, \ldots, b_n)$,

$$c \succ b \Leftrightarrow c_1 \geq \{c_2, b_1\} \geq \{c_3, b_2\} \geq \cdots \geq \{c_{n+1}, b_n\} \geq c_{n+2}.$$

$\{u, v\} \geq \{x, y\}$ means each number in $[u, v]$ is larger than or equal to that in $\{x, y\}$.

Considering

$$S_{(c_1, \ldots, c_{n+2})}(x_1, x_2, y_1, \ldots, y_n)$$

$$= \sum_{c > f} x_1^{|c|-|f|} S_{(f_1, \ldots, f_{n+1})}(x_2, y_1, \ldots, y_n)$$

$$= \sum_{c > f'} x_1^{|c|-|f|} \sum_{f > b} x_2^{|f|-|b|} S_{(b_1, \ldots, b_n)}(y_1, \ldots, y_n)$$

$$= \sum_{c \succ b} S_b(y_1, \ldots, y_n) \sum_{c > f > b} x_1^{|c|-|f|} x_2^{|f|-|b|}, \tag{16.20}$$

we see that the coefficient of $S_b(y_1, \ldots, y_n)$ in (16.20) is not equal to zero only when $b \prec c$. When c, b are fixed, from symmetry, we know that

$$\sum_{c > f > b} x_1^{|c|-|f|} x_2^{|f|-|b|} \tag{16.21}$$

is a symmetric homogeneous polynomial, which can be expressed as the algebraic sum of Schur functions. We want to get this algebraic sum. Recall: $c > f > b$ means

$$c_1 \geq f_1 \geq c_2 \geq f_2 \geq \cdots > c_{n+1} > f_{n+1} \geq c_{n+2},$$

$$f_1 \geq b_1 \geq f_2 \geq \cdots \geq b_n \geq f_{n+1},$$

which is equivalent to

$$c_1 \ge f_1 \ge \{c_2, b_1\} \ge f_2 \ge \{c_3, b_2\} \ge \cdots \ge f_n \ge \{c_{n+1}, b_n\} \ge f_{n+1} \ge c_{n+2}.$$

Let

$$\alpha_1 = c_1, \quad \alpha_2 = \min\{c_2, b_1\}, \ldots, \alpha_{n+1} = \min\{c_{n+1}, b_n\},$$

$$\beta_1 = \max\{c_2, b_1\}, \quad \beta_2 = \max\{c_3, b_2\}, \ldots, \beta_n = \max\{c_{n+1}, b_n\},$$

$$\beta_{n+1} = c_{n+2},$$

and $c > f > b$ may be expressed as

$$\alpha_1 \ge f_1 \ge \beta_1 \ge \alpha_2 \ge f_2 \ge \beta_2 \ge \cdots \ge \beta_n \ge \alpha_{n+1} \ge f_{n+1} \ge \beta_{n+1}. \quad (16.22)$$

Let

$$|\beta| = \beta_1 + \beta_2 + \cdots + \beta_{n+1} = c_{n+2} + \sum_{i=1}^{n} \max\{b_i, c_{i+1}\},$$

$$|\alpha| = \alpha_1 + \alpha_2 + \cdots + \alpha_{n+1} = c_1 + \sum_{i=1}^{n} \min\{b_i, c_{i+1}\},$$

and formula (16.21) becomes

$$\sum x_1^{|c|-|f|} x_2^{|f|-|b|} = \sum_{c>f>b} x_1^{|c|-|\beta|-(|f|-|\beta|)} x_2^{|f|-|\beta|+|\beta|-|b|}$$

$$= x_1^{|c|-|\beta|} x_2^{|\beta|-|b|} \sum_{c>f>b} \frac{x_2}{x_1}. \quad (16.23)$$

For the summation formula $\sum_{c>f>b} t^{|f|-|\beta|}$, we have a lemma.

LEMMA 16.8.

1. $$\sum_{c>f>b} t^{|f|-|\beta|} = \prod_{i=1}^{n+1} (1 + t + t^2 + \cdots + t^{(\alpha_i - \beta_i)}). \quad (16.24)$$

2. *Let* $m = \sum_{i=1}^{n+1} (\alpha_i - \beta_i) = |\alpha| - |\beta|$. *Then* $\exists k, 1 \le k \le m$, *such that*

$$\prod_{i=1}^{n+1} (1 + t + t^2 + \cdots + t^{(\alpha_i - \beta_i)}) = 1 + p_1 t + \cdots + p_k(t^k + \cdots + t^{m-k})$$

$$+ p_{k-1} t^{m-k+1} + \cdots + p_1 t^{m-1} + t^m$$

$$(16.25)$$

and satisfy $p_k \ge p_{k-1} \ge \cdots \ge p_1 \ge 1$.

PROOF. 1. The proof is evident from $f = (f_1, \ldots, f_{n+1})$, $\alpha_i \geq f_i \geq \beta_i$.

2. It can be proved by induction (details omitted). We call the polynomial on the right side of formula (16.25) a polynomial of Newton type; e.g., the binomial $(1 + t)^m$ is of Newton type.

Any polynomial of Newton type may also be written as

$$1 + p_1 t + \cdots + p_k(t^k + \cdots + t^{m-k}) + p_{k-1}t^{m-k-1} + \cdots + p_1 t^{m-1} + t^m$$

$$= (1 + t + \cdots + t^m) + (p_1 - 1)(t + t^2 + \cdots + t^{m-1})$$

$$+ (p_2 - p_1)(t^2 + \cdots + t^{m-2})$$

$$+ \cdots + (p_k - p_{k-1})(t^k + \cdots + t^{m-k}). \tag{16.26}$$

Substitution of Lemma 16.8 and (16.26) in (16.23) yields

$$\sum_{c > f > b} x_1^{|c|-|f|} x_2^{|f|-|b|}$$

$$= x_1^{|c|-|\beta|} x_2^{|\beta|-|b|} \sum_{c > f > b} \left(\frac{x_2}{x_1}\right)^{|f|-|\beta|}$$

$$= \sum_{i=0}^{k} (p_i - p_{i-1}) x_1^{|c|-|\beta|} x_2^{|\beta|-|b|} \left(\left(\frac{x_2}{x_1}\right)^i + \cdots + \left(\frac{x_2}{x_1}\right)^{|\alpha|-|\beta|-i}\right)$$

$$= \sum_{i=0}^{k} (p_i - p_{i-1})(x_1^{|c|-|\beta|-i} x_2^{|\beta|-|b|+i} + \cdots + x_1^{|c|-|\alpha|+i} x_2^{|\alpha|-|b|-i})$$

$$= \sum_{i=0}^{k} (p_i - p_{i-1}) S_{(|c|-|\beta|-i,|\beta|-|b|+i)}(x_1, x_2),$$

and substitution of them in (16.20) gives

$$S_{(c_1,\ldots,c_{n+2})}(x_1, x_2, y_1, \ldots, y_n)$$

$$= \sum_{c > b} S_b(y_1, \ldots, y_n)\left(\sum_{i=0}^{k} (p_i - p_{i+1}) S_{(|c|-|\beta|-i,|\beta|-|b|+i)}(x_1, x_2)\right), \tag{16.27}$$

which shows (cf. formula (16.5))

$$\delta((|c| - |\beta| - i, |\beta| - |b| + i), (b_1, \ldots, b_n); (c_1, \ldots, c_{n+2})) = p_i - p_{i-1}.$$
$$(16.28)$$

The above discussion may be summed up into the following theorem.

THEOREM 16.9. *Given* $c = (c_1, \ldots, c_{n+2})$, $b = (b_1, \ldots, b_n)$.
1. *If* b *does not satisfy* $b \ll c$, *then* $[\sigma_a : \sigma_b] = 0$.
2. *If* $b \ll c$, *then let*

$$\alpha_1 = c_1, \quad \alpha_2 = \min(c_2, b_1), \ldots, \alpha_{n+1} = \min(c_{n+1}, b_n),$$

$$\beta_1 = \max(c_2, b_1), \ldots, \beta_n = \max(c_{n+1}, b_n), \beta_{n+1} = c_{n+2}. \quad (16.29)$$

Set $m = |\alpha| - |\beta| = \sum_{i=1}^{n+1} (\alpha_i - \beta_i)$. *Considering a polynomial of Newton type,*

$$F(t) = \prod_{i=1}^{n+1} (1 + t + t^2 + \cdots + t^{(\alpha_i - \beta_i)}$$

$$= 1 + p_1 t + \cdots + p_{k-1} t^{k-1} + p_k (t^k + \cdots + t^{m-k})$$

$$+ p_{k-1} t^{m-k+1} + \cdots + p_1 t^{m-1} + t^m,$$

we have

$$[\sigma_c : \sigma_b] = \sum_{i=0}^{k} (p_i - p_{i-1}) \sigma_{(|c| - |\beta| - i, |\beta| - |b| + i)}. \quad (16.30)$$

EXAMPLE 16.10. $c = (6, 4, 2, 1, 0)$, $b = (3, 3, 1)$, solve $[\sigma_c : \sigma_b]$.

Solution. From (16.29), we easily see that $\alpha = (6, 3, 2, 1)$, $\beta = (4, 3, 1, 0)$.

$$F(t) = (1 + t + t^2)(1 + t)^2 = 1 + 3t + 4t^2 + 3t^3 + t^4.$$

Therefore, $p_0 = 1$, $p_1 = 3$, $p_2 = 4$. From (16.30) we get

$$[\sigma_c : \sigma_b] = \sigma_{[5,1)} + 2\sigma_{(4,2)} + \sigma_{(3,3)};$$

i.e.,

$$\delta((5, 1), (3, 3, 1); (6, 4, 2, 1, 0)) = 1,$$

$$\delta((4, 2), (3, 3, 1); (6, 4, 2, 1, 0)) = 2,$$

$$\delta((3, 3), (3, 3, 1); (6, 4, 2, 1, 0)) = 1,$$

$$\delta(\text{others}, (3, 3, 1); (6, 4, 2, 1, 0)) = 0.$$

References

1. H. Boerner, *Representations of Groups*, North-Holland, Amsterdam (1963).
2. P. Griffiths and J. Harris, *Principles of Algebraic Geometry*, Wiley, New York (1978).
3. S. Kleiman, *Problem 15, Rigorous Foundation of Schubert's numerative Calculus*, Proc. Sympos. Pure Math., Vol. 28, American Mathematical Society, Providence, RI (1976).
4. I. Macdonald, *Symmetric Functions and Hall Polynomials*, Clarendon Press, Oxford (1979).
5. J.-Q. Zhong, *Proceedings of the 1980 Beijing Symposium on Differential Geometry and Differential Equations*, Vol. III, pp. 1697–1708, Science Press, Beijing, China, 1982.

17

An Expansion in Schur Functions and Its Applications in Enumerative Geometry

Abstract

This chapter presents a method for calculating the coefficient $c_{(\lambda_1,\ldots,\lambda_n)}$ for the expansion of polynomial powers of Schur functions

$$(x_1^k + \cdots + x_n^k) = \sum C_{(\lambda_1,\ldots,\lambda_n)} S_{(\lambda_1,\ldots,\lambda_n)}(x_1,\ldots,x_n)$$

and applies it to the discussion of some problems of enumerative geometry.

17.1. Introduction

The problem of how to express $(x_1^k + \cdots + x_n^k)$ by elementary symmetric polynomials has been investigated using the well-known Newton formula. This chapter addresses the problem of how to express $(x_1^k + \cdots + x_n^k)^m$ by a linear combination of Schur functions; that is, it studies the expansion

$$(x_1^k + \cdots + x_n^k)^m = C_{(\lambda_1,\ldots,\lambda_n)} S_{(\lambda_1,\ldots,\lambda_n)}(x_1,\ldots,x_n). \qquad (17.1)$$

Even for Schur function theory itself, (17.1) has a distinct independent meaning. Based on our previous work [7], we propose a method for calculating the coefficients $C_{(\lambda_1,\ldots,\lambda_n)}$ in Section 17.2. And in Section 17.3 we will apply the calculated results as well as some results in Ref. 6 to some problems of enumerative geometry, including an enumeration problem of plane position in combinatorial mathematics.

17.2. An Expansion

Let us see how to expand $(x_1^k + \cdots + x_n^k)^m$ into a linear sum of Schur functions; i.e., we calculate the coefficients $C_{(\lambda_1,\ldots,\lambda_n)}$ of

$$(x_1^k + \cdots + x_n^k)^m = \sum C_{(\lambda_1,\ldots,\lambda_n)} S_{(\lambda_1,\ldots,\lambda_n)}(x_1,\ldots,x_n).$$

For all k, the method holds. Here we provide details for $k = 2$. For general k we only give results and outlines of proofs.

When $k = 1$, the result is classical. From an identity in Ref. 3, p. 22, Theorem 1.3.5, we know

$$\frac{1}{1 - \sum_{i=1}^n x_i} = \sum_{f_1 \geq \cdots \geq f_n \geq 0} \frac{(\sum f_i)!\, D(\tilde{f}_1,\ldots,\tilde{f}_n)}{\tilde{f}_1! \cdots \tilde{f}_n!} S_{(f_1,\ldots,f_n)}(x_1,\ldots,x_n),$$

where $\tilde{f}_i = f_i + n - i$ $(i = 1, 2, \ldots, n)$, $D(\tilde{f}_1,\ldots,\tilde{f}_n) = \prod_{i<j}(\tilde{f}_i - \tilde{f}_j)$. Hereafter we set

$$\frac{D(\tilde{f}_1,\ldots,\tilde{f}_n)}{\tilde{f}_1! \cdots \tilde{f}_n!} = N(f_1,\ldots,f_n). \tag{17.2}$$

On the other hand, because $(1 - \sum x_i)^{-1} = \sum_{m=0}^{\infty} (\sum x_i)^m$, we obtain

$$\left(\sum_{i=1}^n x_i\right)^m = \sum_{\substack{f_1 \geq \cdots \geq f_n \geq 0 \\ \sum f_i = m}} m!\, N(f_1,\ldots,f_n) S_{(f_1,\ldots,f_n)}(x_1,\ldots,x_n), \tag{17.3}$$

which give the solution to (17.1) for the case of $k = 1$.

Another way leading to (17.3) is to utilize formula (0.4) in Ref. 6; i.e.,

$$\varphi(t_1) \cdots \varphi(t_n) = \sum_{a_1 \geq \cdots \geq a_n \geq 0} S_{(a_1,\ldots,a_n)}(t_1,\ldots,t_n) G_{(\alpha_1,\ldots,\alpha_n)}. \tag{17.4}$$

Let $\varphi(x) = e^x$, and from (17.4) we have

$$\varphi(x_1) \cdots \varphi(x_n) = e^{x_1 + \cdots + x_n}$$

$$= \sum_{m=0}^{\infty} \frac{1}{m!} (\sum x_i)^m$$

$$= \sum_{\substack{f_1 \geq f_2 \geq \cdots \\ \geq f_n \geq 0}} \begin{vmatrix} \dfrac{1}{f_1!} & \dfrac{1}{(f_1+1)!} & \cdots & \dfrac{1}{(f_1+n-1)!} \\[2mm] \dfrac{1}{(f_2-1)!} & \dfrac{1}{f_2!} & \cdots & \dfrac{1}{(f_2+n-2)!} \\[2mm] & & \cdots & \\[2mm] \dfrac{1}{(f_n-n+1)!} & & & \dfrac{1}{f_n!} \end{vmatrix}$$

$$\times S_{(f_1,\ldots,f_n)}(x_1,\ldots,x_n). \tag{17.5}$$

By comparing (17.3) and (17.5), we can prove the next lemma.

LEMMA 17.1. *If* $f_1 \geq f_2 \geq \cdots \geq f_n \geq 0$ *is a sequence of integers, then*

$$\det\left(\frac{1}{(f_i + j - i)!}\right) = N(f_1, \ldots, f_n). \tag{17.6}$$

Here $N(f_1, \ldots, f_n)$ *is defined by* (17.2).

Now let us consider the case of $k = 2$. Put

$$\varphi(x) = e^{x^2} = \sum_{\lambda=0}^{\infty} b_\lambda x^\lambda, \qquad b_\lambda = \begin{cases} 0, & \lambda \text{ is odd,} \\ \left[\left(\frac{\lambda}{2}\right)!\right]^{-1}, & \lambda \text{ is even.} \end{cases}$$

Still from formula (17.4), we have

$$\varphi(x_1) \cdots \varphi(x_n) = e^{\sum_{i=1}^n x_i^2} = \sum_{\lambda_1 \geq \cdots \geq \lambda_n \geq 0} b_{(\lambda_1, \ldots, \lambda_n)} S_{(\lambda_1, \ldots, \lambda_n)}(x_1, \ldots, x_n),$$

$$b_{(\lambda_1, \ldots, \lambda_n)} = \begin{vmatrix} b_{\lambda_1} & b_{\lambda_1+1} & \cdots & b_{\lambda_1+n-1} \\ b_{\lambda_2-1} & b_{\lambda_2} & \cdots & b_{\lambda_2+n-2} \\ & & \ddots & \\ b_{\lambda_n-n+1} & & \cdots & b_{\lambda_n} \end{vmatrix}. \tag{17.7}$$

Label every λ_i in subscript $(\lambda_1, \ldots, \lambda_n)$ as

$$\lambda_i \in \Lambda^+ \Leftrightarrow \lambda_i - i \text{ is odd,}$$

$$\lambda_i \in \Lambda^- \Leftrightarrow \lambda_i - i \text{ is even.} \tag{17.8}$$

Write $\Lambda^+ = \{\lambda_{i_1}, \ldots, \lambda_{i_l} \mid i_1 < i_2 < \cdots < i_l\}$, $l = \#(\Lambda^+)$, $\Lambda^- = \{\lambda_{i_{l+1}}, \ldots, \lambda_{i_n}\}$, $n - 1 = \#(\Lambda^-)$. For example, if $\lambda = (5, 4, 4, 2)$, then $\Lambda^+ = \lceil \lambda_3 \rceil$, $\Lambda^- = \{\lambda_1, \lambda_2, \lambda_n\}$. Evidently, in determinant $b_{(\lambda_1, \ldots, \lambda_n)}$ the row in which λ_i belonging to Λ^+, Λ^- lies takes the form

By putting together the corresponding rows in Λ^+, Λ^- and shifting the rows, the determinant is changed into

$$
l \left\{ \;\; \begin{vmatrix} \overbrace{\begin{matrix} * \cdots * \\ \vdots \quad \vdots \\ * \cdots * \end{matrix}}^{\dfrac{n+1}{2}} & \quad 0 \\[2em] 0 & \begin{matrix} * \cdots * \\ \vdots \quad \vdots \\ * \cdots * \end{matrix} \end{vmatrix} \right\} n-l
\tag{17.9}
$$

Therefore, $b(\lambda_1, \ldots, \lambda_n) \neq 0$, if and only if $l = [(n + 1)/2]$.

When $l = [(n + 1)/2]$, the subdeterminant in (17.9) corresponding to Λ^+ is

$$
\begin{vmatrix}
b_{\lambda_{i_1}-i_1+1} & b_{\lambda_{i_1}-i_1+3} & \cdots & b_{\lambda_{i_1}-i_1+2l-1} \\
b_{\lambda_{i_2}-i_2+1} & b_{\lambda_{i_2}-i_2+3} & \cdots & b_{\lambda_{i_2}-i_2+2l-1} \\
& & \cdot & \\
b_{\lambda_{i_l}-i_l+1} & \cdots & & b_{\lambda_{i_l}-i_l+2l-1}
\end{vmatrix}
\tag{17.10}
$$

because $\lambda_{i_1}, \ldots, \lambda_{i_l} \in \Lambda^+$, the subscript of each element in the above determinant is even; e.g., the first row is actually

$$
b_{\lambda_{i_1}-i_1+1} = \frac{1}{\left(\dfrac{\lambda_{i_1} - i_1 + 1}{2}\right)!}, \ldots, b_{\lambda_{i_1}-i_1+2l-1} = \frac{1}{\left(\dfrac{\lambda_{i_1} - i_1 + 2l - 1}{2}\right)!}
$$

Consequently, subdeterminant (17.10) is also a determinant of a kind similar to (17.6). According to formula (17.6), the value of subdeterminant (17.10) is

$$
N\left(\frac{\lambda_{i_1} - i_1 + 1}{2}, \ldots, \frac{\lambda_{i_l} - i_l + 2l - 1}{2}\right).
\tag{17.11}
$$

In the same way we treat $\Lambda^- = \{\lambda_{i_{l+1}}, \ldots, \lambda_{i_n}\}$, whose subdeterminant value corresponding to the subscript in (17.9) is

$$
N\left(\frac{\lambda_{i_{l+1}} - i_{l+1} + 2}{2}, \ldots, \frac{\lambda_{i_n} - i_n + 2(n - l)}{2}\right).
\tag{17.12}
$$

At the same time, it can be easily checked that in the process of transforming (17.7) into (17.9), the signature is $(-1)^{i_1+i_2+\cdots+i_l-l}$. Note that $e^{\sum x_i^2} = \sum (1/m!)(\sum x_i^2)^m$. Summing up the above discussion, we can get (17.1) for the case of $k = 2$ as follows.

THEOREM 17.2. *Set* $\Lambda^+ = \{\lambda_{i_1}, \ldots, \lambda_{i_l}\}$, $\Lambda^- = \{\lambda_{i_{l+1}}, \ldots, \lambda_{i_n}\}$,

$$(x_1^2 + \cdots + x_n^2)^m = m! \sum_{\lambda_1 \geq \cdots \geq \lambda_n \geq 0} b_{(\lambda_1, \ldots, \lambda_n)} S_{(\lambda_1, \ldots, \lambda_n)}(x_1, \ldots, x_n)$$

and we have

$$
b_{(\lambda_1 \cdots \lambda_n)} =
\begin{cases}
0 & \text{when } l \neq \left[\dfrac{-n+1}{2}\right], \\[4mm]
(-1)^{\sum_{k=1}^{l}(i_k-l)} N\left(\dfrac{\lambda_{i_1}-i_1+1}{2}, \ldots, \dfrac{\lambda_{i_l}-i_l+2l-1}{2}\right) \\[3mm]
\quad \times N\left(\dfrac{\lambda_{i_{l+1}}-i_{l+1}+2}{2}, \ldots, \dfrac{\lambda_{i_n}-i_n+2(n-l)}{2}\right) \\[4mm]
\hspace{4cm} \text{when } l = \left[\dfrac{-n+1}{2}\right], \quad (17.13)
\end{cases}
$$

where $N(*, \ldots, *)$ *is defined by* (17.2).

Now let us discuss the case of a general k. Let

$$\varphi(x) = e^{x^k} = \sum_{\lambda=0}^{\infty} b_\lambda x^\lambda, \qquad b_\lambda = \begin{cases} 0 & \lambda \not\equiv 0 \pmod{k}, \\ 1/(\lambda/k)!, & \lambda \equiv 0 \pmod{k}. \end{cases}$$

Still from (17.4), it follows that

$$\varphi(x_1) \cdots \varphi(x_n) = e^{\sum x_i^k}$$

$$= \sum_{\lambda_1 \geq \cdots \geq \lambda_n \geq 0} b_{(\lambda_1, \ldots, \lambda_n)} S_{(\lambda_1, \ldots, \lambda_n)}(x_1, \ldots, x_n).$$

Hence,

$$\left(\sum_{i=1}^{n} x_i^k\right)^m = m! \sum_{\substack{\lambda_1 \geq \cdots \geq \lambda_n \geq 0 \\ \sum \lambda_i = mk}} b_{(\lambda_1, \ldots, \lambda_n)} S_{(\lambda_1, \ldots, \lambda_n)}(x_1, \ldots, x_n), \qquad (17.14)$$

where $b_{(\lambda_1,\ldots,\lambda_n)}$ is expressed by (17.7), too. Classify every λ_i in $\lambda = (\lambda_1,\ldots,\lambda_n)$. Give an arbitrary positive integer S, $1 \le S \le k$. Define

$$\lambda_i \in \Lambda^{(S)} \Leftrightarrow \lambda_i - i \equiv -S \pmod{k}. \tag{17.15}$$

It can be easily seen that in the determinant $b_{(\lambda_1,\ldots,\lambda_n)}$ the corresponding rows of λ_i that belong to the same $\Lambda^{(S)}$ have the same form; e.g.,

$$\lambda_i \in \Lambda^{(1)} \quad \text{the row} = (*, 0, \ldots, 0, *, 0, \ldots, 0, *, 0, \ldots),$$

$$\lambda_i \in \Lambda^{(2)} \quad \text{the row} = (0, *, \ldots, 0, 0, *, \ldots, 0, 0, *, \ldots).$$

Therefore, after the shift of rows and columns, $b_{(\lambda_1,\ldots,\lambda_n)}$ becomes (differing in a signature)

$$\tag{17.16}$$

where $l_i = \#(\Lambda^{(i)})$ $(i = 1, \ldots, k)$. In consequence,

$$b_{(\lambda_1,\ldots,\lambda_n)} \ne 0 \Leftrightarrow l_i = \left[\frac{n+k-1}{k}\right], \ldots, l_k = \left[\frac{n}{k}\right]. \tag{17.17}$$

Write

$$\Lambda^{(1)} = \{\lambda_{i_{11}}, \ldots, \lambda_{i_{1l_1}}\}, \ldots, \Lambda^{(k)} = \{\lambda_{i_{k1}}, \ldots, \lambda_{i_{kl_k}}\}.$$

Then the value of the subdeterminant corresponding to each $\Lambda^{(s)}$ $(1 < s <$

k), e.g., the value of $\Lambda^{(1)}$, is

$$
\begin{vmatrix}
b_{\lambda_{i_{11}} - i_{11} + 1} & b_{\lambda_{i_{11}} - i_{11} + 1 + k} & \cdots \\
b_{\lambda_{i_{12}} - i_{12} + 1} & b_{\lambda_{i_{12}} - i_{12} + 1 + k} & \cdots \\
\cdots & \cdots & \cdots
\end{vmatrix}
$$

$$
= \begin{vmatrix}
\dfrac{1}{\left(\dfrac{\lambda_{i_{11}} - i_{11} + 1}{k}\right)!} & \dfrac{1}{\left(\dfrac{\lambda_{i_{11}} - i_{11} + 1}{k} + 1\right)!} & \cdots \\
\dfrac{1}{\left(\dfrac{\lambda_{i_{12}} - i_{12} + 1}{k}\right)!} & \dfrac{1}{\left(\dfrac{\lambda_{i_{12}} - i_{12} + 1}{k} + 1\right)!} & \cdots \\
\cdots & \cdots & \cdots
\end{vmatrix}
$$

$$
= N\left(\frac{\lambda_{i_{11}} - i_{11} + 1}{k}, \frac{\lambda_{i_{12}} - i_{12} + 1 + k}{k}, \ldots, \frac{\lambda_{i_{1l_1}} - i_{1l_1} + (l_1 - 1)k + 1}{k}\right).
$$

Similarly, the value of the subdeterminant corresponding to $\Lambda^{(s)}$ in (17.16) is

$$
N\left(\frac{\lambda_{i_{s1}} - i_{s1} + s}{k}, \frac{\lambda_{i_{s2}} - i_{s2} + s + k}{k}, \ldots, \frac{\lambda_{i_{sl_s}} - i_{sl_s} + (l_s - 1)k + s}{k}\right).
$$

Hence,

$$
b_{(\lambda_1, \ldots, \lambda_n)} = (\pm 1) \prod_{s=1}^{k} N\left(\frac{\lambda_{i_{s1}} - i_{s1} + s}{k}, \ldots, \frac{\lambda_{i_{sl_s}} - i_{sl_s} + (l_s - 1)k + s}{k}\right).
$$

$$
\tag{17.18}
$$

What is left to be done is to determine the sign. By simple calculation if $n = kp + q$, $0 < q \le k$, then $b_{(\lambda_1, \ldots, \lambda_n)} = 0$ requires

$$
l_i = \#(\Lambda^{(i)}) = \left[\frac{n + k - i}{k}\right] = \begin{cases} p + 1, & 1 \le i \le q, \\ p, & q < i \le k. \end{cases} \tag{17.19}
$$

Now

$$
\Lambda^{(j)} = \{\lambda_{i_{j1}}, \ldots, \lambda_{i_{jp+1}}\}, \qquad 1 \le j \le q,
$$

$$
\Lambda^{(j)} = \{\lambda_{i_{j1}}, \ldots, \lambda_{i_{jp}}\}, \qquad q < j \le k.
$$

The corresponding signature of (17.18) is

$$\delta(i_{11}, \ldots, i_{k1}, i_{12}, \ldots, i_{k2}, \ldots, i_{1p}, \ldots, i_{kp}, i_{1\,p+1}, \ldots, i_{qp+1}) \quad (17.20)$$

Here $\delta(\ldots, i_{kl_1}, \ldots)$ is the sign for $(\ldots, i_{kl_1}, \ldots)$ being a permutation of $(1, 2, \ldots, n)$.

The combination of (17.14), (17.18), (17.19), and (17.20) yields the following theorem, which provides a complete solution to (17.1).

THEOREM 17.3. *The expansion*

$$\left(\sum_{i=1}^{n} x_i^k \right)^m = m! \sum_{\substack{\lambda_1 \geq \cdots \geq \lambda_n \geq 0 \\ \sum \lambda_i = mk}} b_{(\lambda_1, \ldots, \lambda_n)} S_{(\lambda_1, \ldots, \lambda_n)}(x_1, \ldots, x_n)$$

holds, where the coefficients $b_{(\lambda_1, \ldots, \lambda_n)}$ are determined as follows.

Put $n = kp + q, 0 < q < k$. Classify each part λ_i in $(\lambda_1, \ldots, \lambda_n)$ into $\Lambda^{(s)}$ $(s = 1, \ldots, k)$. Let $l_s = \#(\Lambda^{(s)})$. Then when

$$l_1 = \cdots = l_g = p + 1, \quad l_{g+1} = \cdots = l_k = p,$$

$$b_{(\lambda_1, \ldots, \lambda_n)} = \delta(i_{11}, \ldots, i_{qp+1}) \prod_{s=1}^{q} N\left(\frac{\lambda_{i_{s1}} - i_{s1} + s}{k}, \ldots, \frac{\lambda_{i_{sp+1}} - i_{sp+1} + pk + s}{k} \right)$$

$$\times \prod_{s=q+1}^{k} N\left(\frac{\lambda_{i_{s1}} - i_{s1} + s}{k}, \ldots, \frac{\lambda_{i_{sp}} - i_{sp} + (p-1)k + s}{k} \right) \quad (17.21)$$

In other cases, $b_{(\lambda_1, \ldots, \lambda_n)} = 0$.

17.3. Some Problems Concerning Enumerative Geometry

The Schubert calculus is developed on the basis of enumerative geometry. (For classical enumerative geometry and its modern forms, refer to Ref. 1.) A typical problem of enumerative geometry related to the linear subspace is: How many $\mathbb{P}^d$ are there in $\mathbb{P}^n$ that intersect $h \equiv (d + 1)(n - d) \mathbb{P}^{n-d-1}$ in general position?

The set of all $\mathbb{P}^d$ in $\mathbb{P}^n$ is the set composed of all $(d + 1)$-dimensional subspaces in $\mathbb{C}^{n+1}$, namely, the Grassmann manifold $G(n + 1, d + 1)$. Given a flag, $V = \{V_1 \subset V_2 \subset \cdots \subset V_{n+1} = \mathbb{C}^{n+1} | \dim V_i = i\}$, then by definition of the Schubert cycle, we have

$$\delta_{\underbrace{(1,0,\ldots,0)}_{d+1}}(V) = \left\{ (d+1)\text{-dimensional subspaces } \Lambda \text{ in } \mathbb{C}^{n+1} \right| $$

$$\left. \begin{array}{l} \dim(\Lambda \cap V_{n-\alpha}) \geq 1 \\ \dim(\Lambda \cap V_{n-\alpha+i}) \geq i, \forall i \geq 1 \end{array} \right\}$$

$$= \{\Lambda \mid \dim(\Lambda \cap V_{n-\alpha}) \geq 1\}$$

$$= \text{the number of } \mathbb{P}^d \text{ in } \mathbb{P}^n \text{ that intersect } \mathbb{P}(V_{n-\alpha}) = \mathbb{P}^{n-d-1}$$

This is because $\dim \Lambda = d + 1$. So

$$\dim(\Lambda \cap V_{n-d-i}) = \dim \Lambda + \dim V_{n-d-i} - \dim\{\Lambda, V_{n-d-i}\} \geq n + i - n = i$$

holds naturally. Here $\{\Lambda, V_{n-d-i}\}$ stands for the linear space spanned by Λ and V_{n-d-i}. As a result, the number of $\mathbb{P}^d$ that intersect $h = (n-d)(d+1)$ ($=\dim G(n+1, d+1)$) $\mathbb{P}^{n-d-1}$ in general position is just the intersection number of $\sigma^h_{(1,0,\ldots,0)}$ and $G(n+1, d+1)$ itself (viewed as Schubert cycle $\sigma_{(0,\ldots,0)}$). This intersection number is called $\deg \sigma^h_1$ in algebraic geometry.

According to the Schubert calculus,

$$\sigma^h_1 = \sum_{\substack{n-d \geq \lambda_1 \geq \cdots \geq \lambda_{d+1} \geq 0 \\ \sum \lambda_i = h}} C_{(\lambda_1,\ldots,\lambda_{d+1})} \sigma_{(\lambda_1,\ldots,\lambda_{d+1})}. \tag{17.22}$$

Then by formula (0.1) in Ref. 6, we have

$$\#(\sigma_a \cdot \sigma_b) = \delta^{n-d-b_d}_{a_1} \cdots \delta^{n-d-b_l}_{a_d}.$$

Therefore,

$$\deg \sigma^h_1 = \#(\sigma^h_1 \cdot \sigma_{(0,\ldots,0)}) = C_{(n-d,\ldots,n-d)} \tag{17.23}$$

is the coefficient of $\sigma_{(n-d,\ldots,n-d)}$ in the expansion (17.22).

In Ref. 6 we indicated that for the Schur function the relation

$$S_{(a_1,\ldots,a_n)}(x_1,\ldots,x_n) S_{(b_1,\ldots,b_n)}(x_1,\ldots,x_n) = \sum_c \delta(a,b;c) S_{(c_1,\ldots,c_n)}(x_1,\ldots,x_n) \tag{17.24}$$

holds, which shows that $\deg \sigma^h_1$ is the coefficient of $S_{(n-d,\ldots,n-d)}$ in the expansion of $S^h_{(1,0,\dots,0)}$. But because of

$$S_{(1,0,\dots,0)}(x_1,\dots,x_n) = \sum_{i=1}^{n} x_i,$$

the problem is one of how to expand $(x_1 + \cdots + x_n)^h$ in a linear sum of Schur functions. Equation (17.3) shows

$$\deg \sigma_1^h = h! N(n-d,\dots,n-d) = \frac{h!1!2!\dots d!}{(n-d)!(n-d+1)!\dots n!}, \quad (17.25)$$

where $h = (n-d)(d+1)$. By the same fact, the number of $\sigma_{(\lambda_1,\dots,\lambda_{d+1})}$ in σ_1^h (it can also be interpreted as the number of d-spaces that satisfy certain Schubert conditions) is

$$(\textstyle\sum \lambda_i)! N(\lambda_1,\dots,\lambda_{d+1}) = \frac{(\sum \lambda_i)! \prod_{i<j}(\tilde\lambda_i - \tilde\lambda_j)}{\tilde\lambda_1!\dots\tilde\lambda_{d+1}!}, \quad (17.26)$$

where $\tilde\lambda_1 = \lambda_1 + d,\dots,\tilde\lambda_{d+1} = \lambda_d$. Both (17.25) and (17.26) are classic results. The traditional method for proving them is based on the Pieri formula and the inductive method [2, 4].

A majority of the problems of enumerative geometry have something to do with calculus similar to those which have been mentioned. But as far as I know, no general calculus for $\deg \sigma_k^m$, $\deg \sigma_{k_1} \cdots \sigma_{k_j}$ has been proposed in the literature. This is by no means a simple problem. In this section, with the aid of the expansion in Section 17.2, we will give a partial solution to this problem.

Assuming $2l = dm$, we will calculate $\deg \sigma_2^l$ in $G(m+d, d)$. Based upon the above explanation, we know if

$$\sigma_2^l = \sum_{\substack{\lambda_1 \ge \cdots \ge \lambda_d \ge 0 \\ \sum \lambda_i = 2l}} C_{(\lambda_1,\dots,\lambda_d)} \sigma_{(\lambda_1,\dots,\lambda_d)}, \quad (17.27)$$

then

$$\deg \sigma_2^l = C_{(m,\dots,m)}. \quad (17.28)$$

THEOREM 17.4. Let $2l = md$. Then in $G(d+m, d)$,

$$\deg \sigma_2^l = \frac{1}{2^l} \sum_{k=0}^{l} C_l^k (2k)!(2l-2k)! \left(\sum_{\substack{m \ge \lambda_1 \ge \cdots \ge \lambda_d \ge 0 \\ \sum \lambda_i = 2k}} a_{\tilde\lambda} \cdot b_\lambda \right), \quad (17.29)$$

where $a_{\bar\chi} = N(m - \lambda_d, \ldots, m - \lambda_1)$ *is defined by* (17.2); $b_\lambda = b_{(\lambda_1,\ldots,\lambda_d)}$ *is shown by* (17.13).

PROOF. According to (17.24), $\deg \sigma_2^l$ is the coefficient of $S_{(m,\ldots,m)}(x_1, \ldots, x_d)$ in the case where $S^l_{(2,0,\ldots,0)}(x_1, \ldots, x_d)$ is expanded into Schur functions.

Since

$$S_{(2,0,\ldots,0)}(x_1, \ldots, x_d) = \sum_{i=1}^{d} x_i^2 + \sum_{i<j} x_i x_j$$

$$= \sum x_i^2 + S_{(1,1,0,\ldots,0)}(x_1, \ldots, x_d)$$

and

$$S^2_{(1,0,\ldots,0)}(x_1, \ldots, x_d) = S_{(2,0,\ldots,0)}(x_1, \ldots, x_d) + S_{(1,1,0,\ldots,0)}(x_1, \ldots, x_d),$$

there are

$$2S_{(2,0,\ldots,0)} = S^2_{(1,0,\ldots,0)} + \sum x_i^2$$

$$\frac{1}{1 - 2S_2} = (1 - S_1^2)^{-1}\left(1 - \frac{\sum x_i^2}{1 - S_1^2}\right)^{-1}$$

$$= \sum_{k=0}^{\infty} \frac{(\sum x_i^2)^k}{(1 - S_1^2)^{k+1}}$$

$$= \sum_{k=0}^{\infty} (\sum x_i^2)^k \left(\sum_{n=0}^{\infty} C^n_{n+k} S_1^{2n}\right)$$

$$= \sum_{l=0}^{\infty} \sum_{k=0}^{l} C_l^k (\sum x_i^2)^k S_1^{2l-2k}.$$

Therefore,

$$S_2^l = \frac{1}{2^l} \sum_{k=0}^{l} C_l^k (\sum x_i^2)^k S_1^{2(l-k)}. \tag{17.30}$$

From (17.3) and (17.13), it follows that

$$S^{2(l-k)}_{(1,0,\ldots,0)}(x_1,\ldots,x_d) = (2(l-k))! \sum_{\sum \lambda_i = 2l-2k} a_{(\lambda_1,\ldots,\lambda_d)} S_{(\lambda_1,\ldots,\lambda_d)}(x_1,\ldots,x_d),$$

$$\left(\sum_{i=1}^{d} x_i^2\right)^k = (2k)! \sum_{\sum \mu_i = 2k} b_{(\mu_1,\ldots,\mu_d)} S_{(\mu_1,\ldots,\mu_d)}(x_1,\ldots,x_d),$$

$$S^{2(l-k)}_1 \left(\sum_{i=1}^{d} x_i^2\right)^k = \sum_{\substack{\sum \lambda = 2(l-k) \\ \sum \mu = 2k}} a_\lambda b_\mu S_\lambda S_\mu.$$

Now expand $S_\lambda(x_1,\ldots,x_d) \cdot S_\mu(x_1,\ldots,x_d)$. The coefficient of $S_{(m,\ldots,m)}$ $(x_1,\ldots,x_d)$ in this expansion is equal to

$$\delta(\lambda,\mu_j(m,\ldots,m) = \#(\sigma_{(\lambda_1,\ldots,\lambda_d)} \cdot \sigma_{(\mu_1,\ldots,\mu_d)})$$

$$= \begin{cases} 1, & \text{if } \mu = \tilde{\lambda} = (m - \lambda_d,\ldots,m - \lambda_1), \\ 0, & \text{otherwise.} \end{cases}$$

Thus, we know that the coefficient of $S_{(m,\ldots,m)}(x_1,\ldots,x_d)$ in $S^{2(l-k)}_1 (\sum_{i=1}^{d} x_i^2)^k$ is $(2l-2k)!(2k)!(\sum_{m \geq \lambda_1 \geq \cdots \geq \lambda_\alpha \geq 0; \sum \lambda_i = 2k} b_\lambda a_{\tilde{\lambda}})$. By substituting it in (12.30), we complete the proof.   □

THEOREM 17.5. *Assume* $l \geq n, l \geq k_1 \geq k_2 \geq \cdots \geq k_l \geq 0, \sum_{i=1}^{l} k_i = nl$. *Then in* $G(l + n, n)$.

$$\deg(\sigma_{k_1},\ldots,\sigma_{k_l}) = \text{the coefficient of } x_1^{k_1} \cdots x_l^{k_l} \text{ in the}$$
expansion of $S_{(l,\ldots,l,0,\ldots,0)}(x_1,\ldots,x_l)$ *into monomials.* (17.31)

PROOF. From identity [3], there holds

$$\prod_{i,j=1}^{l} (1 + x_i y_i) = \sum_{l \geq \lambda_1 \geq \cdots \geq \lambda_l \geq 0} S_\lambda(x) S_{\lambda^*}(y), \qquad (17.32)$$

where λ^* denotes the conjugate of λ (cf. Ref. 1, p. 200). On the other hand,

$$\prod_{i=1}^{l} (1 + x_i t) = \sum_{k=0}^{l} e_k(x) t^k,$$

where

$$e_k(x) = S_{\underbrace{(1,1,\ldots,1,0,\ldots,0)}_{k}}(x_1,\ldots,x_l).$$

Therefore,

$$\prod_{i,j=1}^{l} (1 + x_i t_j) = \left(\sum_{k_1=0}^{l} e_{k_1}(x) t_1^{k_1} \right) \cdots \left(\sum_{k_l=0}^{l} e_{k_l}(x) t_l^{k_l} \right)$$

$$= \sum_{k_1,\ldots,k_l=0}^{l} e_{k_1}(x) \cdots e_{k_l}(\lambda) t_1^{k_1} \cdots t_l^{k_l}. \qquad (17.33)$$

Denote by $a(\lambda; k_1, \ldots, k_l)$ the coefficient of the monomial $t_1^{k_1} \cdot t_2^{k_2} \cdots t_l^{k_l}$ in the expansion of $S_\lambda(t_1, \ldots, t_l)$. From (17.32) and (17.33)

$$\sum_{k_1,\ldots,k_l=0}^{l} e_{k_1}(x) \cdots e_{k_l}(x) t_1^{k_1} \cdots t_l^{k_l} = \sum_{l \geq \lambda_1 \geq \cdots \geq \lambda_l \geq 0} S_\lambda(t) S_{\lambda^*}(x)$$

and comparing the coefficients of $t_1^{k_1} \cdots t_l^{k_l}$, we obtain

$$e_{k_1}(x) \cdots e_{k_l}(x) = \sum_{l \geq \lambda_1 \geq \cdots \geq \lambda_l \geq 0} a(\lambda; k_1, \ldots, k_l) S_{\lambda^*}(x). \qquad (17.34)$$

In Ref. 6 we put forth a relation satisfied by the involutive operator T:

$$Te_j = h_j, \qquad T^2 = I \qquad (17.35)$$

and

$$TS_\lambda = S_{\lambda^*}. \qquad (17.36)$$

Now we take the involution on both sides of (17.34). Using (17.35) and (17.36), we have

$$h_{k_1}(x) \cdots h_{k_l}(x) = \sum_{l \geq \lambda_1 \geq \cdots \geq \lambda_l \geq 0} a(\lambda, k_1, \ldots, k_l) S_\lambda(x), \qquad (17.37)$$

where $h_{k_j}(x) = S_{(k_j,0,\ldots,0)}(x_1, \ldots, x_l)$. From the similarity between the Schubert calculus and Schur functions (see Ref. 6)

$$\sigma_a \cdot \sigma_b = \sum_c \delta(a, b; c) \sigma_c$$

and (17.24), we see that (17.37) is equivalent to

$$\sigma_{k_1} \cdots \sigma_{k_l} = \sum_{l \geq \lambda_1 \geq \cdots \geq \lambda_l \geq 0} a(\lambda; k_1, \ldots, k_l) \sigma_\lambda.$$

Therefore,

$$\deg(\sigma_{k_1} \cdots \sigma_{k_l}) = a((\underbrace{l, \ldots, l}_{n}, 0, \ldots, 0); k_1, \ldots, k_l).$$

$$\square \quad (17.38)$$

COROLLARY 17.6. *In $G(l + n, n)$,*

$$\deg \sigma_n^l = \text{the coefficient of } (x_1 \cdots x_l)^n \text{ in the}$$

$$\text{expansion of } S_{(\underbrace{l, \ldots, l}_{n}, 0, \ldots, 0)}. \quad (17.39)$$

By now we have given the special case where $d = 2$, $m = 1$ of Theorem 17.4. By means of Theorem 17.5, this special case corresponds to the solution of a problem in combinatorial analysis.

THEOREM 17.7. *In $G(2 + l, 2)$ let $p_l = \deg \sigma_2^l$. Denote the formal power series $F(t) = \sum_{l=0}^{\infty} p_l t^l$ ($p_0 = 1$). Then*

$$F(t) = \varphi\left(\frac{t}{1+t}\right),$$

$$\quad (17.40)$$

$$\varphi(s) = 1 + \sum_{k=1}^{\infty} \frac{3C_{2k}^{k-1}}{k+2} S^{k+1}.$$

The first terms of $F(t)$ are

$$F(t) = 1 + t^2 + t^3 + 3t^4 + 6t^5 + 16t^6 + 36t^7. \quad (17.41)$$

PROOF. According to Theorem 17.4, p_l is equal to the coefficient of $S_{(l,l)}(x, y)$ after $S_{(2,0)}^l(x, y)$ is expanded into a linear sum of Schur functions. From $S_{(2,0)}(x, y) = x^2 + y^2 + xy$, we can easily get

$$1 - S_2(x, y) = (1 + xy)\left(1 - \frac{(x+y)}{1+xy}\right).$$

Consequently,

$$\sum_{k=0}^{\infty} S_2^k(x, y) = \sum_{j=0}^{\infty} \frac{(x+y)^{2j}}{(1+xy)^{j+1}}.$$

But according to (17.3),

$$(x+y)^{2j} = \sum_{\substack{f_1 \geq f_2 \geq 0 \\ f_1 + f_2 = 2j}} \frac{(2j)!}{(f_1 + 1)!(f_2)!} D(f_1 + 1, f_2) S_{(f_1, f_2)}(x, y)$$

and

$$\frac{1}{(1 + xy)^{j+1}} = \sum_{m=0}^{\infty} (-1)^m C_{j+m}^m (xy)^m,$$

we obtain

$$S_2^k(x, y) = \sum_{\substack{j+m=k \\ f_1+f_2=2j}} (-1)^m C_{j+m}^m \frac{(2j)!}{(f_1 + 1)! f_2!} D(f_1 + 1, f_2) S_{(f_1+m, f_2+m)}(x, y),$$

where the term $S_{(k,k)}(x, y)$ is equivalent to $f_1 = f_2 = j$. So

$$p_k = \sum_{j+m=k} (-1)^m C_k^m \frac{(2j)!}{(j + 1)! j!}$$

$$= \sum_{j=0}^{k} (-1)^{k-j} C_k^j C_{2j}^j \frac{1}{j + 1}.$$

Hence (let $p_0 = 1$),

$$\sum_{n=0}^{\infty} p_n t^n = \sum_{n=0}^{\infty} \left(\sum_{k=0}^{n} (-1)^{n-k} C_n^k \frac{C_{2k}^k}{k + 1} \right) t^n$$

$$= \sum_{k=0}^{\infty} \left(\sum_{n \geq k} (-1)^{n-k} C_n^k t^n \right) \frac{C_{2k}^k}{k + 1}$$

$$= \sum_{k=0}^{\infty} \frac{C_{2k}^k}{k + 1} \frac{t^k}{(1 + t)^{k+1}}$$

$$= \left(1 - \frac{t}{1 + t} \right) \sum_{k=0}^{\infty} \frac{C_{2k}^k}{k + 1} \left(\frac{t}{1 + t} \right)^k$$

$$= 1 + \sum_{k=0}^{\infty} \left(\frac{C_{2k+2}^{k+1}}{k + 2} - \frac{C_{2k}^k}{k + 1} \right) \left(\frac{t}{1 + t} \right)^{k+1}$$

$$= 1 + \sum_{k=0}^{\infty} \frac{3 C_{2k}^{k-1}}{k + 2} \left(\frac{t}{1 + t} \right)^{k+1}$$

$$= \varphi \left(\frac{t}{t + 1} \right).$$

This theorem has significance for the enumerative problem of plane partition in combinatorial theory. The so-called plane partition of shape $\lambda = (\lambda_1, \ldots, \lambda_d)$, $\lambda_1 \geq \cdots \geq \lambda_d \geq 0$ means that one of the arrangements in the following forms (n_{ij} is a positive integer):

$$n_{11}, n_{12}, \ldots, n_{1\lambda_1}$$
$$n_{21}, n_{22}, \ldots, n_{2\lambda_2}$$
$$\cdots\cdots\cdots\cdots$$
$$n_{d1}, n_{d2}, \ldots, n_{d\lambda_d}$$

satisfies the condition that any row or column in the arrangement is nonincreasing; i.e., $n_{ij} \geq n_{ij+1}$, $n_{ij} \geq n_{i+1j}$. In case $n_{ij} > n_{i+1j}$, $\forall i$, it is called strictly column decreasing. The principal problem of plane partitions is: what is the number of plane partitions that satisfy certain conditions? Usually, the condition gives a definite shape $\lambda = (\lambda_1, \ldots, \lambda_d)$ and the sum $N = \sum n_{ij}$ or the set of elements so as to make $n_{ij} \in S$, etc. For the general theory of plane partitions, refer to Ref. 5.

Here we raise a question about plane partitions. For a given shape of $\lambda = (1, 1)$ and element set $S = \{1, 1, 2, 2, \ldots, l, l\}$, how many plane partitions are there which satisfy the condition of strictly column decreasing?

THEOREM 17.8. *Let the number of strictly column-decreasing plane partitions for the shape $\lambda = (1, 1)$ and the element set $S = \{1, 1, 2, 2, \ldots, l, l\}$ be p_l. Then p_l is the very p_l in Theorem 17.7.*

PROOF. Let us recall the definition of combination on Schur functions [5]. This definition shows that if π is the plane partition of shape $(\lambda_1, \ldots, \lambda_r)$ and with $(1, 2, \ldots, l)$ its element set (the number of elements equal to i in π is denoted by a_i), then the monomial corresponding to the definition and its degree are, respectively,

$$M(\pi) = x_1^{a_1} \cdots x_l^{a_l},$$

$$d(\pi) = \text{degree } M(\pi) = a_1 + a_2 + \cdots + a_l.$$

Then the Schur function $S_{(\lambda_1, \ldots, \lambda_r)}(x_1, \ldots, x_l)$ may be defined as

$$S_{(\lambda_1, \ldots, \lambda_r)}(x_1, \ldots, x_l) = \sum_{\substack{\pi = (\lambda_1, \ldots, \lambda_r) \\ d(\pi) = \sum \lambda_i}} M(\pi). \tag{17.42}$$

Thus it is evident that

$$p = \# \left\{ M(\pi) \;\middle|\; \begin{array}{l} \pi \text{ is of shape } (l, l) \\ \text{in } \pi, \; a_i = 2, \; i = 1, 2, \ldots, l \end{array} \right\}$$

$$= \text{the coefficient of } (x_1 \cdots x_l)^2 \text{ in } S_{(l,l,0,\ldots,0)}(x_1, \ldots, x_l)$$

$$= \deg \sigma_2^l \; (\text{in } G(l+2, 2)) \qquad (\text{from } (17.39))$$

$$= p_l \text{ in Theorem 17.7.}$$

When l is relatively small, its value is $p_1 = 0$, $p_2 = p_3 = 1$, $p_4 = 3$, $p_5 = 6$, $p_6 = 16$, $p_7 = 36, \ldots$ (see (17.41)).

References

1. P. Griffiths and J. Harris, *Principles of Algebraic Geometry*, Wiley, New York (1978).
2. W. V. D. Hodge and D. Pedoe, *Methods of Algebraic Geometry*, Vol. 2, Cambridge, University Press (1952).
3. L.-K. Hua, *Harmonic Analysis of Functions of Several Complex Variables in the Classical Domains*, Science Press, Beijing (1965).
4. S. Kleiman, *Problem 15, Rigorous Foundation of Schubert Enumerative Calculus*, Proc. Symp. Pure Math., vol. 27, American Mathematical Society, Providence, RI (1976).
5. R. Stanley, Theory and applications of plane partitions. I, *Stud. Appl. Math.* **50**, 167–188 (1971).
6. J.-Q. Zhong, Schubert calculus and Schur functions, *Zhong Gua Kexue Ser.* (in Chinese) (Chapter 16 of this volume).
7. J.-Q. Zhong, A note on Schubert calculus, in *Proc. 1980 Beijing Symp. Differential Geometry and Differential Equations*, Vol. 3, pp. 1697–1708 (S. S. Chern et al., eds.), Science Press, Beijing; Gordon and Breach, New York (1982).

Index